Carbonyl Compounds

Carbonyl Compounds

Reactants, Catalysts and Products

Feng Shi
Hongli Wang
Xingchao Dai

WILEY-VCH

Authors

Feng Shi
State Key Laboratory for Oxo Synthesis
and Selective Oxidation
Lanzhou Institute of Chemical Physics
Chinese Academy of Sciences
No.18, Tianshui Middle Road
Lanzhou 730000
China

Hongli Wang
State Key Laboratory for Oxo Synthesis
and Selective Oxidation
Lanzhou Institute of Chemical Physics
Chinese Academy of Sciences
No.18, Tianshui Middle Road
Lanzhou 730000
China

Xingchao Dai
State Key Laboratory for Oxo Synthesis
and Selective Oxidation
Lanzhou Institute of Chemical Physics
Chinese Academy of Sciences
No.18, Tianshui Middle Road
Lanzhou 730000
China

Cover: © Andreas Prott/Shutterstock

Library of Congress Card No.:
applied for

British Library Cataloguing-in-Publication Data
A catalogue record for this book is available from the British Library.

Bibliographic information published by the Deutsche Nationalbibliothek
The Deutsche Nationalbibliothek lists this publication in the Deutsche Nationalbibliografie; detailed bibliographic data are available on the Internet at <http://dnb.d-nb.de>.

Print ISBN: 978-3-527-34736-0
ePDF ISBN: 978-3-527-82560-8
ePub ISBN: 978-3-527-82561-5
oBook ISBN: 978-3-527-82562-2

Typesetting Straive, Chennai, India
Printing and binding CPI Group (UK) Ltd, Croydon, CR0 4YY

Printed on acid-free paper

C101029_020821

Contents

Preface

Chemical industry has made a great contribution to the development of human society because it creates a great range of products such as dye, pesticides, paints, pharma drugs, fertilizers, and plastic, etc. Among the different chemicals, the carbonyl compounds represent one of the most basic chemicals in the chemical industry. Carbonyl compounds include CO, CO_2, aldehydes, ketones, carboxyl acids, ester, amides, carbamate, urea, carbonate, and so on. Carbonyl compounds are typical building blocks in myriad of natural products, pharmaceuticals, agrochemicals, and functional materials, etc. In addition, they are indispensable precursors for the preparation of functional molecules and materials. It is considered that about 60% chemical processes are involving the building and transformation of the carbonyl-containing molecules. Thus, the building and transformation of carbonyl-containing molecules hold a very important position in chemical industry process and organic synthesis. Normally, the synthesis of carbonyl-containing molecules and transformation of carbonyl-containing molecules are independently discussed. In fact, many reactions simultaneously involve building and transformation of the carbonyl-containing molecules, for example, borrowing-hydrogen reactions, direct amination of toluene via benzaldehyde intermediate, hydroaminomethylation of olefins, etc. So, the two parts should be looked as a whole. Integration of building and transformation of the carbonyl containing molecules will not only promote exploration of novel catalytic reactions and synthetic methods but also facilitate development of new processes and catalysts and innovation of chemical industry and technology. It eventually will contribute greatly to efficient utilization of resources and the sustainable development of the chemical industry. Nevertheless, we could NOT find any books that look at two individual topics as a whole on the building and transformation of the carbonyl-containing molecules. In this context, we began to prepare a book on this topic in order to fill the gap in this area.

Considering that the building and transformation of the carbonyl-containing molecules involve many contents, we were unable to include every detail involved in these fields in this short volume. This book will highlight the importance of carbonyl-containing molecules, which can be served as a link between reaction, chemicals, and materials. The book is intended to describe and convey the concept of building and transformation of carbonyl-containing molecules and its application in catalysis. The book is divided into four chapters and organized according to the sequence of the synthesis and application of carbonyl compound. Part I focuses on the methods of synthesis of carbonyl compounds. A variety of carbonyl sources, such as CO, carbon dioxide, other C_1 carbonyl molecules, and non-C_1 carbonyl molecules, in situ generated carbonyl molecules, are described for the synthesis of carbonyl molecules. The application of carbonyl compounds as catalysts for the synthesis of fine chemicals is summarized in Part II. Part III presents the applications of carbonyl compounds in the synthesis of functional molecules and materials. The book ends with our personal outlook on this field.

It is a pleasure to be the authors of this book because it provides us with the opportunity to survey the field of the building and transformation of the carbonyl-containing molecules. We gratefully acknowledge the WILEY-VCH editorial staff and extend special thanks to Lifen Yang. We are very grateful to all of our colleagues for their excellent contributions to this book. Contributors include: Xinzhi Wang (Chapter 1), Dongcheng He (Chapters 2 and 3), Kang Zhao (Chapters 4 and 7), and Shujuan Liu (Chapters 5, 6, and 8).

We truly hope this book will be of interest to all those involved in this field, from graduate students to independent catalytic and organic researchers in both academic and industrial sectors. We also hope that it will stimulate a wider general interest and promote further development in the subject.

Lanzhou, China
July 2020

Feng Shi, Hongli Wang, and Xingchao Dai

Part I

Carbonyl Molecules as Reactants

1

Carbon Monoxide

1.1 Hydroformylation of Alkenes and Alkynes

Hydroformylation is one of the most important reactions for the preparation of aldehydes and alcohols from alkenes and synthesis gas [1]. In 1938, Roelen (1897–1993) discovered the reaction between alkenes and an equimolar mixture of carbon monoxide (CO) and hydrogen to form aldehydes [2, 3]. It is called "hydroformylation" and was originally called "oxo-reaction." Nowadays, homogeneous metal complexes commercially based on cobalt and rhodium are used as catalysts. With more than 10 million metric tons of oxo products per year, hydroformylation represents one of the most important industrial applications and achievements of homogeneous catalysis in the chemical industry [4].

The key consideration of hydroformylation is the selectivity of "normal" vs. "iso." For example, the hydroformylation of propylene can afford two isomeric products, butyraldehyde or isobutyraldehyde (Scheme 1.1).

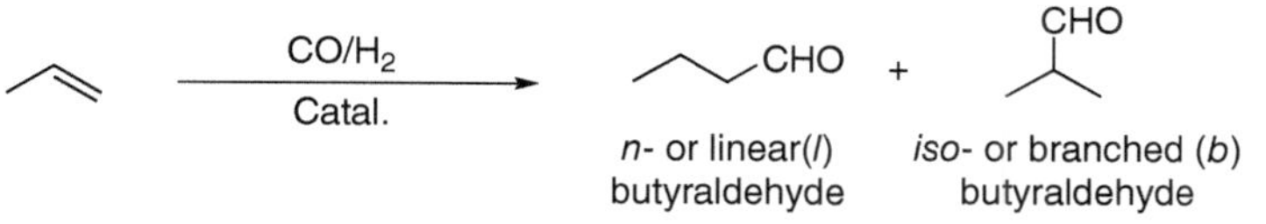

Scheme 1.1 Example of hydroformylation.

These isomers are related to steric hindrance and the rate of CO migration insertion. In addition, they also reflect the regiochemistry of the insertion of alkene into the M–H bond. For example, the reaction mechanism begins with dissociation of CO from hydrido-metal–tetracarbonyl complex (1) to give the 16-electron species $HM(CO)_3$ (2). Then, the alkene starts to coordinate with the $HM(CO)_3$ complex. The π-complex (3) is converted into the corresponding σ-complex (4); the 18 electron species are formed by adding CO (5). In the next step of the reaction cycle, the CO is inserted into the carbon–metal bond (6). Once again, CO is associated to end up in the 18 electron species (7). In the last step of the reaction cycle, the catalytically active hydrido-metal–tetracarbonyl complex (1) is released by adding hydrogen. Moreover, the aldehyde is formed by a final reductive elimination step

Carbonyl Compounds: Reactants, Catalysts and Products, First Edition.
Feng Shi, Hongli Wang and Xingchao Dai.

(Scheme 1.2) [5]. In this section, we will summarize the history and recent advances of catalysts for hydroformylation.

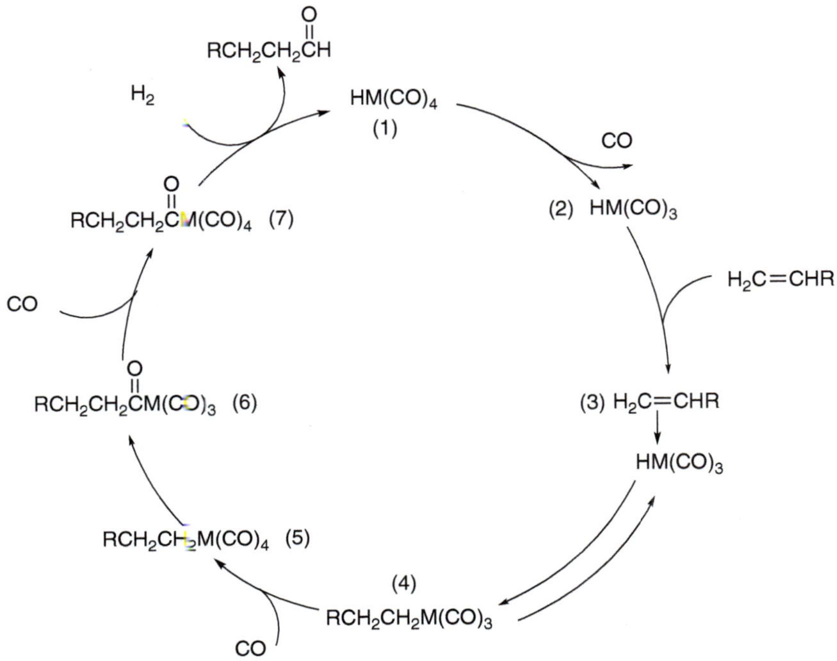

Scheme 1.2 Mechanism of metal-catalyzed hydroformylation. Source: Based on Heck and Breslow [5].

1.1.1 Co Catalysts

A generally accepted rough order of active metals in hydroformylation is given in Table 1.1 [6].

Roelen first discovered and patented hydroformylation of straight-chain 1-olefins under the $HCo(CO)_4$ catalyst system which generally leads to the formation of linear aldehydes and large amounts of branched aldehydes [7]. This laid a good foundation for all hydroformylation studies.

As early as in 1958, the heterogeneous cobalt catalyst for hydroformylation was reported by Aldridge et al. [8] With 0.879 wt% insoluble cobalt as a catalyst system, the conversion of C_7 olefin (mixed isomers) using CO and H_2 (about 1 : 1, 2750–2900 psig) at 177 °C is 83%. But, it was soon recognized that the real active species is the homogeneous complex hydridocobaltcarbonyl. $HCo(CO)_4$ is a yellow liquid and strong acid, which is stable only under CO/H_2 pressure above the melting point (−26 °C) [9]. Therefore, the mechanism of hydroformylation has been extensively studied.

In 1968, the hydroformylation of propene and 1-hexene has been investigated in a tertiary organophosphine–cobalt hydrocarbonyl catalyst system $HCo_2(CO)_8(PBu_3)$

Table 1.1 Activity of metals in hydroformylation.

Active metal	Rh	Co	Ir	Ru	Os	Tc	Mn	Fe	Re
Log(relative activity)	3	0	−1	−2	−3	−3	−4	−6	<−6

Source: Beller [6]. © 2006, Springer Nature.

by Tucci (Scheme 1.3) [10]. Trialkylphosphine complexes yielded 73% less branched isomer formation than the conventional $HCo(CO)_4$ system. The activity of the cobalt hydrocarbonyl-catalyzed branched isomer formation decreases in the order $HCo(CO)_4 > HCo(CO)_3(PAr_3) > HCo(CO)_3(PR_3)$.

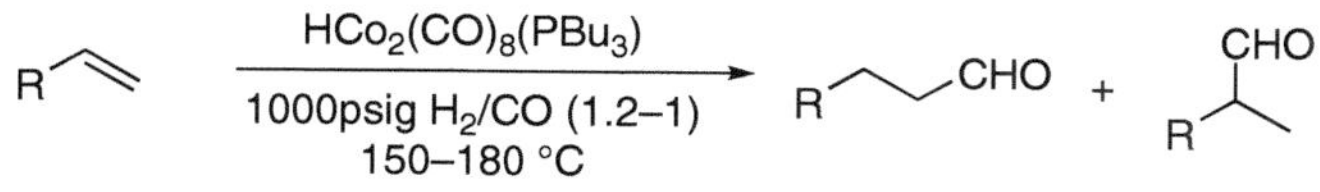

Scheme 1.3 Ligand modification of the catalyst-catalyzed hydroformylation of olefins. Source: Based on Tucci [10].

There has been an important breakthrough in the development of cobalt catalysts recently. Stanley's group found the $[HCo(CO)_n(P_2)]^+$ catalyst showed nice activity at lower pressures for hydroformylation [11]. For example, using $[Co(acac)(depe)](BF_4)$ under standard conditions (1 mM catalyst, 1 M 1-hexene, dimethoxytetraglyme solvent, activate at 140 °C under 34 bar of 1 : 1 H_2 : CO, then reduced to 100 °C and 10 bar), *n*/*iso* aldehyde ratio of 0.8% and 15.1% alkene isomerization can be obtained.

1.1.2 Rh Catalysts

Since the 1970s, most hydroformylation reactions rely on catalysts based on rhodium catalysts [12]. For example, Hanson and coworkers described the Rh/NaX and Rh/NaY catalyst system in a fixed bed reactor consisting of propylene : H_2 : N_2 : CO (3 : 3 : 2 : 1) at 1 atm [13]. The selectivity of *n*-butyraldehyde vs. *iso*-butyraldehyde is 2.0 : 1 and 1.9 : 1 for Rh/NaX and Rh/NaY at 150 °C, respectively (Scheme 1.4).

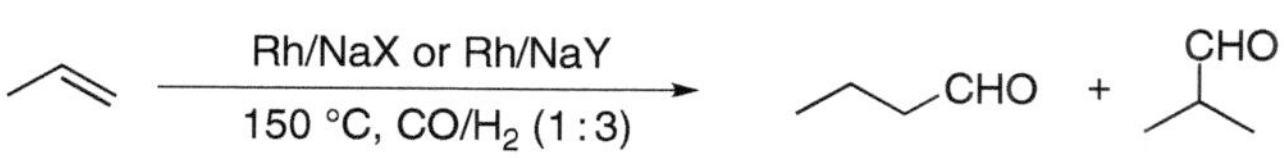

Scheme 1.4 Zeolite-catalyzed hydroformylation of propylene.

Water-soluble catalysts have been developed. They facilitate the separation of the products from the catalyst [14]. For example, Herrmann et al. reported the novel Rh(I)/BISBIS catalyst for hydroformylation of propene and 1-hexene [15]. The high activity (97.7) and productivity (1.26) at low phosphane/rhodium ratios (6.7 : 1) can be obtained in the hydroformylation of propene. Alper and coworker reported the first polymeric water-soluble metal complex Rh/PPA(Na^+)/DPPEA for hydroformylation of aliphatic olefins (Scheme 1.5) [16]. For the hydroformylation of vinyl arenes, the complete conversion, high selectivity (>97%) and *iso*/*n* ratio (7.3–24) can be observed.

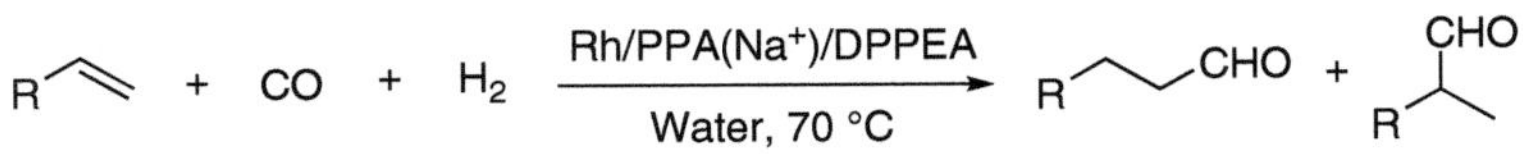

Scheme 1.5 Rh/PPA(Na^+)/DPPEA-catalyzed hydroformylation of olefins. Source: Based on Ajjou and Alper [16].

In 1989, Arhancet et al. described a novel supported aqueous-phase catalyst $HRh(CO)[P(m\text{-}C_6H_4SO_3Na)_3]_3/SiO_2$ (SAPC) for hydroformylation (Scheme 1.6) [17]. The hydroformylation of 1-octene yielded a nonanal/2-methyl octanal (*n*/*iso*) ratio ranging from 1.8 to 2.9 depending on the water content and ligand/rhodium ratio (which varied from 7 to 30).

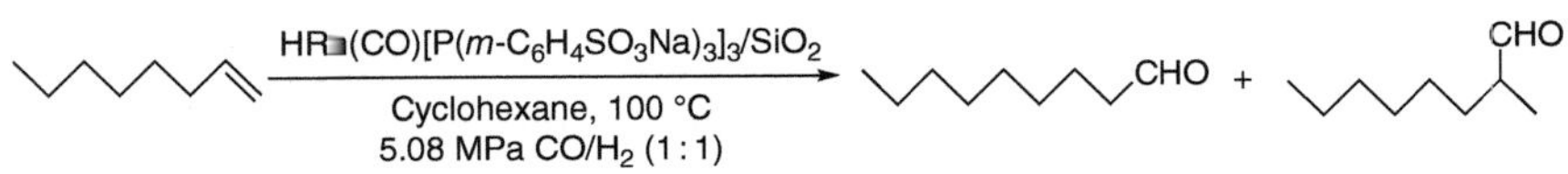

Scheme 1.6 Hydroformylation of 1-octene using SAPC. Source: Based on Arhancet et al. [17].

In 2000, Arya and Alper found a solid-phase synthetic approach to obtain dendritic ligands anchored onto beads for the hydroformylation of several olefins [18]. For example, the complete conversion of styrene (>99%) with a high selectivity for the branched isomer (branched : linear, 16 : 1) was obtained at 65 °C. The catalyst can be recycled five times without deactivation.

Chaudhari and coworkers reported $HRh(CO)(PPh_3)_3$ encapsulated and anchored in NaY, MCM-41 and MCM-48 for hydroformylation of olefins to aldehydes in 2003 (Scheme 1.7) [19]. 99.4% conversion of 1-octene and a 1.7 *n*/*iso* ratio of regioselectivity were obtained when using Rh-MCM-48 at 80 °C. This catalyst is recyclable and can be reused six times without obvious deactivation.

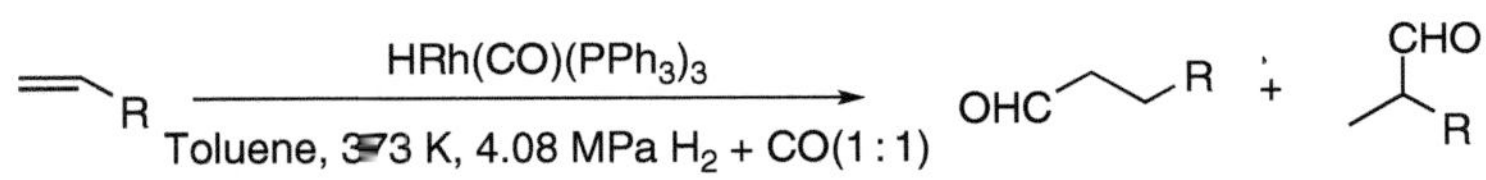

Scheme 1.7 Hydroformylation of olefins using $HRh(CO)(PPh_3)_3$-encapsulated catalysts. Source: Based on Mukhopadhyay et al. [19].

Recently, Shi's group described the simple Rh black as a heterogeneous catalyst for hydroformylation of olefins [20]. This catalyst system has a broad substrate scope including the aliphatic and aromatic olefins, affording the desired aldehydes in good yields (Scheme 1.8). For example, 86% yield of propanal and 200 h^{-1} turnover

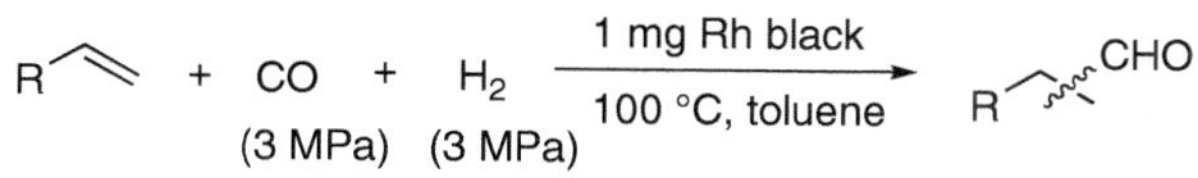

Scheme 1.8 Rh black-catalyzed hydroformylation of olefins.

frequency (TOF) were obtained when ethylene is the reactant. The catalyst could be recycled five times without loss of activity.

For hydroformylation of alkynes, Breit and coworker designed and synthesized a series of new supramolecular ligands containing a functional guanidine group with increasing p-acceptor ability of the phosphine donor ligands in 2018 [21]. The desired aldehydes were obtained in 51–94% yields with regioselectivities up to 25 : 1 (Scheme 1.9).

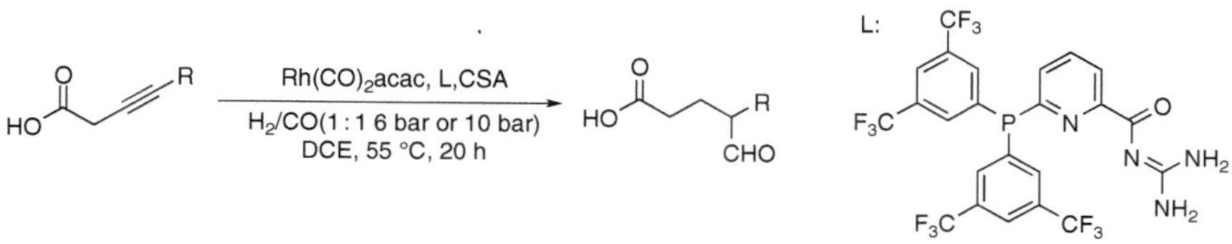

Scheme 1.9 Hydroformylation of alkynes.

1.1.3 Au Catalysts

In 2008, Tokunaga and coworkers presented Au/Co_3O_4 heterogeneous catalysts prepared by coprecipitation for hydroformylation (Scheme 1.10) [22]. Aldehydes were prepared with 99.5% conversion of 1-hexene and 1.2 regioselectivity at 130 °C. The Au/Co_3O_4 catalysts can be recycled by simple decantation with slight decrease in catalytic activity. The role of Au may promote in situ reduction of Co_3O_4 to Co^0 which is the active site for the hydroformylation reaction.

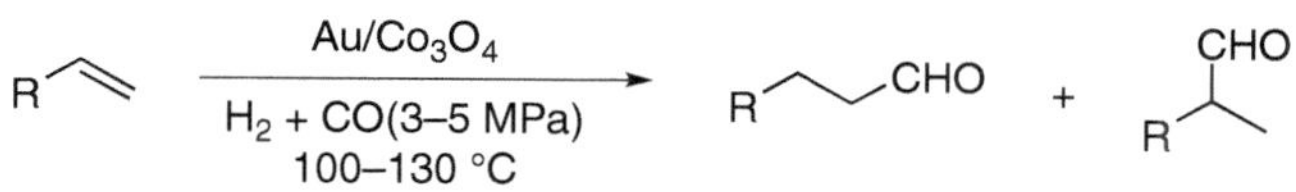

Scheme 1.10 Au/Co_3O_4-catalyzed hydroformylation of 1-olefins. Source: Based on Liu et al. [22].

1.1.4 Ligand-Modified Heterogeneous Catalysts

Subsequent work demonstrated that the ligand modification of the catalyst system can influence the catalytic activity and selectivity of the hydroformylation process. In 2006, Li and coworkers prepared heterogeneous chiral ligand-modified silica-supported rhodium (Rh/SiO_2) with chiral phosphorus ligands [23]. Up to 72% enantiomeric excess (ee) and 100% selectivity of branched aldehyde for the hydroformylation of vinyl acetate at 60 °C were obtained by (*R*)-BINAP-Rh/SiO_2 catalysts

(BINAP = 1,1′-binaphthyl-2,2′-diphenyl phosphine) (Scheme 1.11). Reviews on active cobalt and rhodium complexes, also ligand-modified, and on methods for the necessary spectroscopic in situ methods are reported [24, 25].

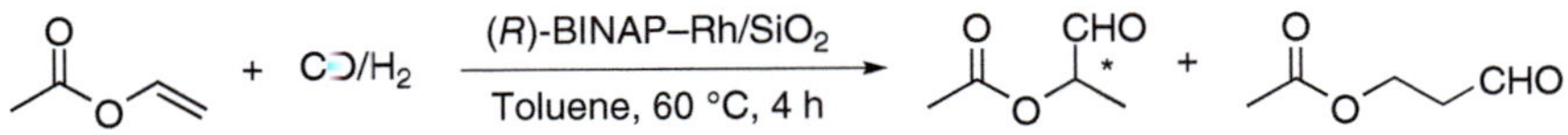

Scheme 1.11 Asymmetric hydroformylation of vinyl acetate.

In 2011, Shukla and coworkers prepared the Rh-TPPTS-hexagonal mesoporous silica (HMS) catalyst that $RhCl(TPPTS)_3$ (TPPTS = *m*-trisulfonato triphenyl phosphine) was in situ encapsulated into the mesopores of HMS for hydroformylation of vinyl esters [26]. The synthesized heterogeneous catalyst evaluated for hydroformylation of vinyl esters gave 100% conversion and 88–94% selectivity to *iso*-aldehyde at 100 °C (Scheme 1.12).

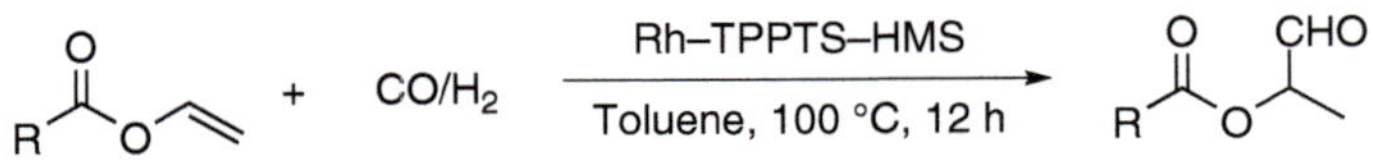

Scheme 1.12 Hydroformylation of vinyl esters.

To enhance the activity and stability of the heterogeneous catalyst, a nice method of organic phosphine ligand functionalized support via ligand immobilization or polymerization was developed and applied in the preparation of heterogeneous hydroformylation catalysts [27–29].

In 2014, Xiao, Ding, Chen, and Meng reported the POL (porous organic ligand) bearing a triphenylphosphine-supported rhodium catalyst (Rh/POL–PPh_3) for hydroformylation of 1-octene [30]. Good conversion (>99%) of 1-octene, regioselectivity (*n*/*iso* = 0.87–1.35), and selectivity (89.0–92.1%) were afforded at 90 °C (Scheme 1.13). The catalyst could be recycled at least six times with negligible loss of activity.

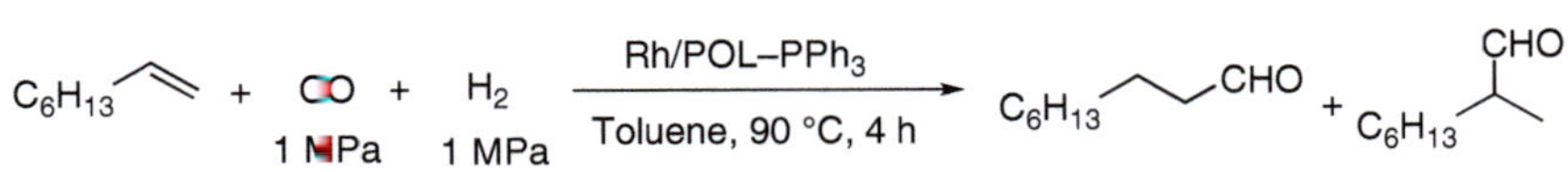

Scheme 1.13 Rh/POL–PPh_3-catalyzed hydroformylation of 1-octene.

Then, Xiao's group synthesized Rh/POL–dppe catalyst for hydroformylation (dppe = 1,2-bis(diphenylphosphino)ethane) [31]. For example, using Rh/POL–dppe as a catalyst, 1-octene in the presence of CO (1 MPa) and H_2 (1 MPa) at 90 °C can give corresponding aldehydes in high conversion of up to 96.9% with regioselectivities of *n*/*iso* up to 2.46 (Scheme 1.14). Meanwhile, the catalyst can be easily separated and recycled five times from the reaction systems without losing any activity and selectivity.

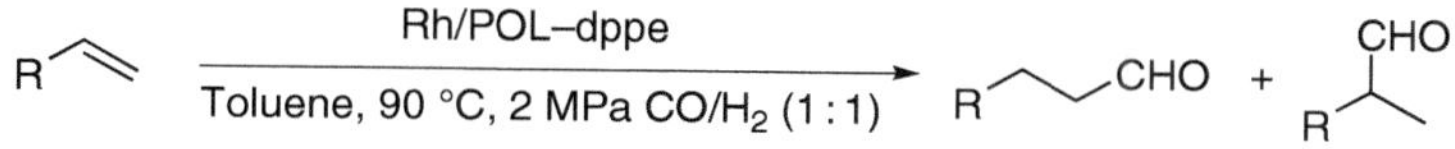

Scheme 1.14 Rh/POL–dppe-catalyzed hydroformylation of olefins.

Ding and Li reported porous organic copolymer (denoted as CPOL–BP&P) which was prepared by copolymerization of vinyl biphephos and tris(4-vinylphenyl) phosphine monomers under solvothermal conditions, and followed with impregnation method provided a highly efficient Rh/CPOL–BP&P catalyst with high activity (TOF = 11 200 h^{-1}) and regioselectivity (the ratio of linear to branched aldehydes, n/iso = 62.2) for hydroformylation of 1-butene (Scheme 1.15) [32]. High regioselectivity was also obtained in the hydroformylation of butene mixture (2-butene: n/iso = 55.8, isomeric mixture of butenes: n/iso = 56.0). Later, they found the hydroformylation of propene to linear butaldehyde by employing single atom dispersed Rh/CPOL–BP&P catalysts in a continuous fixed-bed reactor [33]. In the presence of CO/H_2 (0.5 MPa, 1 : 1) at 70 °C, high regioselectivity (n/iso > 24), activity (TOF > 1200 h^{-1}) and stability (over 1000 h) were obtained.

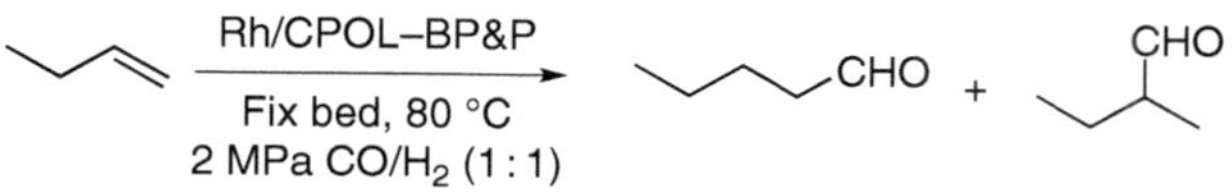

Scheme 1.15 Rh/CPOL–BP&P-catalyzed hydroformylation of 1-butene.

Tsubaki and Yang prepared PPh_3-Rh/GO heterogeneous catalyst by triphenylphosphine (PPh_3)-reduced graphene oxide (rGO) and a rGO-supported Rh–ligand complex catalyst simultaneously [34]. This catalyst system is applied in 1-olefins hydroformylation to form normal aldehyde, exhibiting remarkable catalytic activity and selectivity (Scheme 1.16). For example, the conversion of 1-hexane is 99.9% and the regioselectivities of n/iso is 2.10. This catalyst could be recycled five times without significant decrease in activity.

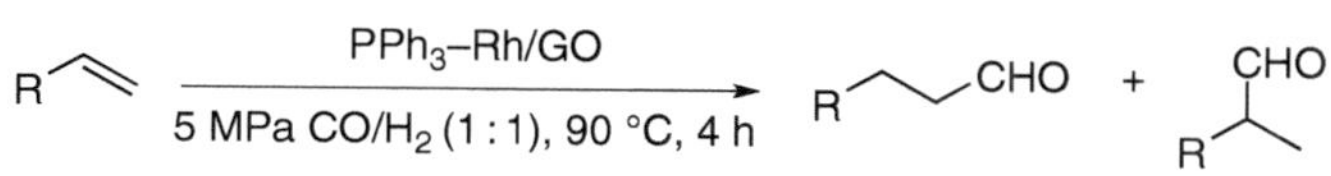

Scheme 1.16 PPh_3–Rh/GO-catalyzed hydroformylation of olefins.

In addition, Yang and coworkers presented an efficient heterogeneous catalytic system about IL-in-oil (IL = ionic liquid) Pickering emulsion constructed with Rh-sulfoxantphos as the catalyst and surface-modified dendritic mesoporous silica nanospheres as the stabilizer for hydroformylation of long-chain alkenes in 2018 [35]. 1-Dodecene could be converted to the corresponding aldehyde with an n/iso ratio of 98 : 2, chemoselectivity of 94% and TOF of 413 h^{-1}, among the highest ever reported for IL-oil biphase hydroformylation of long-chain alkenes. The high efficiency of IL-in-oil Pickering emulsion was primarily attributed to the increased interface area and unique properties of ILs.

Jia and Zong recently developed a novel Rh/POL–BINAPa&PPh_3 catalyst by porous organic polymer-supported rhodium for the hydroformylation of various alkynes (Scheme 1.17) [36]. It can afford the corresponding α,β-unsaturated aldehydes with good to excellent yields (61–89%) with high *E* stereoselectivity (>40 : 1), excellent catalytic activity and good reusability (10 cycles).

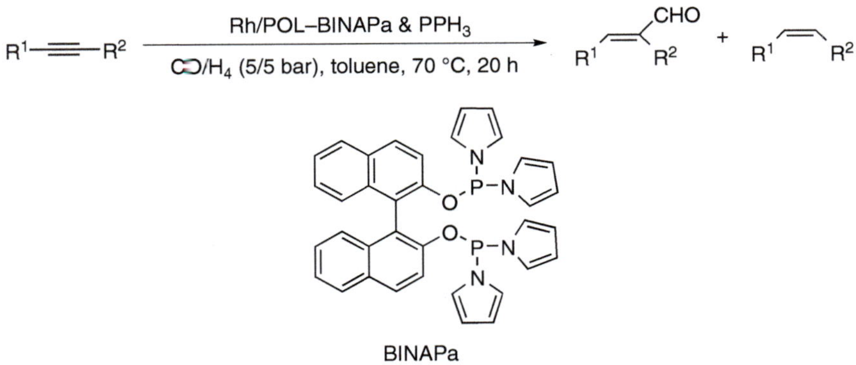

Scheme 1.17 Rh/POLBINAPa&PPh_3-catalyzed hydroformylation of alkynes. Source: Modified from Liang et al. [36].

1.1.5 Single-Atom Catalysts

With the development of single-atom catalysis (SAC), it has also been successfully applied in hydroformylation. Zhang and coworkers synthesized a Rh SAC supported on ZnO nanowires for the hydroformylation reaction (Scheme 1.18). As a result, a complete conversion of styrene, a TON of about 40 000 and 99% selectivity can be obtained [37].

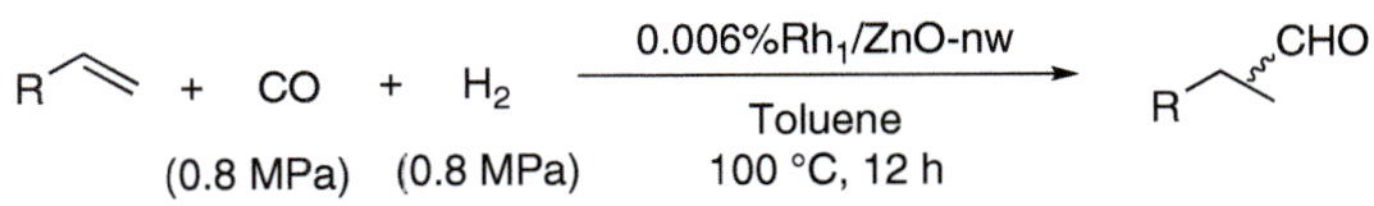

Scheme 1.18 Rh_1/ZnO-catalyzed hydroformylation of olefins.

Wang and Zhang developed CoO-supported Rh single-atom catalysts (Rh_1/CoO) for hydroformylation of propene [38]. The optimal selectivity of 94.4% for butyraldehyde and the highest TOF number of 2065 h^{-1} among the atomic-scale Rh-based catalysts were obtained (Scheme 1.19).

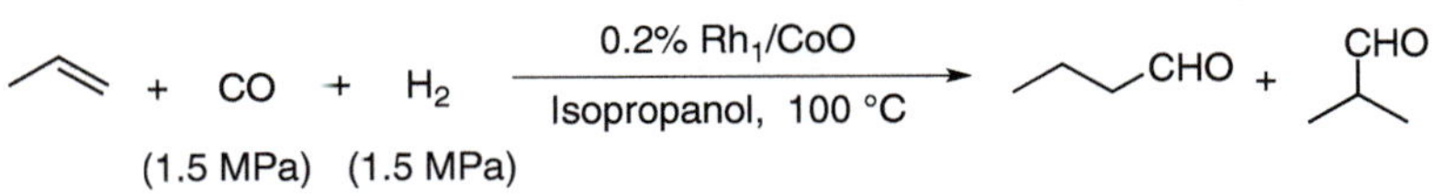

Scheme 1.19 Rh_1/CoO-catalyzed hydroformylation of propene.

1.2 Hydroxy-, Alkoxy-, and Aminocarbonylation of Alkenes and Alkynes

Transition metal-catalyzed carbonylation of alkenes and alkynes with nucleophiles such as water, alcohols or amines are called hydroxycarbonylation, alkoxycarbonylation, or aminocarbonylation, respectively. As the importance of carboxylic acid derivatives, transition metal-catalyzed hydroxyl, alkoxy-, and aminocarbonylation reactions are important transformations in organic synthesis. Several palladium-catalyzed alkoxycarbonylations and aminocarbonylations have also been applied on an industrial scale, such as the carbonylation of 1,2-xylyldichloride to give isochroman-3-one, aminocarbonylation of 2,5-dichloropyridine to give Lazabemide and so on.

1.2.1 Hydroxycarbonylation of Alkenes

In 1969, Von Kutepow et al. patented phosphine-containing palladium complexes for the hydroxycarbonylation of a terminal alkene [39]. Further studies by Fenton [40] showed that palladium–phosphine was a good precursor and the reaction parameters were optimized to combine high conversion and a good selectivity in the linear ester. At pressures ranging from 7 to 55 bar, temperatures near to 150 °C, with an excess of PPh_3, linear to branched ratios as high as 3.5 can be reached. As the conversion is largely improved when a hydrogen partial pressure is introduced, the active catalytic species is presumably $[Pd(H)Cl(PPh_3)_2]$ and the catalytic cycle is shown in Scheme 1.20, in which "Pd" represents $[PdCl(PPh_3)_2]$.

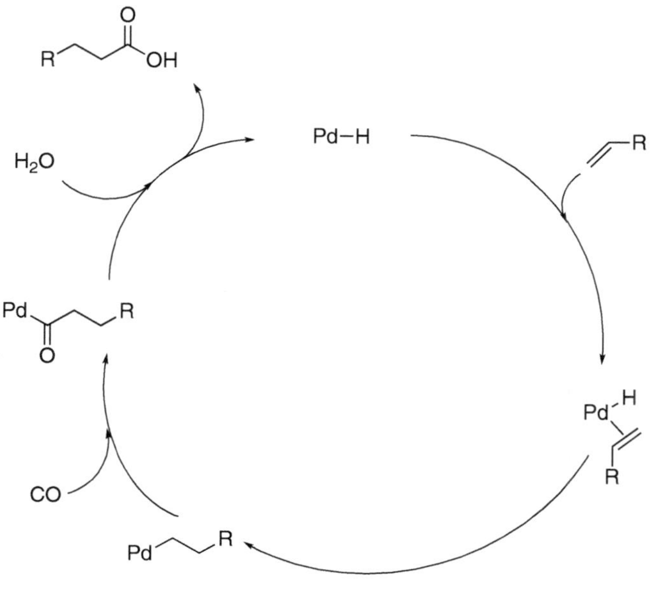

Scheme 1.20 Catalytic cycle of the Pd–H-catalyzed hydroxycarbonylation of alkenes.

In 1983, Alper et al. discovered that the selectivity of the carbonylation reaction can be turned toward the branched acid by adding copper(II) chloride and hydrochloric acid to the palladium catalyst system (Scheme 1.21) [41]. The reaction conditions become very mild since the reaction is performed at room temperature and at 1 bar of CO; the yields in 30–100% can be obtained. By addition of the BNPPA ((*R*)- or (*S*)-1,1-binaphthyl-2,2-diylhydrogenophosphate) chiral ligand, it was possible to reach an ee as high as 91% [42]. It is thus possible to transform *p*-isobutylstyrene into (*S*)-ibuprofen and 2-vinyl-6-methoxynaphthalene into (*S*)-naproxen, which both possess nonsteroidal anti-inflammatory properties [43].

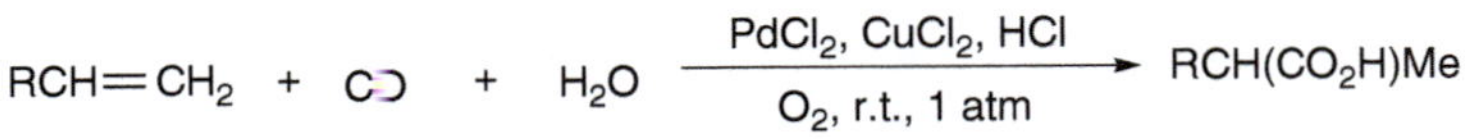

Scheme 1.21 Pd-catalyzed hydroxycarbonylation of alkenes. Source: Alper et al. [41].

Hydroxycarbonylation has been performed in biphasic media, maintaining the catalyst in the aqueous phase using water-soluble mono- or diphosphine ligands. For example, Sheldon and coworkers reported the sodium salt of TPPTS ($P(C_6H_4\text{-}m\text{-}SO_3Na)_3$) with palladium was shown to synthesize carbonylate efficiently propene and light alkenes in acidic media (Scheme 1.22) [44, 45]. By applying propene as a starting material, the TOF was >2800 h^{-1} with 99% selectivity to *n*- and isobutyric acid.

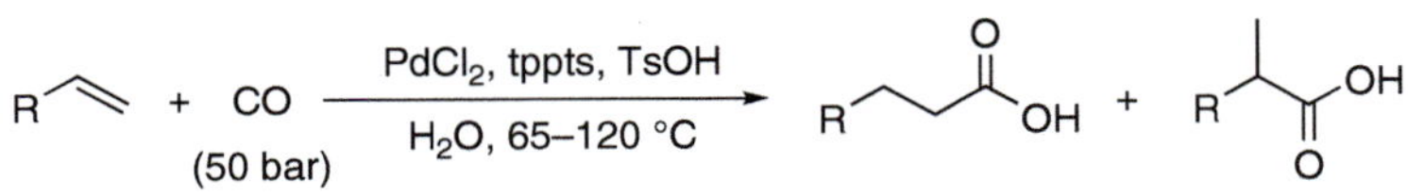

Scheme 1.22 The hydrocarboxylation of olefins in the aqueous phase. Source: Papadogianakis et al. [44]; Verspui et al. [45].

Monflier and coworkers introduced an inverse phase transfer agent particularly dimethyl-β-cyclodextrin to overcome the mass transfer problems between the aqueous and organic phases for heavy alkenes [46, 47]. For instance, the selectivity obtained during 1-decene hydroxycarbonylation reached 90% with per(2,6-di-*O*-methyl)-β-cyclodextrin as ligand.

In 1998, van Leeuwen and coworkers presented a dicationic palladium center coordinated by the bidentate diphosphine ligand 2,7-bis(sulfonato)xantphos catalyzes, in the presence of tolylsulfonic acid, the hydroxycarbonylation of ethylene, propene, and styrene can provide a ca. 0.34 : 0.66 molar ratio for the linear and branched acids [48].

Alper and coworker reported that a mixture of palladium acetate $Pd(OAc)_2$, 1,4-bis(diphenylphosphino butane) (dppb), and formic or oxalic acid is possible to promote high selectivity in linear acid synthesis. For instance, under roughly 7 bar pressure of CO and at 150 °C, α-methylstyrene gives 82% conversion and 100% linearity or 2,4,6-trimethylstyrene gives 98% conversion and 100% linearity, respectively [49]

Recently, Beller and coworkers reported a new state-of-the-art palladium catalyst containing L20 as a ligand for mild and versatile hydroxycarbonylation of olefins (Scheme 1.23) [50]. For industrially relevant olefins, for example, ethylene, excellent catalyst turnover numbers and frequencies are obtained (TON > 350 000, TOF > 15 000 for propionic acid). The catalyst shows outstanding stability and can be easily recycled (>25 runs) without measurable loss of reactivity.

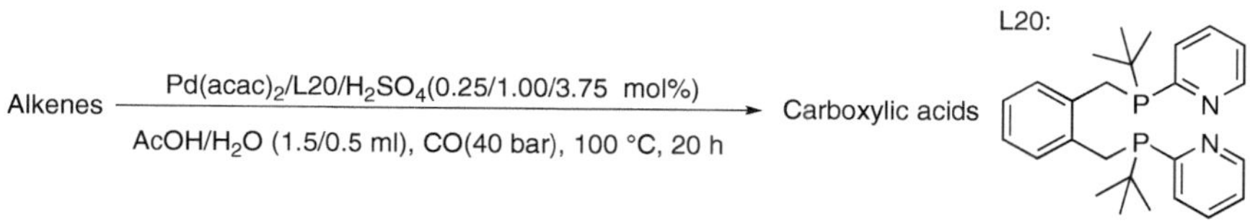

Scheme 1.23 Hydroxycabonylation of olefins. Source: Based on Sang et al. [50].

1.2.2 Hydroxycarbonylation of Alkynes

The pioneering work of Reppe and coworkers reported the industrial preparation of acrylic acid by carbonylation of acetylene as early as 1938 [51]. The reaction was conducted at 200–230 °C and 100 bar of CO and catalyzed by $Ni(CO)_4$ in the presence of a copper halide. Selectivity of 90% and 85% was reached in acrylic acid with regard to acetylene and CO, respectively [52].

Later, Alper and his group reported that hydroxycarbonylation of alkynes can be performed under mild conditions (90 °C, 1 bar of CO), provided that a phase transfer agent is added to the biphasic medium. In the presence of cetyltrimethylammonium bromide, various substituted alkynes or diynes are converted into the corresponding unsaturated carboxylic acids in 50–95% yields by using $Ni(CN)_2$ [53]. When using the bimetallic system $CoCl_2$ and $Ni(CN)_2$ with KCN, saturated carboxylic acids are obtained with a good selectivity for the branched isomer (Scheme 1.24) [54].

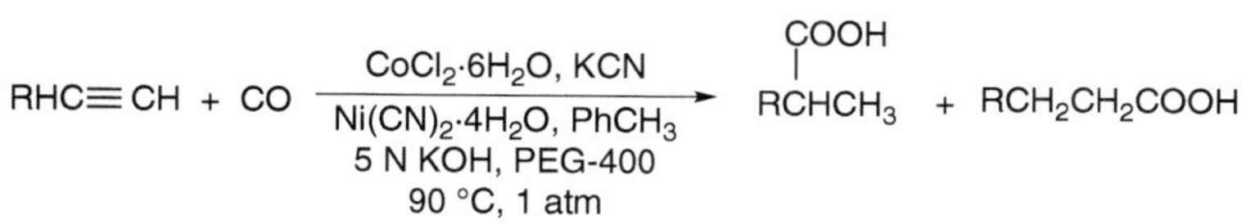

Scheme 1.24 Hydroxycarbonylation of alkynes. Source: Lee and Alper [54].

Similarly, allenes [55] and alkynols [56] were used as starting materials, and their carbonylation provides β,γ-unsaturated acids and unsaturated diacids, respectively.

Gabriele et al. mentioned that some precursors easily catalyze the reductive carbonylation of alkynes from the CO/H_2O couple [57, 58]. The main role of water is to furnish hydrogen through the water–gas-shift (WGS) reaction, as evidenced by the co-production of CO_2. In the presence of PdI_2/KI, terminal alkynes have been selectively converted into furan-2-(5*H*)-ones or anhydrides. Two CO building

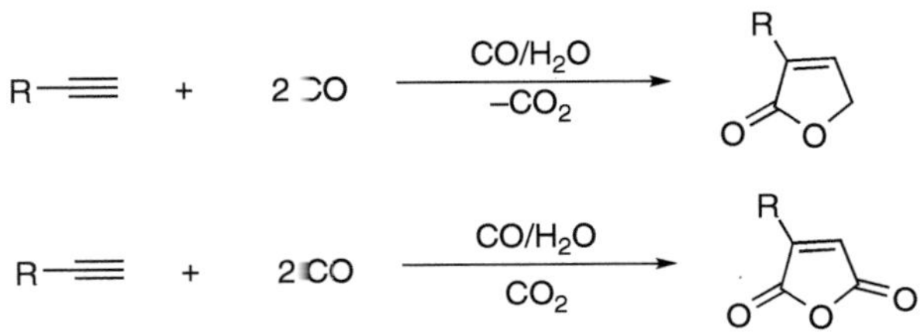

Scheme 1.25 Incorporation of two CO building blocks into alkynes under water–gas-shift conditions.

blocks are incorporated, and the cascade reactions that occur on palladium result in a cyclization together with the formation of an oxygen–carbon bond furnish (Scheme 1.25).

Recently, Liu and coworkers reported an L1-based Pd catalyst for the hydroxycarbonylation of terminal alkynes containing aromatic and aliphatic toward α,β-unsaturated carboxylic acids (Scheme 1.26) [59]. The high conversion (68–91%) and B/L ratio (68 : 32–91 : 9) can be obtained at 120 °C. This catalyst could be recycled for six times in IL [Bmim]NTf_2 without obvious activity loss and detectable metal leaching.

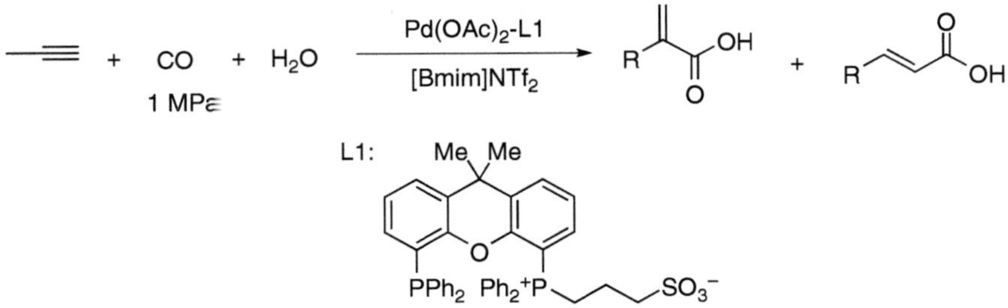

Scheme 1.26 Pd catalyst for the hydroxycarbonylation of terminal alkynes. Source: Based on Yang et al. [59]

1.2.3 Alkoxycarbonylation of Alkenes

Since the first observation by Heck that $Pd(OAc)_2$ can substitute a hydrogen atom in ethylene by a carbomethoxy group [60], Stille and James have discovered that the [Pd–Cu] couple catalyzes the incorporation of a COOMe group arising from CO and methanol [61]. Most of the reactions with an alkene end up with a diester or a methoxyester, copper being used in stoichiometric quantities. For example, in the presence of 3 atm of CO at 28 °C, norbornene yielded 98% diesters products (Scheme 1.27).

Other Pd catalyst systems have been used for the bisalkoxycarbonylation of alkenes; however, the formation of by-products in the Pd-reoxidation process decreases ester yields dramatically [51]. In the absence of an oxidation agent, the reaction is derived to monoalkoxycarbonylation provided the Pd metal center is

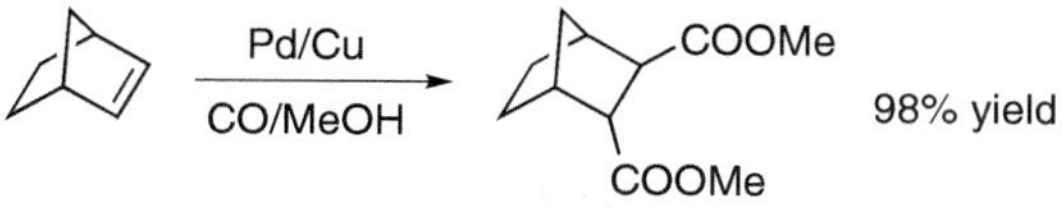

Scheme 1.27 An example of methoxycarbonylation of a cyclic alkene leading to a diester.

stabilized by surrounding ligands. This strategy, first illustrated in 1976 by Knifton with the complex [$PdCl_2(PPh_3)_2$], has been extensively developed since then [62, 63].

In 1984, Semmelhack and coworker reported the first attempts applying mild conditions (25 °C) in intramolecular alkoxycarbonylation of alkenes (Scheme 1.28) [64]. Reactions were under 1 bar of CO in the presence of $PdCl_2$ with $CuCl_2$ as a stoichiometric oxidant; the high yields in 65–88% can be obtained.

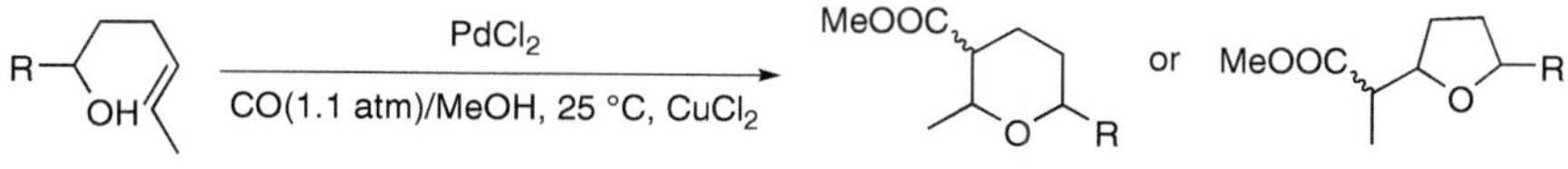

Scheme 1.28 Intramolecular alkoxycarbonylation of alkenes. Source: Based on Semmelhack and Bodurow [64].

The use of heterogeneous catalysts in this reaction has also been achieved: palladium–montmorillonite clays [65] or palladium/activated carbon [66] in the presence of dppb transformed 2-allylphenols into lactones and the regioselectivity of the reaction being largely dependent on the nature of the support.

In 2005, Alper and Touzani used palladium complexes immobilized onto silica-supported (polyaminoamido) dendrimers as catalysts in the presence of dppb for the cyclocarbonylation of 2-allylphenols, 2-allylanilines, 2-vinylphenols, and 2-vinylanilines affording five-, six-, or seven-membered lactones and lactams [67]. Good conversions (up to 100%) are realized at 120 °C, and the catalyst can be recycled 3–5 times.

In 2017, Beller's group designed and synthesized palladium catalyst based on 1,2-bis((*tert*-butyl(pyridin-2-yl)phosphanyl)methyl)benzene L3 (py^{t}bpx) for alkoxycarbonylation of alkenes (Scheme 1.29) [68]. Under 40 bar CO and at 120 °C, the isolated yields in 65–99% can be obtained. In addition, industrially relevant bulk ethylene is functionalized with high activity (TON: >1 425 000; TOF: 44 000 h^{-1} for initial 18 h) and selectivity (>99%).

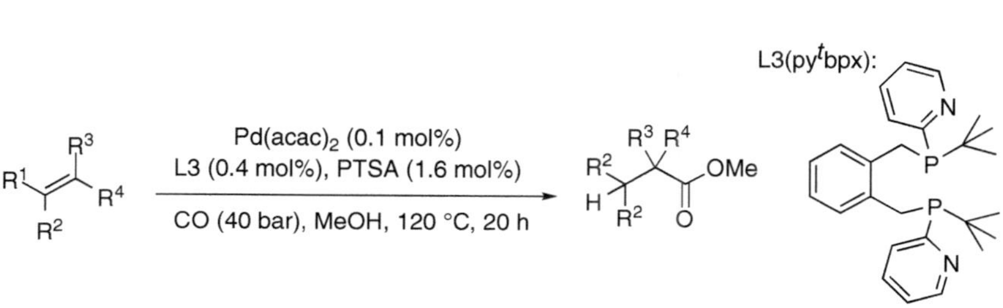

Scheme 1.29 Methoxycarbonylation of various alkenes. Source: Modified from Dong et al. [68].

1.2.4 Alkoxycarbonylation of Alkynes

The first observations, done by Tsuji et al. in 1980 on substituted acetylenes (Scheme 1.30) [69], were reinvestigated by Brandsma and coworkers in 1994 and concern the alkoxycarbonylation of these substrates to product acetylenic esters [70]. The yields in 55–84% can be obtained at room temperature.

$$\mathrm{RC{\equiv}CH + CO + R'OH \xrightarrow[\text{NaOAc, r.t.}]{\text{PdCl}_2\text{(catal.), CuCl}_2} RC{\equiv}C{-}COOR'}$$

Scheme 1.30 Alkoxycarbonylation of alkynes. Source: Modified from Tsuji et al. [69].

In 1997, Alper and coworker reported that CO insertion occurs prior to allene insertion leading to methylene- or vinyl-benzopyranone derivatives [71]. On the contrary, insertion of alkynes precedes insertion of CO, affording coumarine derivatives, as reported by Larock. This unusual selectivity can be explained by the inability of the acyl palladium species to further react with the alkyne; hence, the decarbonylation step occurs preferentially [72–74].

Dialkoxycarbonylation has been reported using a Pd-catalyst/oxidant system on propynols or butynols furnishing, respectively, β- or γ-lactone derivatives with α-(alkoxycarbonyl)ethylene chains (Scheme 1.31) [75–77].

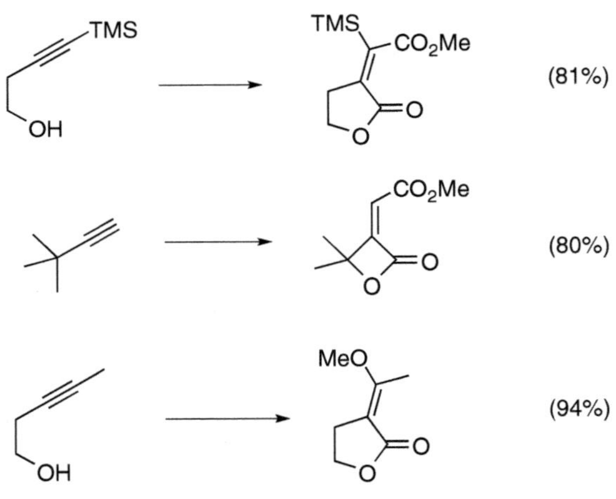

Scheme 1.31 Double carbonylation of various butynols or propynols. Source: Bonardi et al. [75]; Gabriele et al. [76]; Tamaru et al. [77].

This reaction occurs in a stereospecific way leading exclusively to cis-dicarbonylated products in fair to excellent yields (25–97%). Noteworthy, a butynol bearing an alkyl or an aryl substituent instead of a tetramethylsilane one undergoes a different course of reaction under the same conditions that trans-alkoxycarbonylation takes place selectively (Scheme 1.32).

Recently, Liu and coworkers developed ionic tri-dentate phosphine (L2′) enabled Pd-catalyzed alkoxycarbonylation of alkynes with H_2O as an additive instead of acid

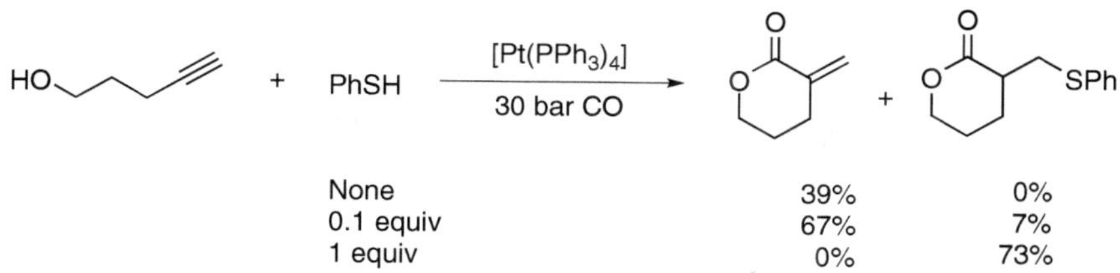

Scheme 1.32 Cyclocarbonylation of an alkynol in the presence of thiols.

(Scheme 1.33) [78]. High conversion (64–90%), complete selectivity, and B/L ratio (83 : 17–95 : 5) can be obtained at 110 °C. This catalyst could be recycled for seven runs without obvious activity loss or metal leaching.

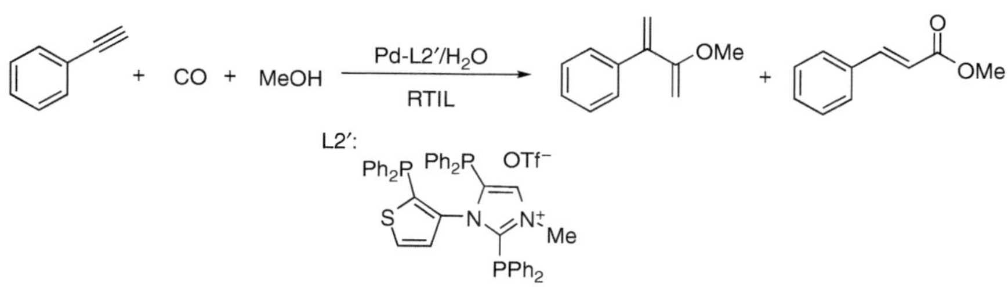

Scheme 1.33 Methoxycarbonylation of phenylacetylene. Source: Yang et al. [78].

1.2.5 Aminocarbonylation of Alkenes

In 2003, Sasai and coworkers reported the first report of an enantioselective intramolecular aminocarbonylation (Scheme 1.34) [79]. The reaction of alkenyl amine derivatives in the presence of Pd(II)-SPRIX catalyst and *p*-benzoquinone (BQ) in methanol under a CO atmosphere affords ester compounds in good yield (47–95%) with moderate enantioselectivity (10–65%).

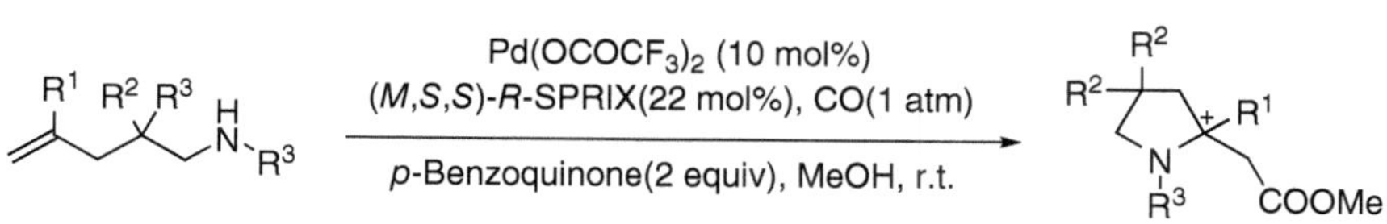

Scheme 1.34 Intramolecular aminocarbonylation of alkenyl amine derivatives promoted by Pd(II) catalyst. Source: Shinohara et al. [79].

In 2013, Beller and coworkers developed $Pd(acac)_2$ catalyst system for aminocarbonylation of alkenes with (hetero)aromatic amines to alkanamides (Scheme 1.35) [80]. For example, aminocarbonylation of ethylene with aniline can get 98% yield of *N*-phenylpropionamide in the presence of 40 bar CO at 100 °C.

In 2014, Liu et al. described a simple, homogeneous PdX_2/tris(2-methoxyphenyl) phosphine system (X = Cl or Br) for the aminocarbonylation of alkenes

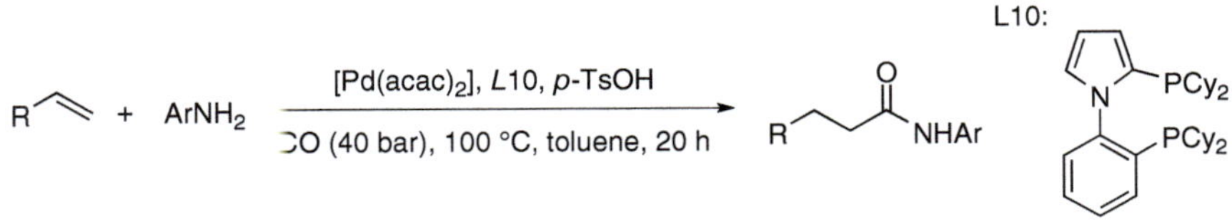

Scheme 1.35 Pd(acac)$_2$ catalyzed aminocarbonylation of alkenes with amines. Source: Fang et al. [80].

(Scheme 1.36) [81]. The wide range of aromatic amines can be efficiently transformed in good yield (27–98%) and usually with high regioselectivity (>99%) for branched carboxamides. Notably, the catalyst does not require acid, base, or any other promoters and employs a commercially available bulky monophosphine ligand thus facilitating reaction scale-up.

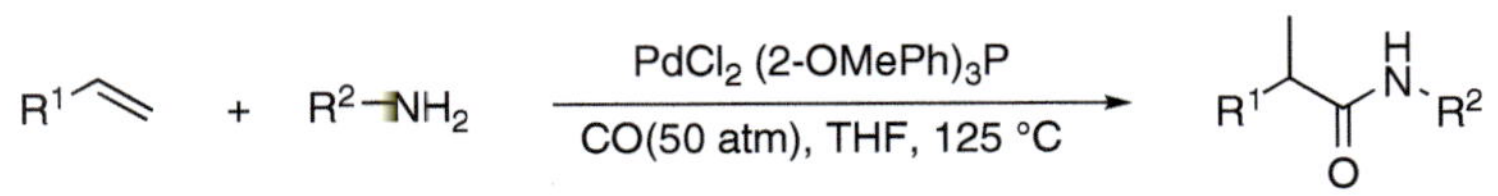

Scheme 1.36 PdCl$_2$ catalyzed aminocarbonylation of alkenes with amines. Source: Liu et al. [81].

Later, a novel convenient methods to generate β-amino acid derivatives from simple alkenes was developed (Scheme 1.37) [82]. The palladium catalyst can be used in the intermolecular aminocarbonylation of broad substrate scope with 1 atm CO at room temperature, providing their corresponding products in excellent yields (41–92%).

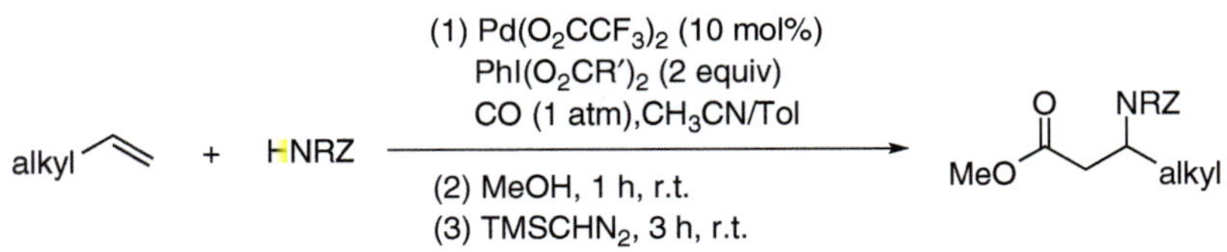

Scheme 1.37 Pd(O$_2$CCF$_3$)$_2$ catalyzed aminocarbonylation of alkenes with amines. Source: Modified from Cheng et al. [82].

In 2015, a cooperative catalytic system combining of Pd(TFA)$_2$/ligand, paraformaldehyde, and acid was established for hydroaminocarbonylation of alkenes by Huang and coworkers (Scheme 1.38) [83]. This catalyst system allows the synthesis of a wide range of *N*-alkyl linear amides in 66–88% yields with high regioselectivity (*n*/*iso* = 80 : 20–97 : 3) and overcomes the difficulty of using aliphatic amines.

In 2018, Shi and Wang described a bulk Pd catalyst for aminocarbonylation of a wide range of olefins with amines under organic ligand-free conditions (Scheme 1.39) [84]. Several kinds of aromatic and aliphatic alkenes with aromatic amines were converted into desired products in 70–99% yields and moderate

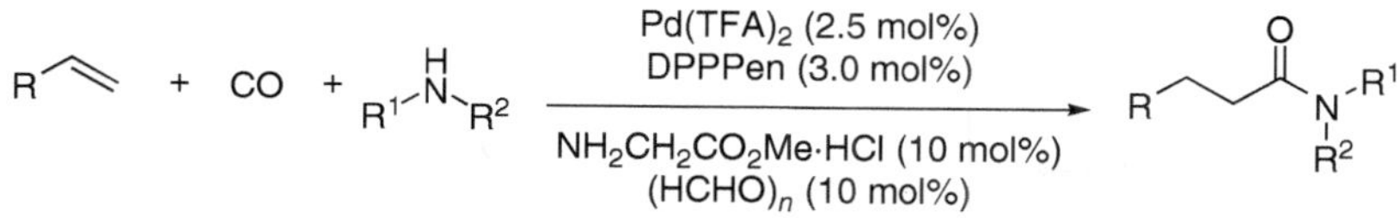

Scheme 1.38 $Pd(TFA)_2$ catalyzed aminocarbonylation of alkenes with amines. Source: Zhang et al. [83].

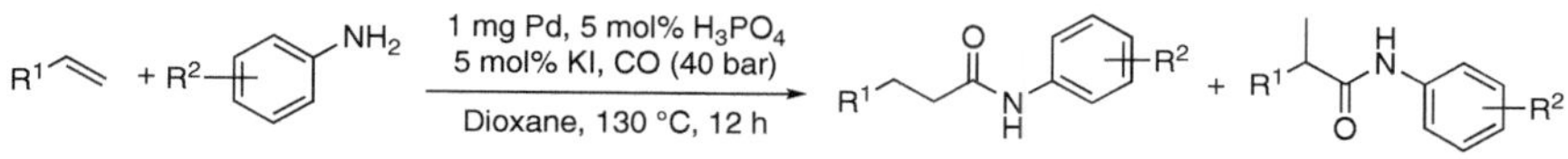

Scheme 1.39 Bulk Pd catalyzed aminocarbonylation of alkenes with amines. Source: Liu et al. [84].

regioselectivities ($n/iso = 50:50–71:29$). The bulk Pd catalyst can be reused at least three times without deactivation. This catalyst can also be applied to the aminocarbonylation reaction of aryl iodides with amines and the oxidative carbonylation reaction of a variety of amines for symmetric urea synthesis.

1.2.6 Aminocarbonylation of Alkynes

Since the pioneering work of Reppe [85], various organometallic catalysts and synthetic procedures have been explored for the carbonylation of alkynes. For example, Yu and Alper reported $Pd(OAc)_2$-catalyzed aminocarbonylation of alkynes in IL [bmim][Tf_2N] without any acid additive under relatively mild conditions in 2006 (Scheme 1.40) [86]. Acrylamides were obtained in 26–85% yields and the catalyst system can be reused five times without loss of catalytic activity. The IL was used as the reaction medium and also acted as a promoter.

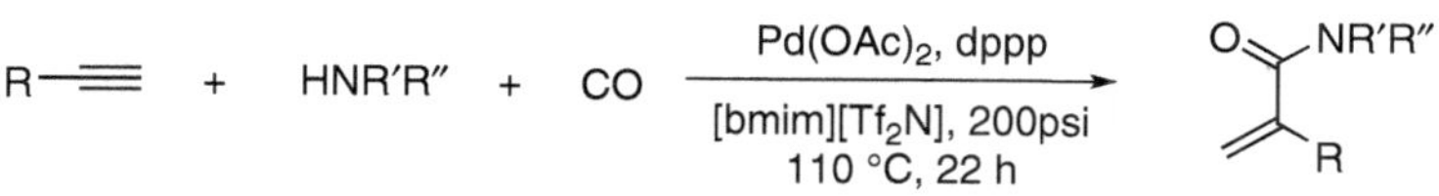

Scheme 1.40 Palladium-catalyzed aminocarbonylation of alkynes in the IL. Source: Li et al. [86].

In 2010, El Ali and coworkers reported the aminocarbonylation reaction of terminal alkynes using a catalyst system $Pd(OAc)_2$–dppb-*p*-TsOH–CH_3CN–CO under relatively mild conditions [87]. Phenylacetylene and various amines can be successfully transformed into the corresponding products with up to 88% yield (Scheme 1.41). The obtained DFT/B3LYP results indicate the formation of trans ester isomer via palladium–hydride cycle, while palladium–amine mechanism is the key for the formation of gem isomer [88]. The results prove the role of acetonitrile solvent in improving the stability of catalytic intermediates.

In 2011, Beller and coworkers developed the first general iron-catalyzed aminocarbonylation of alkynes [89]. Starting from commercially available amines and alkynes, a range of structurally diverse amides are obtained smoothly in the

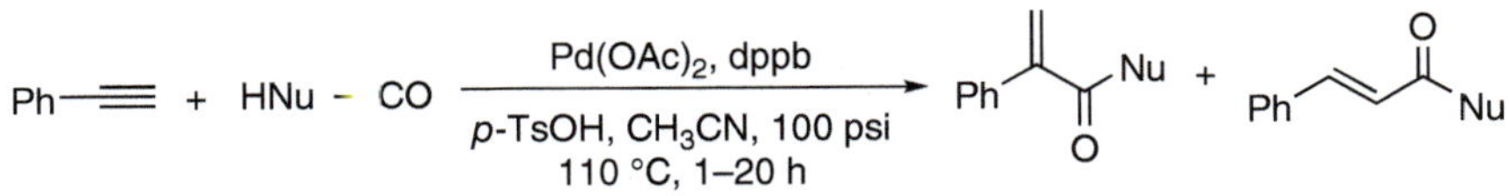

Scheme 1.41 Palladium(II)-catalyzed aminocarbonylation of phenylacetylene using various amines.

presence of catalytic amounts of $[Fe_3(CO)_{12}]$ and L2. For example, mono carbonylation of phenylacetylene with cyclohexylamine can get 66% yield of product at 120 °C (Scheme 1.42). Notably, the method is highly chemo- and regioselective, and no expensive catalyst is required for this novel environmentally friendly reaction.

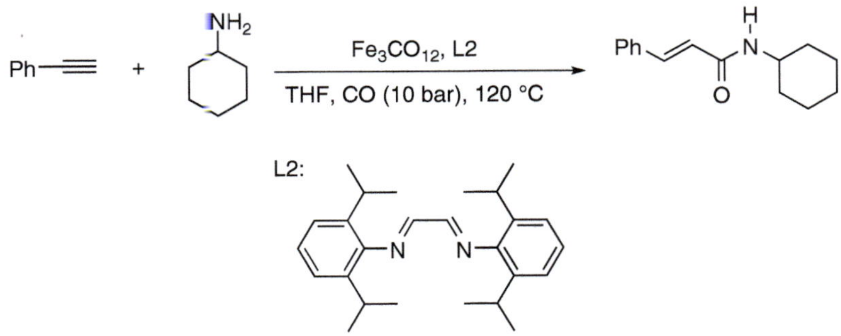

Scheme 1.42 Iron-catalyzed mono carbonylation of phenylacetylene.

In 2017, Alper and coworker reported $Pd(OAc)_2$/dppp-catalyzed aminocarbonylation reactions of alkynes and aminophenols provide direct OH-substituted α,β-unsaturated amides [90]. By a simple change of the ligand and the additives, branched or linear amides were formed in excellent chemo- and regioselectivities (up to b/l > 99 : 1) and in good to excellent yields.

In 2019, Li and coworkers utilized ZrF_4 to promote efficiently iron-catalyzed aminocarbonylation between alkynes and amines without the use of extra ligands (Scheme 1.43) [91]. In the presence of cheap and readily prepared ZrF_4 at 120 °C, various aromatic and aliphatic alkynes including internal alkynes were transformed into the desired α,β-unsaturated amides in 65–93% yields.

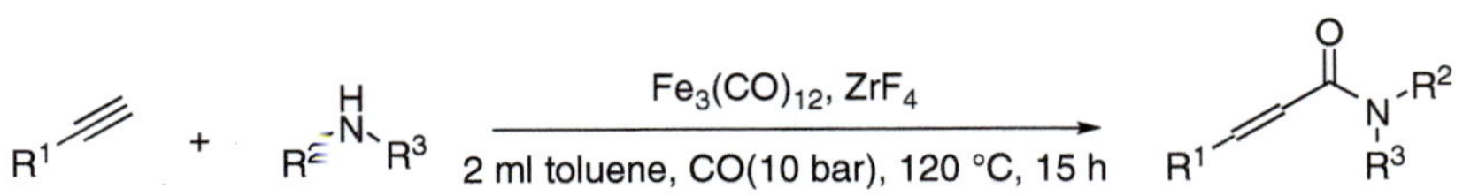

Scheme 1.43 ZrF_4 as co-catalyst promoted iron-catalyzed aminocarbonylation. Source: Modified from Huang et al. [91].

1.3 The Pauson–Khand Reaction

The transition metal-mediated conversion of alkynes, alkenes, and CO in a formal [2 + 2 + 1] cycloaddition process, commonly known as the Pauson–Khand reaction

(PKR), is an elegant method for the construction of cyclopentenone scaffolds. During the last decade, significant improvements have been achieved in this area [92, 93]. For instance, catalytic PKR variants are nowadays possible with different metal sources. In addition, new asymmetric approaches were established, and the reaction has been applied as a key step in various total syntheses. Recent work has also focused on the development of CO-free conditions, incorporating transfer carbonylation reactions. This section attempts to cover the most important developments in this area.

1.3.1 The Catalytic Pauson–Khand Reaction

In 1973, Pauson and Khand reported the first successful example for the conversion of norbornene with the phenylacetylene–hexacarbonyldicobalt complex to give the corresponding cyclopentenone in 45% yield (Scheme 1.44) [94]. With regard to environmental friendliness, catalytic procedures are desirable.

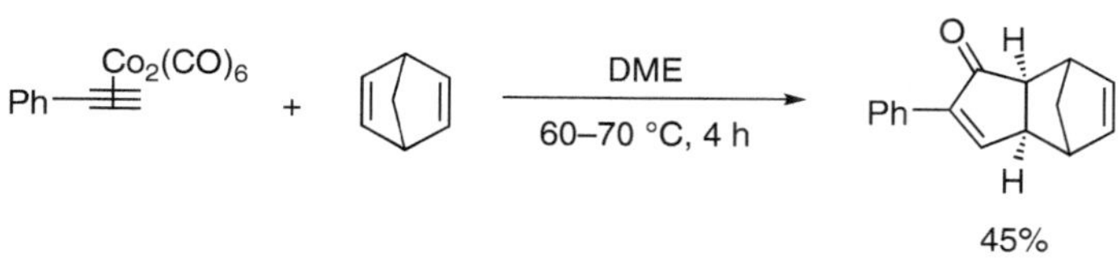

Scheme 1.44 Conversion of norbornene with the phenylacetylene–hexacarbonyldicobalt complex. Source: Modified from Khand et al. [94].

In 1990, Rautenstrauch et al. reported the first catalytic PKR using $Co_2(CO)_8$ as a catalyst [95], who described the conversion of heptyne using a mixture of 40 bar ethylene and 100 bar CO to form the resulting cycloaddition adduct in a yield of 48%. A more practicable procedure was later reported by Jeong et al. [96]. With triphenylphosphite as a ligand, the PKR in 51–94% yields can be obtained at 3 atm of CO and 100 °C by stabilization of the cobalt catalyst.

Apart from $Co_2(CO)_8$, other Co sources, i.e., $Co_3(CO)_9(\mu^3\text{-CH})$, were also successfully applied within the reaction [97]. Lee and Chung [98] examined the use of $Co(acac)_2/NaBH_4$ as a cycloaddition catalyst (Scheme 1.45). In this case, yields were observed in a range of 33–85%. Successful conversions need high CO pressures (40 atm) and long reaction times (up to several days).

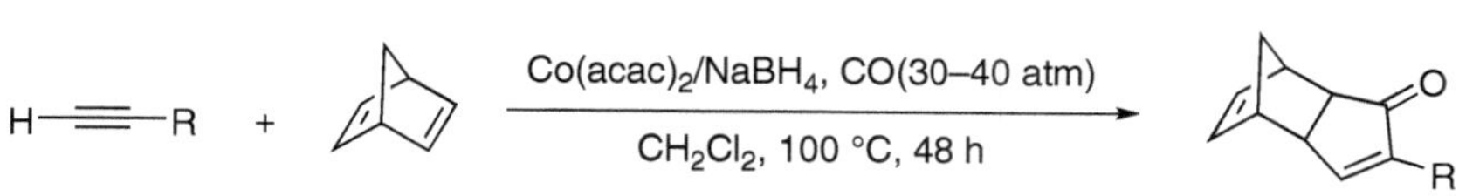

Scheme 1.45 Use of $Co(acac)_2/NaBH_4$.

The first example of a catalytic conversion under atmospheric pressure CO was reported by Rajesh and Periasamy (Scheme 1.46) [99]. A combination of 0.4 equiv

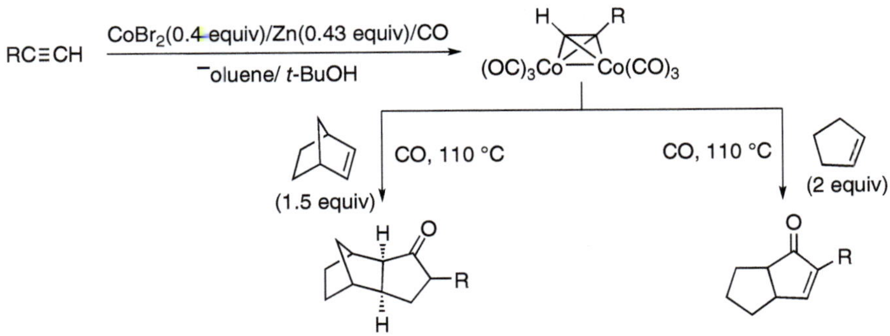

Scheme 1.46 Catalytic conversion under atmospheric pressure of carbon monoxide. Source: Rajesh and Periasamy [99].

$CoBr_2$ and 0.43 equiv Zn was chosen to facilitate the intermolecular PKR. Isolated yields were obtained in the range of 30–88% at 110 °C.

Supplementary to reactions in organic solvents, procedures applying supercritical fluids have become more and more attractive in industry and academic research. Jeong's group reported the use of supercritical CO_2 as an alternative solvent [100]. Using $Co_2(CO)_8$ as a catalyst, 112 atm of CO_2 and 15–30 atm of CO were required at 37 °C to obtain cyclopentenones in yields up to 91%.

In addition to the accelerating effect of sulfides and amines in the stoichiometric $Co_2(CO)_8$-mediated procedure, Hashimoto and coworkers showed that phosphane sulfides are able to promote the $Co_2(CO)_8$-catalyzed reaction [101]. For example, at atmospheric CO pressure and moderate temperature of 70 °C, the desired product was isolated in 92% yield (Scheme 1.47).

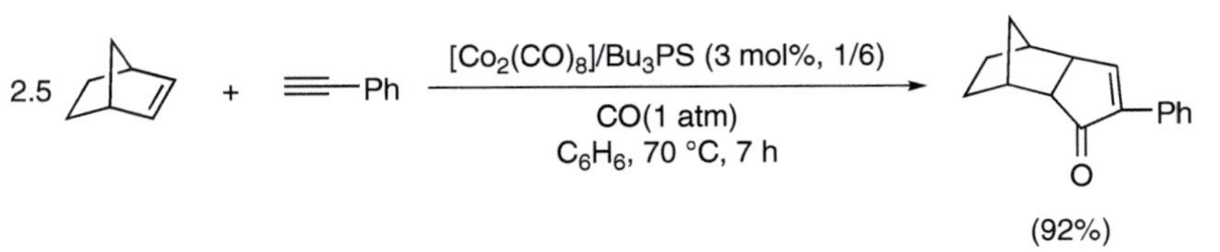

Scheme 1.47 Phosphane sulfides promoted the $Co_2(CO)_8$-catalyzed reaction.

Chung and coworkers developed an easy system that is based on the immobilization of 12 wt% metallic cobalt on commercially available charcoal [102]. The resulting yields were obtained in the range of 61–98%. The catalyst was reused up to 10 times without significant loss of activity after simple filtration.

Although Co carbonyl complexes have been used most often in catalytic PKR, a variety of other transition metal complexes are able to catalyze this reaction. Buchwald and coworkers has shown that 5–20 mol% $Cp_2Ti(CO)_2$ facilitates the PKR at 18 psi CO and 90 °C, giving yields in between 58% and 95% [103].

Mitsudo and coworkers [104] and Murai and coworkers [105] reported independently on the employment of $Ru_3(CO)_{12}$ as an active catalyst. Cyclopentenones were isolated in 41–95% yields. In addition, rhodium catalysts were successfully examined for use in the PKR.

Narasaka and coworkers [106] carried out reactions at atmospheric CO pressure using the dimeric $[RhCl(CO)_2]_2$ complex. Also, in the presence of other rhodium complexes like Wilkinson catalyst $RhCl(PPh_3)_3$ and $[RhCl(CO)(dppp)]_2$ [107] in combination with silver salts, cyclopentenones were obtained in yields in the range of 20–99%.

The pioneering work of Krafft and coworkers revealed that the introduction of coordinating heteroatoms, such as nitrogen or sulfur, attached to the alkene leads to a strong enhancement of reactivity and regioselectivity [108–110]. Hence, further work was dedicated to developing an easily removable directing group, which would open the way to increasing the synthetic potential of the intermolecular PKR.

In this connection, Itami et al. developed an interesting system, which included a 2-pyridyldimethylsilyl group tethered to the alkene part (Scheme 1.48) [111, 112]. As a major advantage, this directing group enabled the regioselective synthesis of cyclopentenones at the 4- and 5-position. The pyridylsilyl group is removed after product formation caused by residual water. $Ru_3(CO)_{12}$ gave the best results even under low pressure of 1 atm CO. Yields were obtained in the range of 40–91%.

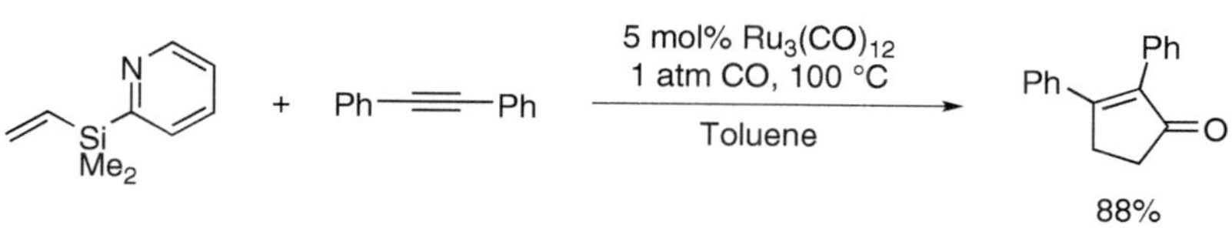

Scheme 1.48 2-Pyridyldimethylsilyl group tethered to the alkene part. Source: Itami et al. [111]; Itami et al. [112].

Because of the remarkable increase of molecular complexity, the PKR serves as a useful methodology for various natural product syntheses. Numerous examples of synthetic applications have been reported. For instance, the PKR was successfully employed in the total syntheses of β-cupraenone [113], loganin [114], hirsutene [115], and (+)-epoxydictymene [116].

1.3.2 Stereoselective Pauson–Khand Reactions

The virtue of performing the PKR in an enantioselective manner has been extensively elaborated. Different powerful procedures were developed, spanning both auxiliary-based approaches and catalytic asymmetric reactions. For instance, the use of chiral N-oxides was reported by Kerr and coworkers, who examined the effect of the chiral brucine N-oxide in the intermolecular PKR of propargylic alcohols and norbornadiene [117]. Under optimized conditions, ee values up to 78% at −60 °C have been obtained (Scheme 1.49). Chiral sparteine N-oxides are also able to induce chirality, but the observed enantioselectivity was comparatively lower [118]. The application of chiral auxiliaries is an alternative route to obtain enantiomerically

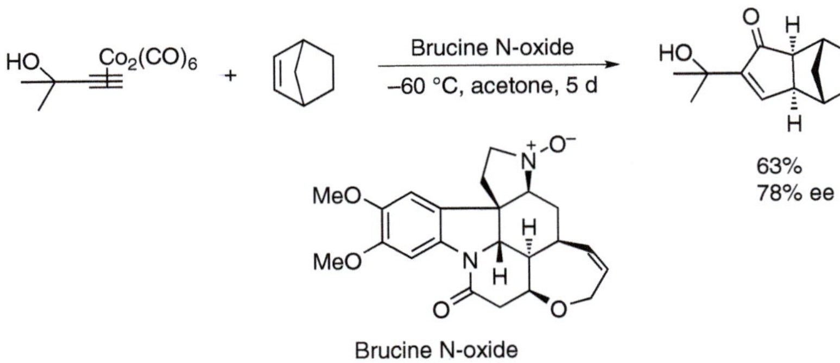

Scheme 1.49 Chiral brucine N-oxide in the intermolecular PKR of propargylic alcohols and norbornadiene.

pure compounds. This approach has been frequently used in the total syntheses of natural products like hirsutene [115] and (+)-15-norpentalene [119].

In 1996, Buchwald and Hicks reported the first successful results that the chiral titanocene complex (*S*,*S*)-(EBTHI)Ti(CO)$_2$ is a useful system for the synthesis of enantiomerically pure cyclopentenones [120, 121]. Various enynes were converted into corresponding product in 72–96% yields with 70–94% ee under 14 psig CO pressure at 90 °C. In addition, RhCl(CO)$_2$ in combination with AgOTf and (*S*)-BINAP was reported to facilitate enantioselective PKR [122]. Values of ee were observed in between 22% and 99%, and isolated yields were in the range of 22–96%.

Moreover, it was shown that chiral iridium diphosphine complexes catalyze such cycloaddition reactions [123]. Using a comparatively large amount of 10 mol% [Ir(cod)Cl]$_2$, reactions afforded enantiomerically enriched cyclopentenones when (*S*)-tolBINAP was employed as a ligand. The corresponding cyclopentenones were obtained in yields up to 85% and ee values ranging from 82% to 98%. Interestingly, $Co_2(CO)_8$ in combination with a chiral bisphosphite also gives access to chiral Pauson–Khand products [124]. Although yields were observed up to 97%, in most cases, the ee was rather low (<20%).

An interesting advance in expanding the scope of the alkene component of the PKR came from Lledo and Riera [125]. The team developed a cobalt-mediated PKR between alkynes and *trans*-cycloalkenes. The scope was established for the reaction between symmetric and terminal alkynes and medium-sized *trans*-cyclooalkenes (Scheme 1.50).

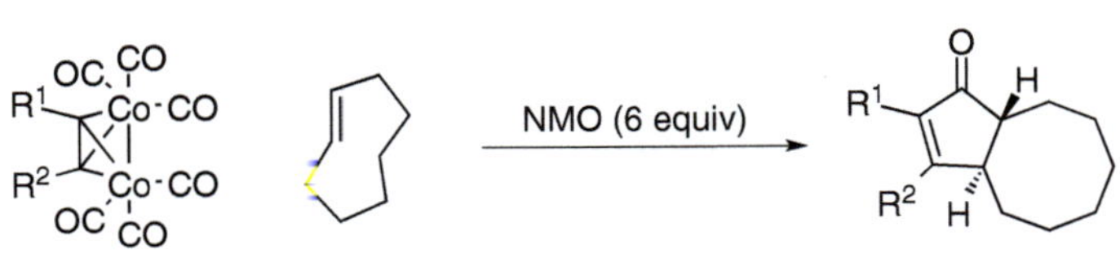

Scheme 1.50 Application of (*E*)-cyclooctene in the PKR.

In 2016, Nordlander and coworkers applied tris-μ carbyne cobalt pre-catalysts to the intermolecular PKR. While they were able to use the catalyst in 2 mol%, yields were modest and enantioselectivity was poor (Scheme 1.51) [126]. Additionally, they used dicobalt octacarbonyl and tetracarbonyl dodecacarbonyl within the PKR and exogenous chiral phosphine ligands and obtained excellent yields, though again no ee.

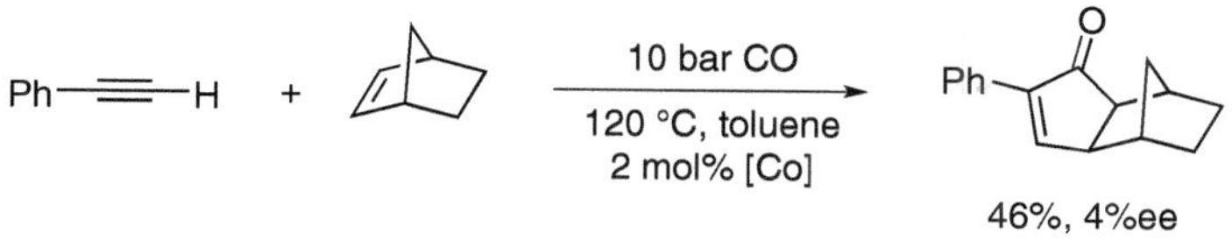

Scheme 1.51 Trinuclear cobalt–WalPHOS-catalyzed coupling of phenylacetylene and norbornene. Source: Modified from Mottalib et al. [126].

1.3.3 Pauson–Khand Transfer Carbonylation Reactions

The PKR has been developed as a straightforward and practicable method for the synthesis of highly substituted cyclopentenones. But for many synthetic chemists, the employment of poisonous CO still represents a disadvantage. Hence, different strategies focused on the replacement of CO within the reaction sequence (Scheme 1.52).

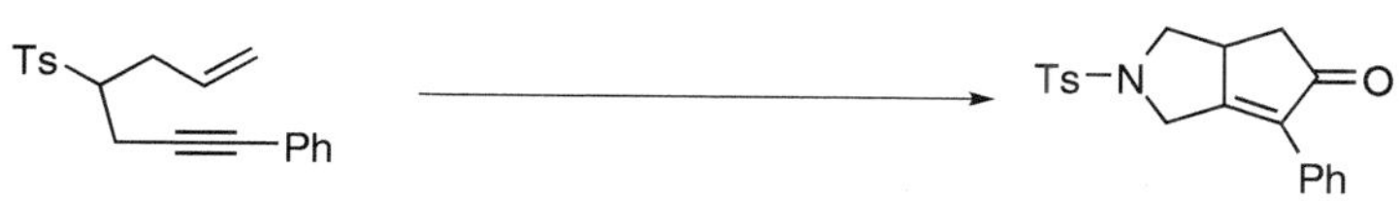

conditions:

a: 5 mol% $[RhCl(cod)]_2$, 11 mol% dppp, 2 equiv C_6F_5CHO, xylene, 4h, 130 °C [127]

b: 5 mol% $Rh(dppp)_2Cl$, 20 equiv cinnamaldehyde, 2 h, 120 °C [128]

c: 5 mol% $[RhCl(cod)]_2$, 10 mol% dppp, 10 mol% TPPTS, 2 equiv SDS, 10 eq HCHO, 2 h, refluxing water [129]

Scheme 1.52 Pauson–Khand transfer carbonylation reactions.

Morimoto, Kakiuchi, and coworkers were the first to show that aldehydes are a useful source of CO in the catalytic PKR [127]. Based on ^{13}C-labeling experiments, it was proposed that after decarbonylation of the aldehyde, an active metal catalyst is formed. The best results were obtained using C_5F_5CHO and cinnamaldehyde as a CO source in combination with $[RhCl(cod)]_2$/1,3-bis(diphenylphosphino)propane (dppp) as the catalyst system. In the presence of an excess of aldehyde, the corresponding products were isolated in the range of 52–97%.

Later, Shibata et al. reported PK-type transfer carbonylation reactions in the absence of solvent and employing a large excess of aldehyde [128]. Also in this case, the reactions went smoothly yielding cyclopentenones in the range of 56–96%. Interestingly, it was also demonstrated that this methodology provides access to chiral products. By applying the chiral catalyst system $[RhCl(cod)]_2$/tolBINAP, ee values of 45–90% were observed.

Another convincing approach to carrying out PK-type transfer carbonylation reactions was again reported by Kakiuchi and coworkers [129]. Formaldehyde was employed as a decarbonylating agent in the presence of a micelle-containing aqueous system. It was proposed that the decarbonylation and carbonylation reactions take place in different phases. More precisely, the water-soluble formaldehyde is decarbonylated in the aqueous phase, enabled by the $[RhCl(cod)]_2$/triphenylphosphine-3,3′,3″-trisulfonic acid trisodium salt (TPPTS) system, while carbonylation takes place in a micelle formed by the surfactant sodium dodecyl sulfate. Also, in this case, the yields obtained were good to quantitative (67–96%).

In 2019, He and coworkers have shown the feasibility of the rhodium-catalyzed PK-type reaction using formic acid for the conversion of CO_2 to CO to prevent the arduous management of gaseous reagents (Scheme 1.53) [130]. A variety of 1,6-enyne substrates bearing electron-withdrawing or electron-donating substituents were smoothly converted to the corresponding cyclopentenones in 49–86% yields.

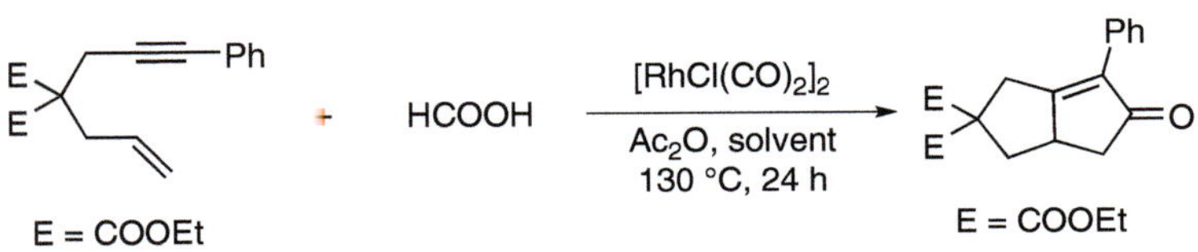

Scheme 1.53 Pauson–Khand transfer carbonylation reactions. Source: Lang et al. [130].

1.4 Synthesis of Acetic Acid

Acetic acid (AA) synthesis via the carbonylation of methanol is one of the most important industrial applications of catalysis using organometallic compounds. The processes based on cobalt, rhodium, and iridium have been developed since the 1960s. Methanol carbonylation comprises more than 60% of the world AA production [131]. The most significant development is the commercialization by BP Chemicals of the iridium-based Cativa™ process. Enhanced stability of iridium catalysts also allows operation at lower water concentration.

1.4.1 Process Considerations

The carbonylation reaction (1) involves formal insertion of CO into the C–O bond of methanol. In a working catalytic system, however, the principal solvent component is AA, so esterification (2) leads to substantial conversion of the substrate into methyl acetate. Methyl acetate is activated by reaction with the iodide co-catalyst (3). Transition metal-catalyzed carbonylation of methyl iodide then gives acetyl iodide (4), which is rapidly hydrolyzed to the product AA (5). The net result of the reactions is described (Scheme 1.54).

As well as playing a role in some of the organic transformations, excess water is also present as a significant component of the solvent mixture, which helps

(1) $MeOH + CO \longrightarrow MeCO_2H$

(2) $MeOH + AcOH \longrightarrow MeOAc + H_2O$

(3) $MeOAc + HI \longrightarrow MeI + AcOH$

(4) $MeI + CO \longrightarrow MeCOI$

(5) $MeCOI + H_2O \longrightarrow AcOH + HI$

(6) $H_2O + CO \longrightarrow CO_2 + H_2$

Scheme 1.54 Process of catalytic carbonylation of methanol for the synthesis of acetic acid.

to maintain catalyst stability. It is found that catalyst solubility is enhanced at high water concentrations, but these results in a more costly separation process to dry the product. The presence of water also results in the occurrence of the WGS reaction (6), which can be catalyzed by Rh and Ir iodocarbonyls, in competition with the desired carbonylation process, resulting in a lower utilization of CO.

From a commercial viewpoint, potential benefits can accrue from operating the methanol carbonylation process at low water concentration, provided that catalyst stability can be maintained. Strategies to achieve this include (i) addition of iodide salts to stabilize the Rh catalyst, (ii) heterogenization of the Rh catalyst on a polymer support to restrict the catalyst to the reactor and (iii) replacement of Rh by a more robust Ir catalyst. These strategies, along with others for improving catalyst activity, will be discussed in the following sections.

1.4.2 Rhodium-Catalyzed Carbonylation

The first process for the carbonylation of methanol to AA commercialized by BASF in 1960 with cobalt/iodide catalyst under a high temperature of 250 °C and 680 bar pressure [132, 133]. The selectivity to AA of the BASF process is about 90% based upon methanol.

In 1968, Paulik and Roth of Monsanto Company announced the discovery of a low-pressure carbonylation of methanol using an iodide-promoted rhodium catalyst [134]. Under CO pressures as low as 1 atm, the selectivity to AA is virtually quantitative (>99%) at 175 °C. Then, scientists began to study the performance and reaction mechanism of rhodium catalysts or iridium catalysts.

In order to overcome the catalyst recycle step, the immobilization of the rhodium complex on a support has been the subject of considerable investigation [135]. These works include activated carbon, inorganic oxides, zeolites, polymeric materials, and so on.

Schultz prepared thermal decomposition of rhodium nitrate impregnated on activated carbon for vapor-phase carbonylation of methanol in the presence of CH_3I promoter (Scheme 1.55) [136]. The 93% conversion of methanol and 91% selectivity to AA can be obtained at 200 °C and 215 psia pressure under this catalyst system (3% rhodium content).

$$CH_3OH + CO \xrightarrow[CH_3I,\ 200\ ^\circ C]{Rh_2O_3/C} CH_3COOH + CH_3COCH_3 + H_2O$$

Scheme 1.55 Rhodium-catalyzed carbonylation of methanol. Source: Modified from Schultz [136].

In 1988, Miessner and Luecke reported the Rh/zeolite (NaX, NaY) catalyst for carbonylation of methanol to AA [137]. The good conversion (about 55%) and the high selectivity (about 90%) can be obtained at 350 °C with a maximum carbonylation activity in the range of the Si/Al ratio of 1.3–1.5.

In 2017, Bae and coworkers investigated a novel heterogeneous Rh-g-C_3N_4 catalyst for a liquid-phase carbonylation of methanol to AA to overcome disadvantages of the commercialized Rh-based homogeneous catalysts [138]. The catalyst system showed a superior catalytic activity in a liquid-phase carbonylation with AA yield above 82% under the reaction conditions of 140 °C and 4.0 MPa CO pressure.

In 2019, Christopher and coworker reported an atomically dispersed Rh catalyst on an acidic support (Al_2O_3, ZrO_2) for methanol carbonylation to AA under halide-free and gas-phase conditions [139]. When operating at 300 °C and 1 : 1 methanol to CO molar feed ratio, 0.2 wt% atomically dispersed Rh loaded on 5%Na–ZrO_2 exhibited stable reactivity over the course of 50 h with 54% selectivity to AA. It is demonstrated that active site pairs consisting of atomically dispersed Rh and support acid sites enable highly selective AA production, whereas Rh clusters drive methanol decomposition to CO and CO_2.

Ding and Lyu prepared single-atom Rh–POL–2BPY catalyst by impregnation of $Rh_2(CO)_4Cl_2$ solution of dichloromethane for methanol carbonylation [140]. Under 195 °C and 2.5 MPa, this catalyst behaved excellent carbonylation activity (TOF > 1400 h^{-1})

The catalytic cycle and mechanism of classic example of homogeneous catalytic rhodium-catalyzed carbonylation was studied (Scheme 1.56) [141]. The Monsanto catalyst system has been the subject of numerous studies [142–145]. The kinetics of the overall carbonylation process are zero order in both reactants (MeOH and CO) but first order in rhodium catalyst and methyl iodide co-catalyst. The active catalyst (identified by in situ high-pressure infrared spectroscopy [146]) is the square-planar Rh(I) complex, *cis*-$[Rh(CO)_2I_2]^-$. Oxidative addition of methyl iodide to this complex is the rate-determining step of the cycle, consistent with the observed kinetics. Rate measurements on the stoichiometric reaction of $[Rh(CO)_2I_2]^-$ with MeI have confirmed this to be second-order, with activation parameters comparable to those for the overall carbonylation process [147].

1.4.3 Iridium-Catalyzed Carbonylation

It was discovered by Monsanto that methanol carbonylation could be promoted by an iridium/iodide catalyst [134]. However, Monsanto chose to commercialize the

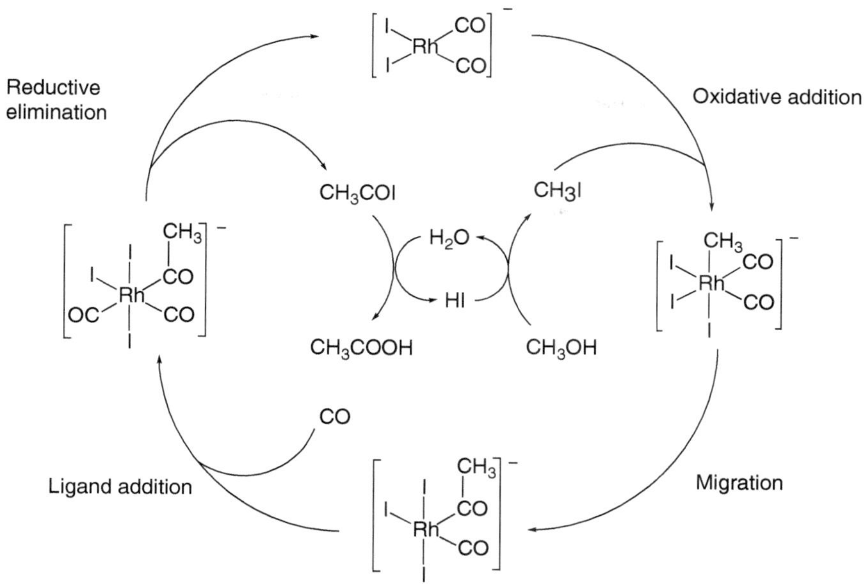

Scheme 1.56 Catalytic cycle of the rhodium-catalyzed methanol carbonylation (Monsanto process). Source: Modified from Forster [141].

rhodium-based process due to its higher activity under the conditions used. In 1996, BP Chemicals announced a new methanol carbonylation process, Cativa™, based upon a promoted iridium/iodide catalyst which now operates on a number of plants worldwide [148–150].

In 2006, Kalck and coworkers discovered $[PtI_2(CO)]_2$ or $[PtI_2(CO)_2]$ has a pronounced synergistic effect for low water content $[Ir(CH_3)I_3(CO)_2]^-$-catalyzed methanol carbonylation to AA [151]. When Pt/Ir ratio is 33 : 67, 35 mol l^{-1} h^{-1} AA can be observed under 30 bar and at 190 °C.

In 2019, Ding and Song investigated the stability of a heterogeneous single-site Ir_1–La_1/AC catalyst for vapor methanol carbonylation [152]. In the presence of CO/H_2 with 240 °C and 2.4 MPa pressure, TOF about 3700 mol_{acetyl}/mol_{Ir} h can be obtained. The La promoter could protect the Ir^+ species from reduction by H_2, favor the single-atom dispersion of Ir metal, and also increase the reaction rate of methanol carbonylation.

Mechanistic studies of iridium/iodide-catalyzed methanol carbonylation have been reported by Forster [153] and others (Scheme 1.57) [154, 155]. The catalytic cycle involves the same fundamental steps as the rhodium system: oxidative addition of MeI to Ir(I), followed by migratory CO insertion to form an Ir(III) acetyl complex. Of course, there are differences between them, the iridium-based cycle involving neutral iridium complexes, and rhodium-based cycle predominantly involving anionic species. Model kinetic studies [156, 157] have shown that the oxidative addition of methyl iodide to the iridium center is about 150 times faster than the analogous reaction with rhodium. The rate-determining step involves

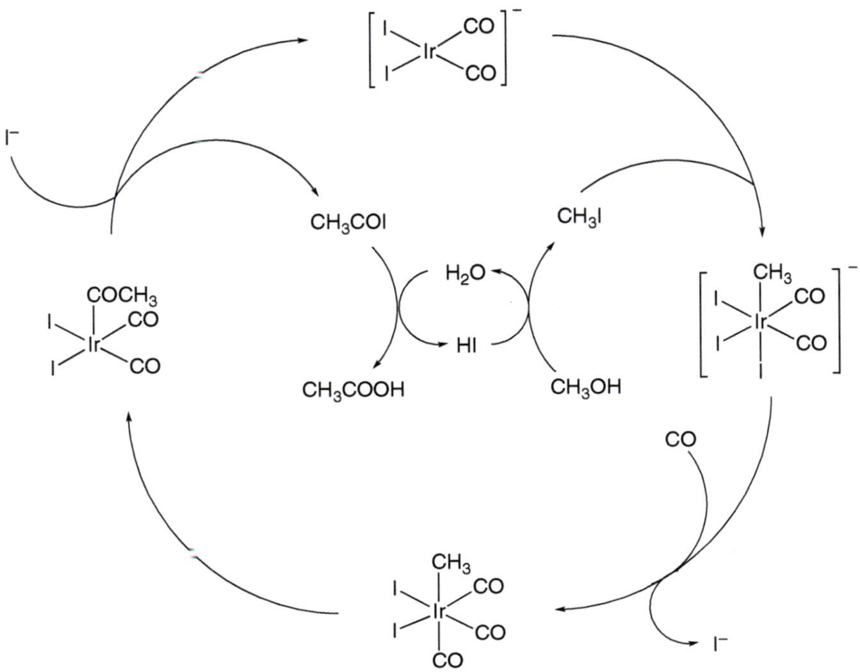

Scheme 1.57 Catalytic cycle of the iridium-catalyzed methanol carbonylation (Cativa process). Source: Matsumoto et al. [154]; Haynes et al. [155].

dissociative substitution of I^- by CO in $[Ir(CO)_2I_3Me]^-$, followed by migratory CO insertion in the tricarbonyl, $[Ir(CO)_3I_2Me]$.

A great advantage of iridium, compared to rhodium, is that a broad range of conditions are accessible for the Ir catalyst without precipitation of IrI_3 occurring. The greater stability of the iridium catalyst can be ascribed to stronger metal–ligand bonding for the third-row metal, which inhibits CO loss from the Ir center.

1.5 Carbonylation of C–X Bonds

Carbonylations of C–X bonds is the insertion of CO into C–X (X = Cl, Br, I) in the presence of nucleophiles. In Scheme 1.58, alcohols, amines, ethers, carboxylic acids, and halides can be converted into acids, amides, esters, ketones, alkynones, alkenones, anhydrides, and acid halides with the assistance of transition metal catalysts in the presence of CO source [158, 159].

1.5.1 Hydroxy-, Alkoxy-, and Aminocarbonylations of C–X Bonds

Transition metal-catalyzed carbonylative activation of C–X bonds with nucleophiles such as water, alcohols, or amines are called hydroxycarbonylation, alkoxycarbonylation, or aminocarbonylation, respectively. From a mechanism point of few, the

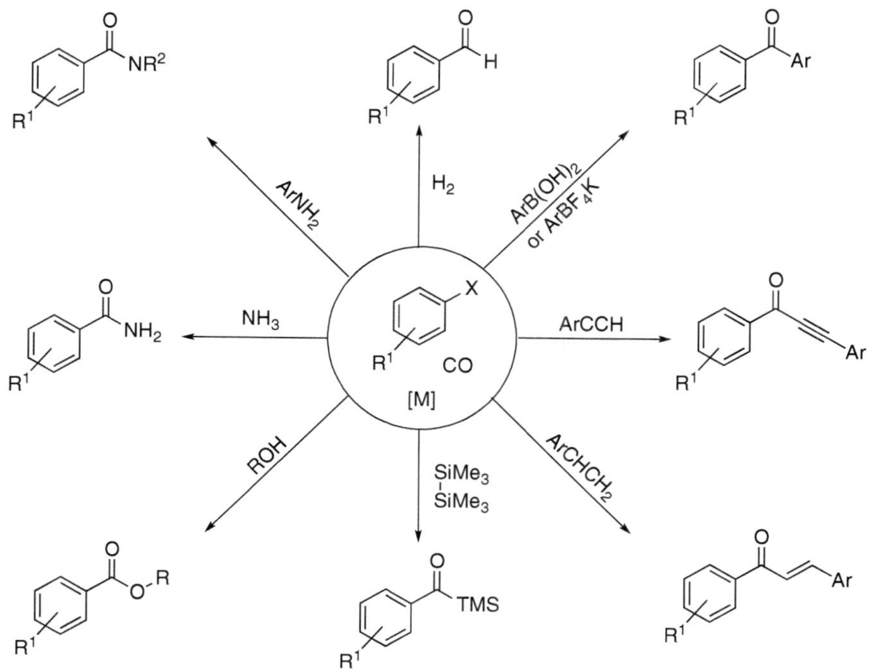

Scheme 1.58 Transition metal-catalyzed carbonylation reactions of C–X bonds.

catalytic cycles for these reactions end with the nucleophilic attack of nucleophiles with an acylpalladium complex and produce carboxylic acids, esters, and amides as their terminal products.

As early as in 1988, Bumagin et al. utilized palladium catalyst ($Pd(OAc)_2$, K_2PdCl_4, $PdCl_2(PPh_3)_2$, and $Pd(NH_3)_4Cl_2$) for hydroxycarbonylation of ArI with CO and H_2O [160]. In the presence of palladium catalyst and base (K_2CO_3 or NaOAc) under 1 atm CO at 25 °C, the product of yield with up to 100%.

In 1999, Uozumi and coworker reported palladium-catalyzed hydroxycarbonylation of aryl or alkenyl halides, which was performed in water in the presence of an amphiphilic solid-supported phosphine–palladium complex (Pd–PEP) [161]. For example, aryl halides can get the corresponding carboxylic acids in 45–100% yields under mild reaction conditions and an atmospheric pressure of CO (Scheme 1.59).

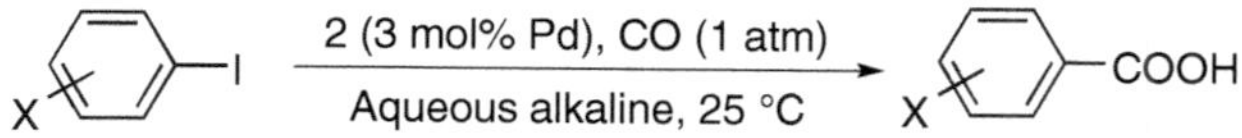

Scheme 1.59 Hydroxycarbonylation of aryl halides in water.

In 2010, Reiser and coworkers developed the reversible immobilization of pyrene-tagged palladium NHC complexes on highly magnetic, graphene-coated cobalt nanoparticles for hydroxycarbonylation of aryl halides in water (100 °C)

[162]. For example, hydroxycarbonylation of 4-iodophenol can yielded 95% product in the presence of K_2CO_3 and 1 atm CO. This catalyst can be recycled 10 times without the significantly decrease amount of nanoparticle-bound palladium.

Heck and colleagues described the first palladium-catalyzed alkoxycarbonylation reaction in 1974 [163]. Carboxylic acid *n*-butyl esters were synthesized from aryl and vinyl iodides and bromides after they reacted with CO (1 bar) in *n*-butanol at 60–100 °C. In the presence of $Pd(OAc)_2$ or $PdX_2(PPh_3)_2$ complexes with a slight excess of tri-*n*-butylamine as a base, 4–83% yields of the corresponding esters were obtained.

Notable progress with regard to catalyst productivity was achieved by Beller's group [164]. Butyl ester was achieved at low pressure (5 bar CO) and 100 °C in the presence of only 0.3 mol% $Pd(PPh_3)_4$ and 3 equiv of Et_3N in *n*-butanol. The optimization resulted in the high turnover (TON up to 7000) for alkoxycarbonylation of aryl halides (Scheme 1.60)

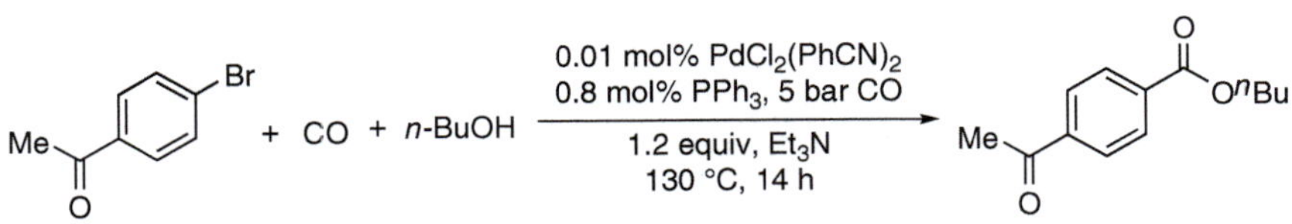

Scheme 1.60 Palladium-catalyzed alkoxycarbonylation of aryl bromides.

Beller's group developed a general palladium-catalyzed carbonylation of aryl and heteroaryl bromides with phenols. The reaction proceeds smoothly in the presence of di-1-adamantyl-*n*-butylphosphine under 2 bar of CO in 1,4-dioxane at 100 °C. Later on, the same group developed a one-pot alkoxycarbonylation of phenols with alcohols and phenols via the in situ formation of ArONf. The reaction proceeds selectively to the desired benzoates in good yields (Scheme 1.61) [165, 166].

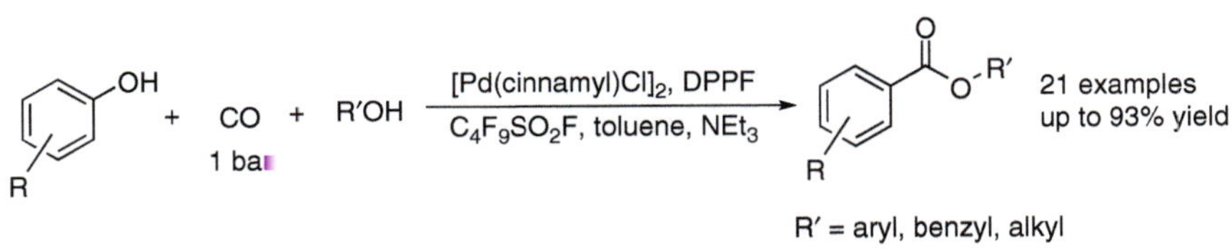

Scheme 1.61 Palladium-catalyzed carbonylation of phenols. Source: Wu et al. [165]; Wu et al. [166].

Chaudhari and colleagues reported a palladium-catalyzed carbonylation–polycondensation reaction of aromatic diiodides and aminohydroxy compounds [167]. With their methodology, alternating polyesteramides were prepared in chlorobenzene with 1,8-diaza-bicyclo[5.4.0]undec-7-ene (DBU) as a base under 3 bar of CO at 120 °C (Scheme 1.62).

Palladium-catalyzed double carbonylation as a more special carbonylation variant usually requires high CO pressures in order to compete with the corresponding monocarbonylation reactions. In 2001, Uozumi et al. reported

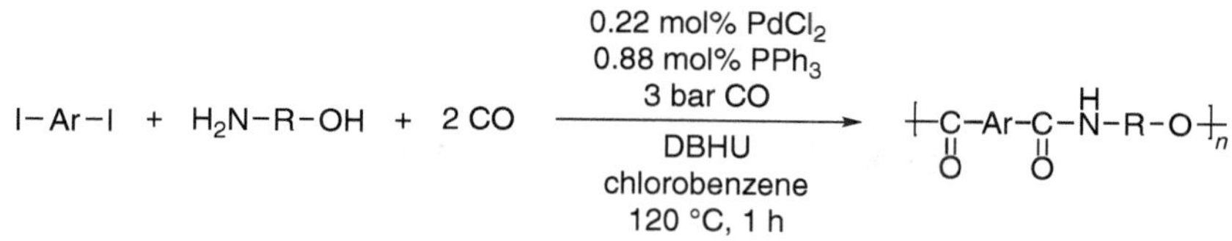

Scheme 1.62 Palladium-catalyzed polycondensations.

a procedure that significantly improved the existing protocols. They found 1,4-diaza-bicyclo[2.2.2]octane to be a superior base for the highly selective double carbonylation of aryl iodides with primary amines (Scheme 1.63) [168]. The desired a-keto amides were prepared under atmospheric pressure of CO at room temperature in the presence of a simple palladium–triphenylphosphine complex.

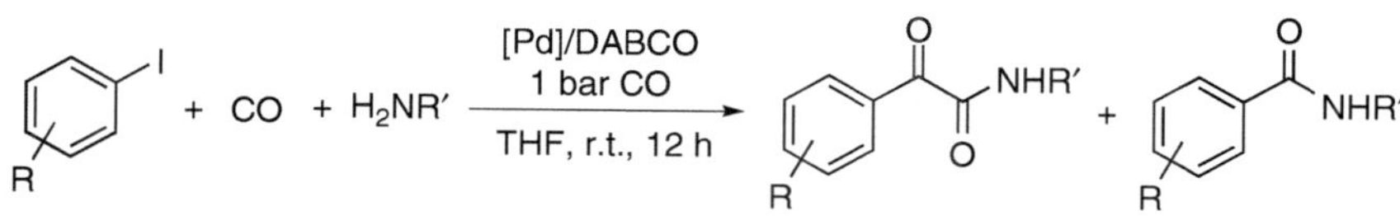

Scheme 1.63 Palladium-catalyzed double carbonylation of aryl iodides. Source: Modified from Uozumi et al. [168].

Schnyder and Indolese proved that the carbonylation of aryl bromides with primary amides or sulfonamides can lead to asymmetrical aroyl acyl imides. When the reactions were carried out under mild conditions, Et_3N was found to be the best base and the desired products were produced in 58–72% yields (Scheme 1.64) [169].

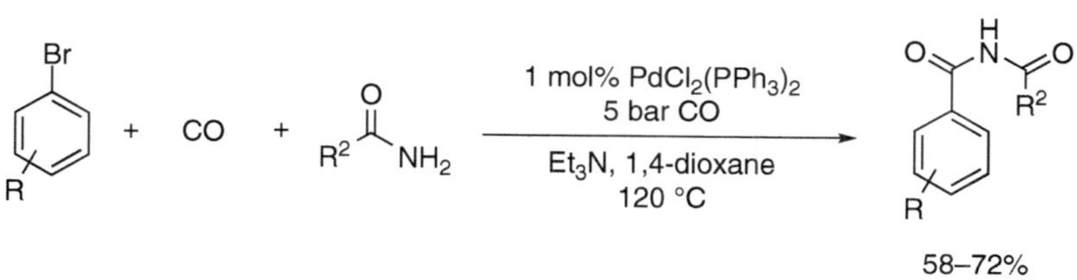

Scheme 1.64 Palladium-catalyzed aminocarbonylation with amides. Source: Modified from Schnyder and Indolese [169].

Beller's group developed several novel methodologies for the primary amides synthesis [170–173]. In the presence of palladium catalysts, aryl halides, phenyl triflates, benzyl chlorides, and even phenols were transformed into the corresponding primary amides in good to excellent yields. Ammonia gas was used directly as an amine source and also as a base. These were the primary reports on using NH_3 and CO for primary amides synthesis (Scheme 1.65).

A microwave-promoted palladium-catalyzed aminocarbonylation of (hetero)aryl halides (X = I, Br, Cl) using $Mo(CO)_6$ and allylamine as a nucleophile was also described [174, 175]. Remarkably, no side products resulting from the competing Heck reaction were detected. Importantly, this was the achievement that

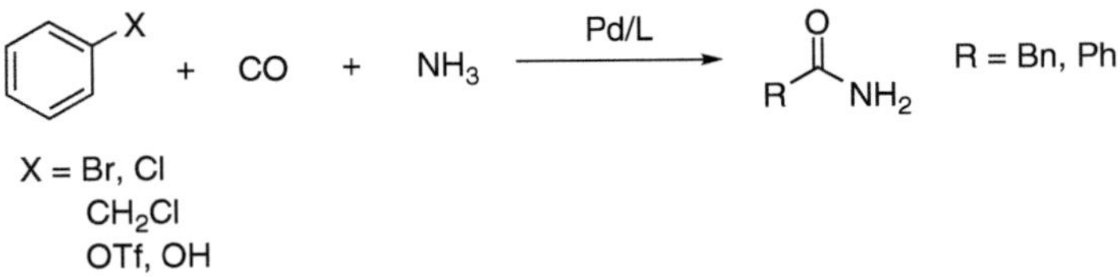

Scheme 1.65 Procedures for carbonylative synthesis of primary amides.

aminocarbonylation was realized on a larger laboratory scale (25 mmol) starting from 4-iodoanisole (Scheme 1.66).

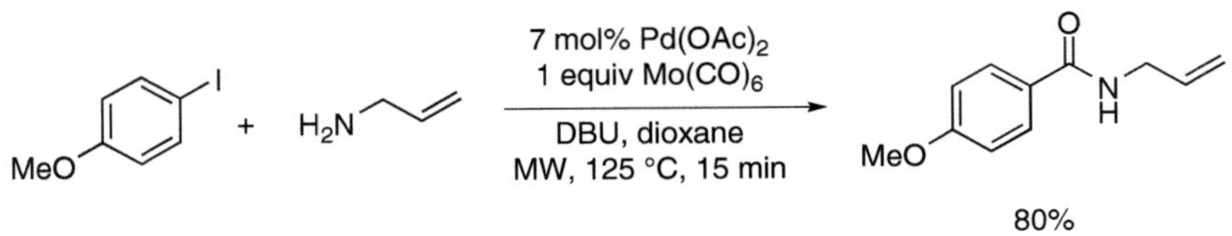

Scheme 1.66 $Mo(CO)_6$-mediated carbonylation of aryl iodides.

1.5.2 Reductive Carbonylations

Reductive carbonylation (also called formylation) catalyzed by transition metal offers a straightforward procedure for aryl aldehyde preparation. Starting from the corresponding aryl–X (X = I, Br, Cl, OTf, etc.), in the presence of catalyst and CO, aromatic aldehydes can be easily prepared (Scheme 1.67).

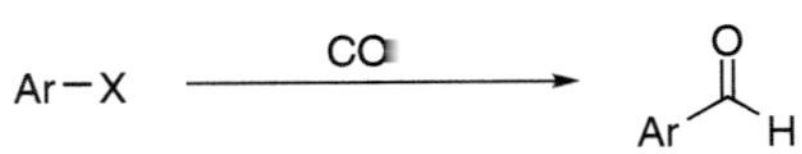

Scheme 1.67 Reductive carbonylation.

The palladium-catalyzed reductive carbonylation reaction was originally discovered by Schoenberg and Heck in 1974 [176]. In the presence of $[PdX_2(PPh_3)_2]$ as a catalyst under 80–100 bar of CO/H_2 and at 80–150 °C, aryl and vinyl bromides or iodides were converted into the corresponding aldehydes in 34–95% yields. This pioneering work was improved by using metal hydrides as reducing agents later.

Baillargeon and Stille [177, 178] established the use of tributyltin hydride (Bu_3SnH) in reductive carbonylation reactions. Under mild conditions (50 °C, 1–3 bar CO), aryl iodides, benzylic halides, vinyl iodides, triflates, and allylic halides were successfully carbonylated in 2.5–3.5 h reaction time. Since then, tin hydrides have been applied for reductive carbonylations in several natural product syntheses. Because of the toxicity and waste generation of tin hydrides, it should no longer be used despite the general application of tin hydrides in the past.

Organosilanes [179–181] are certainly a better choice to be based in conjunction with CO. Ashfield and Barnard recently took up this concept by testing the practicability of various R_3SiH systems for assorted known palladium catalysts [182]. When Et_3SiH was used under mild conditions (3 bar CO, 60–120 °C), the [$PdCl_2$(dppp)]/DMF/Na_2CO_3 system produced good results for most of the substrates (Scheme 1.68). In general, the desired aldehydes were obtained in 79–100% yields.

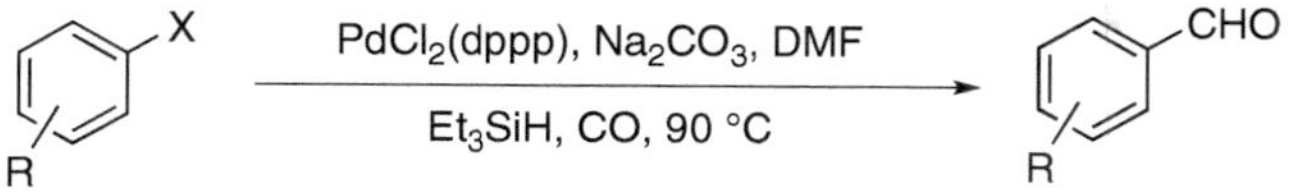

Scheme 1.68 Palladium-catalyzed reductive carbonylation of aryl halides.

The use of readily available and cheap formate salts is an economically attractive variant for performing palladium-catalyzed reductive carbonylations [183, 184]. For example, Cai and his associates developed a silica-supported phosphine palladium complex ("Si"–P–Pd) for the formylation of aryl bromides and iodides with sodium formate (1 bar CO, 90–110 °C) [185]. The polymeric catalyst could be recovered afterward and shown in simple model reactions comparable catalytic activity than homogeneous $PdCl_2(PPh_3)_2$ (Scheme 1.69).

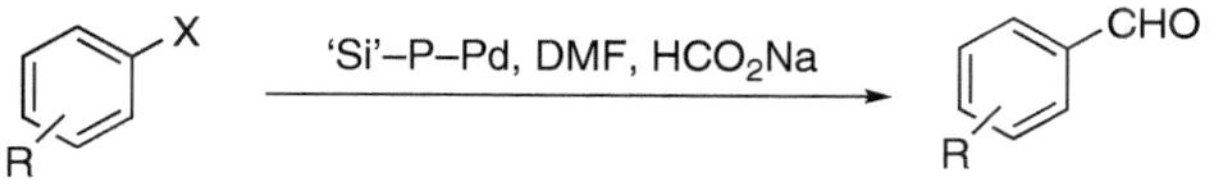

Scheme 1.69 "Si"–Pd-catalyzed reductive carbonylation of aryl halides.

In 1989, Milstein and colleagues realized the reductive carbonylation of aryl chlorides [186]. Under the assistant of $Pd(dippp)_2$ complex, aryl chlorides were transformed into the corresponding aldehydes in good yields in the presence of CO and sodium formate (Scheme 1.70).

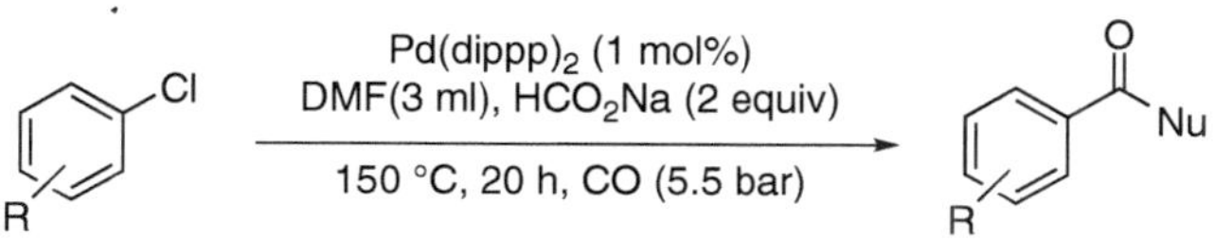

Scheme 1.70 Palladium-catalyzed reductive carbonylation of aryl chlorides.

Hydrogen is an inexpensive, atom-economic, and environmentally friendly reductant, and it is desirable to use syngas as a source of CO and H_2 for reductive carbonylation of C–X. Beller's group has developed the general and efficient palladium-catalyzed formylation procedure for the synthesis of aromatic and heteroaromatic aldehydes [187, 188]. Various (hetero)aryl bromides were successfully carbonylated with the cheap and environmentally benign formyl source, synthesis gas, in the presence of $Pd(OAc)_2$/cata*CX*ium A [189] and *N*,*N*,*N'*,*N'*-

tetramethylethylenediamine at 100 °C (Scheme 1.71). Advantageously, the catalyst system was active at low concentrations (0.25 mol% $Pd(OAc)_2$, 0.75 mol% cata*CX*ium A) and at much lower pressures (5 bar) than those previously reported in the literature. Besides, it was shown that vinyl halides could be formylated under similar conditions to form α, β-unsaturated aldehydes in 41–98% yield [190].

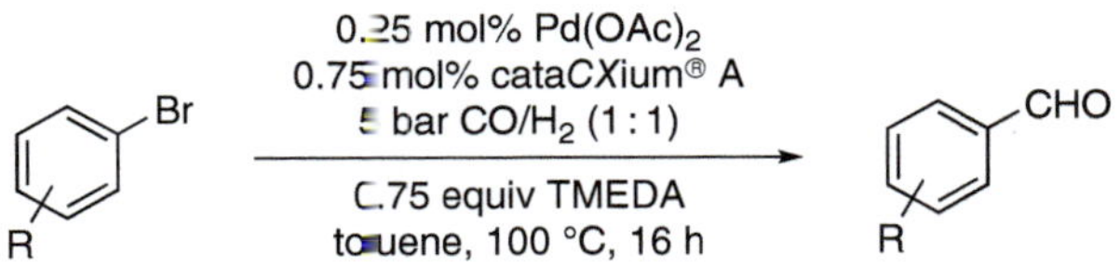

Scheme 1.71 $Pd(OAc)_2$-catalyzed reductive carbonylation.

In 2007, Beller and his colleagues developed the first general palladium-catalyzed carbonylation of aryl triflates with synthesis gas [191]. In contrast to aryl bromides, only the bidentate ligands dppe and dppp led to significant conversion and aldehyde formation. Under mild conditions, various aromatic aldehydes were obtained in a 50–92% yield in the presence of 1.5 mol% $Pd(OAc)_2$, 2.25 mol% dppp, and pyridine in DMF.

Later on, the same group developed an efficient phosphinite ligand for the palladium-catalyzed reductive carbonylation of aryl bromides to aromatic aldehydes based on phosphinite ligands [192]. Several aryl bromides with electron-donating and electron-withdrawing substituents reacted to produce aldehydes in good to excellent yields (Scheme 1.72).

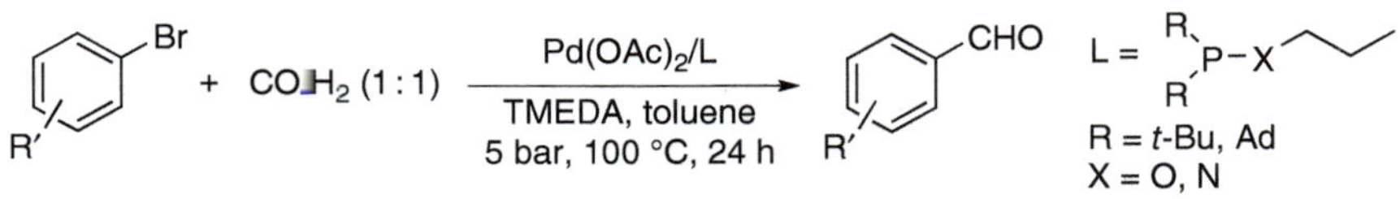

Scheme 1.72 Palladium phosphinite-catalyzed reductive carbonylation of ArBr.

Kappe and coworkers lately developed a continuous-flow protocol for Pd-catalyzed reductive carbonylation of (hetero)aryl bromides to aldehydes, with syngas as an inexpensive, atom-economic, and environmentally friendly source of CO and H_2 [193]. Relatively low catalyst loadings (0.5–1 mol%) and ligand loadings (1.5–3 mol%) provided 55–100% product yields (Scheme 1.73).

Reductive carbonylation offers an interesting pathway for aromatic aldehyde synthesis. Starting from the easily available corresponding parent molecules, aldehydes are selectively produced in good yields. For reductive carbonylation, the choice of hydrogen donor and ligand is important for the success of the transformation.

1.5.3 Carbonylative Coupling Reactions with Organometallic Reagents

In carbonylative coupling with organometallic reagent reactions, transmetalation is normally involved in advance of reductive elimination. As early as 1986, Kojima and

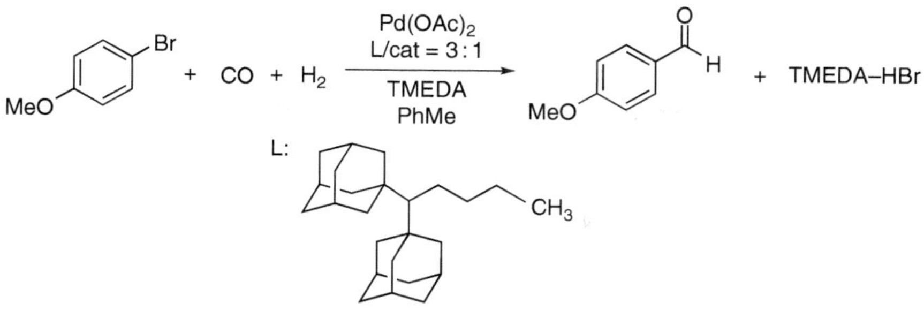

Scheme 1.73 Reductive carbonylation of 4-bromoanisole.

colleagues reported on the carbonylative coupling of aryl iodides or benzyl halides with organoboranes in the presence of dichlorobis(triphenylphosphine)-palladium catalyst [194]. This was the first application of organoboranes in carbonylative coupling reactions mediated by 1.1 equiv of $Zn(acac)_2$ to favor the transmetalation. Various ketones have been produced in good yields starting from aryl iodides and benzyl chloride (Scheme 1.74).

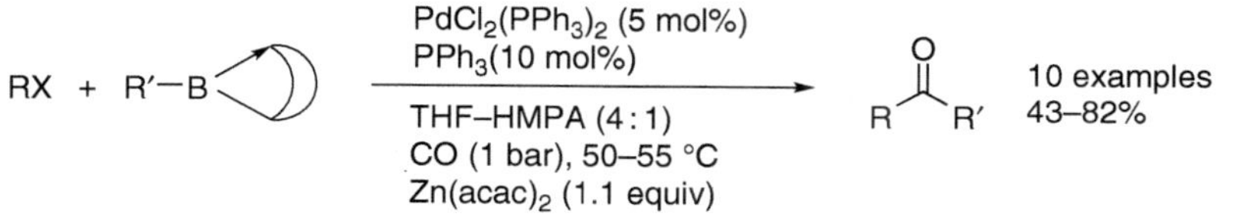

Scheme 1.74 The first Pd-catalyzed carbonylative coupling of organoboranes.

In 1993, Suzuki and colleagues described the palladium-catalyzed carbonylative coupling of aryl iodides with aryl boronic acids [195]. Various diarylketones were produced in high yields (Scheme 1.75). The choice of base and solvent was essential to obtain the desired ketones without biaryl by-products.

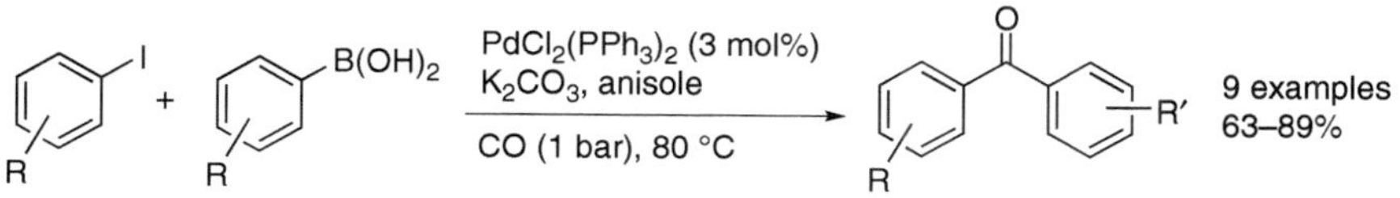

Scheme 1.75 Pd-catalyzed carbonylative Suzuki coupling of aryl iodides with aryl boronic acids.

Zeni and colleagues described the coupling of 2-iodoselenophenes with arylboronic acids and CO in 2006 (Scheme 1.76) [196]. Interestingly, the reaction proceeded with aqueous Na_2CO_3 as base under 1 bar of CO. Concerning the variation of arylboronic acids, strong electron-withdrawing substituted or ortho-substituted arylboronic acids gave low or no yields.

Steroidal phenyl ketones were synthesized by Skoda-Foeldes and colleagues via a related carbonylation pathway [197]. The ketones were produced in high yields by

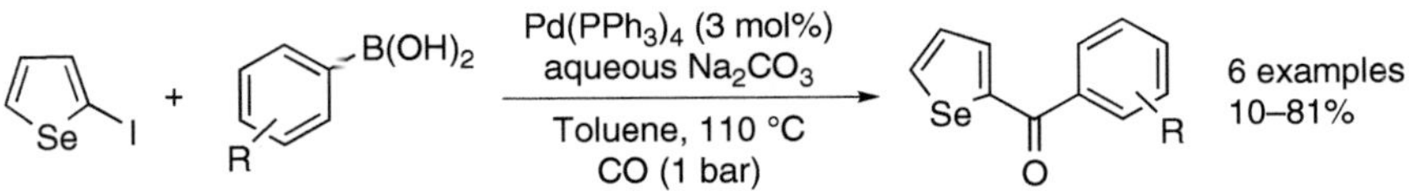

Scheme 1.76 Pd-catalyzed carbonylative coupling of 2-iodoselenophenes. Source: Modified from Prediger et al. [196].

the carbonylation of 17-iodo-androst-16-ene derivatives in the presence of $NaBPh_4$ (Scheme 1.77).

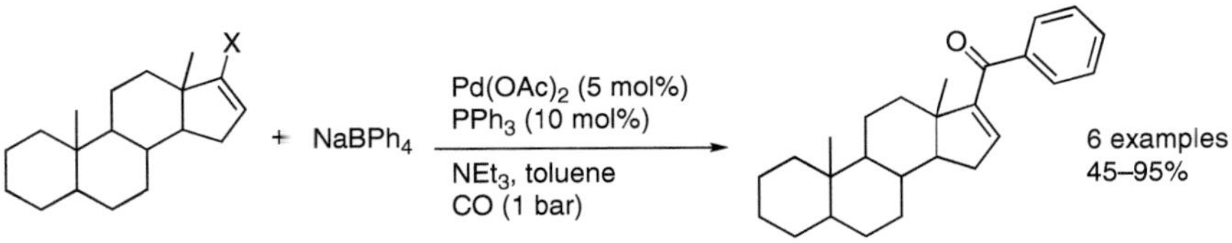

Scheme 1.77 Pd-catalyzed carbonylative synthesis of steroidal ketones.

Castanet and his team demonstrated a palladium-catalyzed carbonylative Suzuki reaction of pyridine halides in 2001. Under their conditions, pyridine halides reacted with aryl boronic acids to 2-pyridyl ketones in 81–95%yields (Scheme 1.78). Later on, they extended this methodology to pyridine chlorides by applying an NHC ligand and Cs_2CO_3 as a base [198–201].

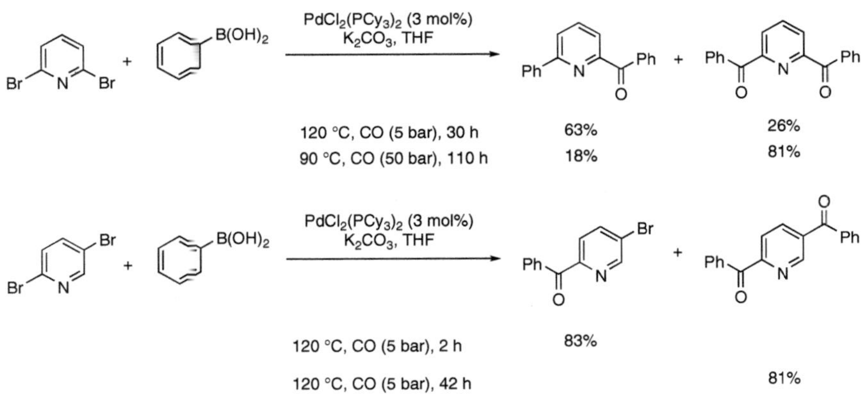

Scheme 1.78 Pd-catalyzed Suzuki carbonylation of halopyridines.

In 2008, Beller's group developed a general method for diaryl ketone synthesis by palladium-catalyzed carbonylative coupling of aryl bromides with arylboronic acids [202]. The combination of $Pd(OAc)_2$ and $BuPAd_2$ allowed the coupling of aryl/heteroaryl bromides with arylboronic acids to produce a wide range of ketones in 60–89% yields (Scheme 1.79).

Schmalz and his colleagues investigated the carbonylative Suzuki reaction of their chloroarene–$Cr(CO)_3$ complexes with phenyl boronic acid [203]. Using

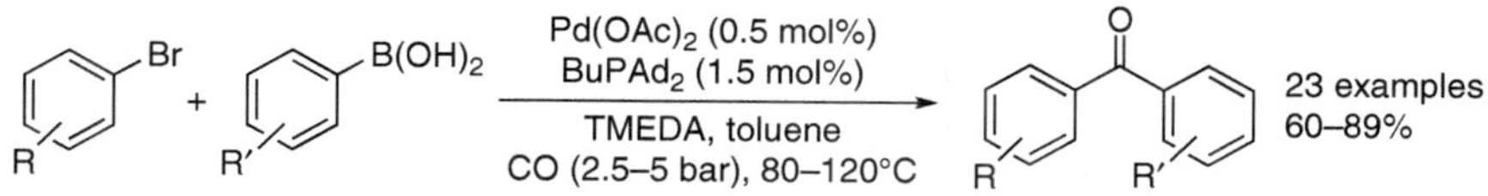

Scheme 1.79 Pd-catalyzed carbonylative Suzuki reaction of aryl bromides.

$PdCl_2(PPh_3)_2$ as a catalyst precursor, benzophenone derivatives were achieved in 48–78% yields (Scheme 1.80).

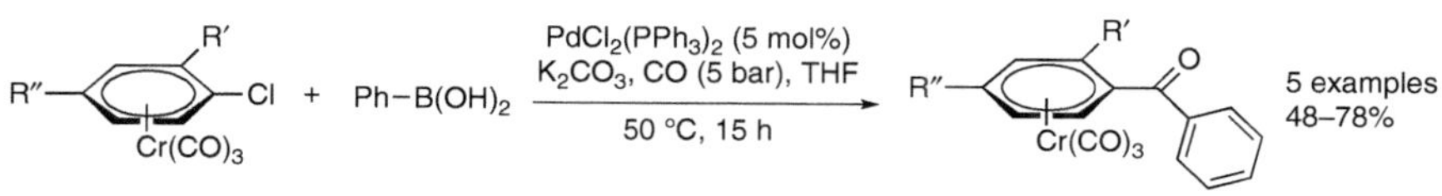

Scheme 1.80 Pd-catalyzed carbonylative coupling of chloroarene–$Cr(CO)_3$ complexes.

Later, Beller and colleagues described a novel carbonylative coupling of benzyl chlorides with aryl boronic acids [204]. This was the first report on carbonylative Suzuki couplings of benzyl chlorides with arylboronic acids (Scheme 1.81). The reaction was carried out using a commercially available $Pd(OAc)_2/PCy_3$ catalyst in the presence of K_2CO_3 and water as the solvent. Twelve ketones have been synthesized in 41–78% yields. Later on, they succeeded in extending their reaction to $ArBF_3K$, a more stable class of borane compounds [205].

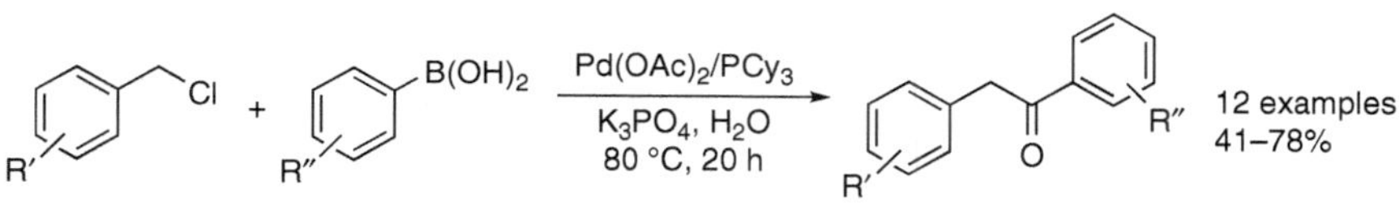

Scheme 1.81 Pd-catalyzed carbonylative Suzuki coupling of aryl boronic acids with hypervalent iodonium salts.

Grushin's team developed a palladium-catalyzed carbonylative coupling of aryl iodides with sodium azide [206]. This catalytic reaction occurs smoothly at temperatures as low as 25–50 °C and 1 bar to cleanly produce aroyl azides from the corresponding aryl iodides, CO, and NaN_3 (Scheme 1.82).

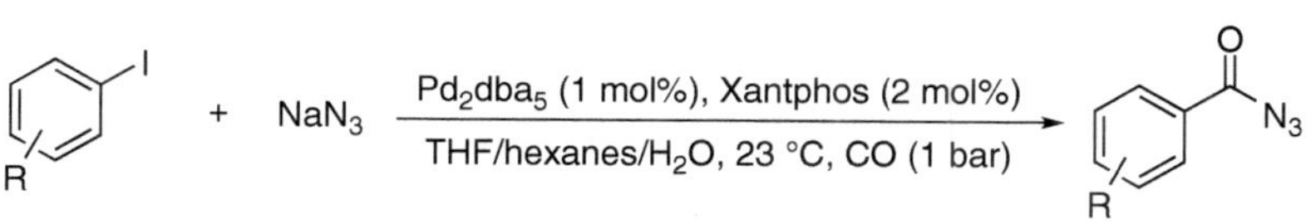

Scheme 1.82 Palladium-catalyzed carbonylative coupling of ArI and NaN_3.

Suzuki and colleagues developed another methodology for the carbonylative coupling of vinyl halides with organoboranes using $Pd(PPh_3)_4$ and K_3PO_4 as a base [207] to synthesize vinyl ketones in moderate to excellent yields (Scheme 1.83).

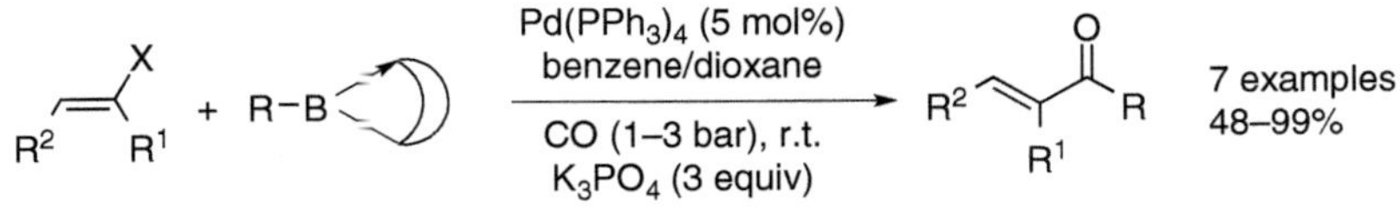

Scheme 1.83 Pd-catalyzed carbonylative Suzuki coupling of organoboranes with vinyl halides.

Jackson and colleagues reported $Pd(PPh_3)_4$ catalyzed an amino acid-derived organozinc reagent with aryl iodides under 1 atm of CO for the synthesis of Kynurenine derivatives [208]. The yields in 13–60% can be obtained at room temperature (Scheme 1.84).

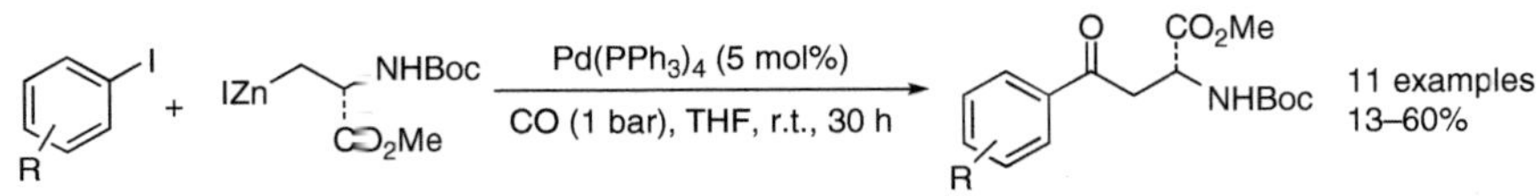

Scheme 1.84 Pd-catalyzed carbonylative coupling of an amino acid-derived organozinc reagent.

In 2008, Martin and colleagues investigated the [1,3-bis(2,6-diisopropylphenyl) imidazol-2-ylidene](3-chloropyridyl)palladium(II) dichloride (PEPPSI-IPr)-catalyzed carbonylative Negishi coupling of ortho-disubstituted aryl iodides with an alkynyl zinc reagent [209]. Alkynones were produced in 67–79% yields under mild conditions (Scheme 1.85).

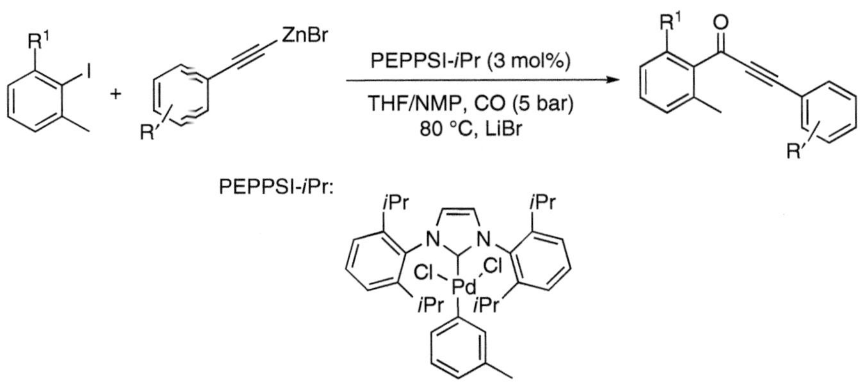

Scheme 1.85 Pd-catalyzed carbonylative Negishi reaction of 2,6-disubstituted aryl iodides.

The first example of Hiyama reactions that palladium-catalyzed C–C bond formation between aryl, alkenyl, or alkyl halides with organosilanes was published in 1989 [210, 211]. In the presence of KF, diaryl ketones were produced in 38–91% yields via the $(C_3H_5PdCl)_2$-catalyzed carbonylative coupling of arylfluorosilanes with aryl iodides (Scheme 1.86).

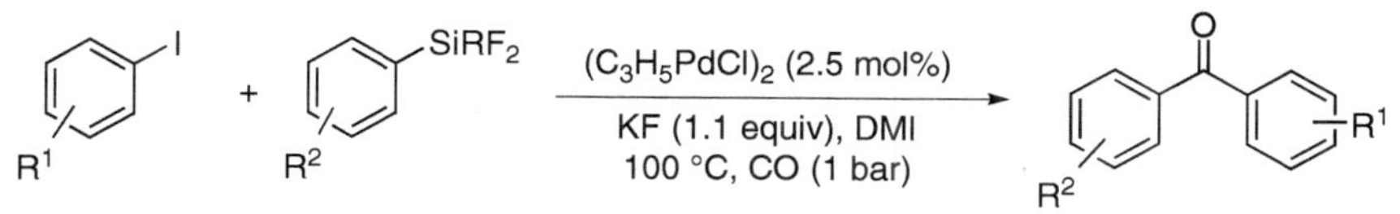

Scheme 1.86 Pd-catalyzed carbonylative Hiyama coupling of aryl iodides.

In 2012, Beller's group reported Pd-catalyzed reaction of aryl iodides and hexamethyldisilane (HMDS) for the synthesis of acyl silanes [212]. In the presence of 10 bar CO, various benzoyl silanes were produced in 41–88% yields (Scheme 1.87).

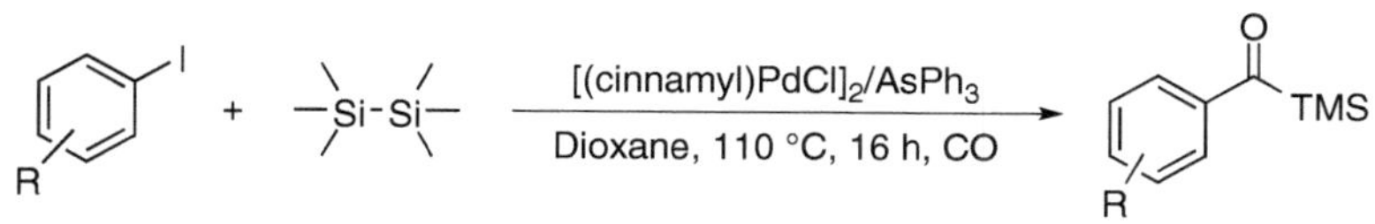

Scheme 1.87 Palladium-catalyzed carbonylative synthesis of acyl silanes.

Aryl iodide, aryl bromide, aryl chloride, and alkenyl halides have been partially discussed. In addition, there are other electrophilic reagents, such as hypervalent iodonium salts [213, 214] and 1-aryltriazene [215], which can also undergo carbonylation with organometallic reagents.

1.5.4 Carbonylative Sonogashira Reactions

The Sonogashira reaction is generally known as a coupling reaction of terminal alkynes with aryl or vinyl halides. This reaction was first reported by Sonogashira and Hagihara in 1975 (Scheme 1.88) [216]. If the Sonogashira reaction is carried out in a CO atmosphere, the reactions are called Carbonylative Sonogashira Reactions, which will give alkynone as an interesting structural motif found in numerous biologically active molecules [217, 218].

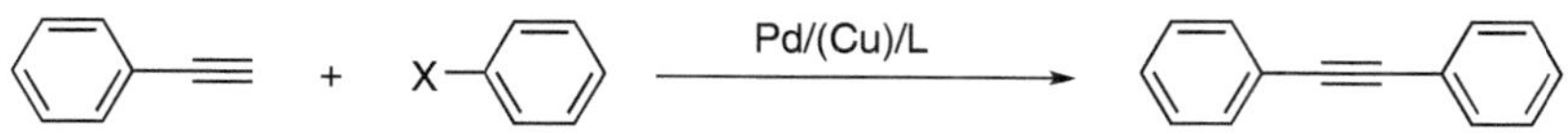

X = I, Br, Cl, OTf, N_2BF_4, etc.

Scheme 1.88 Palladium-catalyzed Sonogashira coupling reaction. Source: Modified from Sonogashira et al. [216].

The first palladium-catalyzed carbonylative Sonogashira coupling was reported in 1981 by Kobayashi and Tanaka [219]. Aryl, heterocyclic, and vinylic halides reacted with CO and terminal acetylenes at 120 °C and 80 bar in the presence of NEt_3 and $PdCl_2$(1,1′-bis(diphenylphosphino)ferrocene, dppf) to form alkynones in 46–93% yields (Scheme 1.89).

In 1991, Alper and Huang interestingly described another type of palladium-catalyzed carbonylative Sonogashira coupling of aryl iodides with benzyl acetylenes

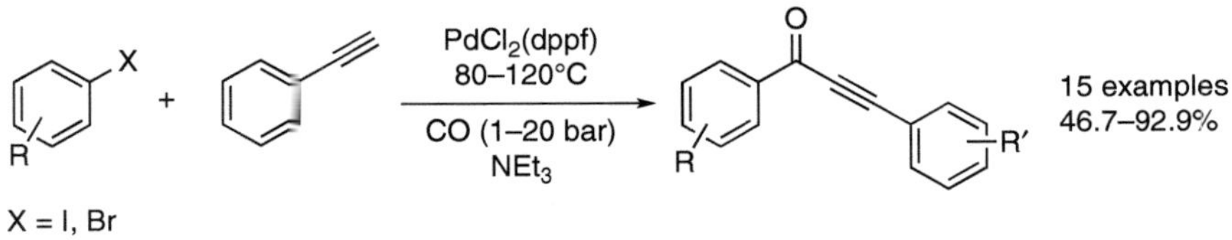

Scheme 1.89 First Pd-catalyzed carbonylative Sonogashira coupling of organic halides.

[220]. In the presence of $Pd(OAc)_2/PPh_3$, aryl iodides and benzyl acetylenes were transformed into furanones in 33–88% yields (Scheme 1.90). Furanones were isolated as the terminal products and not the predicted alkynones.

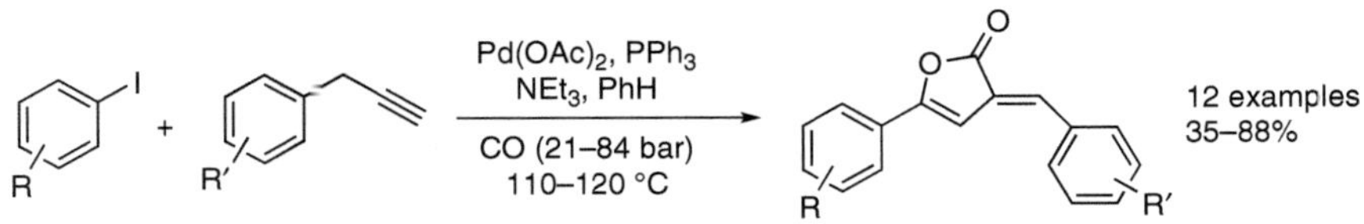

Scheme 1.90 Pd-catalyzed carbonylation of benzyl acetylenes to fuanones.

Ortar and colleagues published a general procedure for the carbonylative Sonogashira couplings of vinyl triflates with terminal acetylenes in 1991 [221]. Various alkynyl ketones were produced in 53–83% yields (Scheme 1.91).

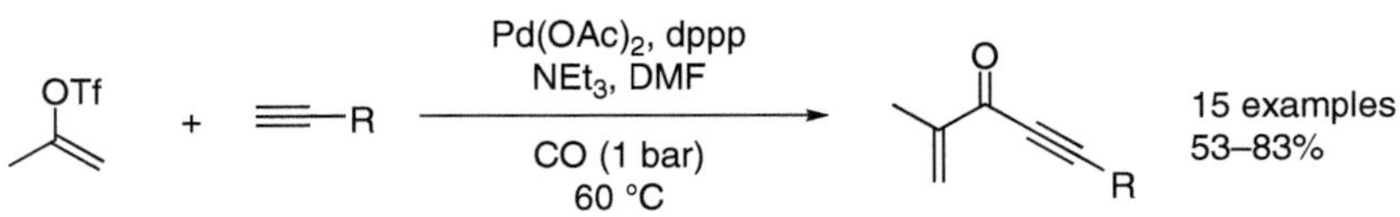

Scheme 1.91 Pd-catalyzed carbonylative Sonogashira coupling of vinyl triflates with 1-alkynes.

The catalytic ability of dimeric palladium hydroxide in carbonylative Sonogashira coupling was demonstrated by Alper and his team in 1994 [222]. Terminal alkynes and alkynols were coupled with aryl iodides in the presence of CO in moderate to good yields.

Kang and colleagues described the carbonylative Sonogashira reaction of iodonium salts with terminal alkynes [223]. Both $Pd(OAc)_2/CuI$ and $Pd(OAc)_2$ systems alone could be used, and various alkynones were synthesized in 61–89% yields with 1 bar CO at room temperature in aqueous media. In 2001, another example of carbonylative Sonogashira coupling reactions with iodinium iodide and 1-alkynes was published by Ma and colleagues [224]. Under atmospheric pressure of CO, iodine-substituted alkynones were produced in 72–94% yields (Scheme 1.92). Aromatic, aliphatic, and heterocyclic terminal acetylenes can be applied as their substrates.

An interesting room-temperature carbonylation using a palladium/copper-catalyst system was published by Mori and Ahmed in 2003 [225–227]. As shown in

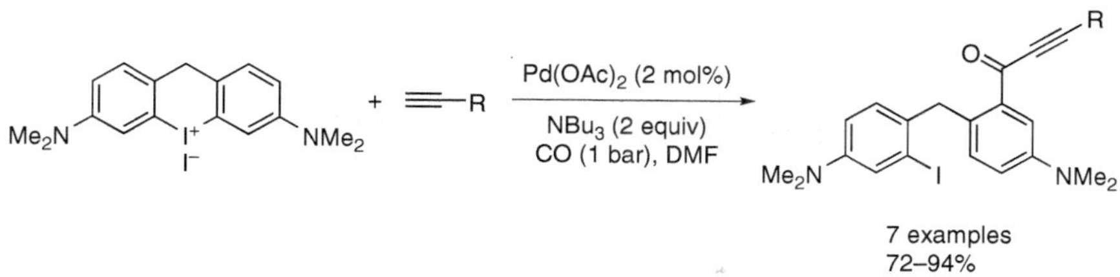

Scheme 1.92 Pd-catalyzed carbonylative Sonogashira coupling of iodinium iodide.

Scheme 1.93, various aromatic alkynones were produced in 47–87% yields using aqueous ammonia as a base.

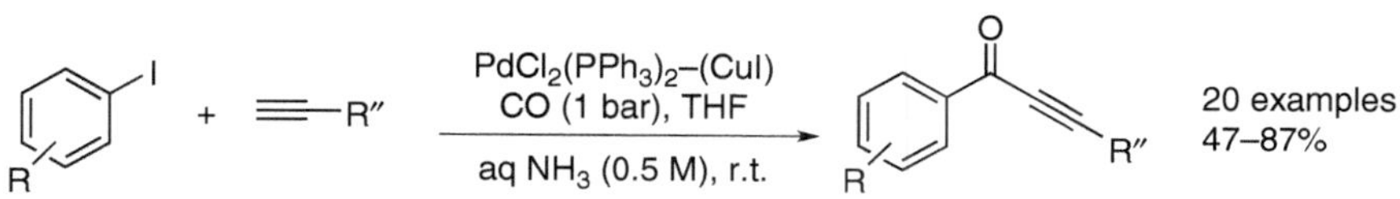

Scheme 1.93 Carbonylative room-temperature Sonogashira reaction in aqueous ammonia.

In 2006, Chen and his colleagues described a convenient, effective method for the carbonylative Sonogashira coupling of aryl iodides with ethynyl ferrocene under one atmosphere of CO [228]. Various aryl ferrocenylethynyl ketones have been synthesized in a 62–88% yields (Scheme 1.94).

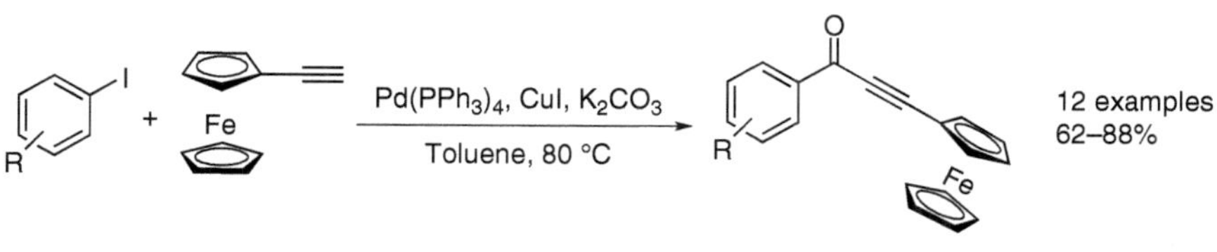

Scheme 1.94 Pd-catalyzed carbonylative Sonogashira coupling of ethynyl ferrocene with aryl iodides.

In 2008, Xia and Chen described a recyclable phosphine-free catalyst system for alkynone synthesis [229]. Using palladium on charcoal (Pd/C) and NEt_3, the carbonylative Sonogashira coupling of aryl iodides with alkynes was smoothly carried out and the desired products were isolated in 63–97% yields (Scheme 1.95). Pd/C could be recycled at least 10 times with only a slight decrease in efficiency.

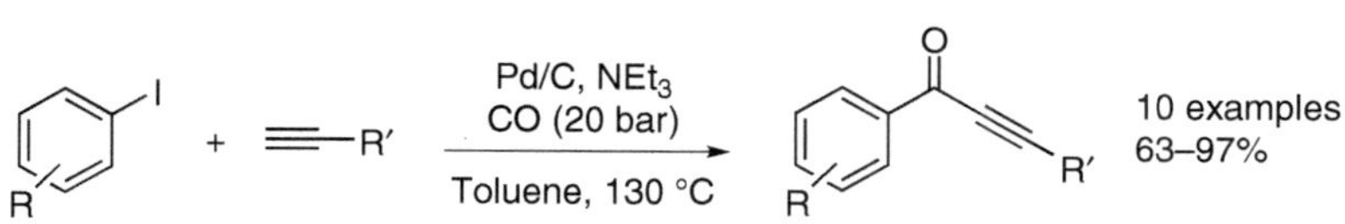

Scheme 1.95 Pd/C-catalyzed carbonylative Sonogashira reaction of aryl iodides.

In 2010, Beller's group discovered a general and convenient palladium-catalyzed carbonylative Sonogashira coupling of aryl bromides [230]. The key to success was the application of BuPAd$_2$ (di-1-adamantyl-*n*-butylphosphine) as a ligand in the presence of K_2CO_3. Alkynones have been generated in 47–88% yields from the corresponding aryl bromides and terminal alkynes (Scheme 1.96).

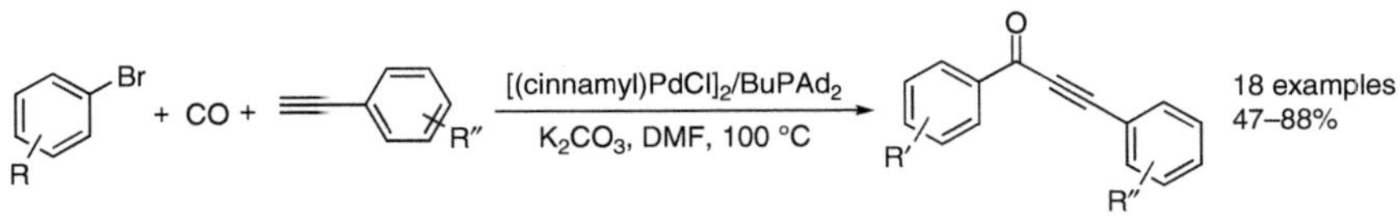

Scheme 1.96 Pd-catalyzed carbonylative Sonogashira coupling of aryl bromides.

In 2019, Li and Zhang reported $PdCl_2(PPh_3)_2$-catalyzed carbonylative Sonogashira coupling in tandem with double annulation reaction to synthesize benzannulated [6,6]-spiroketals [231]. Under mild conditions (under 1 atm pressure of CO at room temperature), a wide range of benzannulated [6,6]-spiroketals in 45–90% yields can be obtained with good functional group tolerance and excellent diastereoselectivities (dr > 20 : 1).

1.5.5 Carbonylative C–H Activation Reactions

Transition metal-catalyzed carbonylation reactions represent an enormous toolbox for CO–X bond formation (X = C, N, O, etc.). While most coupling reactions take place with heteronucleophiles nowadays, carbonylations including C–H activation are attracting more and more attention because the use of stoichiometric amounts of organometallic reagents can be avoided.

In 1986, Kobayashi and Tanaka published the first report of carbonylative C–H activation by $PdCl_2$ [232]. In the presence of NEt_3 under 20 bar of CO, the carbonylation of organic halides with activated methylene compounds produces various ketones in 4–95% yields (Scheme 1.97).

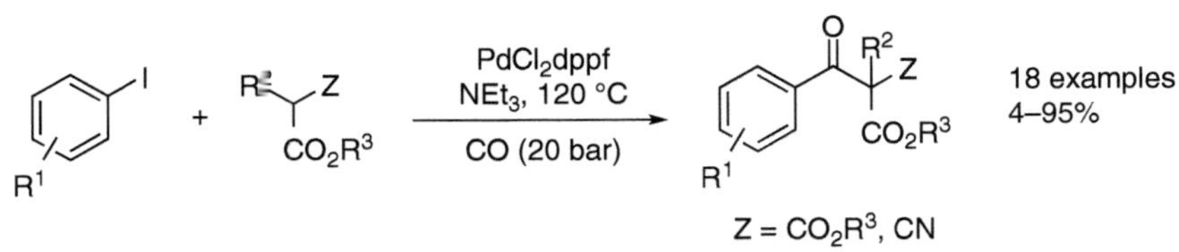

Scheme 1.97 Pd-catalyzed carbonylative coupling with activated methylene compounds.

Later on, Negishi and colleagues reported intramolecular coupling reactions of internal enolates [233] and five- or six-membered rings were synthesized by using the carbonylative C–H activation methodology (Scheme 1.98). In 1998, this group proved that the same reaction can also be catalyzed by $Cl_2Ni(PPh_3)_2$, $Ni(COD)_2$, or Li_2CuCl_4 [234].

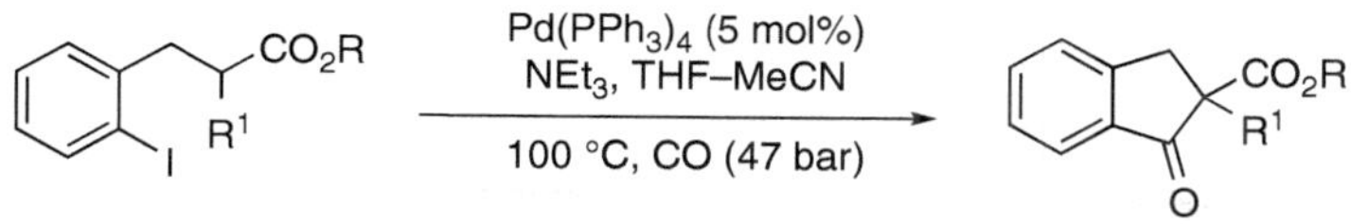

Scheme 1.98 Pd-catalyzed intramolecular carbonylative C–H activations.

In 2002, Larock and Campo reported the palladium-catalyzed cyclocarbonylation of *o*-halobiaryls [235, 236], giving various substituted fluorenones in high yields (Scheme 1.99). The authors succeeded in extending the reaction to polycyclic fluorenones, fused isoquinoline, indole, pyrrole, thiophene, benzothiophene, and benzofuran rings.

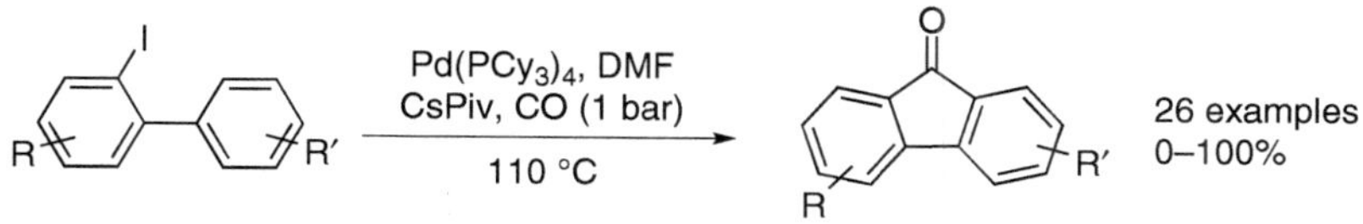

Scheme 1.99 Pd-catalyzed carbonylative synthesis of fluorenones.

In 2007, Wang and his team found the $Pd(PPh_3)_4$-catalyzed coupling of aryl or vinyl iodides with ethyl diazoacetate [237]. Under 1 bar CO, the products yielded in 43–66% at 45 °C. It was the first example of using α-diazocarbonyl compounds as a coupling partner in a palladium-catalyzed carbonylation reaction (Scheme 1.100).

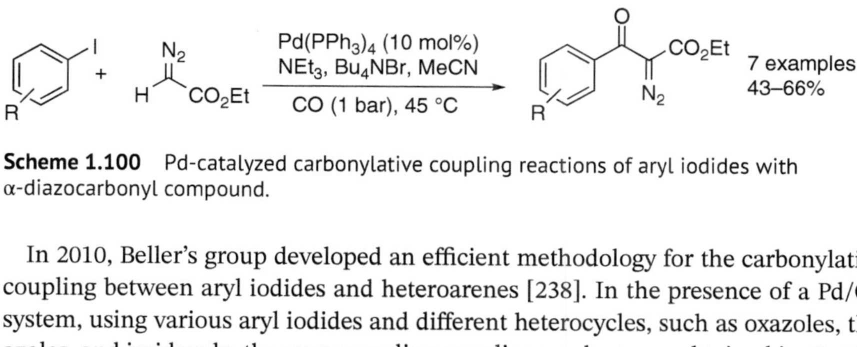

Scheme 1.100 Pd-catalyzed carbonylative coupling reactions of aryl iodides with α-diazocarbonyl compound.

In 2010, Beller's group developed an efficient methodology for the carbonylative coupling between aryl iodides and heteroarenes [238]. In the presence of a Pd/Cu system, using various aryl iodides and different heterocycles, such as oxazoles, thiazoles, and imidazole, the corresponding coupling products are obtained in 40–75% yields at 130 °C (Scheme 1.101). This is the first carbonylative C–H activation reactions of heteroarenes to form diarylketones.

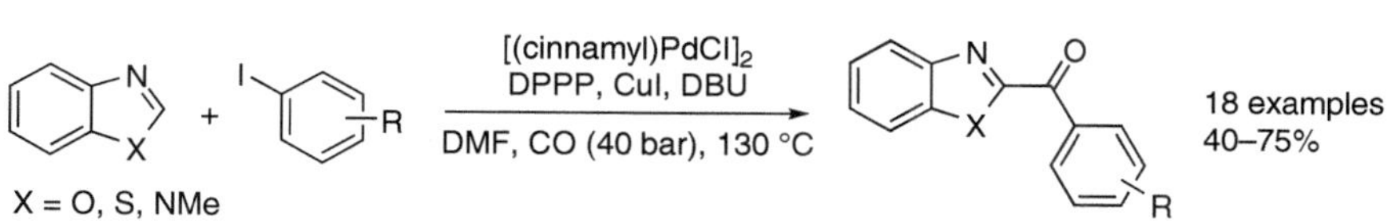

Scheme 1.101 Pd-catalyzed carbonylative coupling of ArI with heteroarenes.

1.5.6 Carbonylative Heck Reactions

The catalytic insertion of olefins into acylpalladium complexes is called a "Carbonylative Heck reaction." The first palladium-catalyzed copolymerization of CO with olefins was described in 1982 [239], and as a consequence, carbonylative coupling reactions with alkenes were reported soon after. In 1983, Negishi and Miller discovered two examples of intramolecular carbonylative Heck reactions of 1-iodopenta-1,4-dienes by applying $Pd(PPh_3)_4$ [240]. 5-Methylenecyclopent-2-enones as the products were produced in moderate yields (Scheme 1.102).

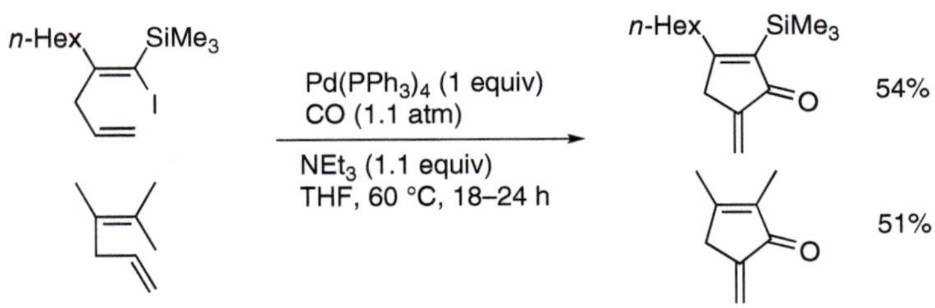

Scheme 1.102 The first examples of palladium-mediated intramolecular carbonylative Heck reactions.

Negishi synthesized various quinones using *o*-iodoaryl cyclohexyl ketones as the starting materials [241]. In the presence of bis(dibenzylideneacetone)palladium ($Pd(dba)_2$) as a catalyst (5 mol%) and under CO pressure (41 bar), quinones were produced in good yields with 100% regioselectivity (Scheme 1.103). In this catalytic system, 58% of furanones were formed instead of quinones if $Pd(OAc)_2/PPh_3$ was used. In 1995, this group published a full account using various vinyliodides [242–244].

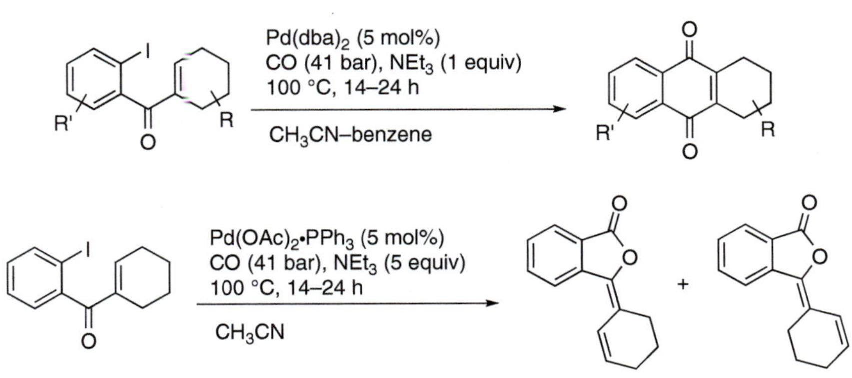

Scheme 1.103 First palladium-catalyzed intramolecular carbonylative Heck reaction.

Notably, Torii and colleagues reported the intramolecular carbonylative Heck coupling of 3-(2-haloarylamino)prop-2-enoates to the corresponding quinolinone

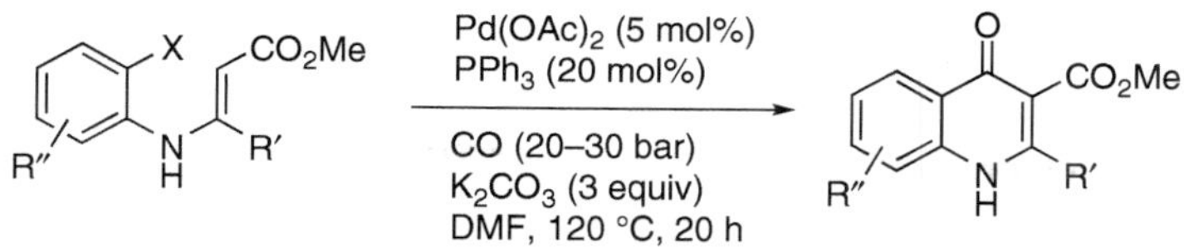

Scheme 1.104 Palladium-catalyzed carbonylative Heck reaction to quinolinones.

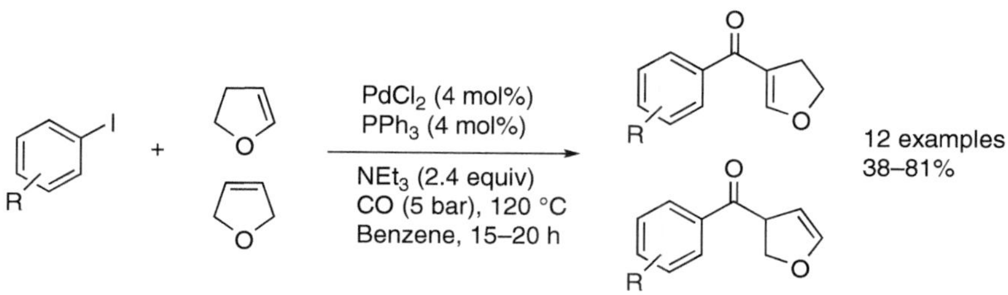

Scheme 1.105 Palladium-catalyzed carbonylative cross-coupling of ArI with cyclic olefins.

derivatives [245]. In the presence of a catalytic amount of $Pd(OAc)_2$ under 20 bar of CO at 120 °C, quinolinones were synthesized in good yields (Scheme 1.104).

In 1995, Miura and colleagues described a palladium-catalyzed carbonylative cross-coupling of aryl iodides with five-membered cyclic olefins [246]. This represents the first palladium-catalyzed intermolecular carbonylative cross-coupling of aryl iodides with olefins. Various benzoylated cyclic olefins were isolated in 38–81% yields (Scheme 1.105).

In 2010, Beller's group reported carbonylative Heck couplings of aryl triflates with styrenes that is a general intermolecular carbonylative coupling of aryl halides or triflates with terminal olefines [247]. The corresponding unsaturated ketones can be obtained in good yields starting from easily available aryl and alkenyl triflates (Scheme 1.106).

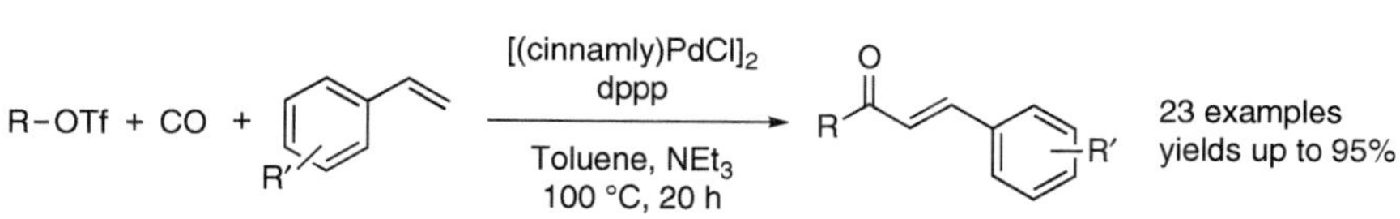

Scheme 1.106 Palladium-catalyzed carbonylative Heck reaction of ArOTf with styrenes.

Then, Beller's group developed a more general $[(cinnamyl)PdCl]_2$-catalyzed carbonylative Heck reaction of aryl halides [248]. Various aromatic and aliphatic alkenes were used successfully in this system for the first time, and 41–90% yields of the corresponding a,b-unsaturated ketones were obtained. When using aryl iodides and bromides, interesting building blocks were obtained at 100 °C (Scheme 1.107).

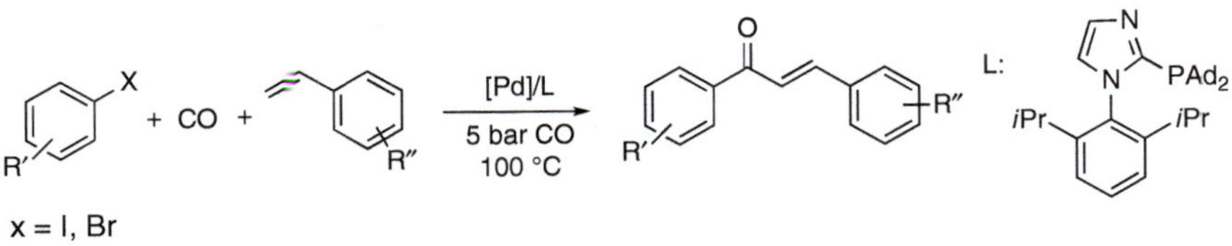

Scheme 1.107 Palladium-catalyzed carbonylative Heck reaction of aryl halides.

1.6 Carbonylation of Epoxides

Carbonylation of epoxides provides direct access to a large variety of hydroxy carbonyl compounds, such as β-lactones, useful synthons in organic synthesis, and poly(3-hydroxyalkanoate), important biodegradable polyesters. Thus, the development of efficient catalysts for such carbonylations has been widely studied.

1.6.1 Ring-expansion Carbonylation of Epoxides

The first successful example of a catalytic ring-expansion carbonylation was reported by Aumann and coworkers in 1970s [249, 250]. Isoprene oxide was allowed to react with CO in the presence of $[Rh(cod)Cl]_2$ to give lactone 67–75% yields (Scheme 1.108).

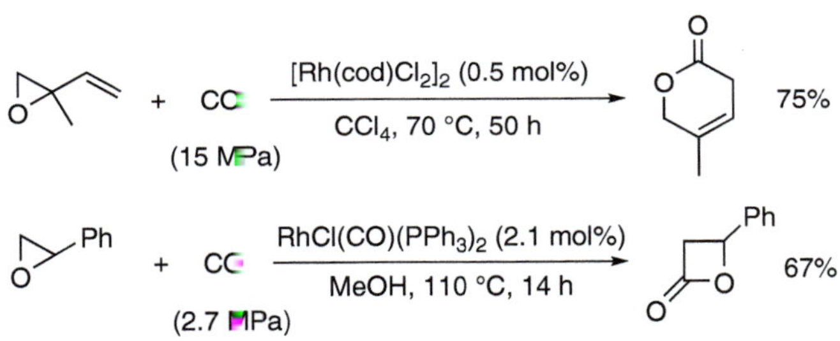

Scheme 1.108 Catalytic ring-expansion carbonylation.

Alper and coworkers reported the regioselective carbonylation of epoxides by using $[PPN][Co(CO)_4]$ [PPN = bis(triphenylphosphine)iminium] in conjugation with a Lewis acid such as $BF_3{\cdot}OEt_2$ or $SnCl_4$ in DME (Scheme 1.109) [251].

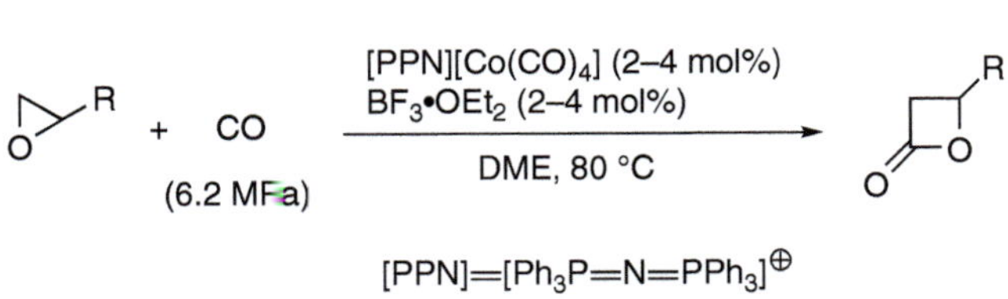

Scheme 1.109 Regioselective carbonylation of epoxides. Source: Modified from Lee et al. [251].

Two molecules of CO were successively incorporated into an epoxide in the presence of a cobalt catalyst and a phase transfer agent [252]. When styrene oxide was treated with CO (0.1 MPa), excess methyl iodide, NaOH (0.50 M), and catalytic amounts of $Co_2(CO)_8$ and hexadecyltrimethylammonium bromide in benzene, 3-hydroxy-4-phenyl-2(5*H*)-furanone was produced in 65% yield (Scheme 1.110).

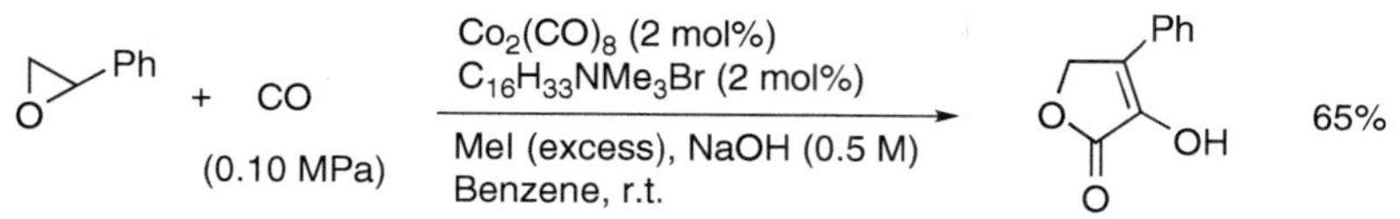

Scheme 1.110 $Co_2(CO)_8$-catalyzed ring-expansion carbonylation.

Alper et al. reported intramolecular cyclization results in the production of 2(5*H*)-furanone and the regeneration of the acylcobalt complex. Furthermore, triple carbonylation proceeded by employing 2-aryl-3-(hydroxymethyl)oxiranes as substrates and tris(3,6-dioxaheptyl)amine as a phase transfer catalyst (Scheme 1.111) [253].

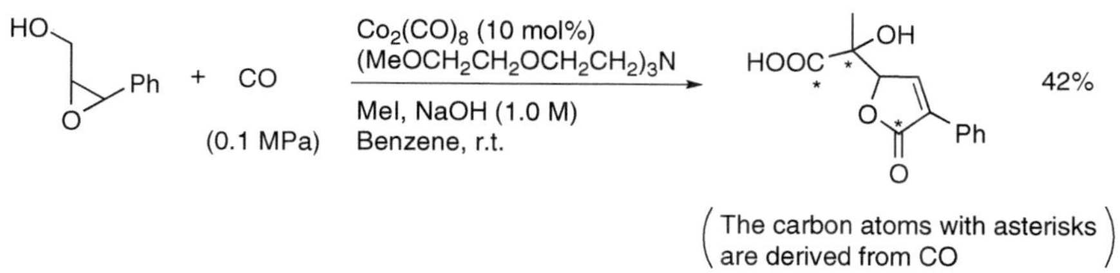

Scheme 1.111 Production of 2(5*H*)-furanone. Source: Modified from Alper et al. [253].

In 2012, Ibrahim and coworker described the first asymmetric ring-expansion carbonylation of *meso*-epoxides by using L*CrCl–$Co_2(CO)_8$ (Scheme 1.112) [254]. Under 500 psi pressure of CO, the corresponding β-lactones in 47–94% yields with up to 56% ee can be obtained at 70 °C.

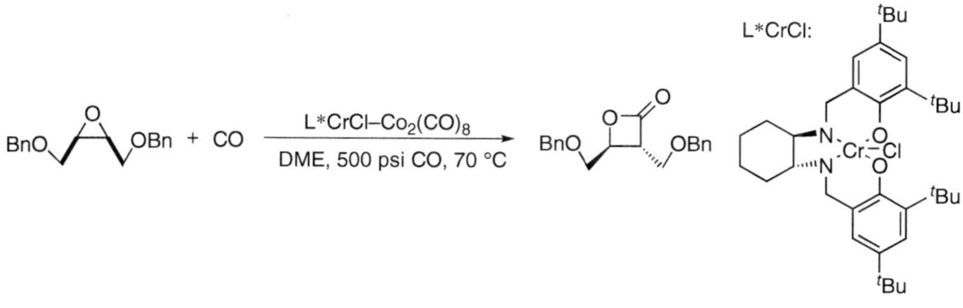

Scheme 1.112 L*CrCl–$Co_2(CO)_8$-catalyzed asymmetric ring-expansion carbonylation. Source: Ganji and Ibrahim [254].

1.6.2 Hydroformylation and Silylformylation of Epoxides

Hydroformylation of ethylene oxide is attractive in industry because hydrogenation of the product 3-hydroxypropanal gives 1,3-propane diol which is an intermediate in the production of polyester fibers and films. To achieve this reaction, many catalysts have been employed [255–257].

A successful example of protection of the resulting formyl group is the cobalt-catalyzed hydroformylation of epoxides in the presence of trimethyl orthoformate reported by Nozaki. The product β-hydroxyaldehydes were converted into its dimethyl acetal forms [258]. Under this catalyst system, hydroformylation–acetalization of epoxides in $HC(OMe)_3$ produces in 13–74% yield (Scheme 1.113).

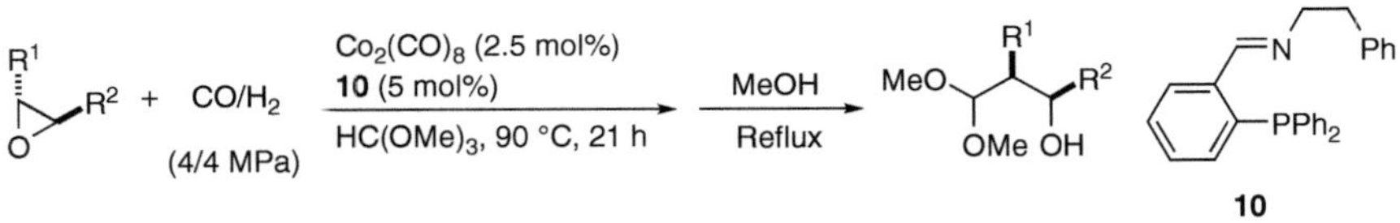

Scheme 1.113 Cobalt-catalyzed hydroformylation of epoxides.

Silylformylation is another successful catalytic formylation of epoxides involving the in situ protection of functional group. As a hydrosilane is deemed to behave much like molecular hydrogen, formylation of epoxides can be carried out using a hydrosilane in place of hydrogen to give β-siloxyaldehydes [259–262].

Murai and coworkers reported silylformylation of epoxides by using rhodium catalysts [261]. Under the optimized conditions ($[RhCl(CO)_2]_2$: 2 mol%, 1-methylpyrazole 40 mol%, CO: 5.1 MPa, $HSiMe_2Ph$: 1.2 equiv, CH_2Cl_2, 50 °C, 20 h), a wide range of epoxides were formylated in good yields (Scheme 1.114).

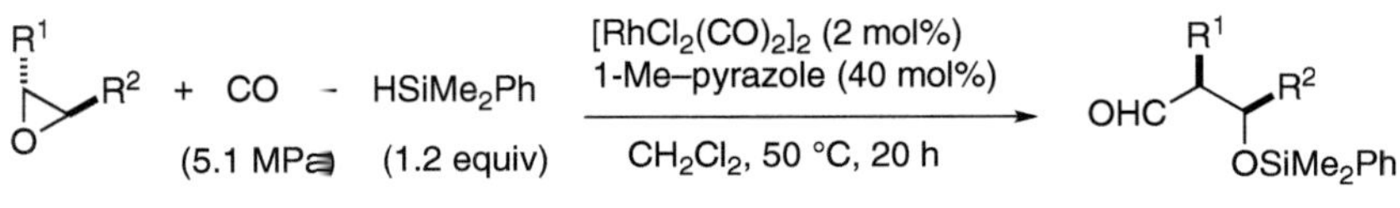

Scheme 1.114 Rhodium-catalyzed silylformylation of epoxides.

1.6.3 Alternating Copolymerization of Epoxides

The alternating copolymerization of epoxides with CO, which was first reported by Furukawa et al. in 1965 [263], should be a more efficient and direct route to produce them.

Rieger and coworkers [264–266] and Osakada and coworkers [267] independently studied the alternating copolymerization of epoxides with CO by using $Co_2(CO)_8$/3-hydroxypyridine catalyst system, which was originally discovered by Drent and coworker [268]. The copolymerization of propylene oxide with CO proceeded with high regioselectivity to give poly(3-hydroxybutyrate). Racemic propylene oxide gave atactic polyester, while enantiomerically pure (*S*)-propylene oxide afforded isotactic polyester with retention of configuration (Scheme 1.115), indicating the ring-opening at the less hindered epoxide carbon through the back-side attack by the tetracarbonyl cobaltate.

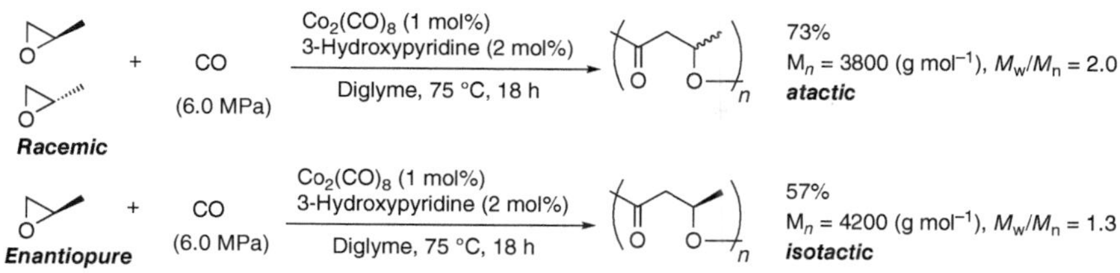

Scheme 1.115 Copolymerization of epoxides with CO.

An equimolar mixture of $Co_2(CO)_8$, benzyl bromide (BnBr), and dihydro-1,10-phenanthroline **9**, which generated $BnCOCo(CO)_4$ under the reaction conditions, copolymerized propylene oxide or 1,2-epoxybutane with CO to give the corresponding atactic polyesters with high regioregularities, perfectly alternating structures, and high molecular weights (Scheme 1.116) [269].

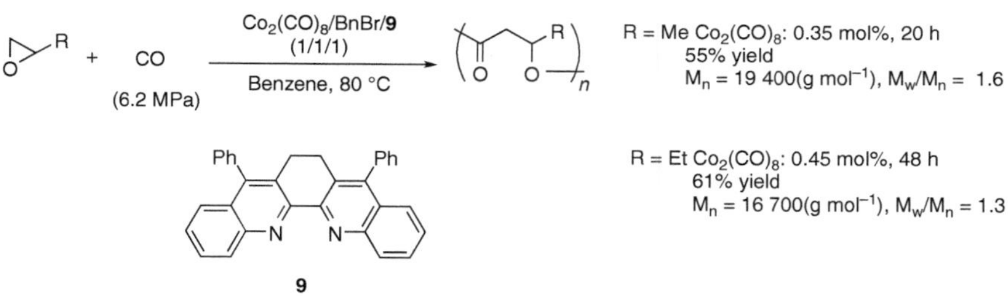

Scheme 1.116 Copolymerized propylene oxide or 1,2-epoxybutane with CO. Source: Modified from Lee and Alper [269].

1.6.4 Alkoxycarbonylation and Aminocarbonylation of Epoxides

Carbonylation of epoxides in the presence of an alcohol gives β-hydroxyesters has been known for a long time [270, 271]. In 1961, Howard prepared methyl/3-hydroxybutyrate by alkoxycarbonylation of propylene oxide with $Co_2(CO)_8$ catalyst. The catalyst system requires 3500 psi pressure, high temperature, and the methyl/3-hydroxybutyrate yields up to 40%.

In 1999, Jacobsen and coworker demonstrated that methoxycarbonylation of epoxides proceeds effectively under relatively mild conditions (55–65 °C) by using $Co_2(CO)_8$/3-hydroxypyridine catalyst system in the presence of methanol (Scheme 1.117) [272]. The high yields in 86–91% and high enantioselectivities (ee > 99%) were obtained.

In 2017, Liu and coworkers reported [1,1-dimethyl-3,3-diethylguanidinium] $[Co(CO)_4]$-catalyzed alkoxycarbonylation of terminal epoxides to β-hydroxy esters

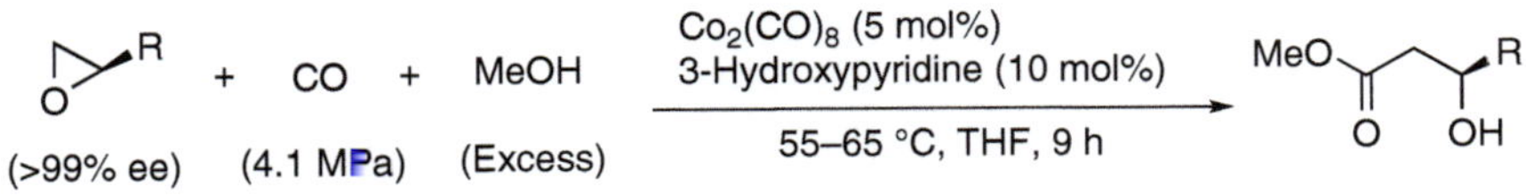

Scheme 1.117 Methoxycarbonylation of epoxides. Source: Modified from Hinterding and Jacobsen [272].

(Scheme 1.118) [273]. Under 6 MPa CO and at 80 °C for 24 h, the corresponding β-hydroxy esters with up to 92% isolated yield can be obtained. This catalyst can be reused for six times without loss of activity.

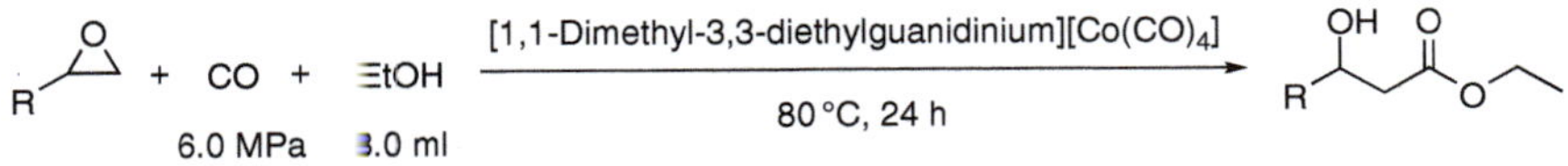

Scheme 1.118 Ethoxycarbonylation of terminal epoxides. Source: Modified from Zhang et al. [273].

Aminocarbonylation can also be carried out by use of CO and a silyl amide. Watanabe and coworkers reported the cobalt-catalyzed aminocarbonylation of epoxides [274]. Some silyl amides such as $PhCH_2NHSiMe_3$ and Et_2NSiMe_3 were applicable to the reaction to give the β-siloxyamide in 30–80% yields, whereas high reaction temperature (100–160 °C) was required. The use of 4-(trimethylsilyl) morpholine was found to be crucial for a milder (25–50 °C) and more efficient carboamination (56–85%): the reaction proceeded at ambient temperature under 0.1 MPa of CO (Scheme 1.119) [275].

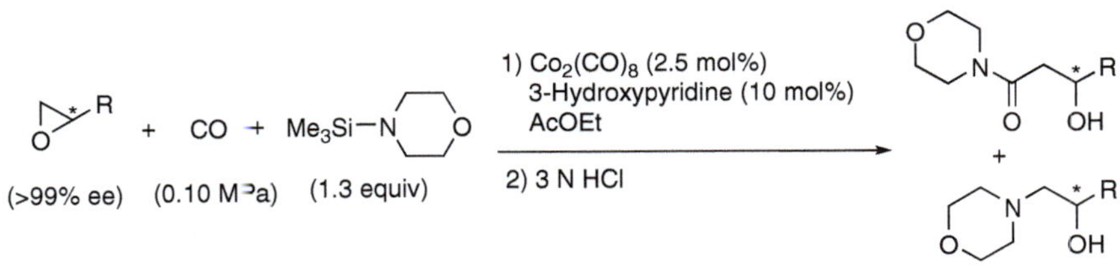

Scheme 1.119 Cobalt-catalyzed aminocarbonylation of epoxides. Source: Modified from Goodman and Jacobsen [275].

1.7 Carbonylation of Aldehydes

Carbonylations of aldehydes incorporate an inexpensive source of carbon, CO, and can convert ubiquitously available aldehydes into various synthetically useful functionalities.

1.7.1 Amidocarbonylations of Aldehydes

As early as in 1971, Wakamatsu et al. described amidocarbonylation reaction of aldehydes and amides in the presence of CO for the synthesis of *N*-acyl-α-amino acids

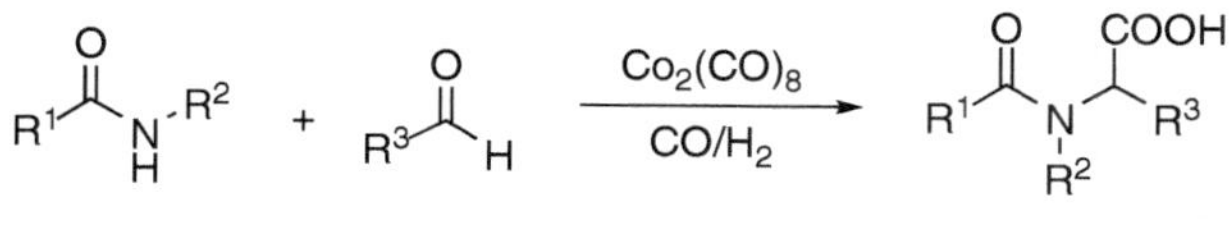

Scheme 1.120 Amidocarbonylation of aldehydes. Source: Wakamatsu et al. [276]; Wakamatsu et al. [277].

(Scheme 1.120) [276, 277]. Under $Co_2(CO)_8$ catalyst system, the products yielded in 26–80% at 110–150 °C.

Lin and coworkers reported the catalytic performance of various cobalt/ligand systems in the synthesis of *N*-acetylglycine. Basic phosphines, such as PBu_3, were shown to allow low-pressure conditions (55 bar). The addition of Ph_2SO or succinonitrile resulted in improved selectivity and facilitated the catalyst recovery [278, 279]. The addition of acid cocatalysts ($pK_a < 3$, e.g., trifluoroacetic acid [TFA]) allowed for low-temperature conditions and absence of hydrogen [280] (Scheme 1.121).

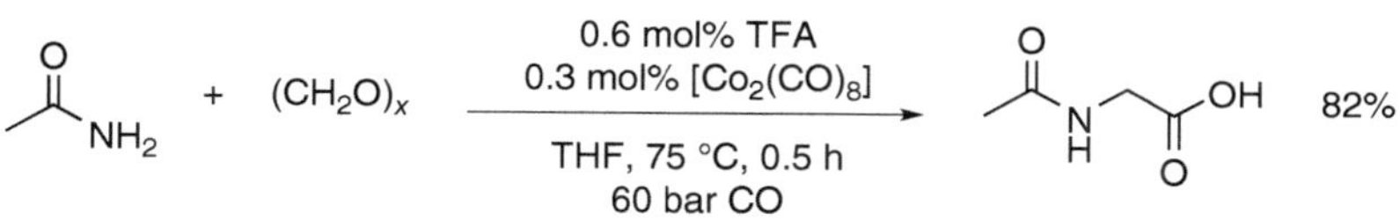

Scheme 1.121 Acid cocatalyst obviates the need for H_2.

The in situ generation of the aldehyde can be achieved by a preceding rearrangement of epoxides and allyl alcohols. Ojima et al. demonstrated the use of styrene oxide and propene oxide in the presence of $[Ti(O\textit{i}Pr)_4]$ or $[Al(O\textit{i}Pr)_3]$ as cocatalysts [281]. The transition metal-mediated isomerization of allylic alcohols ($HRh(CO)(PPh_3)_3$, $Fe_2(CO)_9$, $RuCl_2(PPh_3)_3$, $PdCl_2(PPh_3)_2$) was also shown compatible with amidocarbonylation conditions (Scheme 1.122) [282].

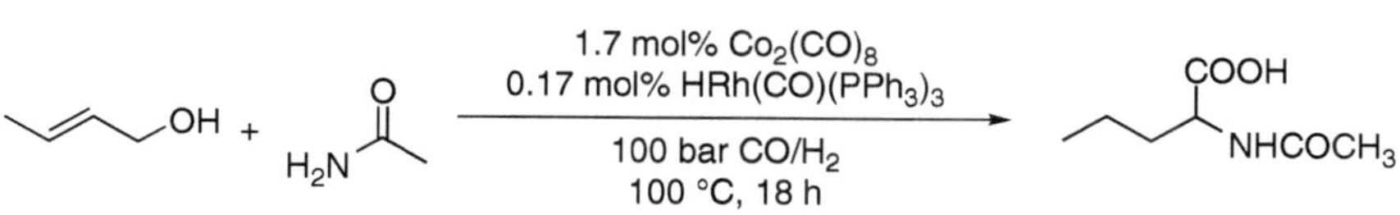

Scheme 1.122 Domino isomerization–amidocarbonylation of allylic alcohols. Source: Modified from Hirai et al. [282].

In 1997, Beller's group found that $PdBr_2/2PPh_3$ is a capable catalysts for the amidocarbonylation of aldehydes. This reaction was typically run at 80–120 °C and 30–60 bar CO. Under optimized conditions, the desired *N*-acyl-α-amino acids could be afforded with TON up to 60 000 (Scheme 1.123) [283, 284].

Later, Beller et al. developed the ureidocarbonylation reaction that provides access to hydantoins containing diverse substituents in the 1-, 3-, and 5-positions (Scheme 1.124) [285]. Under 60 bar CO and 80–130 °C, the isolated products can be obtained in 39–93% yields with TON up to 356.

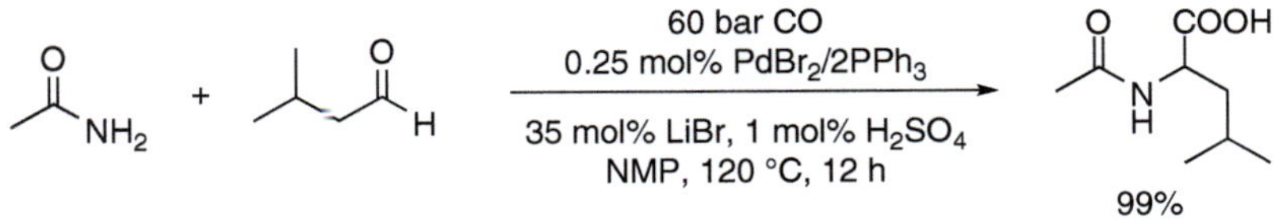

Scheme 1.123 Palladium-catalyzed amidocarbonylation of isovaleraldehyde. Source: Beller et al. [283]; Beller et al. [284].

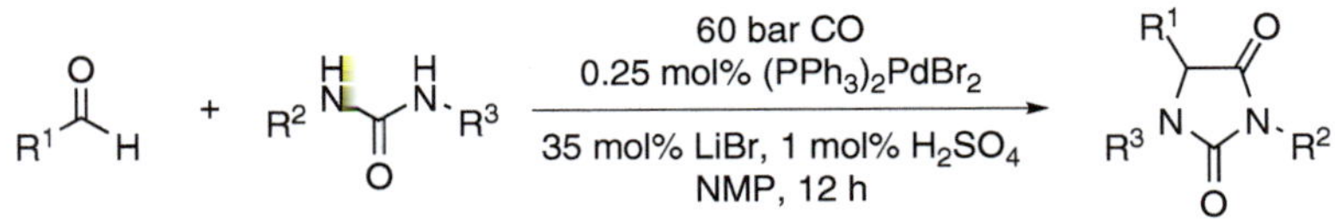

Scheme 1.124 Palladium-catalyzed ureidocarbonylation.

The one-pot amidocarbonylation of commercially more attractive nitriles can be performed via preceding nitrile hydrolysis in HCl/HCO_2H [286, 287]. Systematic studies have shown that amidocarbonylation of benzaldehydes in the presence of palladium catalysts allows for the synthesis of functionalized, racemic *N*-acetyl-α-arylglycines (Scheme 1.125) [288].

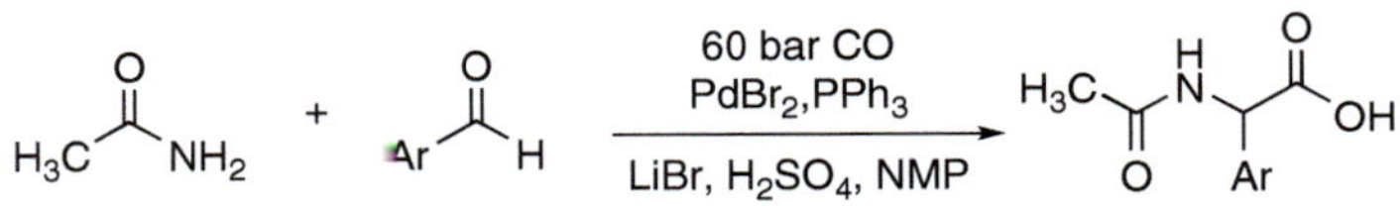

Scheme 1.125 Palladium-catalyzed amidocarbonylation of benzaldehydes. Source: Modified from Beller et al. [285].

1.7.2 Hydroformylation and Silylformylation of Aldehydes

Chan found rhodium/phosphine catalysts and the presence of catalytic amounts of an amine to enhance the catalyst activity most effectively [289]. The nature of the employed ligand (e.g., PPh_3, $(p\text{-}CF_3C_6H_4)_3P$) and base (e.g., NEt_3, py) must be carefully balanced in order to facilitate deprotonation of the hydridorhodium species and prevent base-promoted aldol condensation of glycolaldehyde. Pyridines as solvents greatly enhanced the activity and selectivity of the catalyst, and with pyridines immiscible with water (4-pentylpyridine), 99% of the desired glycolaldehyde could be extracted into the aqueous phase (Scheme 1.126) [290].

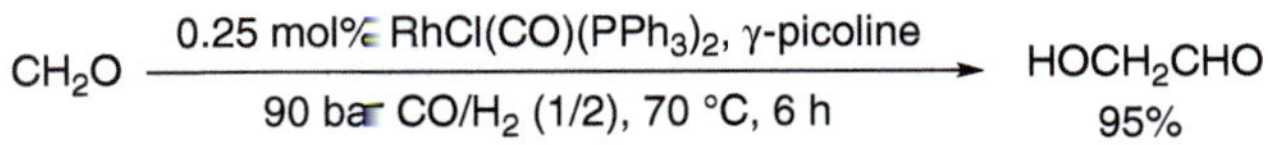

Scheme 1.126 Hydroformylation of formaldehyde. Source: Modified from Okano et al. [290].

Murai et al. reported the direct silylformylation of aliphatic aldehydes (threefold excess) with diethylmethylsilane and CO in the presence of catalytic $Co_2(CO)_8/PPh_3$

to afford α-siloxyaldehydes in moderate yields (100 °C, benzene) [291]. Wright and coworker demonstrated the superior selectivity of a rhodium-catalyzed variant [292]. With 0.5 mol% $[(cod)RhCl]_2$ and Me_2PhSiH, the silylformylation of aliphatic and aromatic aldehydes affords high yields of the α-siloxyaldehydes. The mild reaction conditions (17 bar CO, room temperature) permit discrimination of the starting aldehyde from the newly formed and bulkier siloxyaldehyde.

1.7.3 Hetero Pauson–Khand Reactions of Aldehydes

Hetero PKRs with an aldehyde or ketone component have been shown to afford synthetically versatile γ-butyrolactones. Buchwald and coworkers [293] and Crowe and coworker [294] independently showed that aliphatic enones and enals react with CO under $Cp_2Ti(PMe_3)_2$ mediation (Scheme 1.127). CO insertion and thermal (or oxidative) decomposition gave diastereomerically pure bicyclic γ-butyrolactones and stable $Cp_2Ti(CO)_2$. Imines did not react under the reaction conditions.

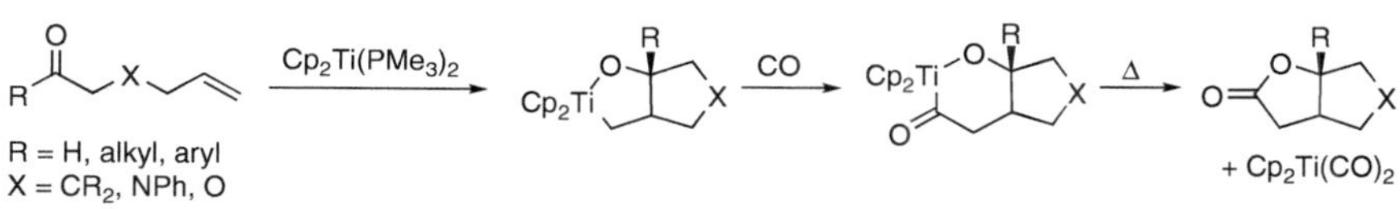

Scheme 1.127 Titanium-mediated hetero Pauson–Khand reaction.

o-Allyl acetophenones have been found capable of displacing CO on $Cp_2Ti(CO)_2$ via electron transfer processes and thus allowing catalytic reactions [295]. While the strong Ti–O bond renders a catalytic reaction difficult in most cases, the use of a late transition metal facilitates the realization of catalytic protocols. Murai reported on the cyclocarbonylation of ynals with CO to give bicyclic lactones in the presence of 2 mol% $Ru_3(CO)_{12}$. Two alternative mechanisms have been discussed, one commencing with oxidative addition of the aldehyde–CH bond to Ru, while the other involves the intermediacy of a metallacycle [296]. Kang et al. developed a related reaction with allenyl aldehydes with CO to afford α-methyl-γ-butyrolactones in good yields (Scheme 1.128) [297].

1.7.4 Reactions of Aldehydes with Acylanions

While various CO-free Umpolung approaches utilize the intermediacy of masked acylanion equivalents, alkyllithium species have been shown to undergo selective carbonylation at low temperatures (<−110 °C) to form acylanions. Such in situ prepared acyllithium species have been demonstrated as versatile reagents for the direct nucleophilic acylation of aldehydes [298, 299]. The resultant acyloins were obtained in good yields from reactions of *s*- and *t*-BuLi with aliphatic and aromatic aldehydes (<95%, Scheme 1.129).

Stable acylzirconocene chloride has been utilized by Hanzawa and coworkers [300] for the synthesis of α-hydroxy ketones. Consecutive hydrozirconation of

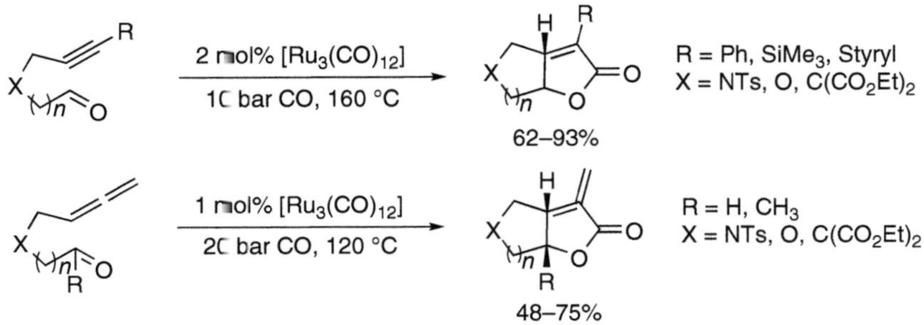

Scheme 1.128 Ruthenium-catalyzed hetero Pauson–Khand reactions with alkynes and allenes. Source: Modified from Kang et al. [297].

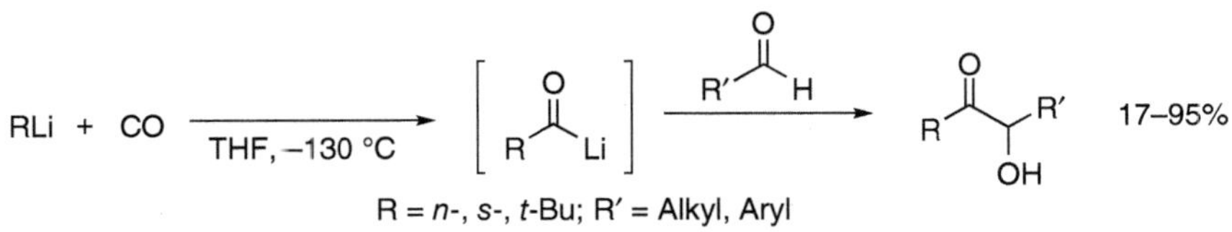

Scheme 1.129 Acyllithium addition to aldehydes.

terminal alkenes or alkynes with Schwartz's reagent and in situ carbonylation with gaseous CO at room temperature afford acylzirconocene chloride. The yields of addition reactions to aldehydes were strongly dependent on the presence of a Lewis acid. Interestingly, the isomeric α-ketol was obtained in some reactions with alkenes when $BF_3{\cdot}OEt_2$ was used (Scheme 1.130).

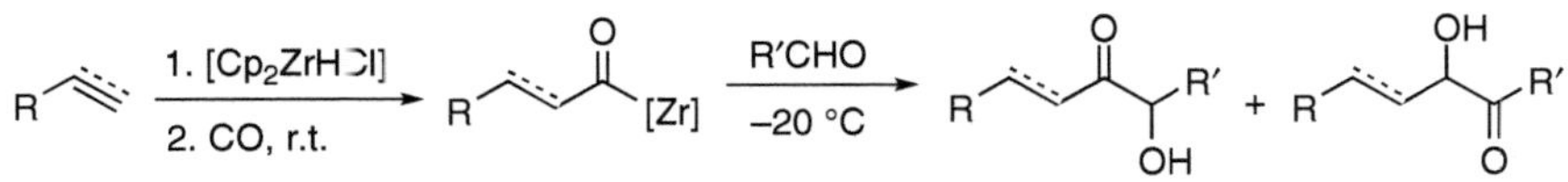

Scheme 1.130 Acylzirconocene addition to aldehydes.

1.7.5 Miscellaneous of Aldehydes

The three-component synthesis of benzo and naphthofuran-2(3*H*)-ones from the corresponding aromatic alcohol (phenols or naphthols) with aldehydes and CO (5 bar) can be performed under palladium catalysis (Scheme 1.131) [301, 302].

Woo showed that α,β-unsaturated aldehydes can be carbonylated in conc. H_2SO_4 to give 3,4-dialkyl-2(5*H*)-furanones (70 bar CO, 50 °C) [303]. Although carbonylation occurs in conjugate fashion, secondary intramolecular carbonylation of the aldehydeoxygen results in the formation of the furanone ring. Simple aldehydes can be converted to α,β-unsaturated aldehydes by rapid aldol condensation under the acidic reaction conditions and give only slightly lower yields (Scheme 1.132).

Toniolo and coworker reported the carbonylation of aromatic aldehydes containing electron-donating substituents with a Pd/PPh_3 catalyst system in the presence of HCl to give phenylacetic acid derivatives [304]. No activity was observed in the

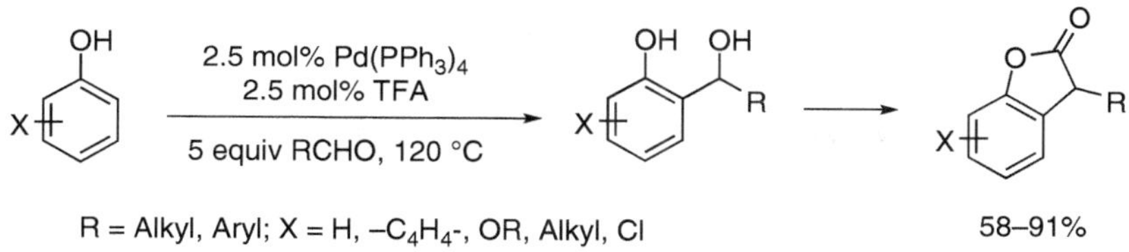

Scheme 1.131 Cyclocarbonylation to furanones. Source: Satoh et al. [301]; Satoh et al. [302].

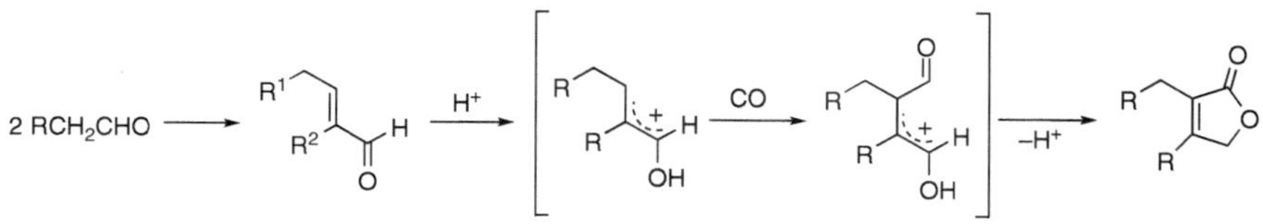

Scheme 1.132 Acid-mediated conjugate carbonylation.

absence of PPh_3 or HCl, and high yields could be achieved with alkanols as a solvent (e.g., EtOH). After decarboxylation of the resultant phenylmalonate, the desired phenylacetic acid derivative in 70–75% yield can be obtained.

1.8 Oxidative Carbonylation Reaction

Oxidative carbonylations have acquired a growing importance during the last few years, owing to the development of new and selective catalytic systems, mainly based on palladium, which are able to promote ordered sequences of transformations under mild conditions with formation of highly functionalized carbonyl compounds in one step starting from simple building blocks [305–307].

1.8.1 Oxidative Carbonylation of Alkenes

In 1963, Tsuji et al. reported the reaction of olefin–palladium chloride complexes with CO and alcohols to produce β-chloroacyl chlorides for the first time [308, 309]. Using internal and terminal aliphatic olefins under 4–10 MPa CO pressure, the corresponding chloroesters in 5–41% yields were transformed.

In 1969, Yukawa and Tsutsumi reported the reaction of a styrene–palladium complex with CO in alcohols [310]. Various cinnamates and phenylsuccinates were synthesized with up to 41% yield.

In 1979, Cometti and Chiusoli published their results on the synthesis of methyl cinnamates from styrene [311]. Using a mixture of $PdCl_2$, $CuCl_2$, $MgCl_2$, and NaOAc, the reaction was run in methanol at room temperature under atmospheric pressure of CO to produce methyl cinnamate with up to 80% selectivity together with dimethyl phenylsuccinate as products (Scheme 1.133).

In 1990, Inomata et al. reported the Pd/C-catalyzed oxidative carbonylation of terminal olefins [312]. The reaction proceeded selectively to mono- or diesters under

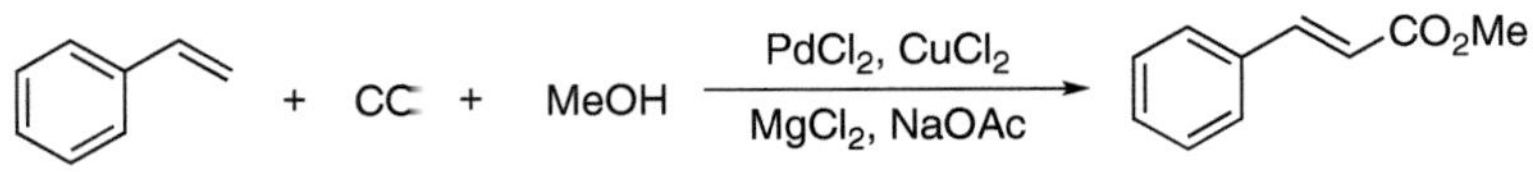

Scheme 1.133 Palladium-catalyzed synthesis of methyl cinnamate.

1 bar of CO at room temperature in good yields in the presence of $CuCl_2$ or CuCl as an additive. Interestingly, mainly monoesters were observed in 82–99% yields if $CuCl_2$ was applied as an oxidant, whereas diesters were formed in 56–99% yields by using CuCl as the oxidizing reagent (Scheme 1.134).

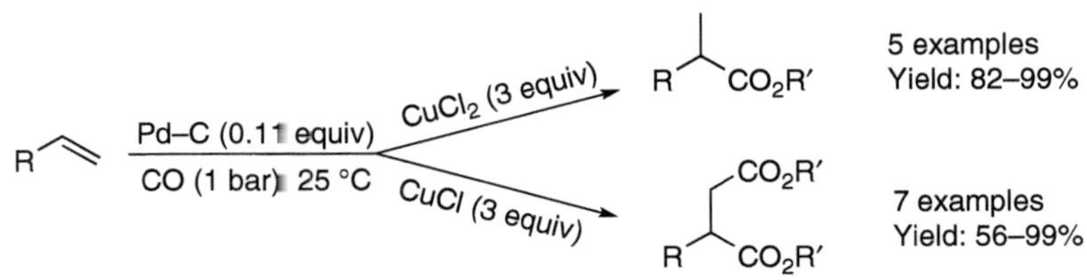

Scheme 1.134 Pd/C-catalyzed oxidative carbonylation of alkenes.

Later, an asymmetric version for the synthesis of diesters was published by Inomata and coworkers [313]. By using a chiral bisoxazoline ligand in the presence of Cu^I triflate at room temperature, terminal olefins were carbonylated to enantiomerically enriched diesters in 35–73% yields with up to 66% ee (Scheme 1.135).

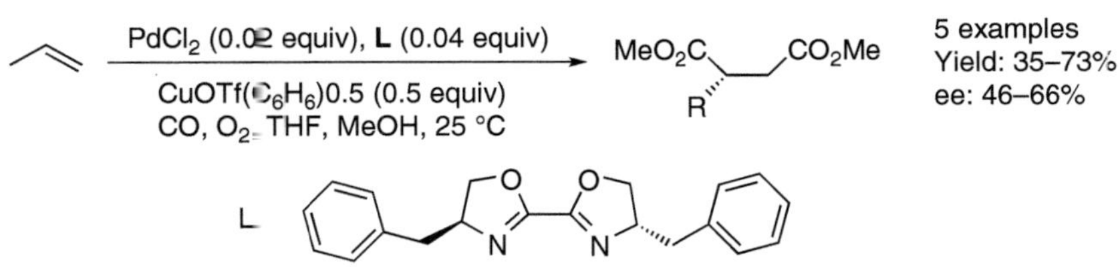

Scheme 1.135 Palladium-catalyzed oxidative carbonylation of olefins. Tf = trifluoromethanesulfonate.

In 1998, Saigo and coworkers reported the use of phosphine sulfides as ligands for the oxidative carbonylation of olefins [314]. Diesters were produced in 35–90% yields starting from the corresponding alkenes at room temperature (Scheme 1.136). Enantioselectivity can be obtained by applying chiral biphosphine sulfides as ligands. Notably, their model system gave rise to 36% and 60% yield in the absence of a ligand and with triphenylphosphine oxide, respectively. However, no reaction occurred in the presence of PPh_3.

Alper and coworkers reported another protocol for the hydroxycarbonylation of alkenes [41]. When $PdCl_2$ and $CuCl_2$ were applied as the catalytic system, alkenes were transformed into branched propionic acids in 30–100% yields in the presence of water, oxygen, and HCl. Later, they extended their protocol to the

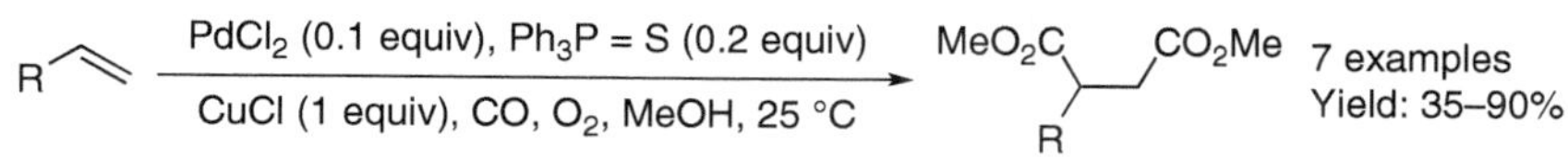

Scheme 1.136 Palladium-catalyzed oxidative carbonylation of olefins using triphenylphosphine sulfide as a ligand.

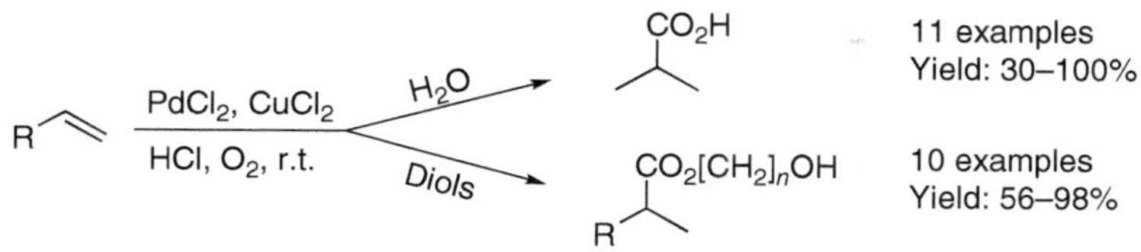

Scheme 1.137 Palladium-catalyzed oxidative carbonylation of alkenes to branched products.

monohydroesterification of diols [315]. Under 1 bar of CO and at room temperature, the products could be yielded in 56–98% (Scheme 1.137).

Sasai and coworkers published the enantioselective oxidative carbonylation of alkenylureas. By using the PE-spiro bis(isoxazoline) system, the desired products were produced in 28–97% yields and with 30–89% ee (Scheme 1.138) [316].

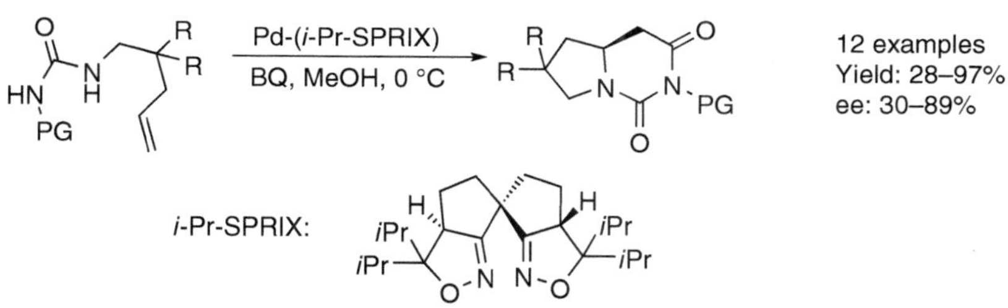

Scheme 1.138 Palladium-catalyzed oxidative carbonylation of alkenylureas. PG = protecting group. Source: Modified from Jamison et al. [116].

The oxidative carbonylation of alkenes has been used in many applications in total synthesis [317–320], but most of the work has focused on ligands and catalyst systems. In general, large amounts of excess copper salts or other organic oxidants are still required. This has an impact on the environment, and the separation of products is difficult. Therefore, the important goal in the future is to utilize air, oxygen, or hydrogen peroxide as "green oxidants" under mild conditions.

1.8.2 Oxidative Carbonylation of Alkynes

The first example of oxidative carbonylations of alkynes was reported in 1964. Tsuji et al. demonstrated the palladium-mediated transformation of acetylene into muconyl, fumaryl, and maleic acid chloride (Scheme 1.139) [321]. Later, they used diphenylacetylene as a substrate for the synthesis of lactones in the presence of alcohol and HCl [322].

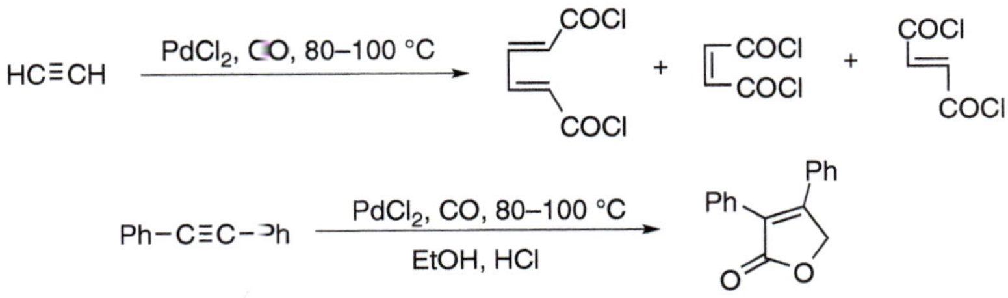

Scheme 1.139 Palladium-mediated oxidative carbonylation of alkynes. Source: Modified from Tsuji et al. [321].

Tsuji et al. further developed the palladium-catalyzed oxidative carbonylation of terminal acetylenes. For example, acetylenecarboxylates were produced in 59–74% yields at room temperature under atmospheric pressure of CO, but stoichiometric amounts of $CuCl_2$ were needed to reoxidize Pd^0 (Scheme 1.140) [69].

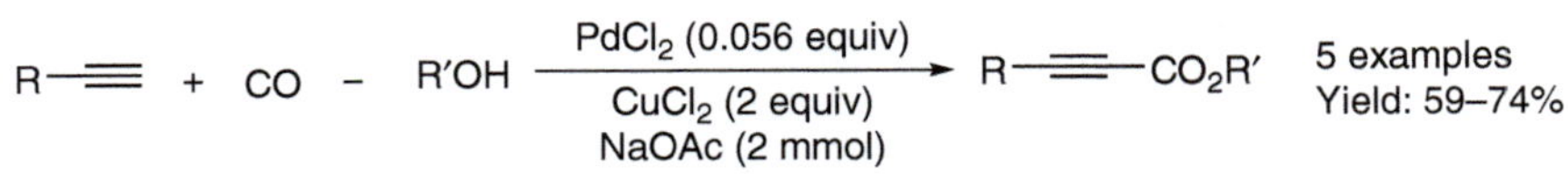

Scheme 1.140 Palladium-catalyzed oxidative carbonylation of acetylenes to acetylenecarboxylates. Source: Modified from Tsuji et al. [69].

Jiang and coworkers described the synthesis of 3-chloroacrylate esters [323]. The reaction proceeded under 1 bar of CO at room temperature yielding 30–72% of the desired products in a highly regio- and stereoselective manner (Scheme 1.141).

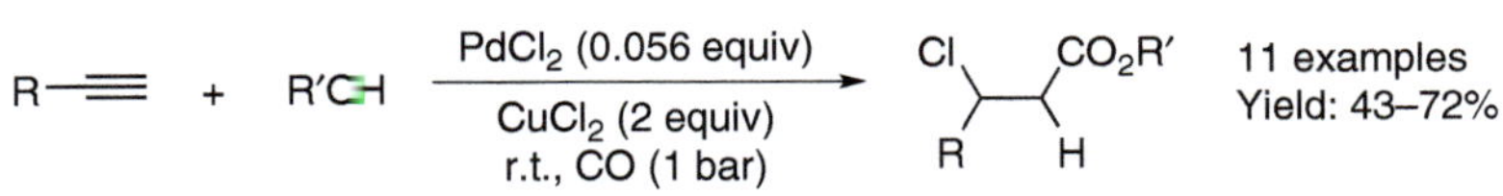

Scheme 1.141 Palladium-catalyzed oxidative carbonylation of acetylenes to chloroacrylates.

The oxidative carbonylation of 1,1-disubstituted propargyl acetates to unsaturated esters was developed by Okumoto et al. in 1999 [324]. Stoichiometric amounts of $CuCl_2$ were used as the oxidant, and the reaction had to be conducted at 0 °C (Scheme 1.142).

A mild and efficient methodology for the palladium-catalyzed carbonylative synthesis of four-membered β-lactones has been published by Ma et al. [325]. In the presence of $PdCl_2$ and $CuCl_2$, β-lactones were produced from 2-alkynols in good yields (Scheme 1.143). Using readily available optically active propargylic alcohols, the corresponding b-lactones were generated with high ee.

Moreover, an improved method for the oxidative carbonylation of hydroxyalkynes was developed by Kato et al. [326]. Applying $Pd(MeCN)_2Cl_2$ (0.05 equiv) and 1.1 equiv of BQ in methanol at 0 °C under 1 bar of CO, the desired products were

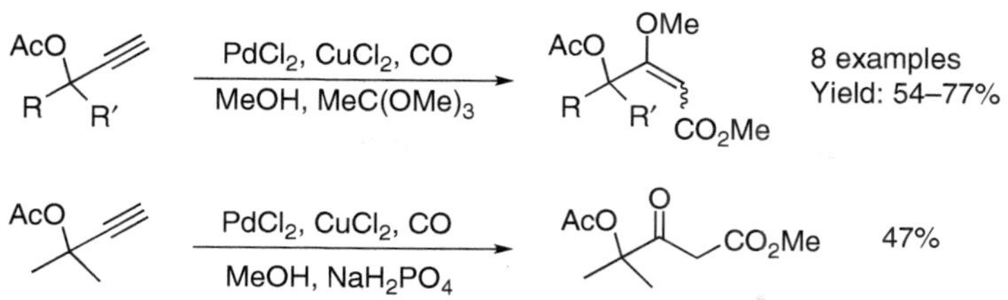

Scheme 1.142 Palladium-catalyzed oxidative carbonylation of propargylic acetates.

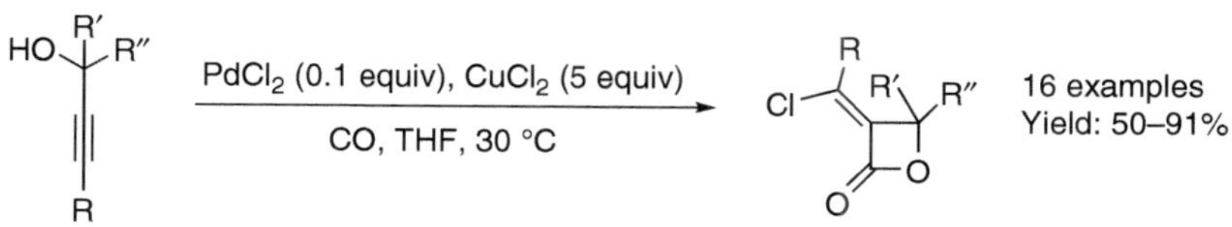

Scheme 1.143 Palladium-catalyzed oxidative carbonylation of alkynes to β-lactones.

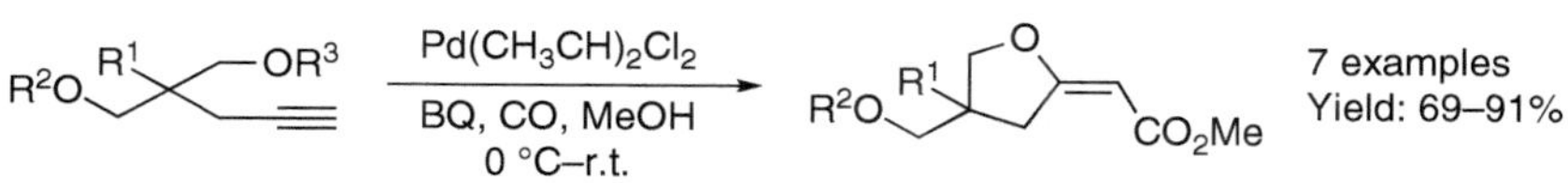

Scheme 1.144 Palladium-catalyzed oxidative carbonylation of alkynes to furans.

obtained from the corresponding cyclic- and acyclic-4-yn-1-ols in 69–91% yields (Scheme 1.144).

In 2009, Kato et al. applied their Pd^{II} bis(oxazoline) complexes in the intermolecular methoxycarbonylation of terminal alkynes. Terminal alkynes were transformed into β-methoxyacrylates in good yields (Scheme 1.145) [327]. In addition, this methodology was applied in the total synthesis of annularin G and annularin H [328].

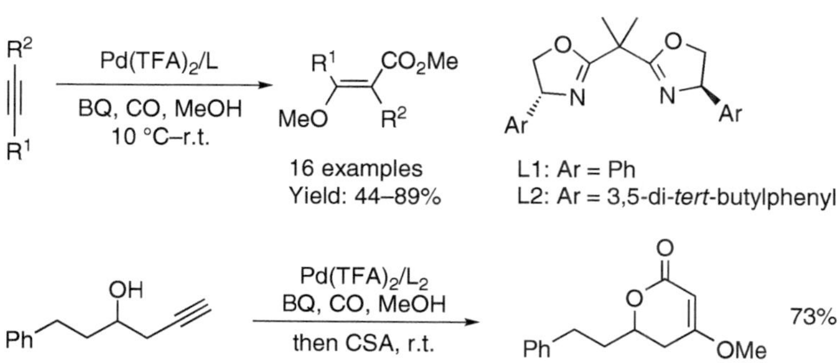

Scheme 1.145 Palladium/BQ-catalyzed oxidative carbonylation of alkynes. CSA = (+)-10-camphorsulfonic acid. Source: Modified from Kato et al. [327].

In 2001, Gabriele and Salerno reported the first catalytic aminocarbonylation of alk-1-ynes [329]. Using a 4 : 1 CO : air mixture (20 atm total pressure), both alkyl- and arylacetylenes could be converted to the corresponding 2-ynamides in 42–80% yields successfully and the latter were more reactive substrates (Scheme 1.146).

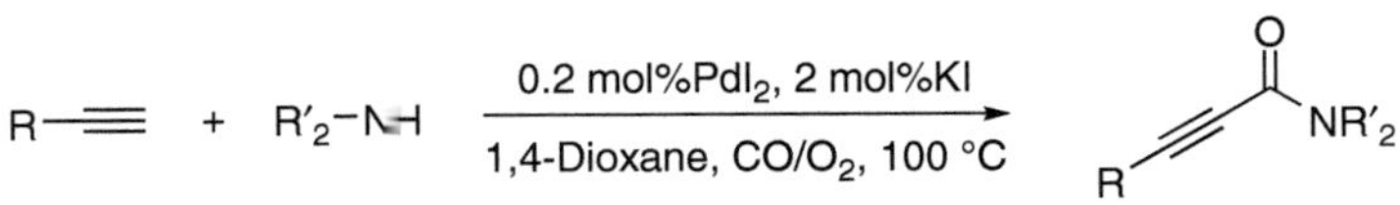

Scheme 1.146 Palladium-catalyzed oxidative aminocarbonylation of alk-1-ynes.

In 2012, Bhanage and coworkers presented Pd/C-catalyzed oxidative aminocarbonylations of alk-1-ynes that is applicable for the synthesis of a wide range of biologically active alk-2-ynamide derivatives (Scheme 1.147). Under the mild conditions, secondary amines provide the corresponding alk-2-ynamides in 55–94% yields [330]. The catalyst can reuse for up to four consecutive cycles.

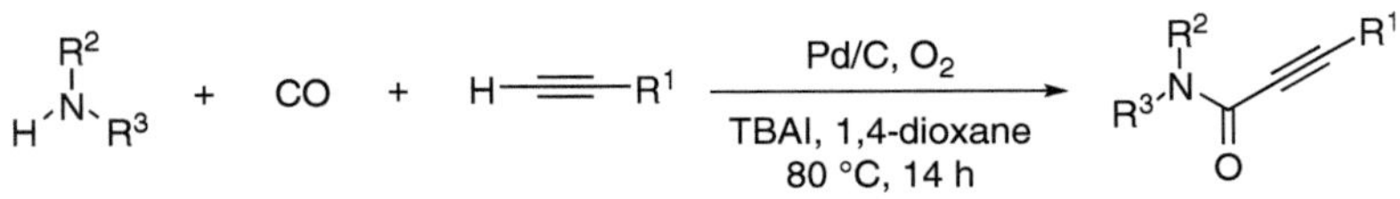

Scheme 1.147 Pd/C-catalyzed oxidative aminocarbonylation of alk-1-ynes.

In 2015, Xia and Liu described palladium–NHC catalytic system that allows the direct formation of a range of 2-ynamides from commercially available alkynes via oxidative aminocarbonylation (Scheme 1.148) [331]. Under 3 MPa CO, 0.5 MPa O_2 and at 100 °C, the corresponding products in 50–95% yields can be obtained. They also prove that amido-Pd could be transformed into amido-Pd-alkynyl in this efficient process.

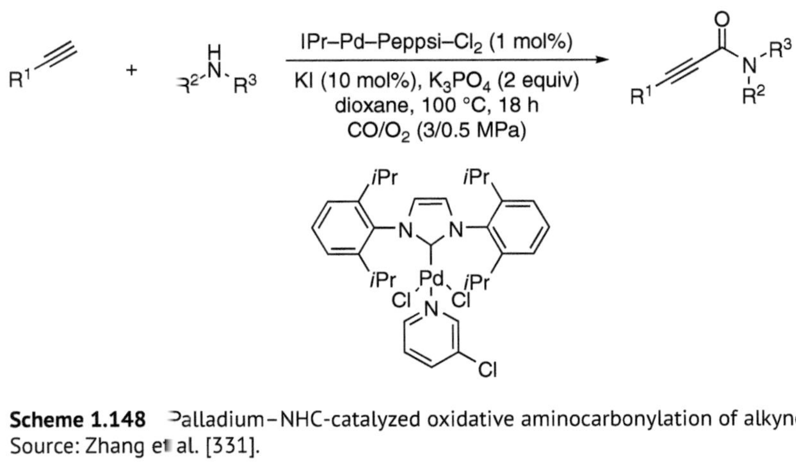

Scheme 1.148 Palladium-NHC-catalyzed oxidative aminocarbonylation of alkynes. Source: Zhang et al. [331].

1.8.3 Oxidative Carbonylation of Organometallic Reagents

In 1974, Stille and Wong published the first oxidative carbonylation based on organomercury compounds [332]. Using stoichiometric amounts of $PdCl_2$, LiCl, and NaOAc, stereoselective carbonylation occurred at room temperature. However, the desired products were obtained only in low yield. Larock improved the methodology by conducting the reaction at −78 °C [333]. Unsaturated carboxylic acids and esters were formed in 72–100% yields starting from the corresponding vinylmercurials. Although stoichiometric amounts of palladium were still necessary, a catalytic version was realized by addition of a large excess of $CuCl_2$ (Scheme 1.149).

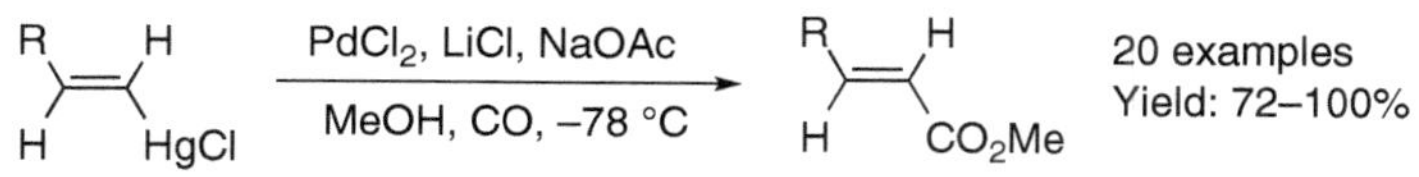

Scheme 1.149 Palladium-promoted carbonylation of organomercuries.

In 1979, Tamao et al. published concerning the oxidative carbonylation of organosilanes [334]. The reaction was carried out at room temperature and under 1 bar of CO, but stoichiometric amounts of $PdCl_2$ were required to form 61–91% of the corresponding vinyl esters (Scheme 1.150).

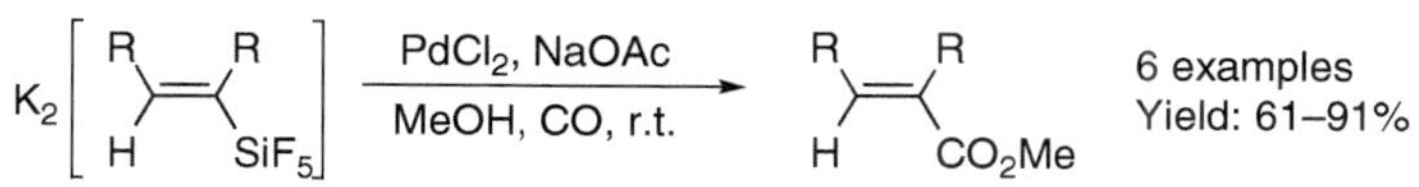

Scheme 1.150 Palladium-promoted carbonylation of organosilanes.

Lei and coworkers developed the first example of palladium-catalyzed oxidative carbonylation of organoindium reagents to produce different esters with desyl chloride as an oxidant [335]. Primary and secondary alkyl indium reagents as well as aryl indium reagents were carbonylated at 60 °C in the presence of catalytic amounts of palladium as a catalyst. The corresponding esters were formed in 60–96% yields, and the methodology showed broad functional group tolerance (Scheme 1.151).

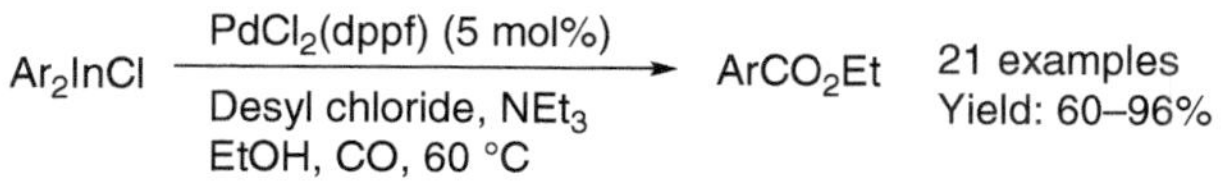

Scheme 1.151 Palladium-catalyzed carbonylation of organoindiums.
DPPF = 1,1′-bis(diphenylphosphino)ferrocene.

In 1995, Knochel and coworker reported the oxidative carbonylation of organozinc reagents to produce symmetric ketones in the presence of stoichiometric cobalt bromide (Scheme 1.152) [336].A series of alkyl and aromatic symmetric ketones were prepared in 56–80% yields at room temperature.

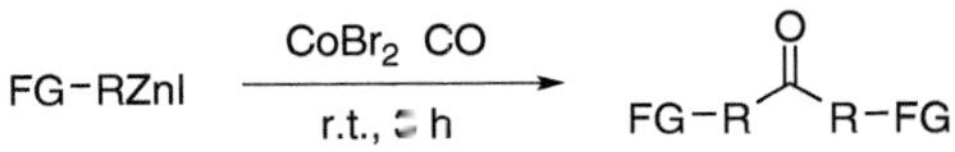

Scheme 1.152 Stoichiometric $CoBr_2$-mediated oxidative carbonylation of organozinc reagents. FG = functional group. Source: Modified from Devasagayaraj and Knochel [336].

Then, Jackson and coworkers reported the palladium-catalyzed oxidative carbonylation of organozinc reagents using O_2 as the oxidant (Scheme 1.153) [208]. Under 1 atm pressure of CO, the corresponding symmetric ketones with up to 84% yield can be obtained at room temperature.

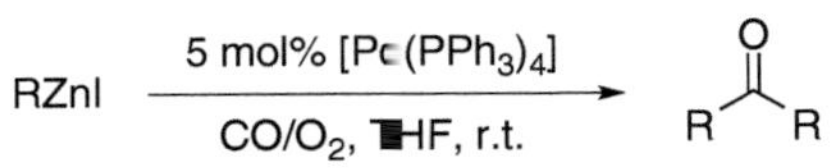

Scheme 1.153 Palladium-catalyzed oxidative carbonylation of organozinc reagents. Source: Modified from Jackson et al. [208].

Kang and coworkers found oxidative carbonylation of organolead reagents was also employed to produce symmetric ketones (Scheme 1.154) [337]. In the presence of $Pd_2(dba)_3{\cdot}CHCl_3$ (5 mol%) and NaOMe (5 equiv) in CH_3CN under atmospheric pressure of CO at room temperature, organolead acetates can be carbonylated to afford symmetrical ketones with wide range of functional groups including heterocycles in 69–85% yields. Notably, the organolead compounds acted as both nucleophiles and oxidants, and the authors proposed that $RPb(OMe)_3$ was formed as a key intermediate.

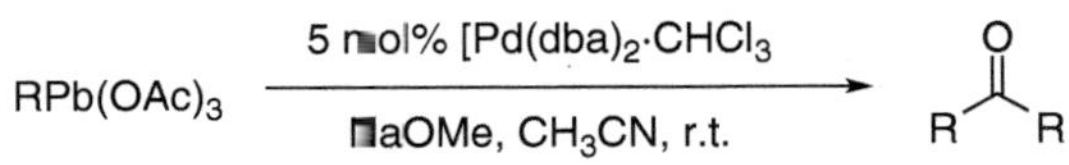

Scheme 1.154 Palladium-catalyzed oxidative carbonylation of organolead reagents. dba=dibenzylideneacetone. Source: Modified from Kang et al. [337].

Suzuki and coworker reported the palladium-catalyzed oxidative carbonylation of alkenylboranes as early as 1981. They prepared 1-alkenylboranes through hydroboration of alkynes, and subsequent oxidative carbonylation mediated by catalytic amounts of $PdCl_2$ in the presence of NaOAc and BQ in methanol provided unsaturated esters in good yields (Scheme 1.155a) [338]. Later, a stereoselective synthesis of β-mono and β,β-disubstituted α,β-unsaturated esters was established through a stepwise cross-coupling alkylation followed by an oxidative carbonylation of 2-bromo-1-alkenylboranes (Scheme 1.155b) [339]. Good yield and excellent stereoselectivity were achieved.

Lei and coworkers discovered a new protocol making use of air as an oxidant at low temperature [340]. Using a balloon pressure of a CO/air mixture, arylboronates were converted into the corresponding esters in 61–95% yields (Scheme 1.156). This was the first example of simply applying air in the oxidative carbonylation of organoboron compounds with alcohols.

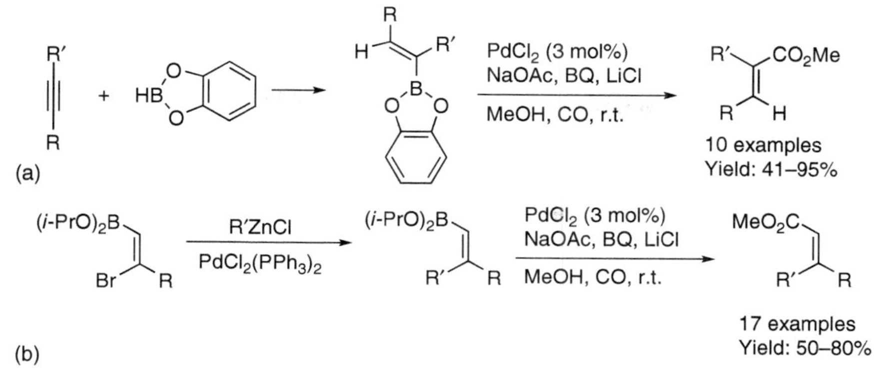

Scheme 1.155 Palladium-catalyzed oxidative carbonylation of alkenylboranes. Source: Modified from Miyaura and Suzuki [338]; Modified from Yamashina et al. [339].

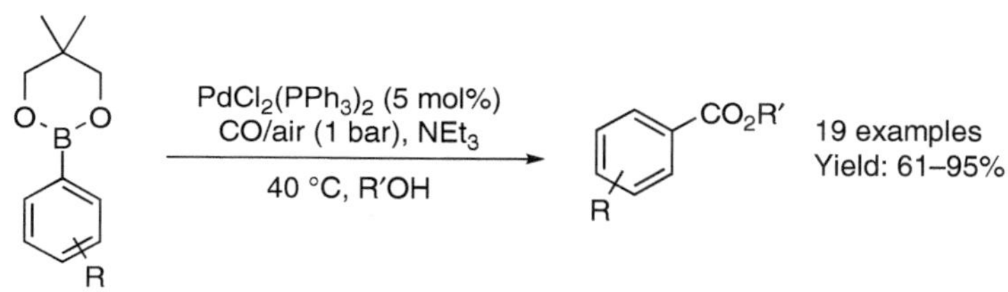

Scheme 1.156 Palladium-catalyzed oxidative carbonylation of arylboronates using air.

1.8.4 Oxidative Carbonylation of Arenes

In 1980, Fujiwara et al. reported for the first time a palladium-mediated oxidative carbonylation of arenes to benzoic acids (Scheme 1.157) [341]. The direct carbonylation of benzene, toluene, anisole, chlorobenzene, furan, and thiophene was performed under CO and in the presence of $Pd(OAc)_2$. 2–43% of the corresponding benzoic acids were formed as the terminal products.

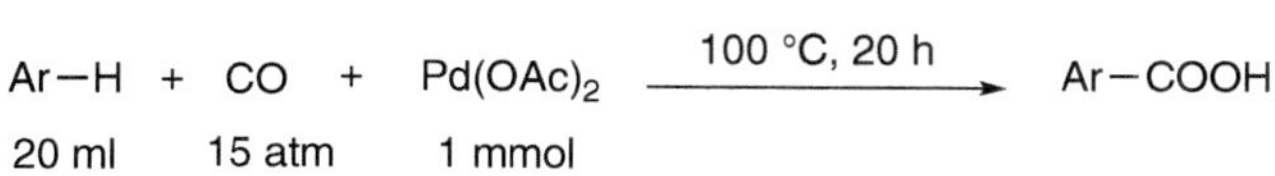

Scheme 1.157 Stoichiometric amounts of $Pd(OAc)_2$ for mediation of the oxidative carbonylation of simple arenes. Source: Modified from Fujiwara et al. [341].

Later, Fujiwara and coworkers developed a more powerful catalytic system (Scheme 1.158) [342, 343]. A series of simple arenes such as benzene, toluene, chlorobenzene, anisole, and naphthalene were oxidatively carboxylated by $Pd(OAc)_2$ in the presence of potassium peroxodisulfate as the oxidant in TFA at room temperature under an atmosphere of CO. The aromatic carboxylic acids were formed in good yields.

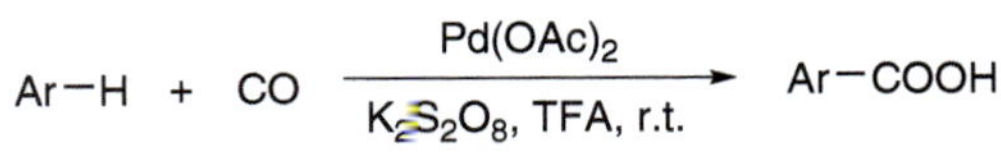

Scheme 1.158 Pd(OAc)$_2$/TFA-catalyzed oxidative carbonylation of simple arenes. Source: Lu et al. [342]. Taniguchi et al. [343].

Widenhoefer and Liu succeeded in palladium-catalyzed oxidative carbonylations for the synthesis of tetrahydrocarbazoles [344, 345]. Starting from alkenyl indoles in the presence of $PdCl_2(CH_3CN)_2$ (5 mol%) and $CuCl_2$ (3 equiv) under 1 bar of CO in THF, the corresponding products were obtained in 45–92% yields with high regioselectivity (Scheme 1.159).

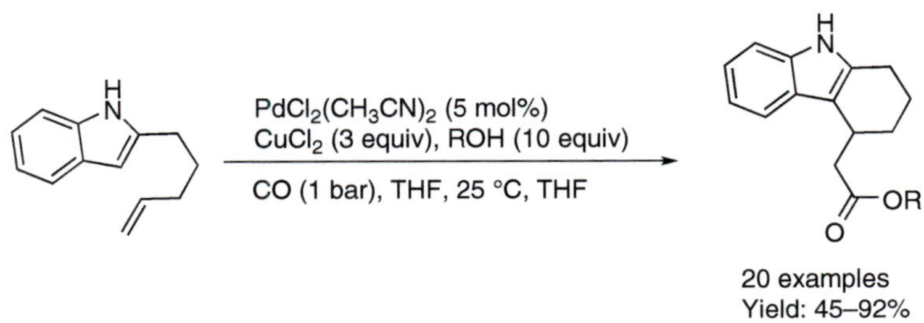

Scheme 1.159 Palladium-catalyzed cyclization/alkoxycarbonylation of alkenyl indoles.

In 2008, Yu and coworker subsequently reported the Pd^{II}-catalyzed direct carboxylation of benzoic and phenylacetic acid derivatives to form dicarboxylic acids (Scheme 1.160) [346]. In the presence of Ag_2CO_3 as an oxidant, 40–93% yields of dicarboxylic acids can be obtained at 130 °C with 1 atm CO. They also characterized the first cyclometalation complex formed from carboxylic acids by X-ray crystallography.

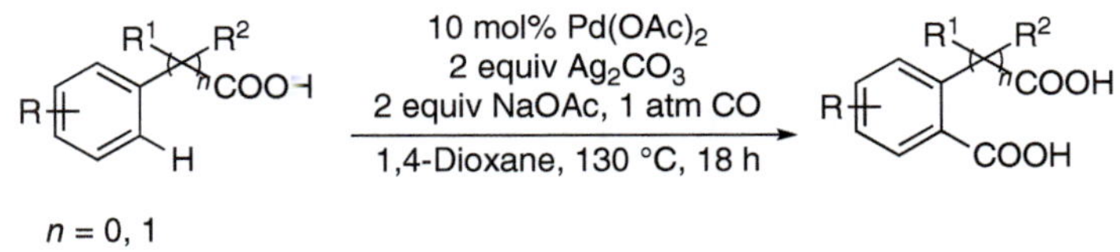

Scheme 1.160 Pd(OAc)$_2$-catalyzed oxidative carbonylation of benzoic and phenylacetic acid. Source: Giri and Yu [346].

Booker-Milburn and Lloyd-Jones published an interesting oxidative carbonylation of aniline derivatives [347]. The reaction proceeded under 1 bar of CO at room temperature with 5 mol% of $[Pd(OTs)_2(MeCN)_2]$ as a precatalyst. Cyclic imidates in 34–85% yields and methyl anthranilates in 30–97% yields can be obtained (Scheme 1.161). Notably, different products can be achieved by simply changing the reaction conditions, such as solvents.

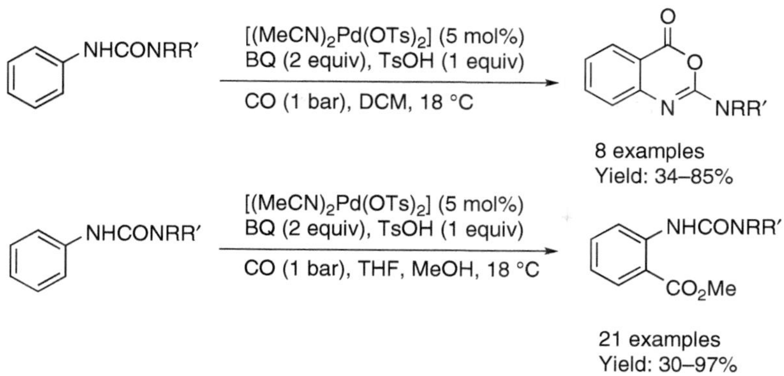

Scheme 1.161 Palladium-catalyzed oxidative carbonylation of aniline derivatives.

1.8.5 Oxidative Carbonylation of Amines

The palladium-catalyzed oxidative carbonylations of amines to ureas were described for the first time by Tsuji and Iwamoto in 1966 [348]. With *n*-decylamine (7.9 g) and palladium chloride (0.5 g) in benzene (30 ml) at 180 °C for 20 hours with shaking, *N*,*N*′-didecyloxamide (3.8 g) and 1,3-didecylurea (2.2 g) were obtained.

In 2001, Deng and Shi reported the first gold(I) complex-catalyzed oxidative carbonylation of amines for the syntheses of carbamates (Scheme 1.162) [349]. At 200 °C and 5 MPa of initial total pressure, 34–100% conversions and 35–93% selectivities can be obtained. Then, they prepared polymer-immobilized gold catalysts ($HAuCl_4{\cdot}4H_2O$) for the oxidative carbonylation of aniline and its derivatives [350]. Carbamates and symmetric ureas with up to 99% selectivity can be obtained at 175 °C with 5.0 MPa mixed gases (CO of 99.99% purity at 4.0 MPa and O_2 of 99.99% purity at 1.0 MPa).

$$R(NH_2)_n + CO + O_2 + R'OH \xrightarrow[200\,°C,\ 3\,h]{[Au(PPh_3)_x]_yZ,\ PPh_3} R(NHCO_2R')_n + H_2O$$

R = Ph, alkyl
$R' = CH_3, CH_3CH_2$
n = 1 or 2; x,y = 1 or 2
Z = Cl, NO_3, S

Scheme 1.162 Gold(I) complex-catalyzed oxidative carbonylation of amines. Source: Shi and Deng [349].

In 2001, Deng and Shi also found a novel $PdCl_2/ZrO_2–SO_4^{2-}$ catalyst for the synthesis of carbamates by oxidative carbonylation of amines [351]. For example, under O_2/CO (MPa) = 0.7/3.3 and at 170 °C, β-naphthylamine was oxidatively carbonylated, almost 100% conversion and 95% selectivity could be achieved as the reaction proceeded in only 15 min. Then, they developed palladium complex–IL catalyst system for the carbonylation of amines to give carbamate or urea [352]. When using $BMImBF_4$ as IL, under 4.0 MPa CO and 1.0 MPa O_2 at 175 °C, 99% conversion of aniline and 98% selectivity of methyl phenylcarbamate with 4540 h^{-1} TOF can be observed. In addition, IL-containing $Pd(phen)Cl_2$ could also be recovered and reused.

In 2005, Deng and Shi prepared silica gel confined IL containing a metal complex ($HRu(PPh_3)_2Cl_2$, $Rh(PPh_3)_3Cl$, $Pd(PPh_3)_2Cl_2$, and $Co(PPh_3)_3Cl_2$) as heterogenized catalysts for the carbonylation of amines and nitrobenzene without molecular oxygen to afford the corresponding ureas (Scheme 1.163) [353]. For example, when using Rh–$DMImBF_4$/silica gel as a catalyst, 92% conversion of aniline and 98% selectivity of disubstituted ureas with 11 548 h^{-1} TOF can be observed.

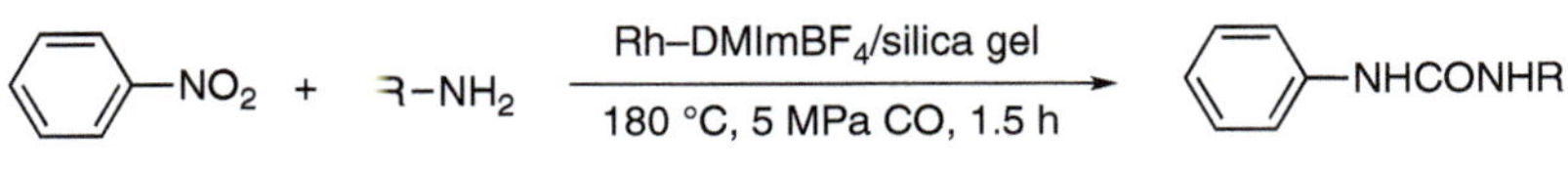

Scheme 1.163 Rh–$DMImBF_4$/silica gel-catalyzed oxidative carbonylation of amines. Source: Modified from Shi et al. [353].

In 2007, Xia and coworkers demonstrated that oxidative carbonylations of amines to ureas are also catalyzed by supported N-heterocyclic carbine (NHC)–palladium complexes (Scheme 1.164) [354, 355]. They showed that the palladium–carbene complex containing both an aniline and an NHC ligand was the active species for this oxidative carbonylation of amines. Without any promoter, various ureas were produced in 46–99% yields. Other immobilized palladium nanoparticles and water-soluble palladium catalysts were also presented [356, 357].

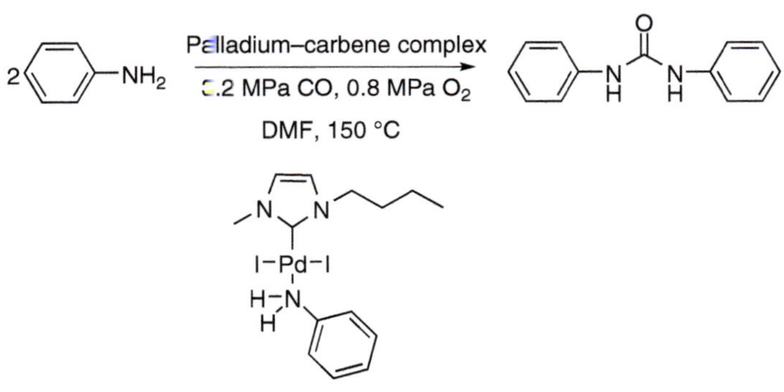

Scheme 1.164 One example of palladium–carbene-catalyzed oxidative carbonylation of amines. Source: Zheng et al. [354]; Zheng et al. [355].

In 2010, Giannoccaro et al. reported $L_nPdCl_2/NEt_3{\cdot}HI$ (L = 2,2′-dipyridine (dipy), 2-(β-diphenylphosphine)ethylpyridine (PN), PPh_3, CH_3CN)-catalyzed oxidative carbonylation of amino alcohols to *N*,*N*′-bis(hydroxyalkyl)ureas under mild conditions (30–60 °C, $P(CO/O_2) = 0.1$ MPa) using molecular oxygen as the oxidant [358]. 74–96% substrate conversions and 67–90% urea isolated yields can be obtained. The catalytic system can be easily recovered and recycled at least six times. Interestingly, the related catalytic carbonylation of amino alcohols to 2-oxazolidinones was also published previously [359]. 53–100% yields of the corresponding products were obtained through direct PdI_2/KI-catalyzed oxidative carbonylation of readily available 2-amino-1-alkanols (Scheme 1.165).

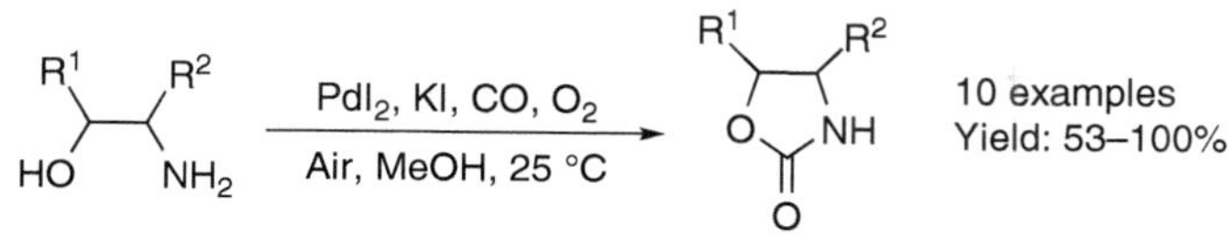

Scheme 1.165 Palladium-catalyzed oxidative carbonylation of 2-amino-alcohols.

In 2019, Shi's group reported Pd/TiO_2 catalyst system for the oxidative carbonylation of amines to ureas with ultra-low Pd content under organic ligand- and solvent-free conditions (Scheme 1.166) [360]. Under 35 bar CO and 5 bar O_2, the corresponding ureas can be obtained in 30–88% yields with up to 250 000 h^{-1} catalytic TOFs (moles of amines converted per mole of Pd per hour) at 122 °C. This is the reported highest TOF numbers so far. In addition, Pd/TiO_2 catalyst can be recycled at least 10 runs without loss in catalytic activity.

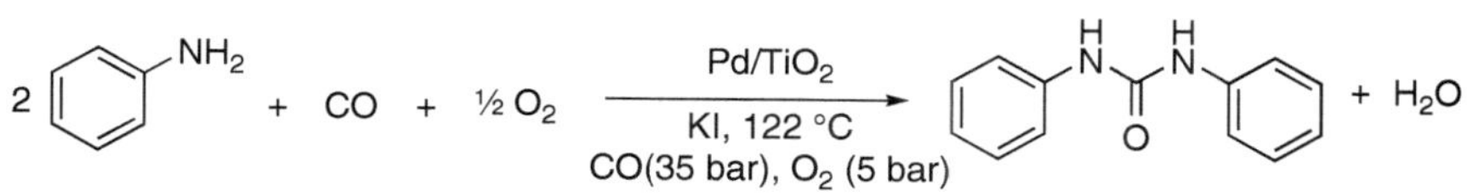

Scheme 1.166 One example of Pd/TiO_2-catalyzed oxidative carbonylation of amines. Source: Modified from Liu et al. [360].

Some other metal-catalyzed oxidative carbonylation of amines with homogeneous transition metal/organic ligand complexes [361–367] and heterogeneous catalysts [368–370] were also studied.

1.9 Other Reactions

To explore other ways to use CO, the replacement of the diazo group in diazoalkanes by CO to afford the corresponding ketenes (Scheme 1.167) might be a synthetically useful reaction since ketenes are versatile intermediates in organic synthesis [371, 372].

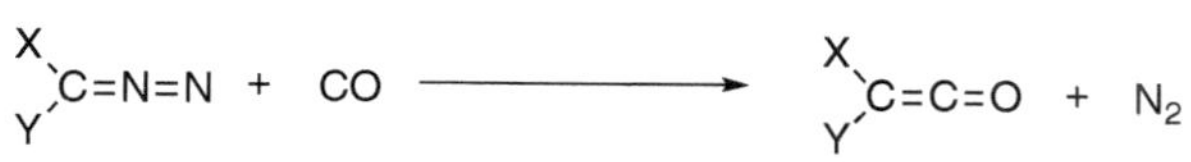

(X, Y = H, alkyl, aryl, alkylthio, acyl, alkoxycarbonyl, alkylamino, F, Cl, trialkylsilyl, etc.)

Scheme 1.167 Replacement of the diazo group in diazoalkanes by carbon monoxide.

1.9.1 Reactions of Diazoalkanes with Carbon Monoxide

In 1994, Lyashchuk et al. reported the reactivity of stable imidazol-2-ylidenes toward CO. Under atmospheric pressure of CO, the synthesis of 1,3-di-1-adamantylimidazol-2-carbonyl from 1,3-di-1-adamantylimidazol-2-ylidene in 20–30% yield at 10–15 °C in THF solutions (Scheme 1.168) [373].

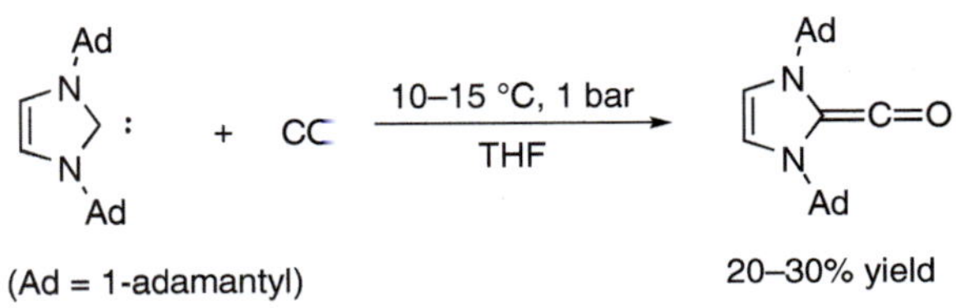

Scheme 1.168 Synthesis of 1,3-di-1-adamantylimidazol-2-carbonyl. Source: Modified from Lyashchuk and Skrypnik [373].

In 2006, Bertrand and coworkers described a smooth addition of CO at 1 bar pressure to the stable amino-substituted acyclic carbenes and a cyclic carbene at room temperature in THF, which resulted in the corresponding stable ketenes in 80–82% and 65% yields, respectively (Scheme 1.169) [374].

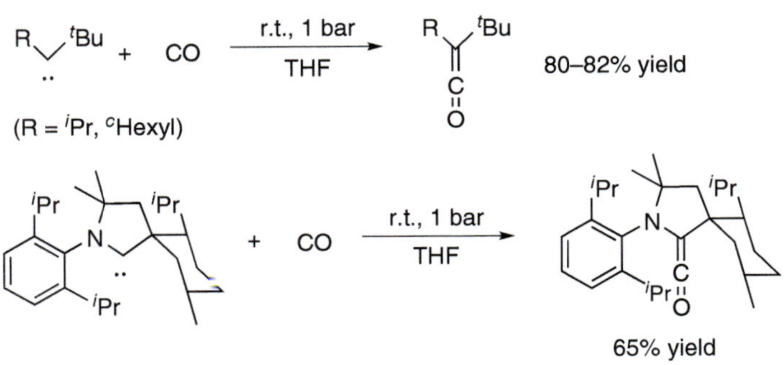

Scheme 1.169 Reactions of amino-substituted acyclic carbenes and a cyclic carbene. Source: Modified from Lavallo et al. [374].

In 1989, Grubbs and coworkers found heterobinuclear complexes with *m*-methylene ligands $Cp_2Ti(Cl)CH_2Pt(CH_3)L_2$ (L = $P(CH_3)_2Ph$, PCH_3Ph_2) to react with CO at atmospheric pressure rapidly to give *m*-(C,O)-ketene species $Cp_2Ti(Cl)–OC(=CH_2)–Pt(CH_3)L_2$ in 80–95% yields (Scheme 1.170) [375].

Kron et al. reported ketene $Ph_3Si(EtO)C=C=O$ prepared in situ from the carbene complexes $(CO)_5M=C(OEt)SiPh_3$ (M = Cr, Mo, W) with 50 bar CO to react at 80–100 °C, among others, with *N*-methylbenzimine to give the corresponding β-lactam in 90% isolated yield (Scheme 1.171) [376].

Geoffroy and coworkers reported the interconversion of methylene and ketene ligands on a triosmium cluster. It was found that a triosmium methylene complex in

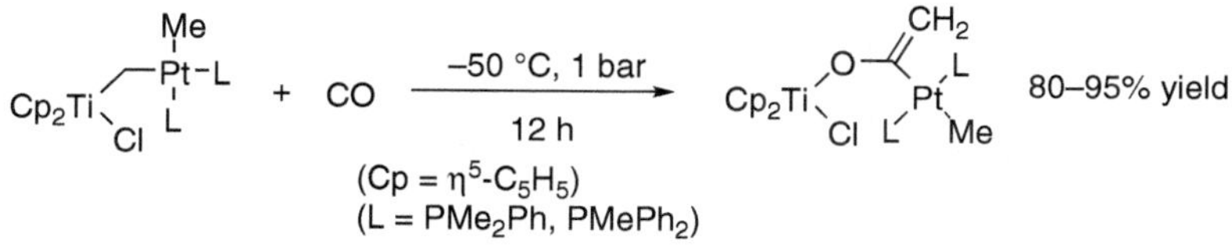

Scheme 1.170 Heterobinuclear complexes with *m*-methylene ligands. Source: Modified from Ozawa et al. [375].

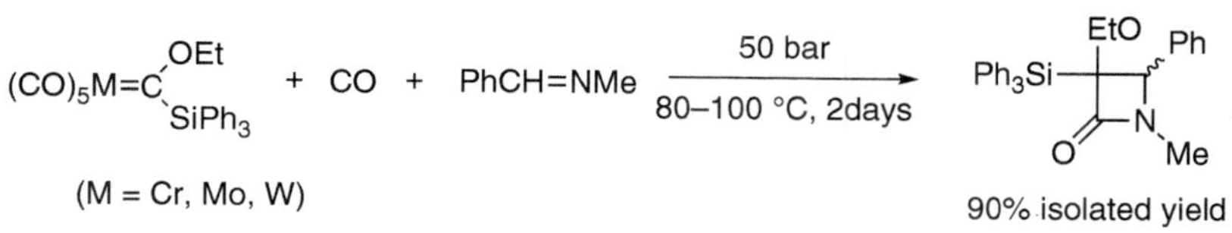

Scheme 1.171 Preparation of $Ph_3Si(EtO)C{=}C{=}O$. Source: Based on Kron and Schubert [376].

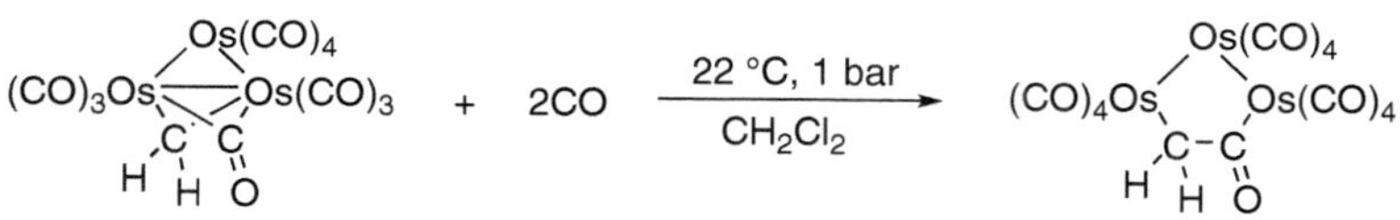

Scheme 1.172 Interconversion of methylene and ketene ligands. Source: Morrison et al. [377]; Morrison et al. [378].

CH_2Cl_2 solution readily adds two moles of CO at 22 °C to yield 16–20% the ketene derivative (Scheme 1.172) [377, 378].

In 2000, Werner et al. reported carbonylation of the rhodium diphenylcarbene complexes **1** and **2** resulted in the formation of diphenylketene in 83–89% yields (Scheme 1.173). It was assumed that the diphenylketene is formed by carbon–carbon coupling in the coordination sphere of rhodium [379].

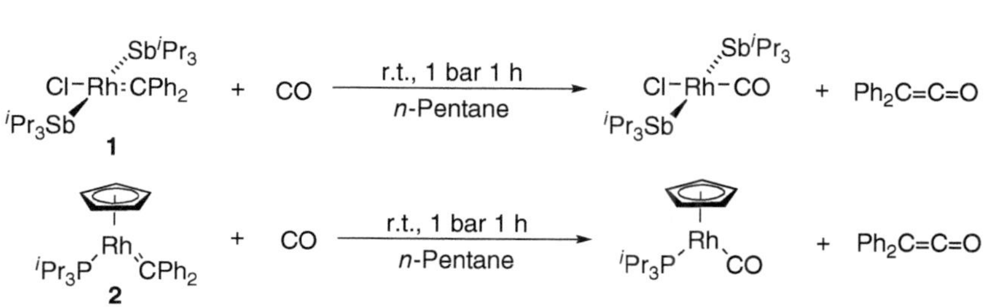

Scheme 1.173 Carbonylation of the rhodium diphenylcarbene complexes.

Miyashita et al. found the η^2-(C,C)-ketene complex of platinum could be isolated in 23% yield from the reaction of $Pt(PPh_3)_4$ and CH_2Br_2 in the presence of zinc powder under 3 bar CO pressure (Scheme 1.174) [380].

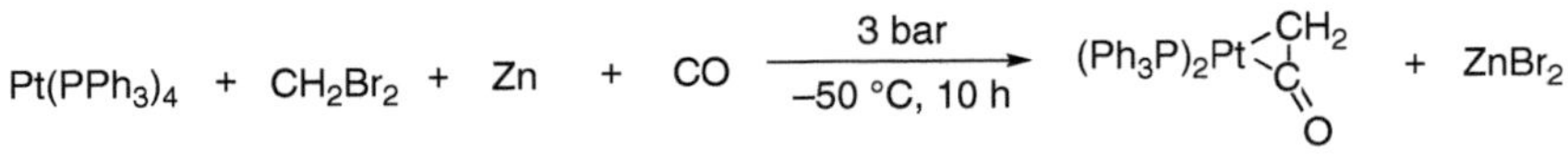

Scheme 1.174 Synthesis of the η^2-(C,C)-ketene complex of platinum. Source: Modified from Miyashita et al. [380].

Since 2003, Ungváry and coworkers have carried out systematic studies on $CO_2(CO)_8$-catalyzed carbonylation of ethyl diazoacetate [381, 382]. In the presence of 1 mol% $CO_2(CO)_8$ and under 50 bar pressure of CO at room temperature, ethyl diazoacetate (EDA) was converted to diethyl 2-diazo-3-oxoglutarate in 91% yield (Scheme 1.175).

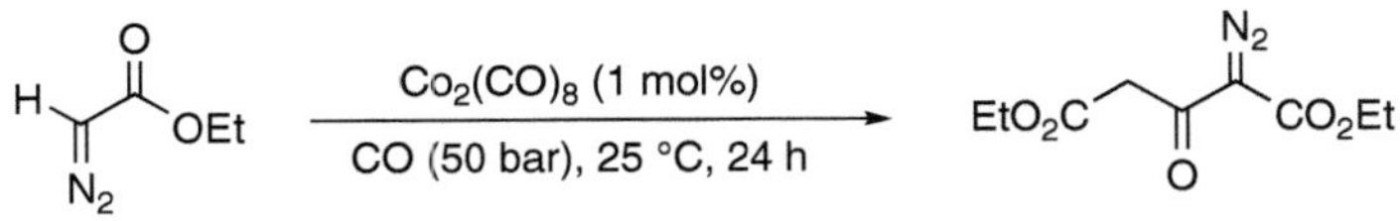

Scheme 1.175 $Co_2(CO)_8$-catalyzed carbonylation of ethyl diazoacetate.

In 2007, Wang and coworkers reported $Pd(PPh_3)_4$-catalyzed cross-coupling of aryl iodides with EDA [237]. Under atmospheric CO pressure, the carbonylation occurred to afford β-keto α-diazocarbonyl compounds in 43–66% yields at 45 °C (Scheme 1.176).

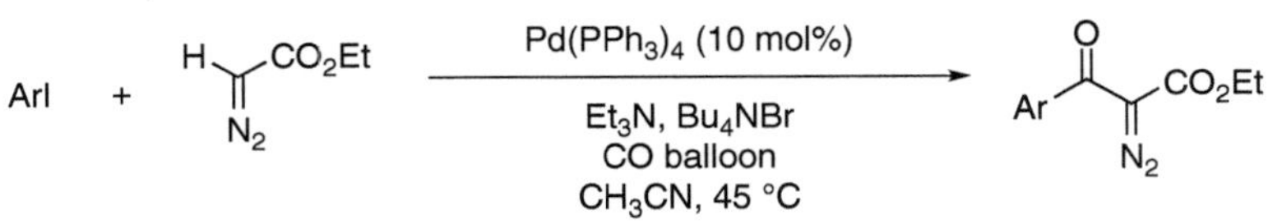

Scheme 1.176 $Pd(PPh_3)_4$-catalyzed cross-coupling of aryl iodides with EDA.

Later, Wang and Li found a comprehensive study of $Pd_2(dba)_3$-catalyzed carbonylation of diazo compounds [383]. In the initial study with EDA, similar results as in the $CO_2(CO)_8$-catalyzed system were observed with Pd(0) catalyst under atmospheric CO pressure (Scheme 1.177).

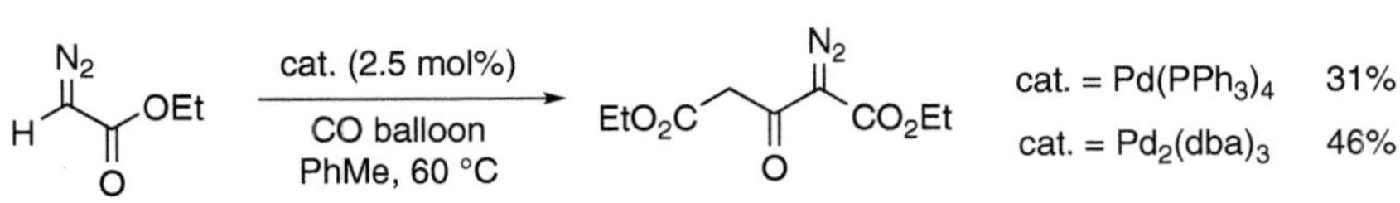

Scheme 1.177 $Pd_2(dba)_3$-catalyzed carbonylation of diazo compounds.

In 2013, de Bruin and coworkers reported cobalt(II)-porphyrin-catalyzed carbene carbonylation reactions (Scheme 1.178) [384]. Under 10 bar CO and at 50 °C, the β-ketoester products can be obtained in 53–75% yields when using CO, a-diazocarbonyl compounds and different nucleophiles.

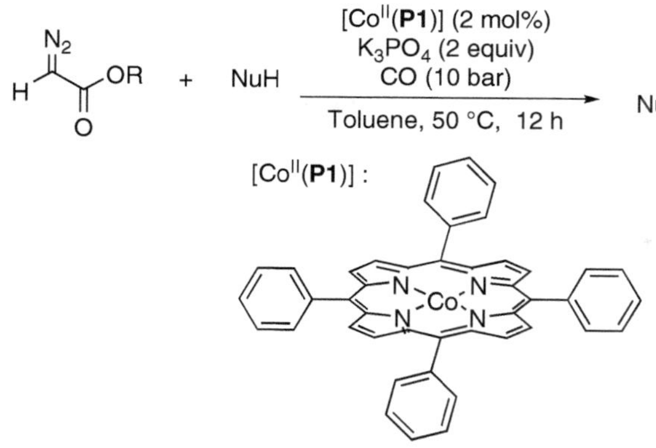

Scheme 1.178 [Co^{II}(Por)]-catalyzed β-ketoester synthesis. Source: Paul et al. [384].

In 2017, Sivasankar and Ramakrishna described that octacarbonyldicobalt as a nongaseous CO source catalyzed carbonylation of diazo compounds for the synthesis of amido esters and amido phosphonates [385]. When using 0.5 equiv $Co_2(CO)_8$ at room temperature, amido esters in 67–95% yields and amido phosphonates in 65–83% yields can be obtained.

1.9.2 Reaction of C–NO_2 with CO

The catalytic carbonylation of nitro compounds is a field of great interest, as a number of important industrial products can be produced (Scheme 1.179) [386–388], such as the synthesis of isocyanates, carbamates, ureas, etc., which normally needed toxic phosgene as the reaction reagent and can be replaced by cheap CO.

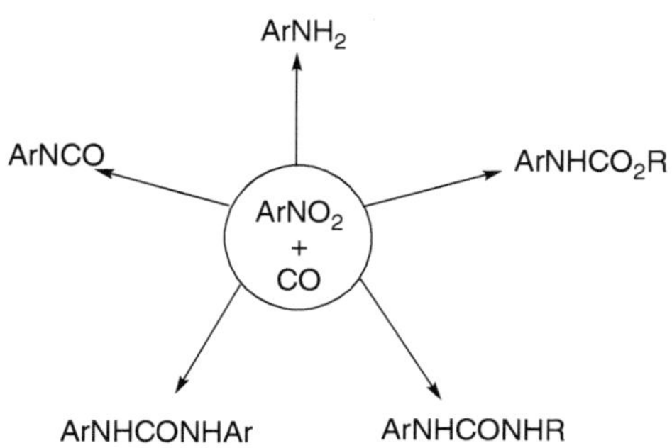

Scheme 1.179 Carbonylation of nitro compounds. Source: Paul [386]; Tafesh and Weiguny [387].

In 1990, the first isolation and characterization of a metallacyclic complex were performed [389, 390]. By combining $Pd(OAc)_2$, 1,10-phenanthroline (3 equiv), and $PhNO_2$ (40 equiv) and heating it in ethanol under CO (30 bar) at 80 °C, a yellow complex was precipitated, with 80% yield. PhNCO, $(PhNH)_2CO$, $(PhNHCO)_2NPh$, and CO_2 were produced by heating the complex in 1,2-dichlorobenzene under CO

(20 bar) at 170 °C. The addition of 2,4,6-trimethylbenzoic acid can increase the production of PhNCO. Finally, a mixture of the complex, $PhNO_2$, EtOH, under CO atmosphere, together with 2,4,6-trimethylbenzoic acid heated at 135 °C, 100% of the $PhNO_2$ was converted into $PhNHCO_2Et$ in 91% selectivity. Based on these studies, a reaction mechanism was given that was modified afterward for different product syntheses (Scheme 1.180).

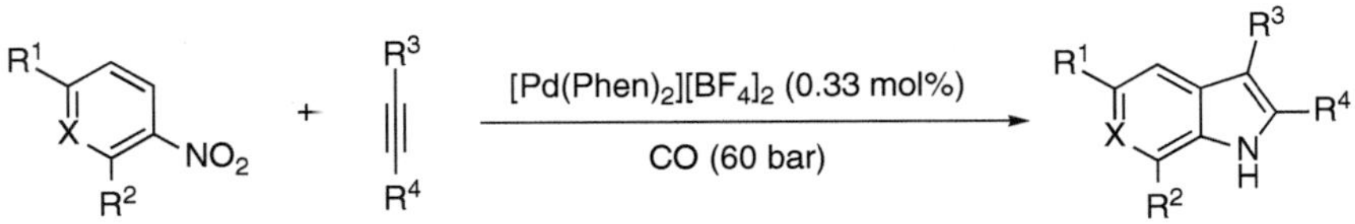

Scheme 1.180 Reaction mechanism for Pd-catalyzed carbonylation of RNO_2.

Yu, Lu, and colleagues found that selenium can catalyze the carbonylation of nitroarenes to symmetrical 1,3-diarylureas under atmospheric pressure of CO [391]. In the presence of KOH or NaOAc as base, various ureas were produced with up to 94% yield in DMF at 95 °C (Scheme 1.181). They also reported a selenium-catalyzed reaction of nitroarenes and amides in the presence CO to *N*-arylamides [392]. Under the assistant of mixed organic bases NEt_3 and DBU (1,8-diazabicyclo[5.4.0]undec-7-ene), *N*-arylamides were produced in 41–88% yields by a selenium catalyst. Other catalytic carbonylation of organic nitro compounds can be found in reviews [386–388, 393].

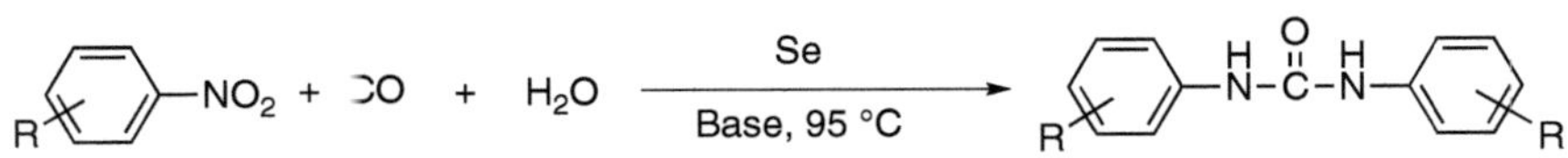

Scheme 1.181 Selenium-catalyzed amide synthesis.

2

Carbon Dioxide

2.1 Synthesis of Urea Derivatives

Ureas and derivatives are an important class of carbonyl compounds and useful chemical intermediates in the synthesis of pharmaceuticals, agricultural chemicals and dyes as well as antioxidants in gasoline and additives in plastics [394, 395]. The reductive CO_2 conversion is one of the transformation methods of CO_2, but the reductive conversion needs a larger energy input using highly reactive reducing agents such as H_2. Non-reductive conversion could obtain carbonates, carbamates, ureas, et al. and is promising as an alternative method (Scheme 2.1).

2.1.1 Metal-free Catalyst Systems

Urea can produced from CO_2 and ammonia under high pressure and high temperature (Scheme 2.2) [396].

The mechanism of the formation of the carbon dioxide compound of primary amine including its constitution and the mechanism of the formation of *N*,*N*′-di-substituted urea from primary amine–carbon dioxide compound were discussed by Hayashi and coworker in 1951 [397].

In 1974, Yamazaki et al. presented a simple method for the preparation of ureas from carbon dioxide by means of diphenyl and triaryl phosphites in the presence of tertiary amines (Scheme 2.3) [398].

2.1.2 Ph_3SbO as Catalyst

Nomura group found that triphenylstibine oxide (Ph, SbO) with assistance from tetraphosphorus decasulfide (P_4S_{10}) was a good catalytic system for the carbonylation reaction [399]. The Ph_3SbO/P_4S_{10} system can promote the carbonylation of several types of amines by 4.9 MPa CO_2 at 80–120 °C for 12–48 h with up to 94% yield (Scheme 2.4).

Carbonyl Compounds: Reactants, Catalysts and Products, First Edition.
Feng Shi, Hongli Wang and Xingchao Dai.

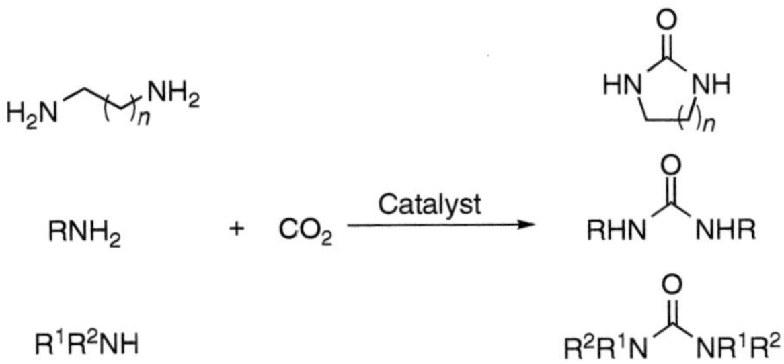

Scheme 2.1 Reactions of amines and CO_2 to form ureas and derivatives.

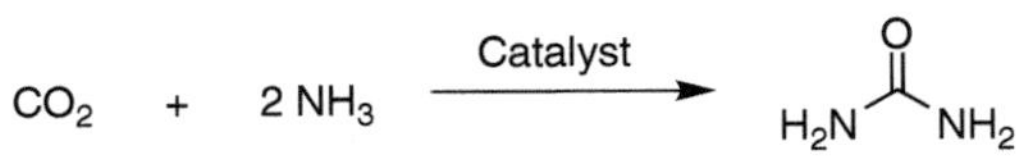

Scheme 2.2 Production of urea. Source: Pagani [396].

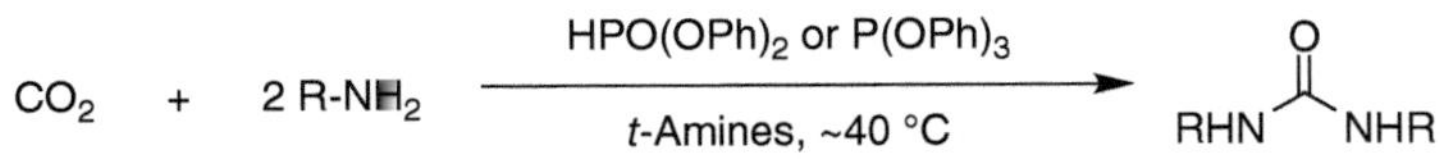

Scheme 2.3 Preparation of ureas in the presence of diphenyl and triaryl phosphites. Source: Modified from Yamazaki et al. [398].

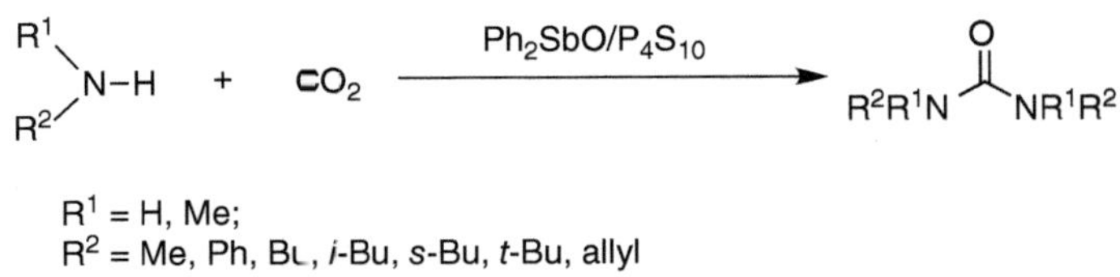

Scheme 2.4 Carbonylation of amines by Ph_3SbO/P_4S_{10} system.

2.1.3 Pd Catalyst Systems

Literature studies also describe dialkylureas from CO_2 and a series of amines including primary amines, secondary amines, and tertiary amine. However, the preparation of tetraalkylureas from CO_2 and secondary amines has been more difficult [400]. Tai et al. reported that the synthesis of tetraalkylureas from dialkylamines and CO_2 can be performed at 60 °C with CCl_4 up to 99% yield (Scheme 2.5) [401]. A two-step procedure, in which dialkylammonium dialkylcarbamate was produced in situ and then reacted with CCl_4 and free dialkylamine, gave greater yields of urea than a simple single-stage procedure.

2.1.4 Ionic Liquids as Catalyst

In order to obtain substituted ureas, dehydration is a key step. In 2003, Deng and coworkers [402] reported that a series of aliphatic amines, and even aromatic amines, react with CO_2 to afford corresponding *N*,*N*′-dialkylurea in the CsOH/ILs

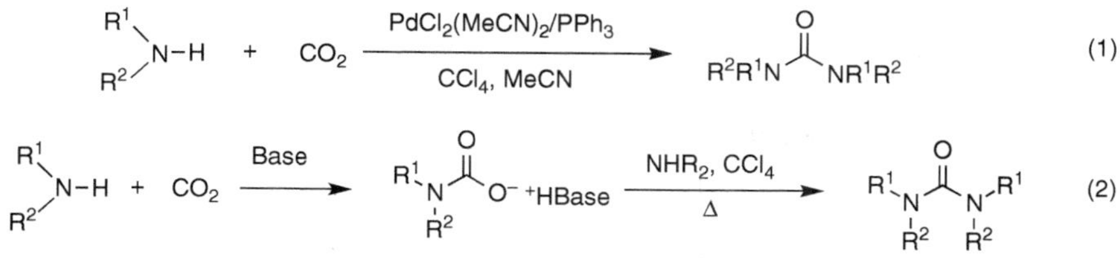

Scheme 2.5 Preparation of dialkylureas by $PdCl_2(MeCN)_2/PPh_3$. Source: Based on Tai et al. [401].

catalyst system without organic solvent and dehydrating agent with moderate to 27–98% yields for 4 h (Scheme 2.6). Furthermore, the concept of reaction and separation has been fulfilled with CsOH/IL catalyst system with the advantage of the nonvolatility and selective solubility of the IL.

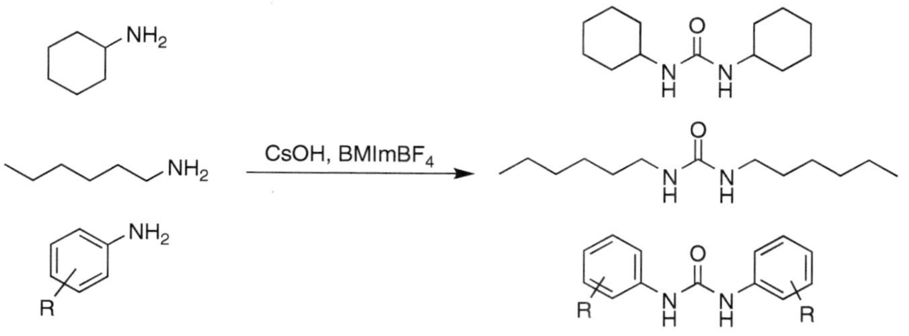

Scheme 2.6 CsOH/ionic liquid catalyst system for the synthesis of ureas.

Polyethylene glycol-supported potassium hydroxide (KOH/PEG_{1000}) [403] was reported for the synthesis of cyclic ureas or urea derivatives at 8 MPa CO_2 pressure and 150 °C for 6 h with 82% yield in a wide range of substrates, including primary aliphatic amines, secondary aliphatic amines, and diamines.

2.1.5 CeO_2 as Catalyst

Nanoparticulate CeO_2 was prepared by a biopolymer template process using mesoporous alginate aerogel at 0.7 MPa CO_2 and 160 °C with low yields of cyclic ureas (37%) [404].

Tomishige group [405] prepared that the catalysis of pure CeO_2 using 2-propanol as a solvent directly synthesized cyclic ureas from CO_2 and diamines under mild reaction conditions (0.3–0.5 MPa CO_2 and 160 °C) for 1 h. As the author report that 2-propanol was the best solvent to achieve 78–98% yields and >99% selectivity of cyclic ureas, the alkyl group of 2-propanol plays an important role in the suppression of N-alkylation of amine, a main side reaction. The catalyst system can be applied to various diamines to afford the corresponding cyclic ureas, particularly six-membered ring ureas; however, the reaction temperature was not changed.

Primary alkylamines were converted to the corresponding ureas from 5 MPa CO_2 over CeO_2 in NMP with 70–87% yield at room temperature for 24 h [406], but secondary amine and aniline were not converted. After adding 2-cyanopyridine, the systems of CeO_2 in NMP with 2-cyanopyridine promoted the converted unreactive amines, like *tert*-butylamine and aniline, to the corresponding ureas in 80% yields which are much higher than those reported in the previous literatures.

2.2 Synthesis of Carbamate Derivatives

Carbamates as one of the products of the reactions from amines and CO_2 are important raw materials for the manufacture of a variety of polymers used in foams, coating, adhesives, plastics, and fibers. They are also used as herbicides, fungicides, and pesticides in agrochemical industry and drug intermediates in pharmaceutical industry [386, 407, 408]. Usually they are obtained in a reaction between amines, phosgene, and alcohols, but because of a highly toxic nature of phosgene, safer routes toward carbamates are in high demand. The most notable examples include the urea alcoholysis, the carbonate aminolysis or the urea–carbonate reaction and direct paths from CO_2 toward carbamate are available. Secondary or primary amines can react with CO_2 to form carbamic acids and/or ammonium carbamates (Scheme 2.7) [407, 409, 410].

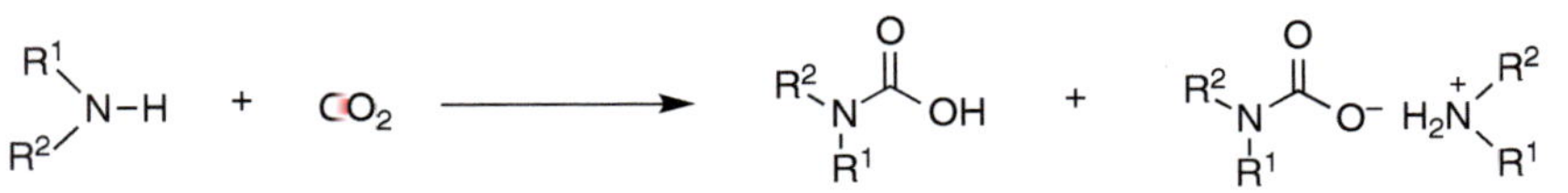

Scheme 2.7 Formation of carbamic acid and ammonium carbamate between amines and CO_2. Source: Dell'Amico et al. [407]; Aresta et al. [409]; Dibenedetto et al. [410].

Reversible transformation of diethylamine and supercritical carbon dioxide into *N,N'*-diethylcarbamic acid was confirmed by Ikariya group in 2008 [411]. The result showed that low temperature favors formation of carbamic acid, whereas high temperature causes decarboxylation.

2.2.1 Ru Catalyst Systems

Basis on carbamic acid formation under $scCO_2$ conditions, the formation of alkenyl carbamates from terminal alkynes, diethylamine and CO_2 was catalyzed by *trans*-$[RuCl_2\{P(OC_2H_5)_3\}_4]$ with 85% yield at 80 °C, 8 MPa CO_2 for 15 h (Scheme 2.8). It was investigated to demonstrate the transformation from carbamic acids to *Z*-alkenyl carbamates with high stereoselectivity [411].

Reactions of in situ generated carbamic acids with electrophiles such as organic halides result in the formation of urethanes (carbamates) (Scheme 2.9). Various homogeneous and heterogeneous catalysts are applicable to synthesize urethane.

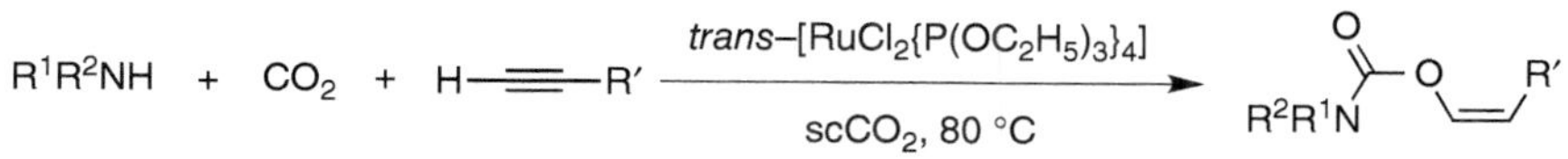

Scheme 2.8 Synthesis of enol carbamates catalyzed by *trans*-$[RuCl_2\{P(OC_2H_5)_3\}_4]$.

$$RNH_2 + CO_2 + R'X + Base \longrightarrow RNHCO_2R' + HX(Base)$$

Scheme 2.9 Urethane synthesis with an organic halide.

Aresta and coworker [412] reported the synthesis of some alkylammonium carbamates RNHC(O)OR from primary amines, CO_2 and alkyl halides (R′X, R = benzyl, allyl, *tert*-butyl, cyclohexyl, etc.), in the presence of a suitable macrocyclic polyether under ambient conditions with 9–57% yield for 28–72 h. The role of the macrocyclic ligand in modifying the reactivity of monoalkylammonium carbamates toward R′X has also been rationalized in article.

If electrophilic "dehydrating agents," like $POCl_3$ or P_4O_{10}, were added, carbamate anions can be transformed to corresponding isocyanates [413].

In 2000, Yoshida et al. founded that some common onium salts also catalyzed selective urethane production at 100 °C for 72–91% yield from amines and alkyl halides in supercritical CO_2 (Scheme 2.10) [414].

$$R^1R^2NH + R^3X + K_2CO_3 \xrightarrow[scCO_2]{\text{Onium salt}} R^1R^2NCOOR^3$$

Scheme 2.10 Urethane production catalyzed by onium salts. Source: Based on Yoshida et al. [414].

2.2.2 Sn or Ni Catalyst Systems

CO_2 was converted to urethanes by reaction with amine and alcohol, in the presence of Sn complexes or Ni complexes as a catalyst (Scheme 2.11) [415, 416].

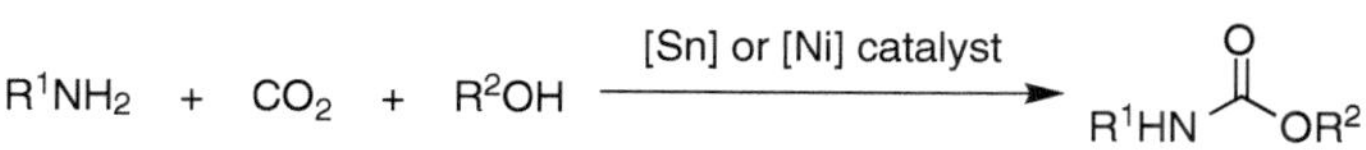

Scheme 2.11 Urethane synthesis with [Sn] or [Ni] catalyst. Source: Abla et al. [415]; Abla et al. [416].

In 2003, the synthesis of cyclic ureas and urethanes was achieved with catalyst-free from diamine or amino alcohols at 6 MPa CO_2 and 150 °C (Scheme 2.12) [417].

2.2.3 Zeolite as Catalyst

In 2004, Srinivas and Ratnasamy and coworkers [418] studied the synthesis of alkyl and aryl carbamates in high yields (80%) under mild reaction conditions (0.34 MPa,

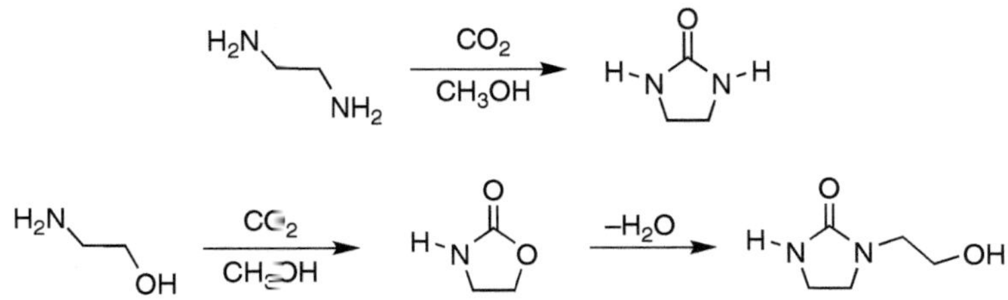

Scheme 2.12 Reactions of diamines or amino alcohols with CO_2 in methanol. Source: Bhanage et al. [417].

80 °C, 3 h) by reaction of the amines, CO_2, and *n*-BuBr over either titanosilicate molecular sieves (TS-1) or metal phthalocyanine complexes encapsulated in zeolite-Y (MPc-Y). These reactions used zeolite catalysts, without onium salts, crown ether, or organic bases as additional co-catalysts, different from the current processes including CO, phosgene, and isocyanate.

Apart from zeolite, polymers, carbon nanomaterials, metal oxides, and many others are possible supports. In 2016, a series of tetraalkylammonium catalysts immobilized on polystyrene or carbon nanotubes were investigated by Krawczyk et al. for the same reaction [419]. The yield of carbamates was up to 85% at 25 °C for 1 h and another 3 h at 100 °C, all with 1 MPa CO_2 in the presence of Cs_2CO_3.

Carbamates can also be synthesized through nucleophilic addition of carbamic acids to alkenes (Scheme 2.13) [420].

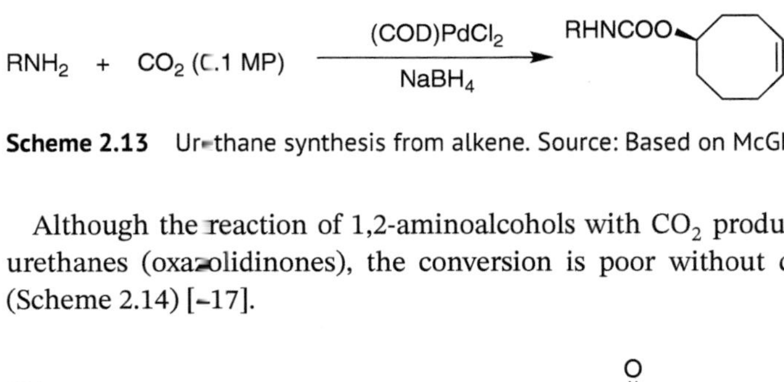

Scheme 2.13 Urethane synthesis from alkene. Source: Based on McGhee and Riley [420].

Although the reaction of 1,2-aminoalcohols with CO_2 produces five-membered urethanes (oxazolidinones), the conversion is poor without dehydrating agents (Scheme 2.14) [417].

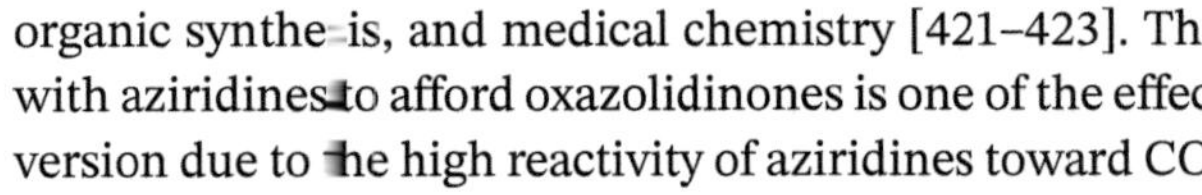

Scheme 2.14 Oxazolidinone from aminoalcohol. Source: Based on Bhanage et al. [417].

Oxazolidinones have important applications as chiral auxiliaries, intermediates in organic synthesis, and medical chemistry [421–423]. The coupling reaction of CO_2 with aziridines to afford oxazolidinones is one of the effective strategies for CO_2 conversion due to the high reactivity of aziridines toward CO_2 and ideal atom efficiency (Scheme 2.15).

Like cyclic carbonate synthesis, a nitrogen-containing three-membered ring (aziridine) reacts with CO_2, which leads to a nitrogen-containing five-membered ring

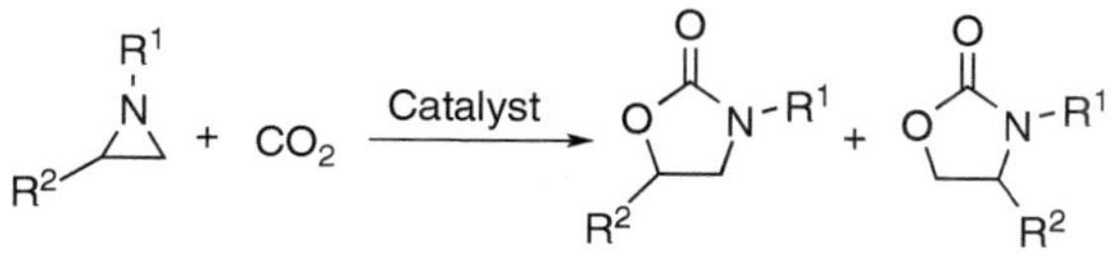

Scheme 2.15 Reactions of aziridines and CO_2 to form oxazolidinones.

(oxazolidinone) (Scheme 2.16) [424–428]. Although the reaction proceeds without a catalyst, the presence of various compounds promotes the reaction; typical catalysts are alkali metal salts, quaternary ammonium salts, chromium salen complexes and so on [429–431]. Ring-opening polymerization under supercritical conditions to yield polyurethanes has also been reported (Scheme 2.17).

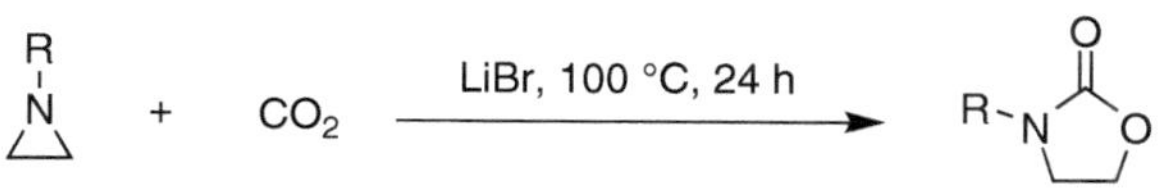

Scheme 2.16 Oxazolidinone synthesis. Source: Kawanami et al. [424]; Miller and Nguyen [425]; Sudo et al. [426]; Sudo et al. [427]; Shi et al. [428].

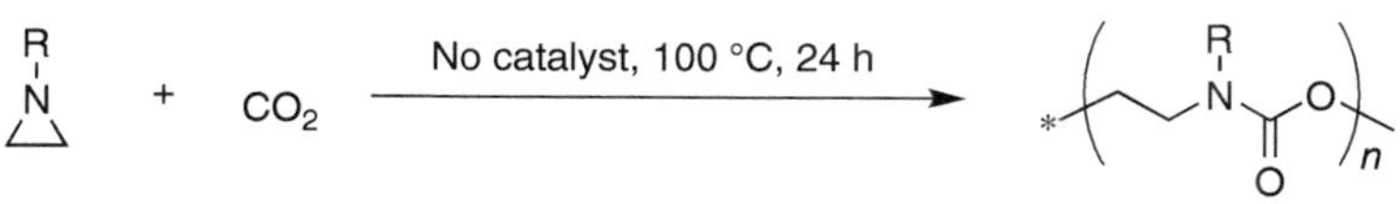

Scheme 2.17 Polyurethane from CO_2.

Previous research about aziridine/CO_2 coupling reaction has employed several catalyst systems lithium iodide, tin, ammonium, and antimony salts and nickel complexes albeit with limited success [432–437]. While these methods suffer from either the use of high pressure or low selectivity and multiple product isomers.

2.2.4 Other Catalyst Systems

(Salen) chromium(III)/4-(*N*,*N*-dimethylamino) pyridine (DMAP) was found to be an active catalyst system for the coupling of CO_2 and aziridines with 82–91% yield [425]. This catalyst system worked for a wide variety of monosubstituted *N*-aryl and *N*-alkyl aziridines as well as a 2,3-disubstituted *N*-alkyl aziridine (Scheme 2.18).

In 2005, Deng and coworkers reported that the reactions of propargylic alcohols, aliphatic primary amines, and CO_2 were conducted in the [BMIm]BF_4 system in the presence of CuCl to produce the corresponding 5-methylene-1,3-oxazolidin-2-ones under relatively mild conditions (2.5 MPa CO_2, 100 °C, 10–15 h). (Scheme 2.19) [438].

Due to the advantages of COF, high reactivity of aziridines toward CO_2, Zhao and coworkers first reported a Cu/COF–TBAB catalytic system for converting CO_2 into

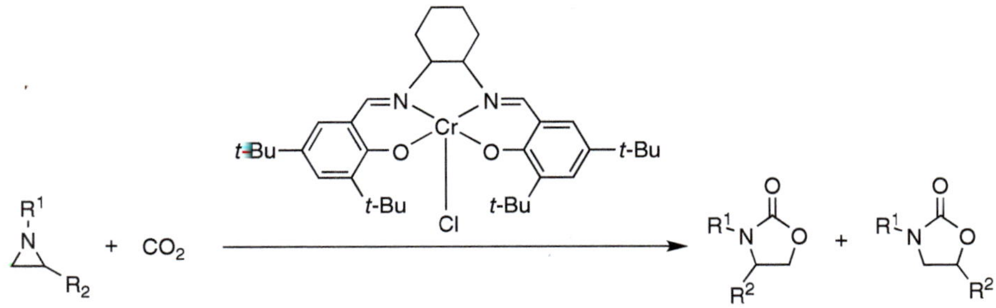

Scheme 2.18 The coupling of CO_2 and aziridines with (Salen) chromium(III)/DMAP.

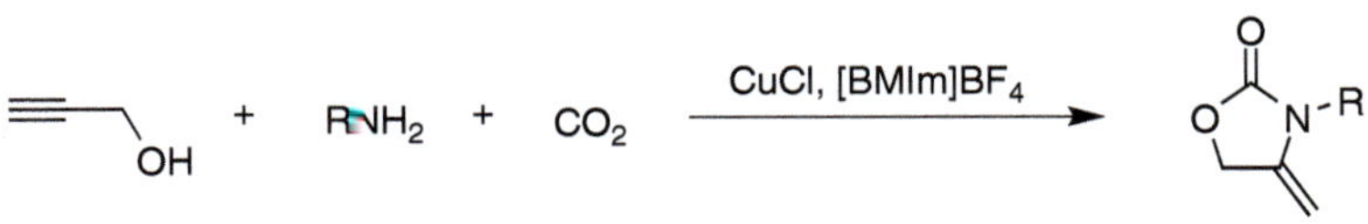

Scheme 2.19 Oxazolidinone from propargyl alcohol. Source: Based on Gu et al. [438].

oxazolidinones under 2 MPa CO_2 at 100 °C for 12 h and got high yields up to 99% [439].

In 2018, CO_2 adducts of carbodicarbenes were first synthesized [440] and applied for the carboxylative cyclization of CO_2 (2 MPa) at 80 °C with aziridines, affording 2-oxazolidinones in good to excellent yields (51–99%) and high functional group tolerance.

The resulting poly(hydroxyurethane) which prepared from bisepoxide CO_2 and diamine could be converted in situ to polyurethanes derivatives bearing ester or silyl ether groups (Scheme 2.20) [430].

Therefore, intramolecular nucleophilic attack of carbamic acid to nitrile results in quinazoline formation in addition of DBU under supercritical CO_2 for 97% yield (Scheme 2.21) [441–443].

2.3 Synthesis of Carboxyl Acid Derivatives

The alkynyl carboxylic acid products can serve as important synthetic intermediates for further applications in medical chemistry as well as organic synthesis to give coumarins, flavones, aminoalkynes, alkynylarenes, and arylidene oxindoles [444–446].

Carboxylic acids are counted suitable synthetic targets starting from CO_2. Carbon nucleophiles such as Grignard reagents, alkyl lithiums, active methylene compounds, and metal enolates attack CO_2 under relatively mild conditions to generate carboxylic acids (Scheme 2.22) [447–450].

In 1997, palladium-catalyzed carboxylation of allyl stannanes was studied by Shi and coworker [451] and rhodium(I) complexes catalyzed the insertion of CO_2 into boron–carbon bonds form aryl- and alkenylboronic esters in 2006 by Ukai group (Schemes 2.23 and 2.24) [451, 452].

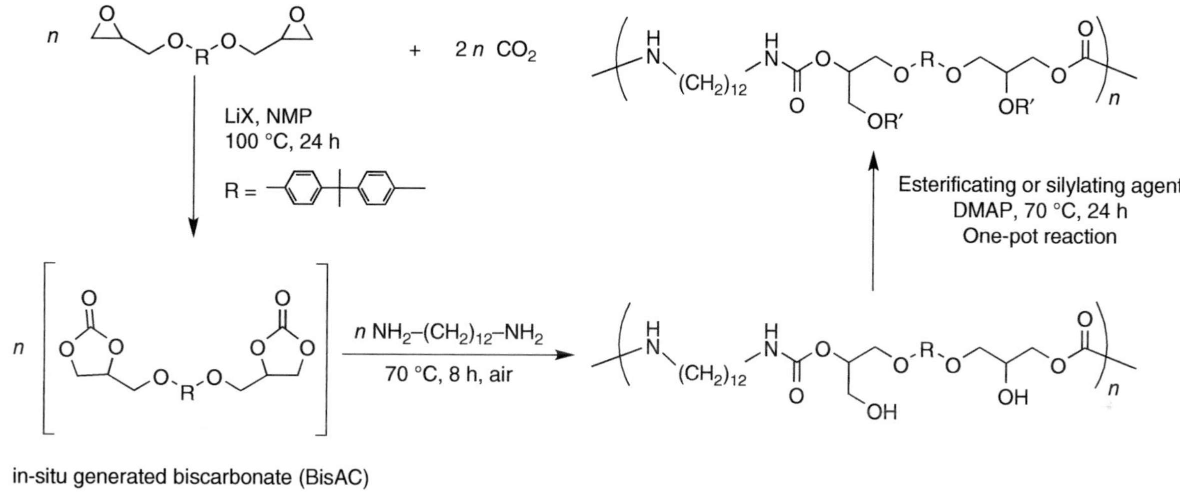

Scheme 2.20 Polyaddition of in situ generated BisAC and diamine and direct conversion to polyurethane derivatives. Source: Ochiai et al. [430].

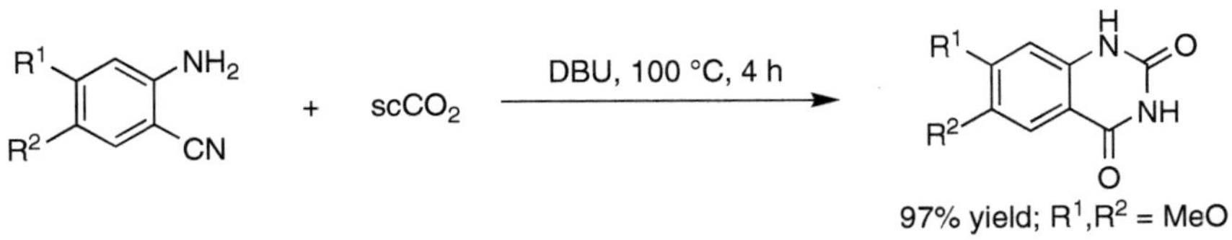

Scheme 2.21 Quinazoline synthesis from CO_2. Source: Connolly et al. [441]; Mizuno and Ishino [442]; Mizuno et al. [443].

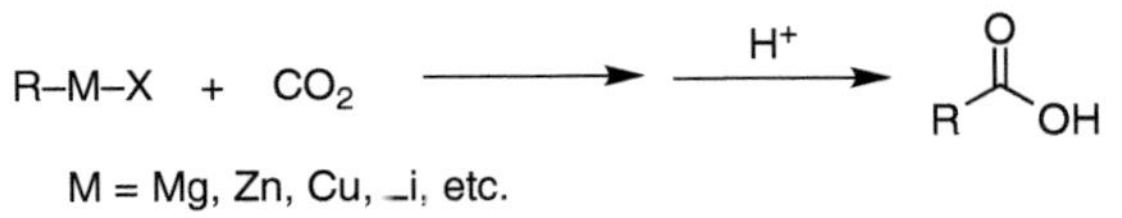

Scheme 2.22 Carboxylic acid from organometallic. Source: Chiba et al. [444]; Mankad et al. [445]; Abstracts of Papers, t. A. N. M [446]; Abe and Inoue [446]; Quirk et al. [447].

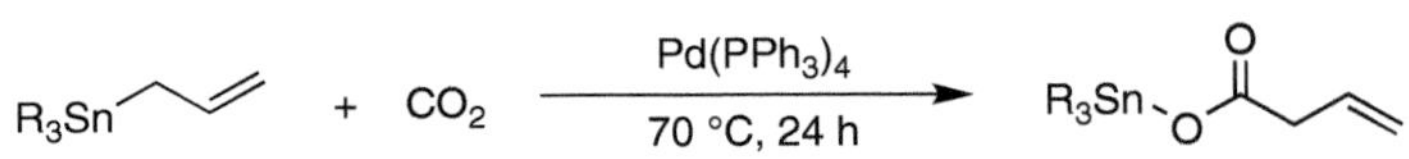

Scheme 2.23 CO_2 insertion into a tin–carbon bond. Source: Based on Shi and Nicholas [451].

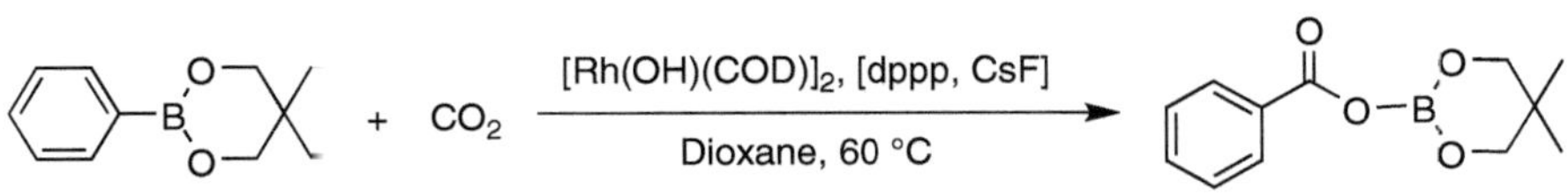

Scheme 2.24 CO_2 insertion into a boron–carbon bond. Source: Modified from Ukai et al. [452].

Acrylic acid, a typical unsaturated carboxylic acid, is currently consumed on a large industrial scale as a raw material for water-absorbing polymers. Usually, the acid is produced by oxidation of propylene via acrolein. It was attractive that alkene and CO_2 can directly synthesize to corresponding acid (Scheme 2.25).

Because the reaction is thermodynamically possible, the key is to develop an efficient catalyst. The stoichiometric formation of acrylic acid on a molybdenum complex has been reported (Scheme 2.26) [453–455].

Then, desired unsaturated carboxylic acids are selectively synthesized via metallacycles (Schemes 2.27–2.31) [456–460], although a stoichiometric amount of metal was consumed and DBU was used to promote the reactions, presumably by trapping CO_2 as a carbamate or bicarbonate form [461, 462].

The introduction of alkyl in reaction system was achieved by the addition of a stoichiometric amount of dialkyl zincs (Schemes 2.32–2.34) [463, 464, 466]. And, asymmetric synthesis can be conducted using optically active phosphine ligands (Scheme 2.33) [464].

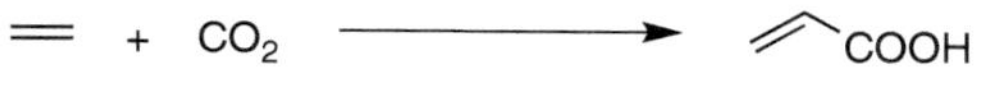

Scheme 2.25 Acrylic acid from CO_2.

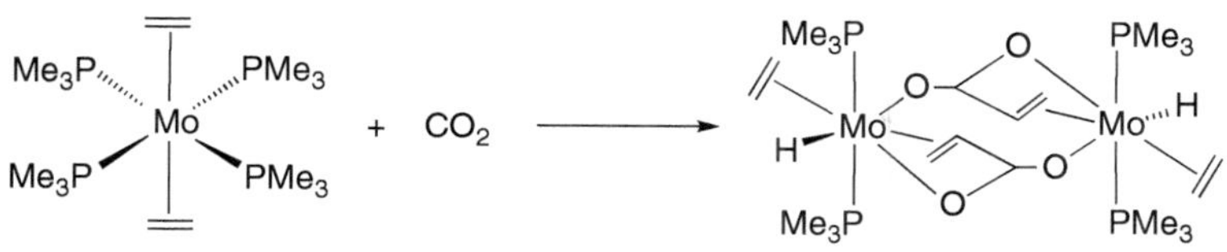

Scheme 2.26 Acrylic acid from CO_2 by metal complexation.

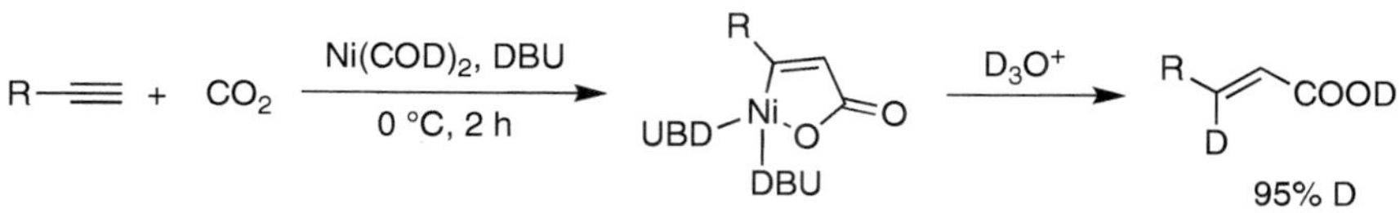

Scheme 2.27 Acrylic acid derivative from acetylene. Source: Modified from Saito et al. [456].

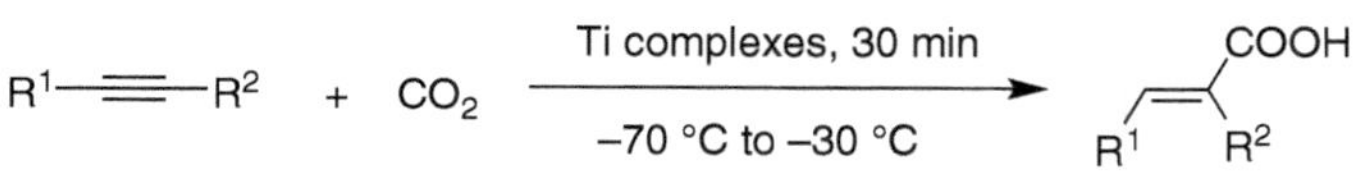

Scheme 2.28 Acrylic acid derivative from acetylene by Ti-complexes. Source: Based on Takimoto and Mori [457].

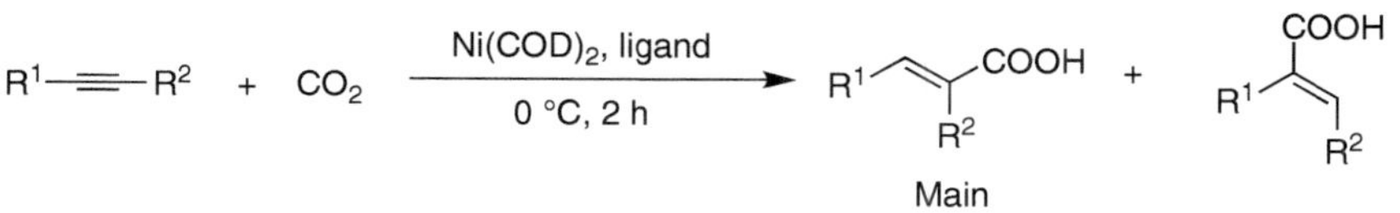

Scheme 2.29 Acrylic acid derivative from acetylene by $Ni(COD)_2$ and ligand. Source: Based on Six [458].

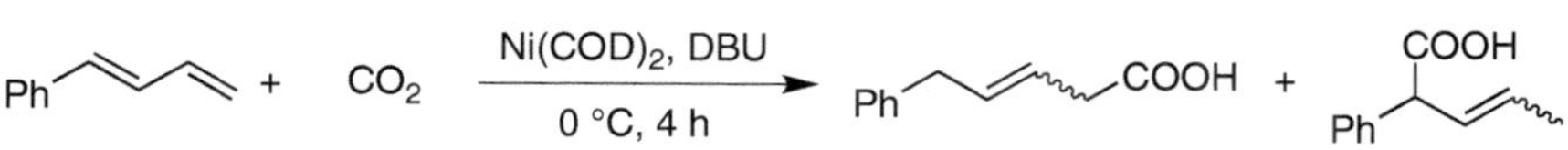

Scheme 2.30 Acrylic acid derivative from 1,3-butadiene. Source: Based on Takimoto et al. [459].

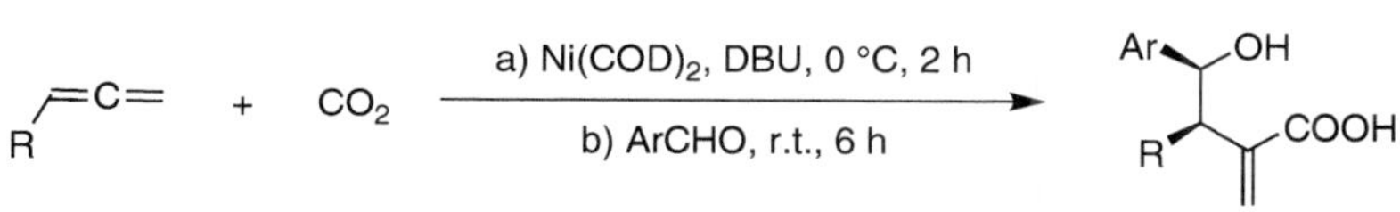

Scheme 2.31 Acrylic acid derivative from allene. Source: Based on Aoki et al. [460].

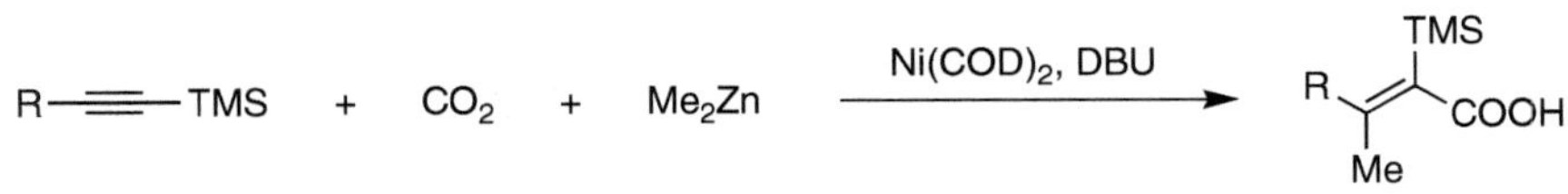

Scheme 2.32 Acrylic acid derivative from acetylene by $Ni(COD)_2$ and DBU. Source: Based on Takimoto and Mori [463].

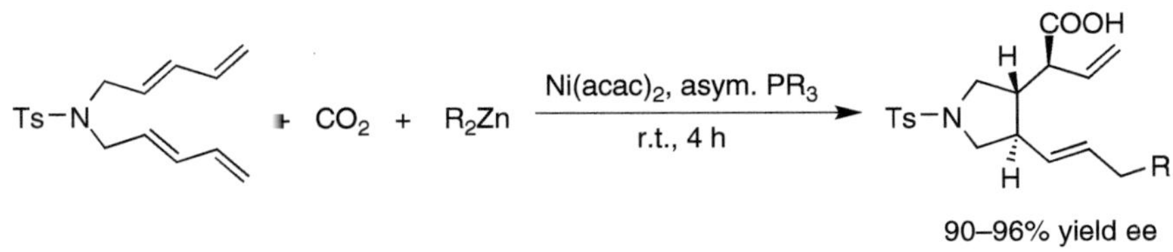

Scheme 2.33 Asymmetric cyclization–carboxylation with CO_2. Source: Modified from Takimoto et al. [464].

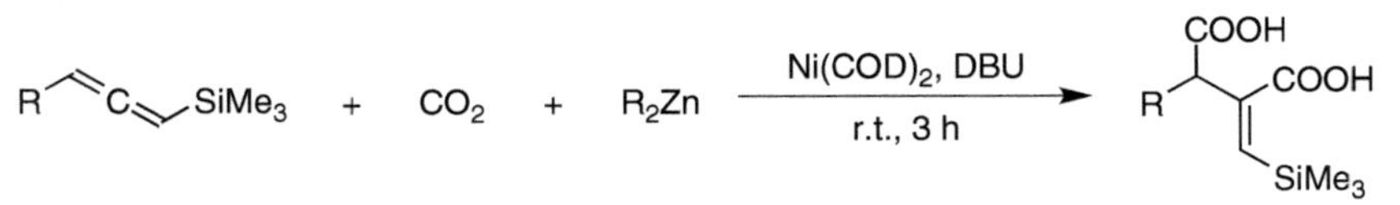

Scheme 2.34 Dicarboxylation of allene with CO_2. Source: Based on Kosugi et al. [465].

Carboxylation with CO_2 can be carried out under either basic or acidic conditions. Hydroxybenzoic acid is one of the few chemicals that is industrially produced from CO_2 (Kolbe–Schmidt reaction) (Scheme 2.35) [465, 467, 468].

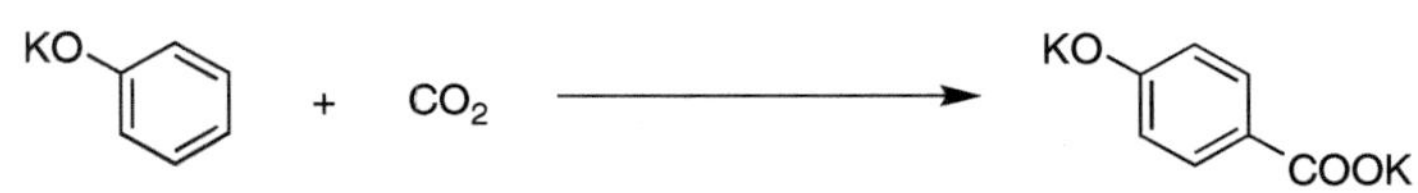

Scheme 2.35 Kolbe–Schmidt reaction. Source: Kosugi et al. [465]; Rahim et al. [467]; Sclafani et al. [468].

For example, the reaction with metal carbonates (Henkel-type reaction) (Scheme 2.36), and the vanadium- or palladium-catalyzed reaction with CO_2 in the presence of $K_2S_2O_8$ as an oxidizing reagent with only 7% yield (Scheme 2.37) [469, 470].

C–H bond activation by the $PdK_2S_2O_8$–CF_3CO_2H system is widely applicable to various reactions [471]. In 2002, the reaction of CO_2 (5.8 MPa) under electrophilic conditions using $AlCl_3$ as a Lewis acid and the direct activation of aromatic C–H bonds have been demonstrated by Prakash and coworkers at 40 °C for 18 h with 85% yield (Scheme 2.38) [472]. Such reaction of silylated aromatic compounds was studied by Miyano and coworkers at room temperature with 3 MPa CO_2 and 3 h for 73% yield (Scheme 2.39) [473, 474].

Metal-catalyzed reactions of organic halides, CO_2, and hydrogen have also produced carboxylic acids (Scheme 2.40) [475, 476].

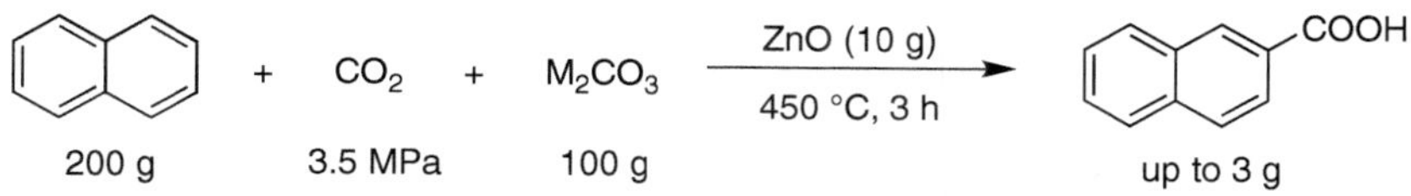

Scheme 2.36 Carboxylic acid formation by C–H bond activation.

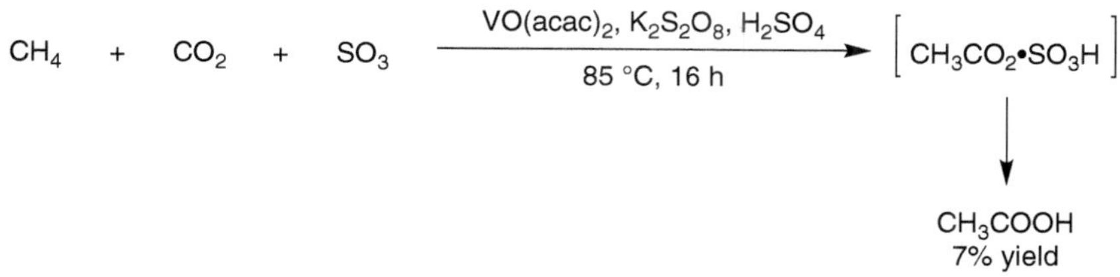

Scheme 2.37 Carboxylic acid formation by C–H bond activation from CH_4. Source: Zerella et al. [469]; Fujiwara et al. [470].

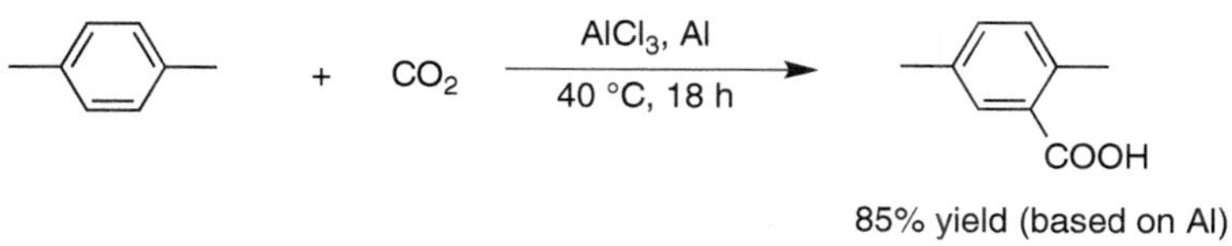

Scheme 2.38 Carboxylic acid formation by C–H bond activation. Source: Modified from Olah et al. [472].

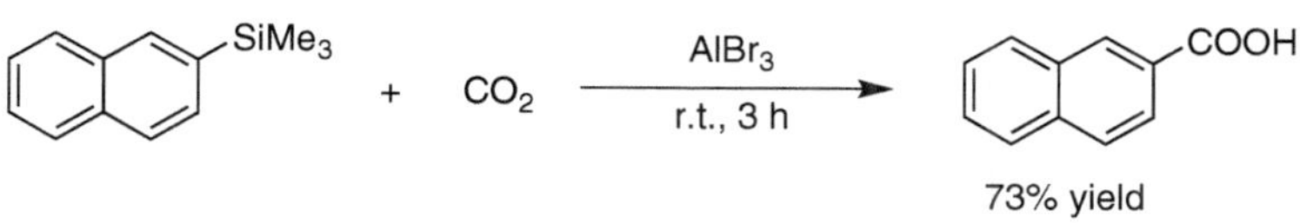

Scheme 2.39 Carboxylation with CO_2. Source: Hattori et al. [473]; Suzuki et al. [474].

In heterogeneous field, one of the best strategies for CO_2 conversion is the synthesis of propiolic acids through the C–H bond activation of terminal alkynes with CO_2 as a C1 building block (Scheme 2.41).

In 2015, Cheng and coworkers [477] synthesized an Ag@MIL-101 catalyst by a simple liquid impregnation at zeolite-type metal–organic framework (MOF) MIL-101, which was found to have an excellent catalytic performance for the conversion of terminal alkynes and CO_2 into propiolic acids with up to 98% yield at atmospheric pressure and low temperature (50 °C) for 15 h. This work showed for the first time the use of MOF-derived materials as catalysts in the synthesis of propiolic acids from CO_2.

Then, Trivedi and Rath [478] prepared a Pd-Cu/MIL-101 catalyst using a double solvent impregnating method. The catalytic activities of Pd-Cu/MIL-101 toward the conversion of terminal alkynes with CO_2 into propiolic acids showed up to 98% yield at room temperature and ambient pressure.

Other catalysts, like AgNPs/Co-MOF [479], Ag/KAPs-P [480], and Ag/CeO_2 [481], also have been reported with 85–98% yields at 60–80 °C. The influences on catalytic

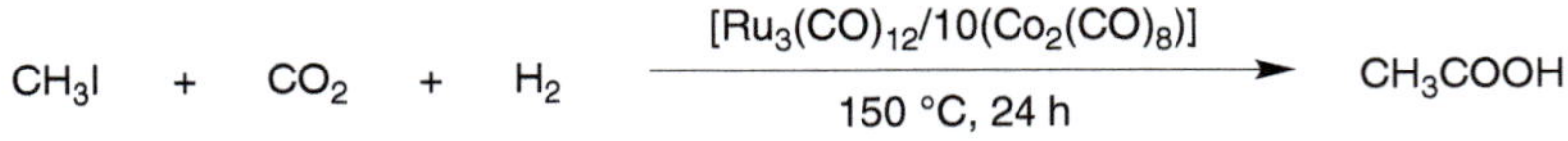

Scheme 2.40 Carboxylic with organic halide. Source: Fukuoka et al. [475]; Osakada et al. [476].

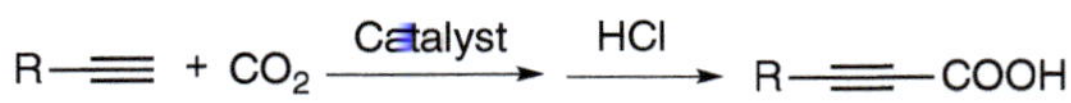

Scheme 2.41 Reactions of terminal alkynes and CO_2 to form alkynyl carboxylic acid.

activity, such as MOF surface areas, thermal and chemical stability, especially the acid–base characteristics of the pores, are compared and discussed systematically at Ag@MOF systems [482].

Huang and coworkers [483] prepared a general route to synthesize a novel Schiff base-modified silver catalyst (Ag/Schiff-SiO_2) by the in situ reduction of a silver precursor under ultrasonic conditions. The reaction for the carboxylation of terminal alkynes with CO_2 into alkynyl carboxylic acids and got 86–98% yields under 1.0 atm CO_2 at 60 °C for 24 h.

Controlled encapsulation of atomically precise nanoclusters (APNCs) into MOFs was an efficient way to create new types of multifunctional crystalline porous materials, owing to the well-defined compositions of both components. In various APNCs@MOF catalysts [484], the as-obtained $Au_{12}Ag_{32}(SR)_{30}$@ZIF-8 composite shows 79–99% yield in capturing CO_2 and converting phenylacetylene into phenylpropiolate under 50 °C and 0.1 MPa CO_2 pressure with a TON as high as 18 164, which was far exceed most known catalysts.

2.4 Cycloaddition of Epoxide with CO_2

Among all CO_2 fixation reactions, the cycloaddition of epoxides with CO_2 to form cyclic carbonates has been proposed as a promising strategy for recycling carbon resource (Scheme 2.42), not only because of its 100% atom economy but also on account of numerous applications of cyclic carbonates, which have been widely used as green solvents, polycarbonates, synthetic organic intermediates, electrolytes, precursors for biomedical applications, raw materials for engineering plastics, etc. [485–487]. Therefore, much attention has been paid to the development of efficient catalyst system for cyclic addition of epoxides with CO_2.

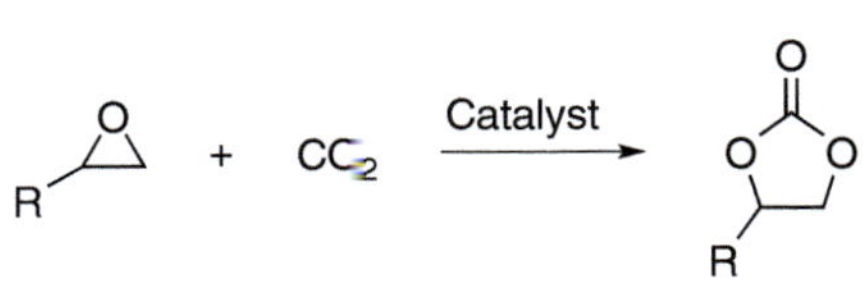

Scheme 2.42 Cycloaddition of epoxides with CO_2 to form cyclic carbonates.

The carboxylation of epoxides is known since time (IG Farben, 1943). A typical process of CO_2 cycloaddition with oxiranes uses the produced carbonate as a solvent in the presence of an onium salt such as tetraethyl ammonium bromide as a catalyst [488].

In addition to onium salts, the reaction proceeds in the presence of a wide range of catalysts, including metal halides. Supporting halide salts on solid supports such as ZnO, SiO_2, carbon, and zeolite results in high catalytic activities [489–492].

Among metal complexes, salen complexes of aluminum [493–495], chromium [496, 497], cobalt [498, 499], zinc [498], and tin [500] exhibit high catalytic activities. Aluminum complexes of phthalocyanines or porphyrins are also highly active [501–504]. Utilizing ILs in organic synthesis has attracted increasing attention [505]. Cyclic carbonate synthesis using ILs (i.e., imidazolium salts) as catalysts [506, 507], or by combining ILs and metal salts, has been reported [508–510]. In addition, a reaction in liquid quaternary ammonium salts, like Bu_4NBr, has been proposed [511].

Furthermore, various metal complexes of nickel [512], ruthenium [513], zinc [514], copper [515], manganese [515, 516], and so on have been reported to be active. Combining with organic compounds such as crown ethers, onium salts, and phosphines has significantly improved the catalytic activities of metal halides [517–520]. Cobalt- or manganese-substituted polyoxometalates have been proposed as catalysts with non-halide counteranions [521]. These polyoxometalates work as homogeneous catalysts. In addition, fully organic catalysts based on phenols and amines have also been investigated [522, 523].

Oxiranes and oxetanes undergo cycloaddition with CO_2 to give five- and six-membered ring carbonates (Scheme 2.43) [488, 524]. Baba et al. have reported a selective synthesis of trimethylenecarbonate from oxetane and CO_2 in the presence of Ph_4SbI in medium condition (5 MPa CO_2, 100 °C, 4 h) for 96% yield [524].

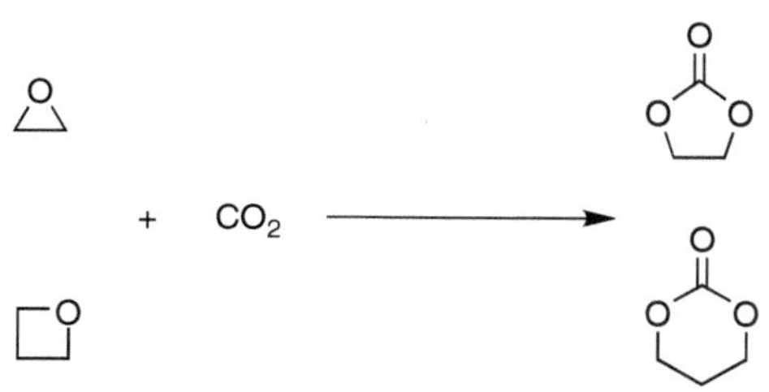

Scheme 2.43 Synthesis of a cyclic carbonate from an oxirane. Source: Darensbourg et al. [488]; Baba et al. [524].

Polyalkylenecarbonates are formed by the direct copolymerization of epoxides with CO_2 (Scheme 2.44). Also, the terpolymerization with cyclic acid anhydrides is feasible [525]. Fundamentally, one differentiates between two distinct catalyst classes: in the case of cobalt(III)-catalysts [526–529], chromium(III)-salen and -salan complexes [506, 530], aluminum(III)-salen complexes, or Lewis acidic compounds [507, 531] such as zinc glutarate [532, 533], zinc diiminate-ethylsulfinate [534], or rare-earth complexes [535] as catalysts, polyalkylenecarbonates with alternating repetition units are obtained. In contrast, if one uses double metal cyanide (DMC) catalysts, one acquires polyalkylenecarbonates, in which ether units occur besides the carbonate linkages.

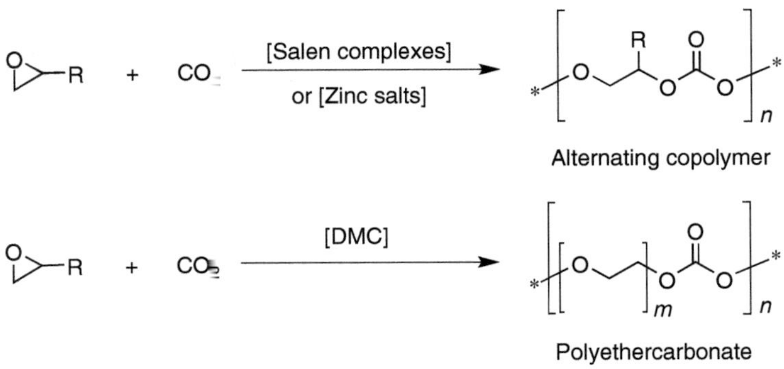

Scheme 2.44 Copolymerization of epoxides and CO_2.

The alternating copolymerization shown in Scheme 2.45 comprises the same reactants as the synthesis of five-membered cyclic carbonates; polymers are kinetic products, while cyclic carbonates are thermodynamic ones [536].

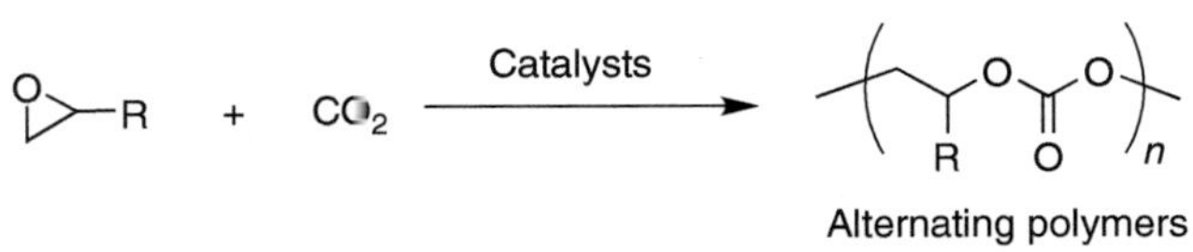

Scheme 2.45 Alternating polymerization of oxirane and CO_2.

The direct oxidative carboxylation of olefins for the synthesis of carbonates that combines both the epoxidation and the carbonation processes. Nb-catalyzed five-membered carbonate synthesis starting from 4.6 MPa CO_2, 0.5 MPa O_2, and olefins has been reported by Aresta (Scheme 2.46) [537].

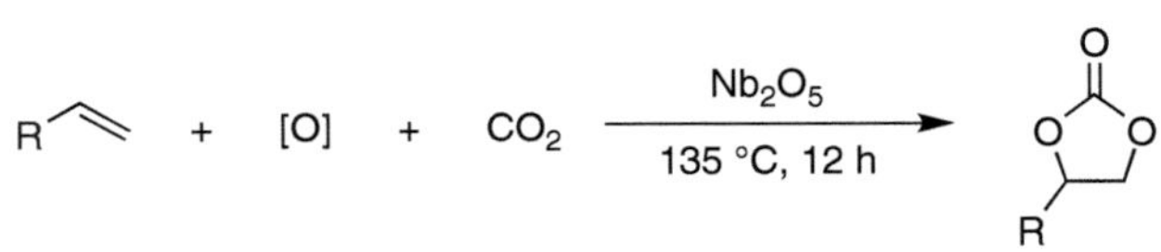

Scheme 2.46 Synthesis of cyclic carbonates from olefins. Source: Based on Aresta and Dibenedetto [337].

On the other hand, the dehydrative reaction of 1,2-diols to give cyclic carbonates is catalyzed by CeO_2–ZrO_2 or Bu_2SnO with 15 MPa CO_2 (Scheme 2.47) [538, 539].

An iron- or copper-catalyzed reaction of a cyclic ketal with CO_2 has also been proposed (Scheme 2.48) [540].

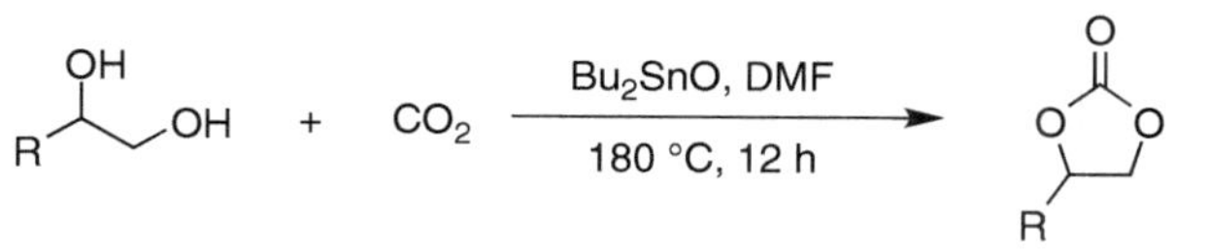

Scheme 2.47 Synthesis of a cyclic carbonates from diols. Source: Tomishige et al. [538]; Du et al. [539].

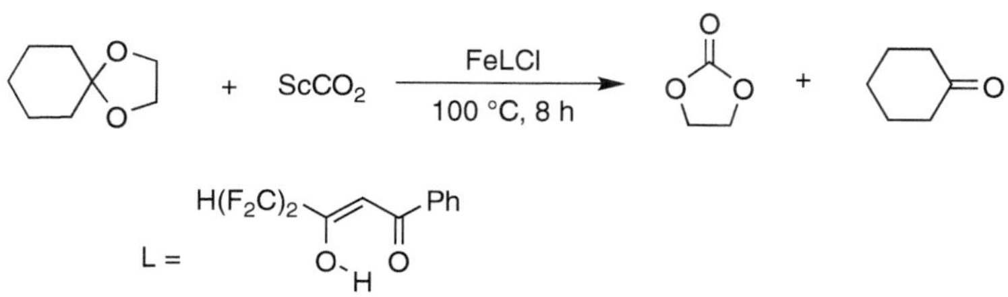

Scheme 2.48 Synthesis of cyclic carbonates from cyclic ketal. Source: Modified from Aresta et al. [540].

The reaction of propargyl carbonate with phenol proceeds in the presence of a palladium catalyst, which leads to phenoxy-substituted cyclic carbonates via CO_2 evolution (Scheme 2.49) [541, 542].

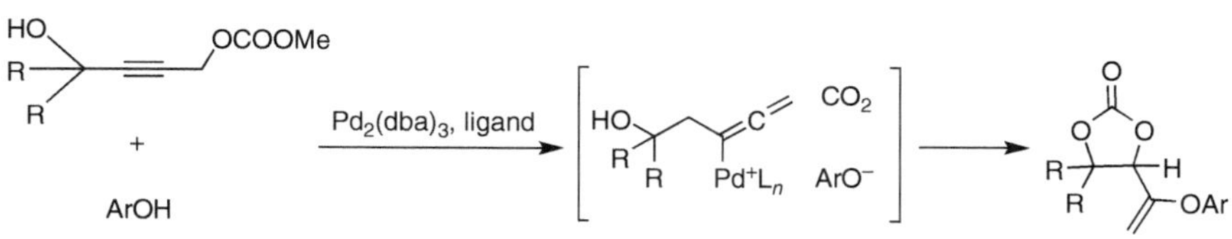

Scheme 2.49 Synthesis of cyclic carbonates from propargyl carbonates.

What is more, solid acid–base catalysts such as zirconia catalyze DMC formation from methanol and CO_2 [543, 544].

Heteropoly acids supported on zirconia are also active for the DMC formation from methanol and CO_2 [545]. In addition to zirconium-based catalysts, vanadium-based catalysts such as H_3PO_4/V_2O_5 [546] and Cu–Ni/V_2O_5–SiO_2 [547] have been applied to gas-phase reactions. Most of the heterogeneously catalyzed reactions seem less productive compared to the homogeneous reaction using ketals (Scheme 2.50).

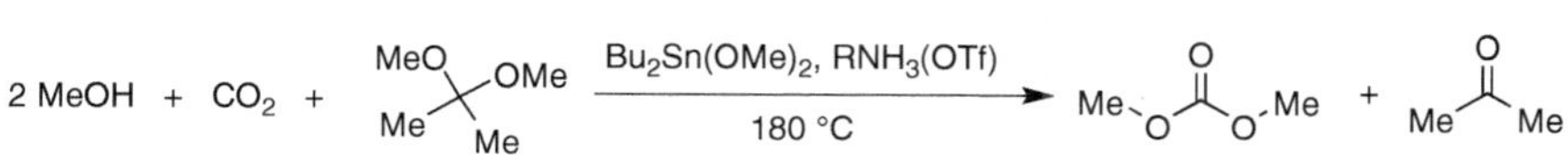

Scheme 2.50 DMC synthesis with dimethyl ketal.

Dialkylcarbonates, in particular dimethylcarbonate, are widely used intermediates for organic synthesis and polymer production. Currently, dialkylcarbonates are commercially produced mostly from the corresponding alcohol and phosgene [548]. Because of the toxic of phosgene, the oxidative carbonylation of aliphatic alcohols has been studied as one of the alternative routes (Scheme 2.51).

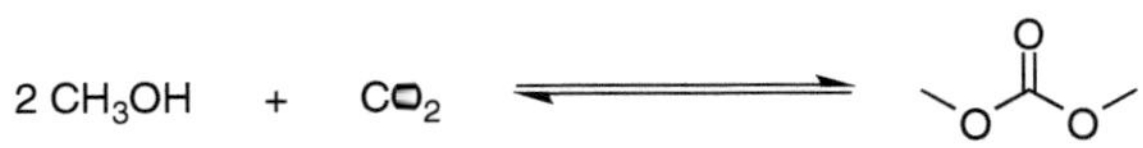

Scheme 2.51 Dimethylcarbonate from methanol and CO_2.

The carboxylation of aliphatic alcohols [549] with tin- and titanium-based [429, 550] catalysts has been described. As the reaction is equilibrium-limited, the resulting water has to be removed from the process to achieve reasonable conversions [551]. As yet, this has been a major obstacle for implementation of the carboxylation reaction.

Combining various unsaturated compounds and CO_2 results in the formation of esters and lactones. For example, polymerization of vinyl ethers under CO_2 in the presence of aluminum compounds affords polymers that incorporate CO_2 into the main chain (Scheme 2.52). However, the yields of these polymers are very low and further improvement is required [552–554]. A similar CO_2 incorporation has also been reported for the polymerization of a three-component system, which contains a phosphorus compound (Scheme 2.53) [555].

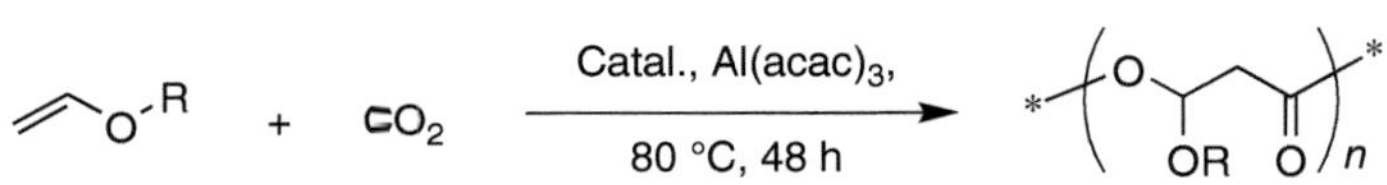

Scheme 2.52 Copolymerization of vinyl ether with CO_2.

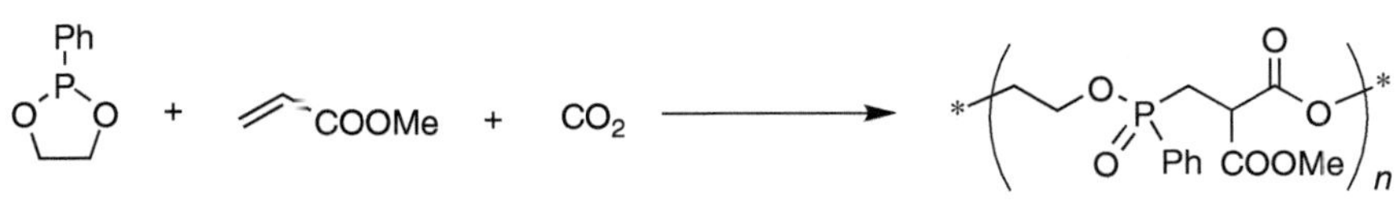

Scheme 2.53 Three-component polymerization. Source: Based on Saegusa et al. [555].

Unsaturated compounds used for these reactions are dienes [556, 557], allenes [558], acetylenes [559], diynes [560, 561], benzynes [562], and so forth.

Nickel complexes with bulky carbene ligands efficiently catalyzed cycloaddition between the diyne parts and CO_2 under very mild reaction conditions (60 °C, 0.1 MPa) to yield six-membered cyclic lactones (Scheme 2.54) [563].

A five-membered ring can be obtained through the oxidative cyclization of CO_2 and O_2 from air with a crystalline palladium complex bearing NHC ligands by Yamashita et al. (Scheme 2.55) [564]. A similar rhodium compound has also been reported [565].

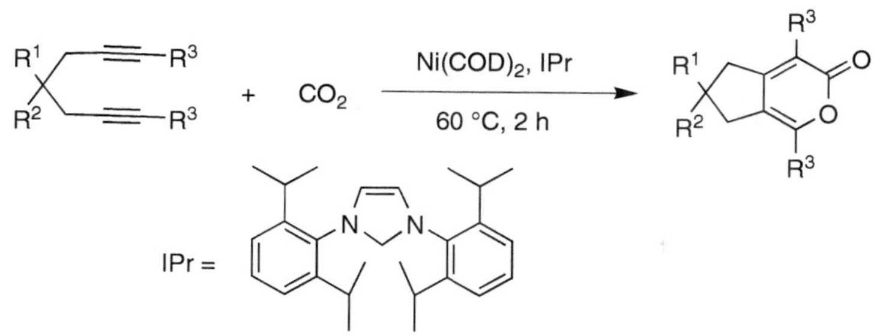

Scheme 2.54 Lactone from diacetylene. Source: Modified from Louie et al. [563].

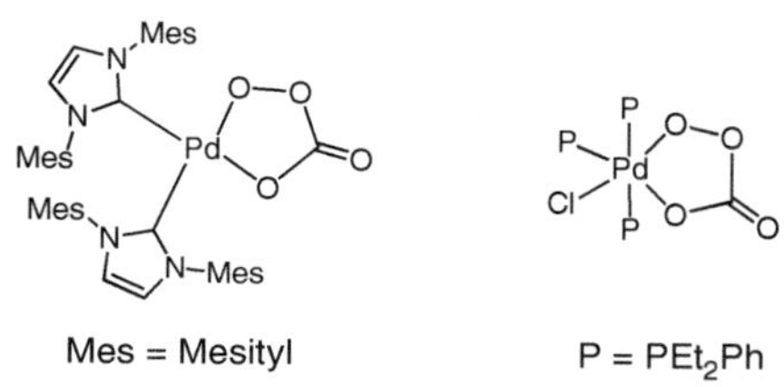

Scheme 2.55 O_2 activation by a metal complex. Source: Based on Yamashita et al. [564].

Applying these complexes to catalytic oxidation reactions with molecular oxygen seems interesting. Palladium-catalyzed C–C cross-coupling under a CO_2 atmosphere can give esters. The first report was a palladium-catalyzed C–C coupling between allyl chloride and allyl(tributyl)tin (Scheme 2.56), but the reactant is limited to allyl compounds [566].

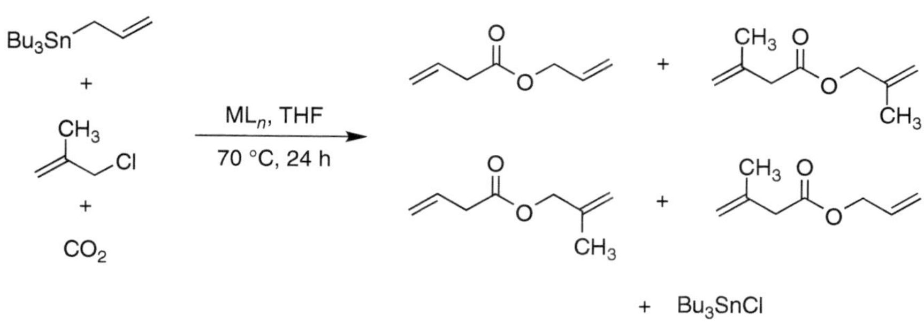

Scheme 2.56 Palladium-catalyzed C–C cross-coupling with CO_2.

2.4.1 Oxides Catalysts

In 1997, Maeshima and coworkers [567] presented for the first time that magnesium oxide is an efficient catalyst for the cycloaddition of carbon dioxide to epoxides under 2 MPa CO_2 at 135 °C. The reaction of carbon dioxide with (*R*)-styrene oxide in the presence of magnesium oxide gives (*R*)-phenyl carbonate in 72% yield with 97% ee.

Following this work, Mg–Al mixed oxides obtained by the calcination of hydrotalcites acted as efficient catalysts for the addition of CO_2 with various epoxides gave yields of up to 99% under 0.5 MPa CO_2 at 100 °C [568]. It was proposed that Mg^{2+} ions are isomorphically substituted by Al^{3+} ions to form Mg–O–Al bonds to disorder the small portion of the Mg–Mg shell, which would bring about the formation of acid and strong basic sites on the surface of the mixed oxides.

In 2015, a hydric microsphere Fe_3O_4@carbon@poly(In^{III}-carboxylate) consisting of a cluster of Fe_3O_4 nanoparticles as the core, a carbon layer as the inner shell, and a porous In^{III}-carboxylate coordination polymer as the outer shell was prepared by template-mediated coordination polymerization and showed high catalytic performance for the cycloaddition of CO_2 with propylene oxide, giving a conversion of 96.5% under the optimized reaction conditions (40 °C, 3.0 MPa) [569].

Recently, a highly porous quaterphenyl-containing covalent triazine polymer incorporated with iron oxide nanoparticles (Fe_3O_4@QP-CTP) was prepared for the cycloaddition of CO_2 and epoxides [570]. After improvement, conversion of epoxides reached 95% and the selectivity to cyclic carbonates is up to 98%. This catalyst exhibited high catalytic conversion even after eight consecutive cycles with high stability.

2.4.2 Zeolite Catalysts

In 1999, Davis and coworkers [571] reported the first example of zeolite-catalyzed cycloaddition reactions of epoxides with CO_2 to form cyclic carbonates. A cesium-exchanged zeolite CsX was prepared by triply-ion-exchanging NaX in 1 M solution of cesium acetate, washing with an aqueous solution of cesium hydroxide (pH = 13.3), and then contacting with deionized water to remove excess alkali metal occluded in the zeolite cages during ion exchange. The catalyst showed catalytic performance for the cycloaddition of CO_2 with ethylene oxide, giving a yield of 26.6% under the optimized reaction conditions (150 °C, 10.3 MPa). Incorporation of the occluded cesium oxide species in the ion-exchanged zeolite (CsO_X/CsY) via decomposition of impregnated cesium acetate further promoted their catalytic activity with yield of 33.4% under the same condition.

Subsequent investigation indicates that Ti-SBA-15 possesses improved reactivity in this coupling reaction. Ratnasamy and coworkers [572] prepared an adenine-modified Ti-SBA-15 solid catalyst by anchoring adenine on Ti-SBA-15, which was found to exhibit excellent catalytic performance for the conversion of epoxides (epichlorohydrin, propene oxide, and styrene oxide) into corresponding cyclic carbonates with conversions of up to 98% and product selectivity of 94.6% under atmospheric pressure of CO_2 (0.69 MPa) at 120 °C. Besides adenine, DMAP was also used to modify zeolite for the reaction of CO_2 to epoxides. In 2007, a SBA-15-immobilized DMAP analog was synthesized and produced propylene carbonate yields of 81% under 1.7 MPa pressure of CO_2 at 120 °C [573].

In addition to modification of base compound, graft of acids could promote the catalytic activity of zeolite. A sulfonic acid-tethered mesoporous silica material (SBA-15-SO_3H) with TBAB co-catalyst was developed for the cycloaddition of CO_2 and styrene oxide at low CO_2 pressure of 0.1 MPa at 80 °C and resulted in 89% styrene carbonate yield [574]. The high catalytic activity of SBA-15-SO_3H/TBAB was attributed to the hydrogen bonding interaction between acidic hydrogen atom (from $-SO_3H$) and the oxygen atom of the epoxide, and nucleophilic effect of bromide anion (from TBAB).

Recently, Islam and coworkers [575] reported a carboxylic acid-grafted SBA-15 material by directly bonding 3-aminopropyltriethoxysilane (3-APTES) on the surface of SBA-15 to form 3-aminopropyl-functionalized SBA-15 and then grafted with 4-formyl benzoic acid to form AFS 1. The formation of cyclic carbonates using different kind of epoxides in the presence of carbon dioxide at atmospheric pressure and room temperature was performed by using TBAB as a co-catalyst and up to 96% yields were achieved. It was proposed that Brønsted acidic sites due to the presence of $-CO_2H$ sites facilitate the protonation of epoxide oxygen, whereas basic sites due to the imine–N facilitate the activation of CO_2 molecules.

2.4.3 Supported Nanoparticle and Lewis Acid Catalysts

In 2005, Deng and coworkers [526] described the efficient synthesis of polymer-immobilized nanogold catalyst (Au/poly) via pretreating polymers with NaOH solution, then ion-exchanging with $HAuCl_4$ (aq), and drying at 150 °C for 6 h. Au/poly showed remarkable catalytic activity for activation of carbon dioxide, with TOF > 50 000 h^{-1} for the synthesis of cyclic carbonate under 3 MPa CO_2 at 150 °C. It was speculated that the synergism between the nanogold species and the peculiar microenvironment of the polymer surface resulted in the exclusive catalytic activity.

In addition to Au catalyst, copper catalyst was also used for cyclization of epoxides to carbonates. In 2017, Xiao and coworkers [576] demonstrated that copper metalated hierarchically porous bipyridine (bpy)-constructed polymer (Cu/POP–Bpy) in the synthesis of cyclic carbonates through the cycloaddition of CO_2 to epoxides under ambient conditions. For cycloaddition of CO_2 to epichlorohydrin, with addition of 3.5 mol% nBu$_4$NBr, a yield of up to 99% was obtained with TON of 237 in 24 h. It was proved that the cooperative effects of fully accessible active sites and enriched CO_2 concentration in the nanopores were responsible for the high activities of the catalysts for CO_2 transformation.

Subsequently, copper-based magnetic nanocatalyst 2-acetylbenzofuran (ABF)@ASMNPs were prepared by in situ synthesis of silica-coated magnetic nanoparticles (SMNPs) from TEOS and Fe_3O_4, anchored APTES on the surface of silica, grafted with ABF, and then stirred with a solution of copper acetate [577]. The performance of the catalyst was investigated in the reaction of CO_2 with propylene oxide to produce propylene carbonate. The magnetic catalyst showed high catalytic performance for the cycloaddition of CO_2 with various epoxides, giving a yield of at least 92% under the optimized reaction conditions (4 mol% DBU, 80 °C, 1 atm).

In another significant progress, novel silica-supported Nb species were prepared by reacting a molecular niobium precursor, $[NbCl_5{\cdot}OEt_2]$, with silica dehydroxylated at 700 °C (SiO_2-700) or at 200 °C (SiO_2-200) to generate diverse surface complexes [578]. The reactions of SiO_2-200 with the niobium precursor generated surface mostly monopodal and mostly bipodal complexes. According to the SSNMR/DFT study, species vicinal monopodal complexes, in which the Nb centers are placed in close proximity, displayed a significant higher catalytic activity than isolated monopodal and bipodal complexes, for the synthesis of propylene carbonate from CO_2 and propylene oxide under mild catalytic conditions (60 °C, 1 MPa). Based on DFT calculations, the cooperative interaction between two Nb centers may contribute to reduce the cyclization barrier during the catalytic cycle (Scheme 2.57).

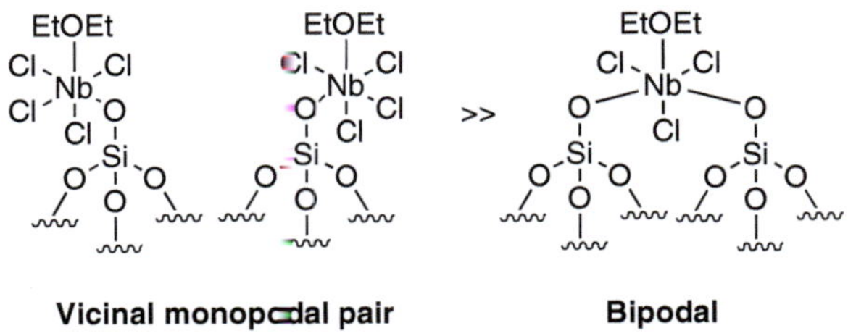

Scheme 2.57 Proposed structures for silica-supported species.

Following this work, a reusable zirconium-based catalyst, $ZrCl_4@SiO_2$, was investigated by the same group in the reaction of CO_2 with propylene oxide to produce propylene carbonate [579]. Comparing with silica-supported Nb species, the isolated monopodal, vicinal monopodal, and binopodal lined Zr species all performed as active catalysts for the cycloaddition of pure CO_2 to propylene oxide (96%, 90%, and 93%) under the optimized reaction conditions (60 °C, 1 MPa) with the addition of 1 mol% TBAB. According to systematic DFT calculation on the $ZrCl_4$-catalyzed cycloaddition reaction, the high catalytic activity of isolated monopodal zirconium complexes is justified by the fact that CO_2 activation proceeds with a lower barrier on $ZrCl_4$ than on $NbCl_5$ and a bimetallic mechanism is not required to lower the barrier for the rate-determining step of the cycloaddition reaction.

Moreover, palladium nanocatalyst $Pd@MTiO_2$ have been designed and prepared by embedding Pd nanoparticles at the surface of a mesoporous TiO_2 material [580]. Synthesis of carbonate over $Pd@MTiO_2$/TBAB gave >90% yield with >98% selectivity at 80 °C, under 1 atm of CO_2.

In addition, supported Zn heterogeneous was also developed for this reaction. In 2016, Liu and coworkers [581] synthesized a series of M/HAzo-POPs catalysts through diazo-coupling reaction between diazonium salts and multihydroxy benzene (Scheme 2.58), followed by chelating metal ions by azo and phenolic –OH groups. The Zn/HAzo-POP-2 showed excellent performance for catalyzing the reaction of CO_2 with epoxide and affording a TOF of 3330 h^{-1} under 3 MPa CO_2 at 100 °C in the presence of TBAB.

Recently, another polymer support heterogeneous [PS-Zn(II)L] catalyst was prepared by reaction of chloromethylated polystyrene with Schiff base ligand, followed by chelating zinc chloride (Scheme 2.59) [582]. With TBAB as a source of Br^-, the formation of organic cyclic carbonates from epoxides via CO_2 fixation was realized with 99% selectivity, 94–100% conversion, and TOF > 20 h^{-1} at room

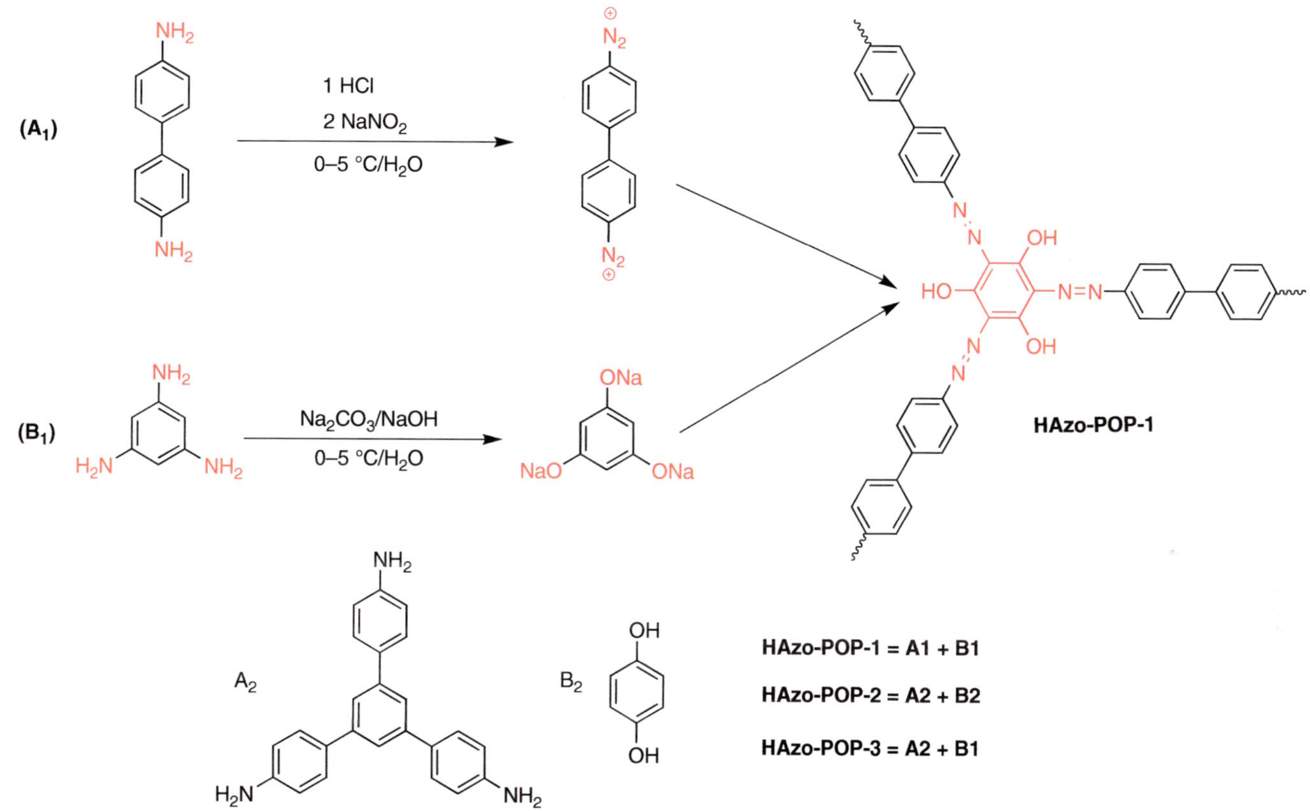

Scheme 2.58 Synthetic process of *o*-hydroxy azo-hierarchical porous organic polymers.

temperature under 1 atm CO_2 pressure in solvent-free condition. After that, a similar polystyrene-functionalized zinc anthranilic acid complex catalyst was prepared by this group and used in the synthesis of cyclic carbonates in the same reaction conditions [583]. A plausible mechanistic pathway proposed that the active sites of PS–Zn anthra complex activate the oxiranes through non-covalent interaction with the oxiranes O atoms, whereas the basic site present at the surface owing to amine can activate the molecules of CO_2 gas.

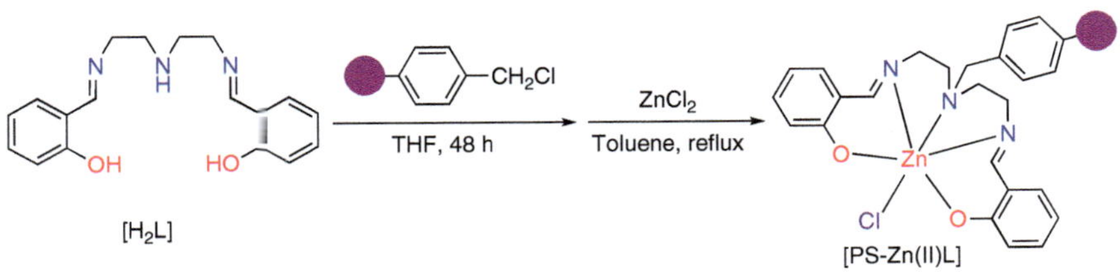

Scheme 2.59 Schematic presentation of synthesis of [PS-Zn(II)L] catalyst. Source: Biswas et al. [582].

2.4.4 Carbon Catalysts

In 2011, Park and coworkers [584] synthesized a mesoporous carbon nitride (MCN) materials by a hard templating method, followed by evaluation of its catalytic performance for the cyclization of CO_2 to propylene oxide. Under optimized reaction conditions, 90% selectivity to propylene carbonate was obtained with up to 16% conversion using MS-MCN catalyst under 0.55 MPa CO_2 at 100 °C.

Following this report, a g-C_3N_4 and n-$Bu_4N^+Br^-$ combination is used for the conversion of epoxides to cyclic carbonates under a CO_2 filled balloon at 105 °C [585]. This combination was found to be very effective for the efficient conversion of a wide number of epoxides.

Except C_3N_4, as-received graphene oxides (GOs) alone were also used as efficient and green heterogeneous catalysts for the cycloaddition between epoxides and CO_2 at 1.0 atm (140 °C) [586]. As received GOs enabled the styrene oxide conversion of 97.8% with a 97.4% chemoselectivity to phenylethylene carbonate in the presence of DMF. Kinetic and X-ray photoelectron spectroscopy analyzes revealed that as-received GOs with the highest amount of the oxygen-containing groups exhibited the highest reaction rate. The activation of epoxide was achieved by the oxygen-containing groups in the GO catalysts as the surface-active sites.

2.4.5 Salen, Porphyrins, and Phthalocyanines Catalysts

The metal–salen complexes, metal–porphyrins, and metal–phthalocyanines are widely recognized in catalysis and proven to be effective homogeneous catalysts for the cycloaddition of CO_2 to epoxides [587–589]. However, these soluble complexes are not easily separated from the organic constituents of the reaction mixture. Therefore, heterogenization of these catalysts by a variety of routes for coupling of epoxides with carbon dioxide was explored. Different immobilization strategies have been developed, such as coordination of the metal center to a modified support surface or covalent-bound of these complexes on support surface, polymerization of these complexes to form porous organic polymer.

In 2002, He and coworkers first described immobilization of aluminum phthalocyanine complex on MCM-41 molecular sieve for the cycloaddition of CO_2 and epoxides to produce cyclic carbonates [590]. The catalyst was synthesized by reaction of $ClAlPc(SO_2Cl)_2$ with 3-aminopropyltriethoxysilane, followed by grafting of obtained $ClAlPc\text{-}(SO_2\text{-}NHCH_2CH_2CH_2Si(OEt)_3)_2$ with a MCM-41 molecular sieve (Scheme 2.60). For the addition of epichlorohydrin with CO_2, a TOF of 452 h^{-1} was achieved under 110 °C, 4 MPa CO_2 using ClAlPc-MCM-41/TBAB as a catalyst system.

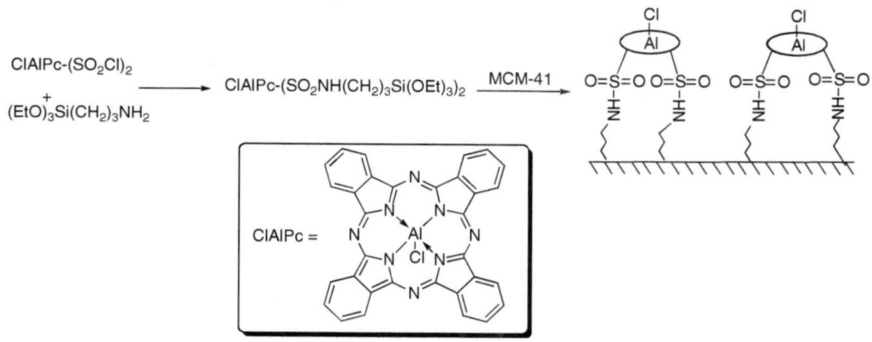

Scheme 2.60 Synthetic routes of monomeric ClAlPc catalyst supported on MCM-41.

Following this report, chromium salen catalysts were covalently anchoring to silica support through nucleophilic substitution between aminopropylsilyl-modified silica and chloromethyl-substituted chromium salen complex (Scheme 2.61) [534]. The synthesized catalyst was active in reaction of styrene oxide with CO_2 under supercritical conditions (10 MPa, 80 °C), yielding 74% conversion of styrene oxide with 100% selectivity after 6 h.

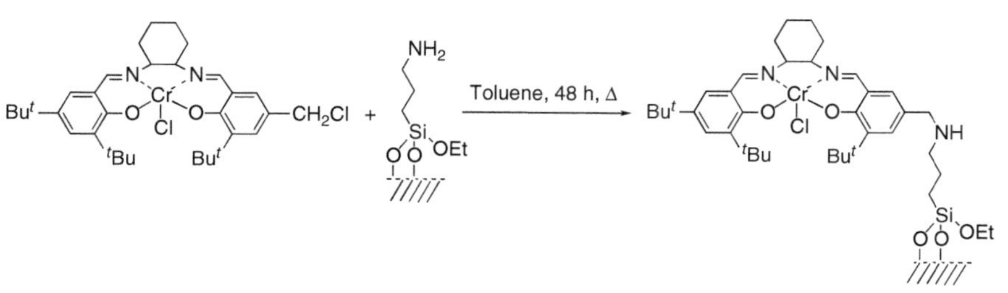

Scheme 2.61 The preparation of Cr–salen/SiO_2. Source: Modified from Alvaro et al. [534].

In addition to immobilization, polymerization of metal–salen complexes, metal–porphyrins, and metal–phthalocyanines complexes was also used to fabricate a heterogeneous catalyst. In 2013, Deng and coworkers successfully synthesized microporous polymers containing metal–salen species by Sonogashira coupling of dibromo-functionalized salen–Co/Al complexes and 1,3,5-triethynylbenzene (Scheme 2.62) [591]. Remarkably, the high activity of Co–salen-based heterogeneous catalyst allows facile cyclization of propylene oxide with 81.5% yield at room temperature and atmospheric pressure in the presence of a co-catalyst TBAB.

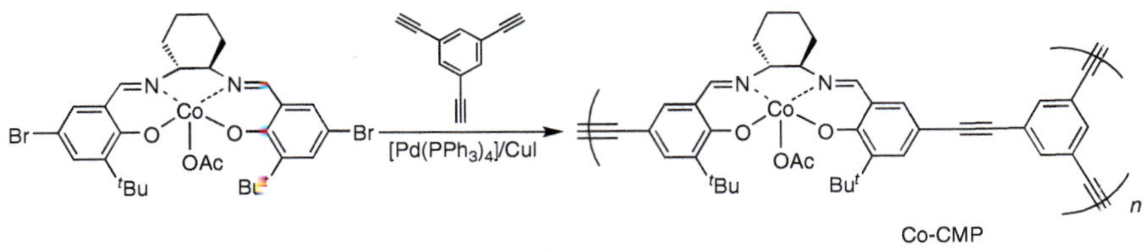

Scheme 2.62 Synthesis of Co–CMP. Source: Modified from Xie et al. [591].

Using similar Sonogashira coupling strategy, a hollow and microporous Cr and Zn porphyrin networks (H–MCrPN and H–MZnPN) for the room-temperature CO_2 fixation to cyclic carbonates was also established [592]. The Cr–porphyrin networks were prepared via the Sonogashira coupling of Cr-tetra (4-ethynylphenyl) porphyrins and 1,4-diiodobenzene on the surface of silica templates followed by silica etching. By using Cr(III)–F porphyrin catalyst, H–MCrPNs and TBAB as catalyst system, serious epoxides can be cyclized with high yields at room temperature under ambient pressure.

Another good method is polymerization of vinyl-functionalized salen, porphyrins, and phthalocyanines. A porphyrin-based porous organic polymer (POP-TPP) is successfully synthesized from polymerization of vinyl-functionalized porphyrin monomer, followed by metalation (Scheme 2.63) [593]. These porous heterogeneous catalysts are really active in cycloaddition of epoxides with CO_2 to cyclic carbonates under ambient conditions. Notably, Co/POP-TPP exhibits the conversion at 95.6% with 99% selectivity under ambient conditions (29 °C and 1 atm) with TBAB as an additive.

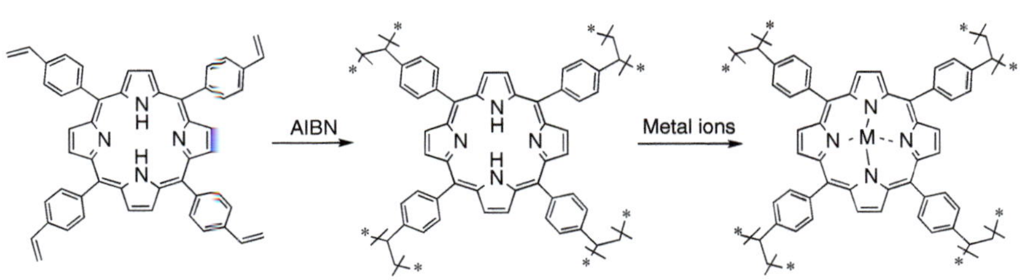

Scheme 2.63 Synthesis of POP–TPP and M/POP–TPP. Source: Dai et al. [593].

Further investigation shows that metalporphyrin-based microporous organic polymer can be successfully prepared by a solvent knitting hyper-crosslinked polymers method using dichloromethane as an external crosslinker and 5,10,15,20-tetraphenylporphyrin (TPP) as the building block (Scheme 2.64) [594]. The as-synthesized catalyst HUST-1-Co showed good catalytic efficiency for the conversion of epoxides into cyclic carbonates with excellent yields (>93%) at room temperature and atmospheric pressure in the presence of the co-catalyst TBAB.

2.4.6 Ionic Liquid Catalysts

ILs have attracted increasing attention in the fields of CO_2 capture and conversion because of their unique properties, including low vapor pressure, high thermal stability, good solubility, and tunable functionalization [595]. Furthermore, ILs are able to generate strong van der Waals and electrostatic interactions with CO_2 and activate CO_2, thus showing amazing activity for CO_2 capture and/or conversion [596, 597]. Chemical immobilization onto solid supports or polymerization was usually used to construct heterogeneous IL catalysts.

In 1993, Nishikubo et al. first described the use of quaternary ammonium or phosphonium salts grafted polystyrene as a catalyst to promote the addition of epoxides with CO_2 under atmospheric pressure [530]. Utilizing polymer-supported quaternary phosphonium salts as the catalyst, the reaction proceeded smoothly to give the corresponding cyclic carbonates in 74–92% yields at 80–90 °C. Since then, considerable advancements have been achieved in the cycloaddition of the epoxides with CO_2 using heterogeneous IL catalysts.

In 2006, Xia and coworkers developed an elegant method for immobilization of IL catalyst and used it as a catalyst for the cycloaddition of carbon dioxide to epoxides [598]. The catalyst was prepared by the coupling of 1-(triethoxysilylpropyl)-3-*n*-butylimidazolium bromide with TEOS using the sol–gel protocol (Scheme 2.65). The combination of the immobilized IL with zinc chloride afforded active catalyst system for the conversion of propylene oxide into propylene carbonates with TOF of up to 2712 h^{-1} and yield of 95% under 1.5 MPa pressure of CO_2 at 110 °C. Following this work, grafted IL catalyst [599] OH/SiO_2 was prepared by the covalent bond of 1-(triethoxysilyl)propyl-3-methylimidazolium hydroxide [599] OH with silica gel [599]. Without $ZnCl_2$ as an additive [599], the OH/SiO_2 catalyst showed high catalytic performance for the cycloaddition of CO_2 with propylene oxide, giving high conversion of 97% and TOF of 1347 h^{-1} under 2.5 MPa pressure of CO_2 at 120 °C for 4 h.

Subsequently, Liu and coworkers described the efficient synthesis of covalent immobilization of glycidyl-group-containing ILs onto various supports through reactions between the glycidyl group in the IL and the polar groups (i.e. amino and carboxyl group) on the support surfaces, followed by investigation of their catalytic performances for the synthesis of cyclic carbonates from CO_2 and epoxides [600]. With the amino-group-functionalized polymer-supported ILs (PNH_2–IL) as a catalyst, coupling of propene oxides with CO_2 proceeded smoothly to afford the desired PC in 95% yield and the TOF could reach about 1693 h^{-1} under 2.0 MPa pressure of CO_2 at 130 °C for 8 h.

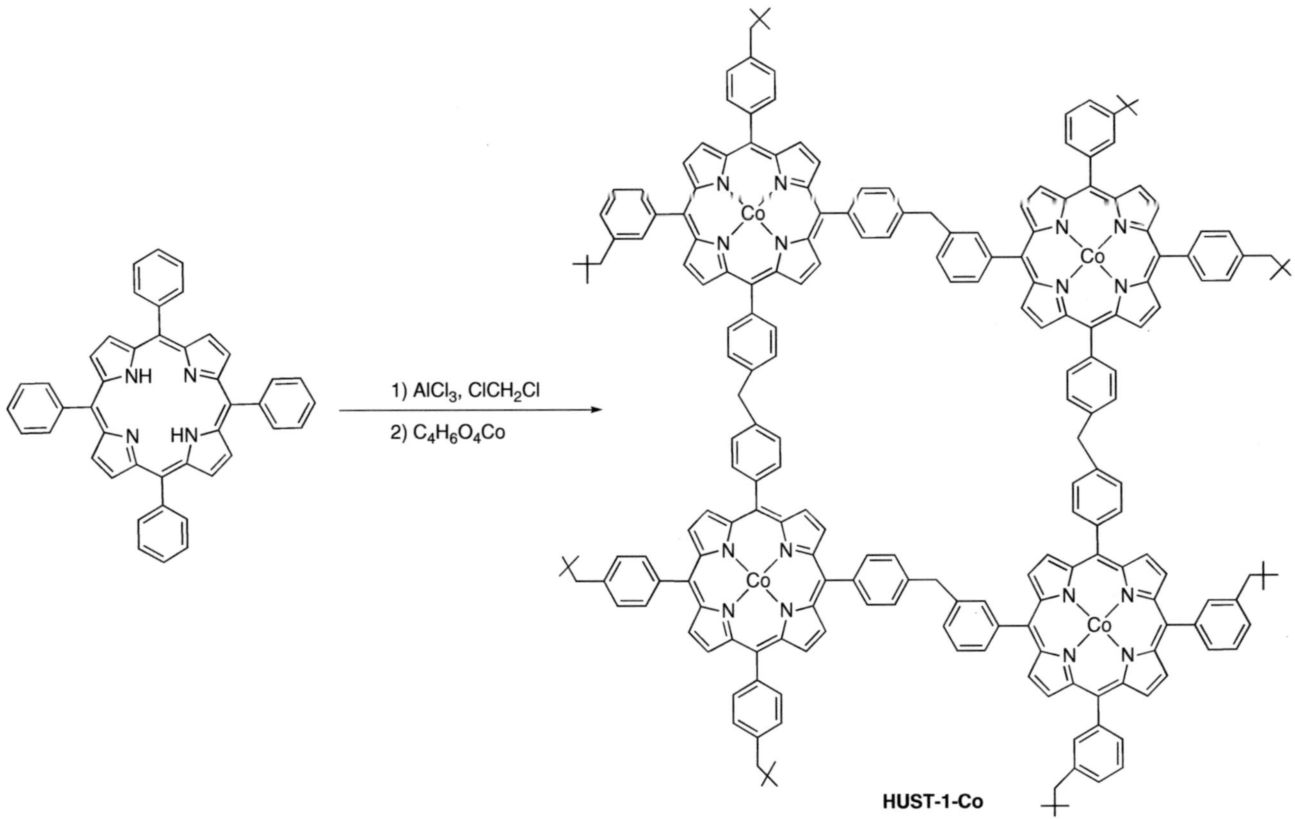

HUST-1-Co

Scheme 2.64 The synthetic pathway to synthesize the network structure. Source: Wang et al. [594].

Scheme 2.65 Preparation of the immobilized ionic liquid via sol–gel method.

Several years later, a chitosan-grafted quaternary ammonium salt catalyst was designed for the incorporation of CO_2 into epoxides [601]. The chitosan was modified with allyl glycidyl ether (AGE) and glycidyl trimethylammonium (GTA) chloride, and followed by thiol–ene addition with 3-mercaptopropionic acid (Scheme 2.66). Cyclic carbonates were obtained with high yields in the presence of KI under mild conditions (0.2–0.7 MPa CO_2, 80 °C).

Further investigation shows that polystyrene-supported resorcinarenes could allow facile cycloaddition of propene oxides with CO_2 in a quantitative yield under 0.5 MPa CO_2 pressures at 80 °C [602]. The immobilized catalyst was synthesized by reaction of chlorinated polystyrene resin with hydroxyl-substituted tetrabenzoxazine derivative, followed by treatment with MeI (Scheme 2.67).

Recently, Li and coworkers constructed SBA-16-grafted IL catalysts through the coupling of an silica-based amino-functionalized imidazolium IL (Si–IM–NH_2) and silica-based quaternary ammonium salt (Si–TBAI) on SBA-16 for the cycloaddition of CO_2 with epoxides under mild conditions (50 °C and 0.5 MPa CO_2) [603].

Moreover, covalent immobilization of ILs onto a series of supports, such as carbon nanotubes [604, 605] and tristriazine [606], has also been studied for cycloaddition of CO_2 with epoxides.

Besides chemical immobilization of ILs onto solid supports, polymerization was also extensively used to construct heterogeneous IL catalysts. In 2007, Han and coworkers first studied the use of polymer-supported IL (PSIL) as a catalyst for the cycloaddition of CO_2 to epoxides [607]. The highly cross-linked PSIL was prepared by copolymerizing of 3-butyl-1-vinylimidazolium chloride ([VBIM]Cl) with the cross-linker divinylbenzene (DVB) (Scheme 2.68). Using PSIL as the catalyst, cyclic carbonates were obtained in good yields at 110 °C and 6 MPa.

The catalytic activity was substantially improved by introducing $ZnBr_2$ onto the polymer-supported IL [608]. The Zn/PS-IL[Br] catalyst gave a 97.5% yield of styrene carbonates with a TOF value that can be up to about 3800 h^{-1} at 120 °C under 3 MPa CO_2.

An important breakthrough for the cyclization of epoxides with CO_2 is achieved when multifunctional sites heterogeneous catalysts comprising IL, zinc halide, and triphenylphosphine are employed [609, 610]. The active heterogeneous catalysts were prepared through free radical polymerization of vinyl-functionalized PPh_3 and vinyl-functionalized IL (imidazolium salt or phosphonium salt) under solvothermal conditions, and subsequently impregnated by ZnX_2 salts. The phosphonium-based PPh_3–ILBr–$ZnBr_2$@POPs catalyst provided the excellent catalytic activity, achieving TOF up to 6022 h^{-1} in cycloaddition of propylene oxide under 3 MPa CO_2 at 140 °C [609].

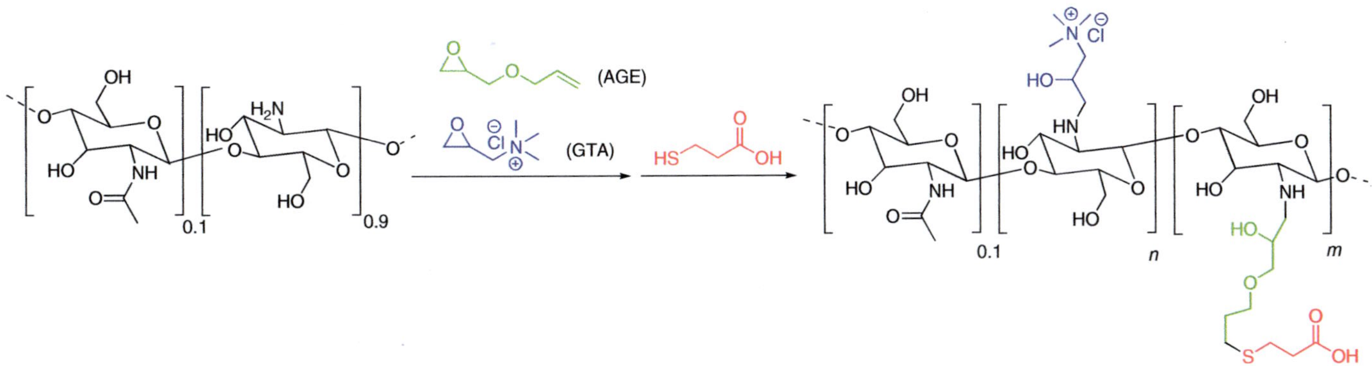

Scheme 2.66 Modification of chitosan with AGE and GTA followed by thiol–ene addition with 3-mercaptopropionic acid.

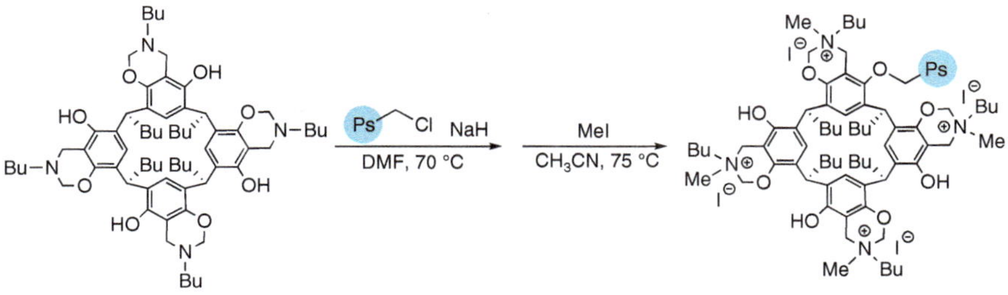

Scheme 2.67 Synthesis of resorcinarene derivatives.

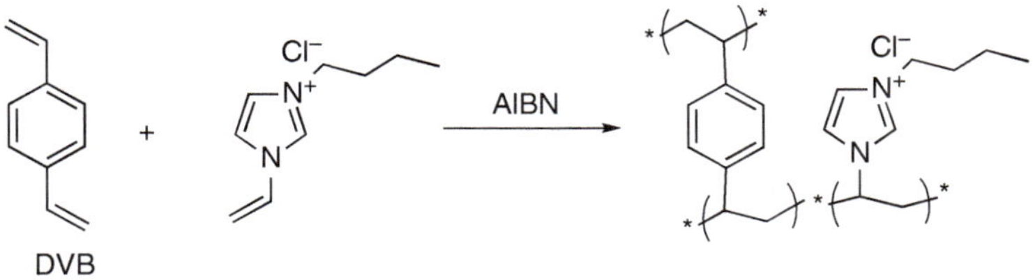

Scheme 2.68 Synthesis of the cross-linked-polymer-supported ionic liquid.

In addition, vicinal dual hydroxyl functional poly(IL)s was constructed through the copolymerization of epoxy-containing IL monomers and divinylbenzene, followed by ring opening in water. The resulting PGDBr-5-2OH catalyst facilitates the cycloaddition reaction with high yields under mild conditions (1 atm CO_2, 70 °C) when n-Bu_4NI was employed as a co-catalyst [611].

In another significant contribution, Ma and coworkers designed ionic polymer encapsulated within the channels of a COF catalysts which was prepared by in situ radical polymerization of vinyl-functionalized phosphonium bromide monomers in channels of COF-TpBpy, followed by metalation with $Cu(OAc)_2$ [612]. The flexibility and movability of the catalytic moieties on the linear polymer enable them to cooperate with the active sites anchored on the COF pore walls. In addition, the essential catalytic components enriched in the confined space are beneficial to boosting the cooperation and promoting the catalytic efficiency. As results, much higher TON of up to 28 764 was observed in cycloaddition of epichlorohydrin with CO_2 (1 atm) at 120 °C for 48 h with PPS⊂COF–TpBpy–Cu as a catalyst. Notably, propene oxide can be cyclized at very mild conditions (1 atm CO_2, 25 °C) to give the propylene carbonate in 94% yield.

In addition to the vinyl polymerization, Sonogashira coupling of tetrakis(4-ethynylphenyl)methane and diiodoimidazolium salts [613], reaction of bisimidazole and silicon tetrachloride [614], reaction of arene-bridged bis- and tris-alkyl halides with trimethylsilylimidazole [615], and phenol–formaldehyde condensation reaction between phloroglucinol and IL-modified dialdehydes [616] have also been utilized for the synthesis of the poly-IL catalysts. The catalytic performance of as-synthesized materials has been explored for fixation of CO_2 with epoxides into cyclic carbonates.

Moreover, silica-supported ILs [617, 618], montmorillonite clay ion-exchanged with ILs [619], carboxymethyl cellulose-supported IL [620], and MCM-22-supported ILs [621] have also been used for the carbonation of epoxides.

2.4.7 Metal–Organic Framework (MOF) Catalysts

MOFs represent a special class of porous materials consisting of easily constructed organic linkers and metal clusters [622]. Owing to their extra-large internal surface area, structural flexibility, tailorable pores and chemical environment, interesting host–guest interactions, and existence of confined nanospace, they have been explored as a new type of functional materials for heterogeneous catalysts over the past decade [623–625]. Therefore, much effort has been devoted to use MOFs as catalyst for the cyclic addition of epoxides with CO_2 [626].

MOF-5 is one of the representative MOFs having a BET surface area ranging from 260 to 4400 $m^2 g^{-1}$ based on the synthetic method and has a good thermal stability of up to 400 °C. It was prepared by reaction of $Zn(NO_3)_2$ with benzene-1,4-dicarboxylic acid (H_2bdc) (Scheme 2.69). In 2009, Han and coworkers presented for the first time that MOF-5 and quaternary ammonium salts had excellent synergetic effect in promoting the coupling reaction of CO_2 with epoxides [627]. By applying MOF-5 and n-Bu_4NBr as a catalyst system, propene carbonates can be produced in 95% yield under mild conditions (0.2 MPa CO_2, 50 °C, 6 h).

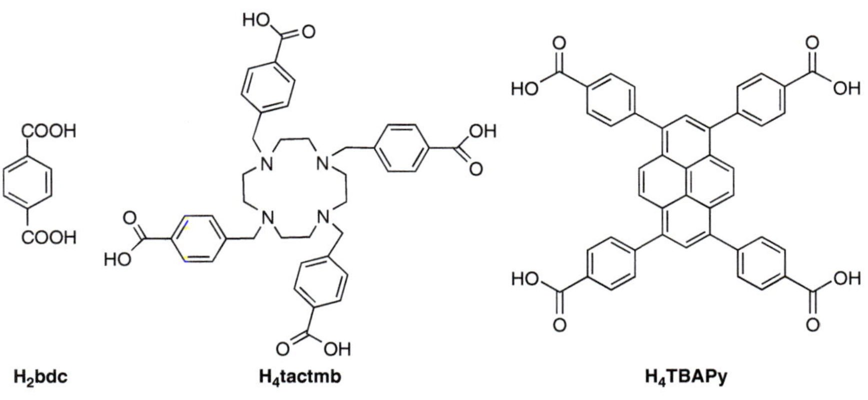

Scheme 2.69 The structures of H_2bdc, $H_4tactmb$, and H_4TBAPy.

Besides MOF-5, ZIF-8 which was synthesized from 2-methylimidazole and a Zn precursor is another famous MOF. The catalytic activity of ZIF-8 was also evaluated in cycloaddition of CO_2 to epichlorohydrin reaction [628]. The ethylene diamine-functionalized ZIF-8 displayed good catalytic performance, affording chloropropene carbonate with 73.1% yield at 80 °C under 0.7 MPa of CO_2 for 4 h.

Sooner after, the catalytic performance of the MOF Cr-MIL-101 in solvent-free coupling of CO_2 and epoxides has been explored [629]. The Cr-MIL-101 with the formula Cr_3–$F_{0.8}(NO_3)_{0.2}(H_2O)_2O[(O_2C)$–$C_6H_4$–$(CO_2)]_3$ $n$$H_2O$ ($n = 25$) was synthesized by reaction of $Cr(NO_3)_3$with terephthalic acid (H_2bdc) in the presence of HF. With MIL-101/TBABr as a catalyst system, the coupling reaction of epoxides with CO_2 proceeds effectively at milder conditions (25 °C, 0.8 MPa) and propylene and styrene carbonates can be obtained with the yields of 82% and 95%, respectively.

Subsequently, an MMCF-2 (empirical formula: $[Cu_2(Cu\text{-tactmb})(H_2O)_3(NO_3)_2]$) was synthesized by the self-assembly of the azamacrocyclic tetracarboxylate ligand 1,4,7,10-tetrazazcyclododecane-*N*,*N*′,*N*″,*N*‴-tetra-*p*-methylbenzoic acid (H_4tactmb) (Scheme 2.69) with $Cu(NO_3)_2$ under solvothermal conditions [630]. This MOF exhibited high catalytic activity for the chemical fixation of CO_2 into cyclic carbonates at room temperature under 1 atm pressure. For examples, 95.4% yield of propylene carbonate was achieved by using MMCF-2 after 48 h at room temperature under 1 atm CO_2 pressure in the presence of nBu$_4$NBr.

In another significant contribution, a new Hf-based MOF (Hf-NU-1000) incorporating Hf_6 clusters is prepared by self-assembly of 1,3,6,8-tetrakis(*p*-benzoic acid)pyrene (H_4TBAPy) (Scheme 2.69) with $HfOCl_2 \cdot 8H_2O$ under solvothermal conditions [631]. The Hf-NU-1000 demonstrated good catalytic activity for the quantitative cycloaddition of styrene oxide and propene oxide to corresponding carbonates at room temperature under 1 atm pressure CO_2 in the presence of nBu$_4$NBr.

Furthermore, a discrete single-walled metal–organic nanotube was synthesized by incorporating a tetraphenyl-ethylene moiety as the four-point connected node and subsequently used as a heterogeneous catalyst for the cycloaddition of carbon dioxide to epoxides [632]. The solvothermal reaction of $Ni(NO_3)_2 \cdot 6H_2O$, tetrakis(4-carboxyphenyl)ethylene (H_4TCPE) (Scheme 2.70), and L-proline in a mixture of DMF and H_2O at 100°C for 3 days gave the compound Ni–TCPE1. High PC yield of 99% with TON of up to 2000 was obtained in the reaction of CO_2 and styrene oxide over Ni–TCPE1 with TBABr as co-catalysts for 12 h (100 °C, 1 MPa). The Ni–TCPE1 catalyst also features the good stability with a TON reach to 35 000 after 20 times (70 h).

Subsequently, a sulfonate-based MOF was also developed for the coupling of epoxides with CO_2 [633]. It was constructed by hydrothermal reaction of $Cu(NO_3)_2$, 4,4′-bpy, and disodium 1,2-ethanedisulfonate (EDSNa$_2$) to afford blue cubic crystals of $[Cu(bpy)_2(EDS)]_n$, which we denote as TMOF-1. In conjunction with TBAB as a cocatalyst under ambient conditions (i.e., room temperature, 1 bar), TMOF-1 exhibited a nearly quantitative yield in converting propene oxides to the corresponding cyclic carbonate.

One year later, a Co(II)-based MOF having formula $\{[CO_2(tzpa)(OH)(H_2O)_2] \cdot DMF\}_n$ was synthesized by employing a tetrazolyl-carboxyl ligand 5-(4-(tetrazol-5-yl)phenyl)isophthalic acid (H_3tzpa) (Scheme 2.70) [594, 634]. The $\{[CO_2(tzpa)(OH)(H_2O)_2] \cdot DMF\}$n facilitates the chemical fixation of CO_2 coupling with epoxides into cyclic carbonates with up to 99% yield under very mild conditions (25 °C, 1 atm CO_2) in the presence of TBAB.

Shortly afterward, Ma and coworkers designed and constructed a stable copper paddlewheel-based MOF (JUC-1000) using a custom-designed buffer behaving ligand 2,4-bis(3,5-dicarboxyphenylamino)-6-ol triazine (H_4BDPO) (Scheme 2.70) that features both weak acidic and basic functional groups [635]. Benefiting from the weak acid–base pairs that facilitate both the activation and interactions of epoxide and CO_2 molecules, the MOF, JUC-1000, exhibited excellent performance in the fixation of carbon dioxide to form cyclic carbonate under ambient conditions.

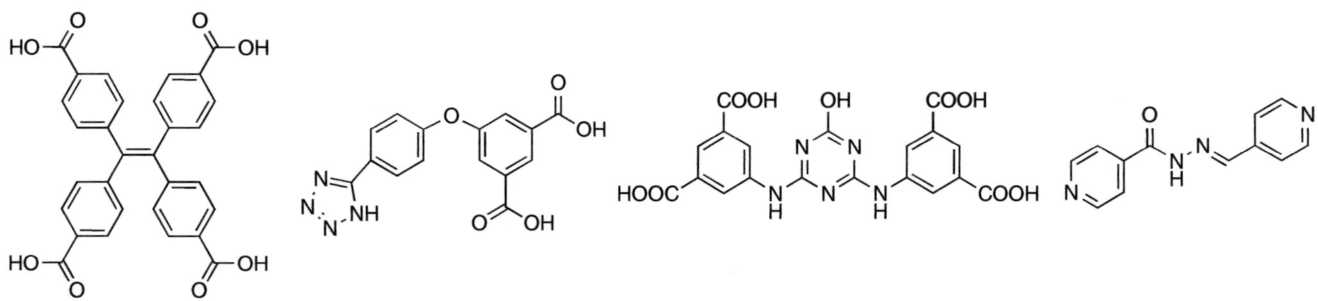

H_4TCPE **H_3tzpa** **H_4BDPO** **PCINH**

Scheme 2.70 The structures of H_4TCPE, H_3tzpa, and H_4BDPO.

JUC-1000/TBAB was capable of the cycloaddition of propylene oxide and CO_2 to form propylene carbonate with a yield of 96% at room temperature under 1 atm CO_2 for 48 hours.

Very recently, mixed-ligand 3D/2D ZnMOFs were synthesized and exploited for their potential as a catalyst in CO_2 fixation [636]. The ZnMOF-1 having formula $((Zn(bdc)(PCINH))\cdot Xg)_n$ (H_2BDC = benzene-1,4-dicarboxylic acid, PCINH = 4-pyridyl carboxaldehyde isonicotinoylhydrazone, G = lattice guests) (Scheme 2.70) was able to catalyze PO to PC under 1 atm CO_2 pressure at room temperature in the presence of the $n Bu_4NBr$ with 86% yield, which is amplified up to 99% upon increase in temperature to 40 °C within 8 h.

In addition, a number of other acid ligands, such as, 4,4′,4″-*s*-triazine-1,3,5-triyl-tri-*p*-aminobenzoic acid (H_3TATAB) [637], L-glutamic acid [638], resorcin arene-functionalized dodecacarboxylic acid ($H_{12}L$) [639], 1′,2′,3′,4′,5′,6′-hexakis (4-carboxyphenyl)benzene (H_6CPB) [640], were also developed to prepare MOFs for the coupling of CO_2 with epoxides (Scheme 2.71).

2.4.8 Bifunctional Catalysts

The rational design and integration of multiple functional components, such as Salen (porphyrin or phthalocyanine), IL, and MOF, into a composite material could combine advantages of each component and bridge the gap between each component, thus resulting in enhanced activity tailored for highly efficient production of cyclic carbonates from CO_2 and epoxides. Therefore, plenty of effort has been devoted to the development of efficient bifunctional catalyst systems for cyclic addition of CO_2 with epoxides.

In 2016, the catalytic behavior of salen–IL bifunctional heterogeneous catalyst TBB-Bpy@Salen-Co (TBB = 1,2,4,5-tetrakis(bromomethyl)benzene, Bpy = 4,4′-bipyridine) was exploited for cycloaddition of epoxides under 1 MPa CO_2 at 60 °C [641]. The catalyst was prepared by polymerization of TBB with Bpy in a one-pot solvothermal method to give TBB–Bpy, followed by reaction of TBB–Bpy with amino-based Salen–Co Schiff (Scheme 2.72). The synergistic effect of the active sites functionalized with Co atoms and Br^- anions afforded excellent conversion and selectivity to give cyclic carbonates at 60 °C, even at low CO_2 pressure (0.2 MPa). Interestingly, the binary catalyst system TBB–Bpy@Salen–Co led to a 99.2% propene epoxide conversion and a turnover number (TON) of 496 under 1 MPa CO_2 at 60 °C for 6 h.

Subsequently, a salen–IL bifunctional cationic porous organic polymer based on a Salen–Al metalloligand (Al–CPOP) containing imidazolium functionality was prepared from the chloromethyl-functionalized Salen–Al and 2,4,6-tris(imidazol-1-yl)-1,3,5-*s*-triazine (TIST) (Scheme 2.73) [642]. The obtained catalyst exhibited good activity in the cycloaddition of carbon dioxide with epichlorohydrin to give the product with about 95% yield under 1 atm CO_2 at 120 °C.

Furthermore, a porphyrin–IL bifunctional catalyst (denoted as Al–iPOPs) were synthesized by Yamamoto–Ullmann coupling reaction between 5,10,15,20-tetrakis (4-bromophenyl)porphyrin-aluminum(III) chloride (Al–TBPP) and brominated ILs

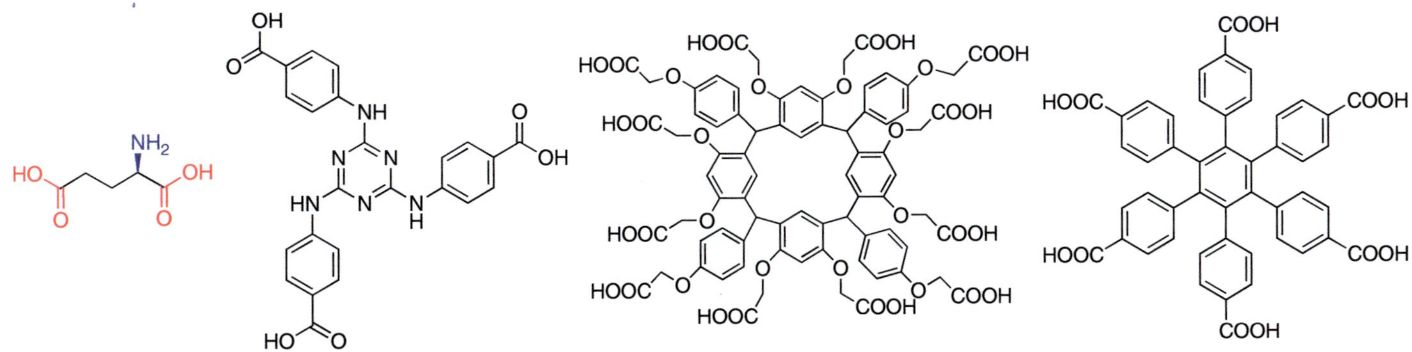

L-Glutamic acid　　H_3TATAB　　$H_{12}L$　　H_6CPB

Scheme 2.71 The structures of H_3TATAB, L-glutamic acid, H12L, and H_6CPB.

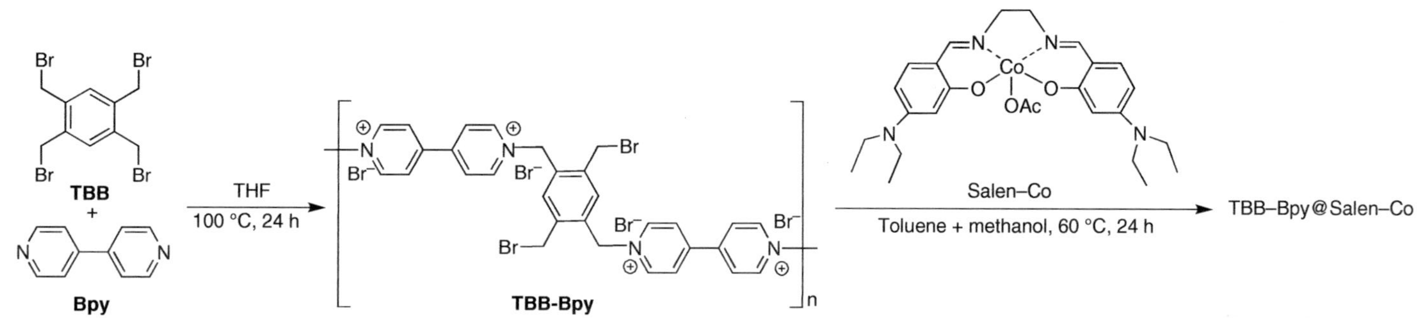

Scheme 2.72 Schematic of the synthesis of bifunctional catalyst TBB-Bpy@Salen-Co.

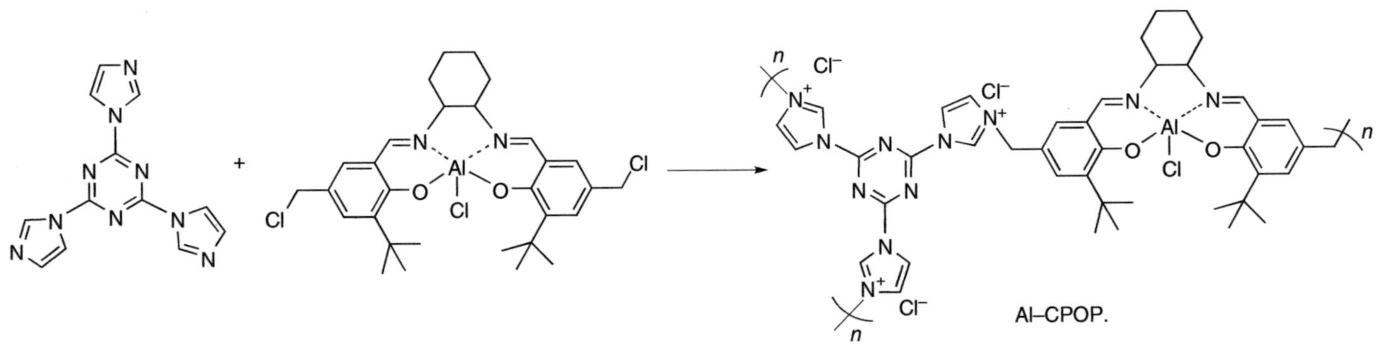

Scheme 2.73 Synthesis of Al–CPOP. Source: Liu et al. [642].

(Scheme 2.74) [643]. Notably, a high TOF of 7600 h^{-1} was obtained with propylene epoxide at 100 °C and 1.0 MPa. Remarkably, Al–iPOP-2 could promote the PO/CO_2 cycloaddition reaction within 8 h to achieve an excellent yield of 99% at 25 °C and atmospheric CO_2 pressure (0.1 MPa).

In addition, a porphyrin–IL bifunctional catalyst (denoted as Mg-por/pho@POP) based on a magnesium porphyrin and phosphonium salt-integrated porous organic polymer (POP) was designed for cyclic carbonate production which uses epoxides and CO_2 [644]. The Mg-por/pho@POP catalyst can be obtained from the copolymerization of a vinyl-functionalized Mg–porphyrin complex and vinyl-functionalized phosphonium salt under solvothermal conditions (Scheme 2.75). This catalyst showed excellent catalytic activity in cyclic addition of propene with CO_2, achieving an ultrahigh TOF value of 15 600 h^{-1} (1 h, 78% yield) at 140 °C under 3 MPa CO_2 without the addition of co-catalysts.

Very recently, a series of novel zinc(II) porphyrin-based ionic polymers (SYSU-Zn@ILs) have been synthesized by imine condensation reaction of zinc(II) tetrakis(4-aminophenyl)porphyrin (Zn–TAPP) and various IL-functionalized dialdehyde compounds (Scheme 2.76) [645]. Owing to the intramolecularly cooperative behavior between Lewis acidic metal center and nucleophilic bromide ion as well as the polar CO_2-philic nature, SYSU-Zn@IL_2 was capable of insertion of CO_2 to propene epoxide with 80% yield at ambient temperature (25 °C) and atmospheric CO_2 pressure (0.1 MPa).

Besides the combination of ILs with salen and porphyrin, integration of MOF with salen and porphyrin could also be used to construct bifunctional catalyst.

In 2013, a series of highly stable MOFs namely PCN-224 have been prepared by self-assembly of $ZrCl_4$ and MTCPP (TCPP, tetrakis(4-carboxyphenyl)-porphyrin) (Scheme 2.77) and tested as catalysts for the cycloaddition of CO_2 and propylene oxide to propylene carbonate [646]. By applying PCN-224(Co)/Bu_4NCl as a catalyst system, a 39% propylene oxide conversion with TOF of 129 h^{-1} was obtained under 2 MPa CO_2 at 100 °C for 4 h.

In addition, a new heterogeneous catalyst was synthesized by nucleophilic substitution of amino-based porphyrin [$Sn^{IV}(TNH_2PP)(Cl)_2$] with chloromethylated MIL-101(Cr), followed by exchange of anion (Scheme 2.78) [647]. The obtained catalyst showed good catalytic performance for the CO_2 cycloaddition to different epoxides with excellent yields at 50 °C and 1 atm CO_2 in the presence of TBPB.

Moreover, Salen–Cu(II)@MIL-101(Cr) catalyst was developed for the synthesis of propylene carbonate from CO_2 and propylene oxide under room temperature and ambient pressure with a yield of 87.8% over 60 h with TBAB co-catalyst [648]. Salen–Cu(II)@MIL-101(Cr) was in situ assembled in the MOFs by the bonding of metal ions with Salen ligands according to the ship in a bottle method, that made Salen compounds encapsulated inside the cavities of MOFs and difficult to pass through the windows of MOFs.

Very recently, a new porphyrinic MOF having formula $[Cu_2(C_{40}H_{24}N_8)_{0.5}(C_{14}H_8O_5)(DMA)](DMA)(H_2O)_6$ has been prepared by solvothermal reaction TPyP, $Zn(NO_3)_2{\cdot}6H_2O$, and 4,4′-oxybisbenzoic acid (H_2OBA), followed by exchanging of Zn^{2+} by Cu^{2+} ions [649]. The combination of this Cu catalyst with TBAB as

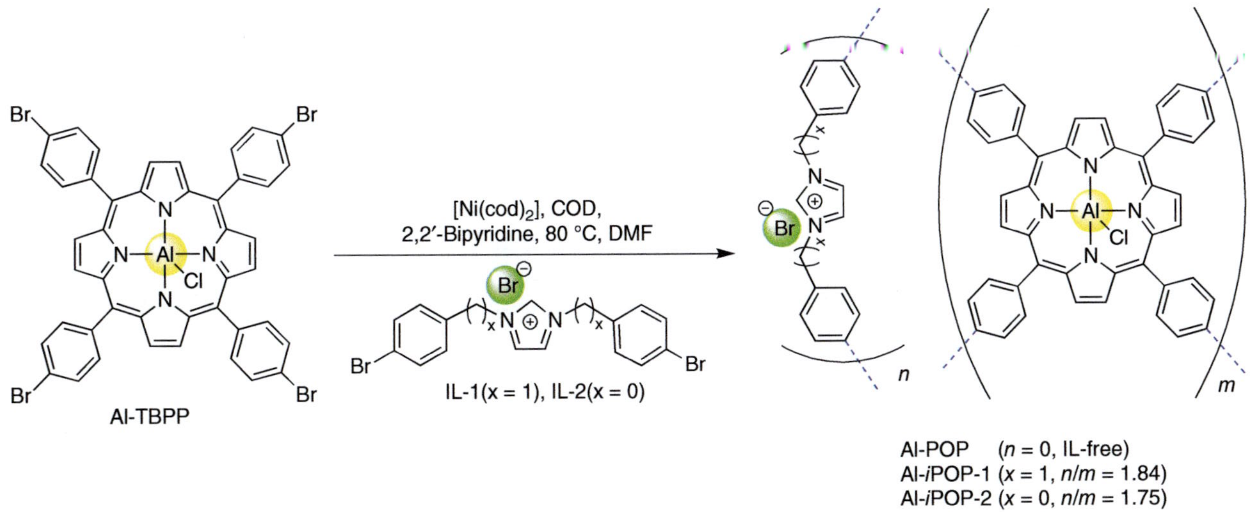

Scheme 2.74 Synthesis of Al–POP, Al–iPOP-1, and Al–iPOP-2. Source: Chen et al. [643].

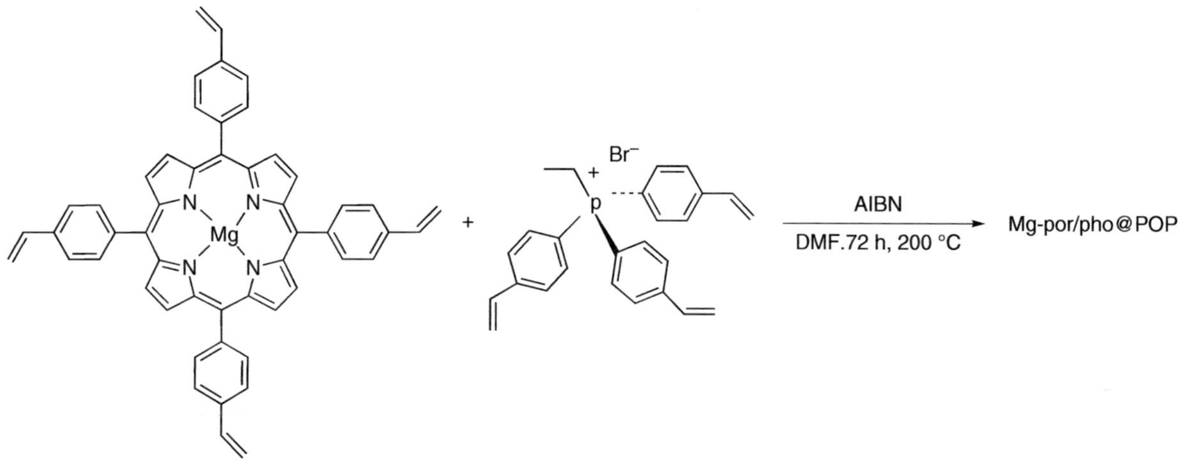

Scheme 2.75 Synthesis of Mg-por/pho@POP.

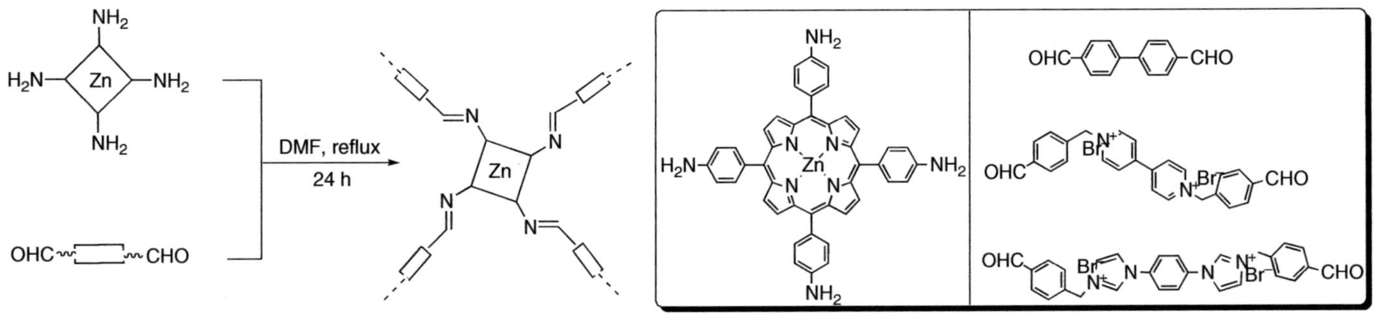

Scheme 2.76 Synthesis route to SYSU-Zn, SYSU-Zn@IL1, and SYSU-Zn@IL2. Source: Based on Chen et al. [645].

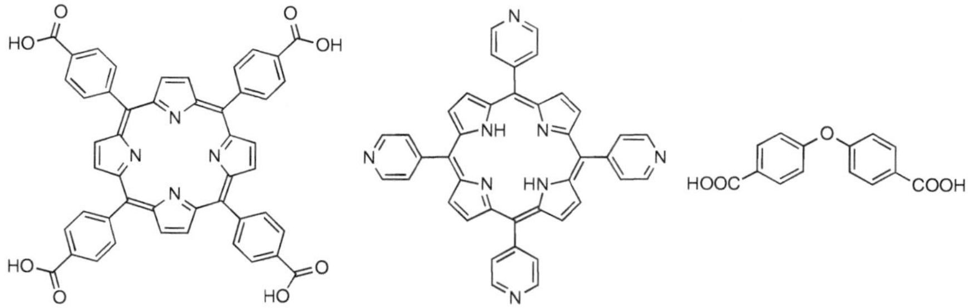

Scheme 2.77 The structures of TCPP, TPyP, and H_2OBA.

a co-catalyst was found to be a good catalyst for the conversion of CO_2 to cyclic carbonate with excellent yields under ambient conditions (1 atm CO_2, 25 °C)

The combination of MOF and ILs is another strategy to fabricate bifunctional catalyst. Recently, imidazolium-based poly(IL)s have been confined into the MOF material MIL-101 via in-situ polymerization of encapsulated monomers [650]. The resultant composite polyILs@MIL-101 exhibits good CO_2 capture capability that is beneficial for the catalysis of the cycloaddition of CO_2 with epoxides to form cyclic carbonates at subatmospheric pressure in the absence of any co-catalyst. The significantly enhanced activity of polyILs@MIL-101 is attributed to the synergistic effect among the good CO_2 enrichment capacity, the Lewis acid sites in the MOF, as well as the Lewis base sites in the polyILs.

In addition, an acid–base-mediated self-assembly approach was developed to fabricate a heterogeneous catalyst composed of MOF-supporting ILs (Scheme 2.79) [651]. The obtained IL@MIL-101–SO_3H catalyst exhibits a high catalytic activity for cyclizing CO_2 with epichlorohydrin with 98% yield under atmospheric pressure at 90 °C without the addition of a cocatalyst such as Bu_4NBr.

2.4.9 Other Catalysts

In 2015, a nanoporous polymer incorporating sterically confined air-stable NHCs (NP-NHCs) was first synthesized and exhibited excellent catalytic performance for the conversion of epoxides into cyclic carbonates with yields of up to 98% at 120 °C under atmospheric pressure of CO_2 [652].

Recently, two triazine-based covalent organic frameworks (COF-JLU6 and COF-JLU7), with high crystallinity and good stability, were successfully synthesized under solvothermal conditions. Using TBAB as the co-catalyst, the COF-JLU7 as a metal-free organocatalyst displayed an efficient catalytic performance (yields of 58–99%) for the cycloaddition of CO_2 with epoxides under mild conditions with a broad substrate scope [653].

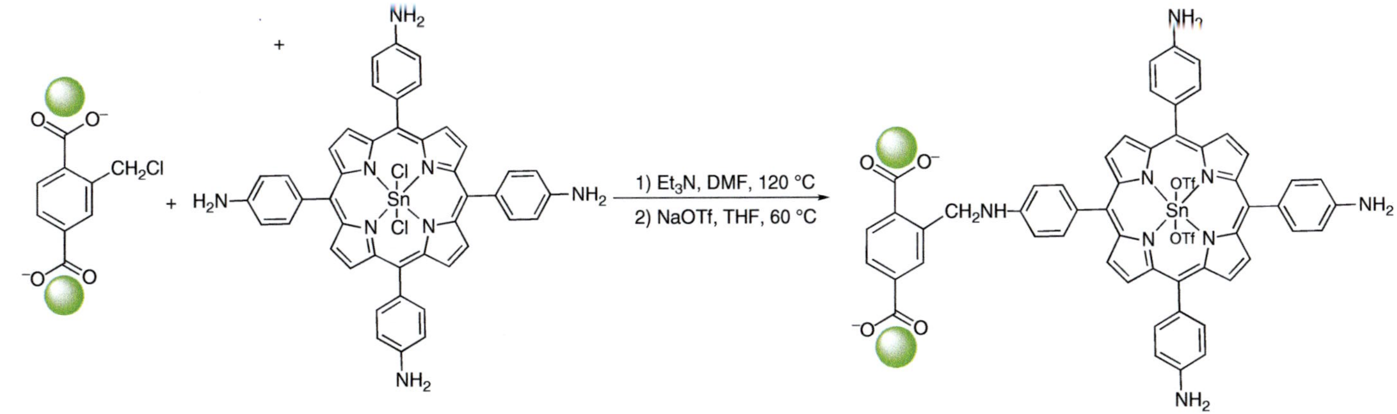

Chloromethylated MIL-101(Cr) **$Sn^{IV}(TNH_2PP)Cl_2$** **$[Sn^{IV}(TNH_2PP)OTf_2]$/CM-MIL-101**

Scheme 2.78 Preparation of $[Sn^{IV}(TNH_2PP)OTf_2]$/CM-MIL-101 catalyst. Source: Modified from Zadehahmadi et al. [647].

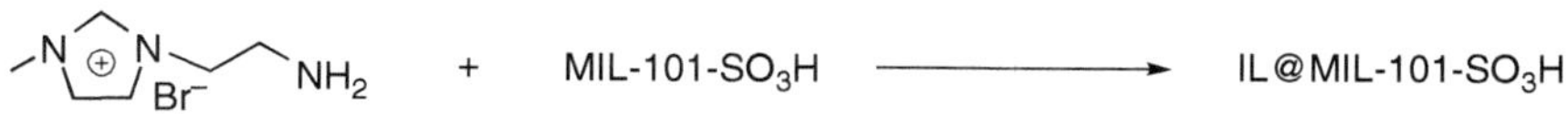

Scheme 2.79 Fabrication procedure of self-assembled MOF-supported IL heterogeneous catalyst. Source: Based on Sun et al. [651].

2.5 Reaction of Polyalcohols/Olefins with CO_2

Cyclic carbonates provide an application as a raw material in the synthesis of polycarbonates and glycols as well as fuel additives, electrolyte solvents for lithium batteries, and substitutes for non-environmental [654–657]. Synthesis of cyclic carbonates can be accomplished by coupling carbon dioxide with epoxides, polyalcohols and oxidative carboxylation of olefins. The process of obtaining cyclic carbonates via the direct reaction between glycerol and CO_2 is more sustainable and ecological than the reaction with epoxides or use of phosgene ($COCl_2$) (Scheme 2.80) [658–660].

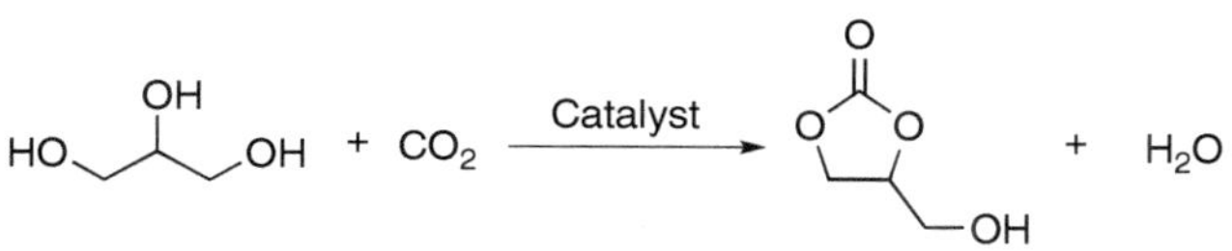

Scheme 2.80 Reactions of polyalcohols and CO_2 to form cyclic carbonates. Source: Sonnati et al. [658]; Ma et al. [659]; Dai et al. [660].

In 2017, different metal oxide materials (ZnO, SnO_2, Fe_2O_3, CeO_2, La_2O_3) were produced by sol–gel method using polyvinyl alcohol (PVAl) as a template and ZnO exhibited the best performance for direct carbonation of glycerol with CO_2, affording glycerol carbonate with a yield of 8.1% at 180 °C, 1.5 MPa after 12 h [661]. The proposed mechanism suggests that zinc oxide surface might avail to form an adsorbed alkoxide and then attacks the CO_2 molecule to form a carboxylate intermediate. In the sequence, the secondary hydroxyl of the glycerol may attack the carboxyl group to produce the cyclic carbonate.

Light olefins are the most inexpensive and readily available chemical building blocks; oxidative carboxylation of olefins are potentially significant approaches for both chemical utilization of CO_2 and production of cyclic carbonates. One of the catalytic systems contains two-step about epoxidation step and cycloaddition step (Scheme 2.81).

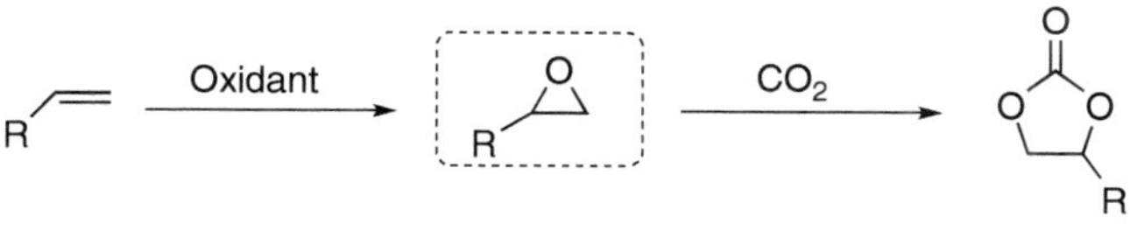

Scheme 2.81 Reactions of olefins and CO_2 with two-step to form cyclic carbonates.

Metal oxides, like Ag_2O, have been used both in the epoxidation of olefins and the carbonation of epoxides after the patent of 1941. Oxidative carboxylation with olefins and CO_2 using metal oxides or transition metal complexes has been studied. Commercial Nb_2O_5 as a heterogeneous catalyst was reported to synthesis cyclic carbonate obtaining 27% conversion and 16.6% selectivity with the process of epoxy intermediate from styrene. 4.5 MPa CO_2 and 0.5 MPa O_2 in 2000 [662]. Then, catalytic system of two-step about Au/CNT for the epoxidation step and [bmim]Br–$ZnBr_2$ for cycloaddition step achieved 60% yield of styrene carbonate [663]. The gold nanoparticles immobilized on multi-walled carbon nanotubes (Au/CNT) by impregnation method for the first step at 80 °C for 4 h using anhydrous *tert*-butyl hydroperoxide (TBHP) as an oxidant. The reaction of the resulting epoxide and CO_2 was catalyzed by the 1-butyl-3-methylimidazolium bromide ([bmim]Br^-ZnBr_2) system under the CO_2 pressure (1.2 MPa) at 120 °C. The illustration is shown in Scheme 2.82 [663].

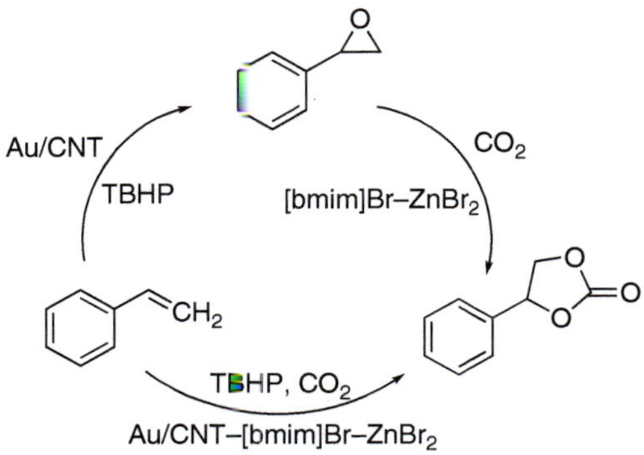

Scheme 2.82 The illustration of reactions about olefins and CO_2 with Au/CNT-[bmim]Br-$ZnBr_2$. Source: Jasiak et al. [663].

Recently, the synthesis of heterobimetallic dual catalyst [664] (MNP@SiO_2–8Mn and MNP@SiO_2–4Cr) with hybrid Mn(III)- and Cr(III)-halogenated porphyrin magnetic nanocomposites has been studied for the same reaction by using O_2 or H_2O_2 as the oxidant showed conversion of 99% and selectivity of 94% under room temperature with 45 min for epoxidation step and using 1 MPa CO_2 for cycloaddition reactions with 24 h at 80 °C.

Apart from two-step of epoxidation and cycloaddition reactions, direct synthesis of olefins carbonate has been reported (Scheme 2.83). The Au/SiO_2–$ZnBr_2$/Bu_4NBr catalyst system [665] was studied for direct synthesis of styrene carbonate from styrene, TBHP and CO_2 with 42% yield under mild reaction conditions (80 °C, 1 MPa CO_2) for 4 h. Au/SiO_2 is active for the epoxidation of styrene, and $ZnBr_2$ and Bu_4NBr cooperatively catalyze the subsequent CO_2 cycloaddition to epoxide. Similar catalyst system, like Au/$Fe(OH)_3$–$ZnBr_2$/Bu_4NBr system [666], nanogold catalyst-supported R-201 resin (Au/R-201) system [667] enhanced 10% yield under similar condition.

The use of TBHP as an oxidant instead of O_2 and ILs as catalysts increases carbonate selectivity, but still requires high CO_2 pressure (4–8 MPa). Catalytic properties of the MOF Cr-MIL-101 prepared by impregnation method have been explored under

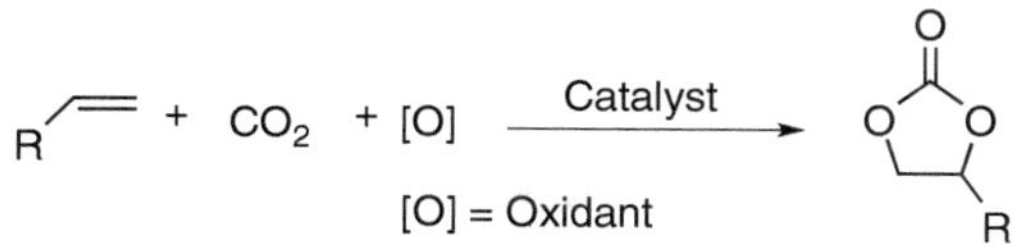

Scheme 2.83 Reactions of olefins and CO_2 direct synthesis of cyclic carbonates.

mild reaction conditions (0.8 MPa CO_2, 25 °C, 48 h) in the presence of both TBHP and H_2O_2 with up to 44% selectivity of styrene carbonate and 57% conversion of styrene [668].

Chiral POMOFs (e.g. ZnW-PYI) [669] containing pyrrolidine moiety as a chiral organocatalyst and polyoxometalate as an oxidation catalyst were designed for conversion of light olefins into cyclic carbonates with TBHP as an oxidant in a one-pot procedure under 0.5 MPa CO_2, 50 °C, 96 h with 92% conversion and 80% selectivity. In this process, captured CO_2 was fixed and activated synergistically by well-positioned pyrrolidine and amine groups, providing further compatibility with epoxidation intermediate and driving the tandem catalytic process.

The nonmetallated nature of the porphyrin ligand was exploited for implantation of coordinatively unsaturated Fe(III) ions to generate a Fe(III)@MOF-1 framework, which catalyzed for the oxidation of styrenes to the corresponding epoxides at room temperature. Moreover, the one-pot direct synthesis of styrene carbonates with up to 99% conversion and 90% selectivity from styrenes and 0.8 MPa CO_2 was also achieved using Fe(III) @MOF-1 as a catalyst in the presence of PhIO and TBAB as an oxidant at 50 °C for 48 h [670].

Recently, a series of catalysts like Ti-MMM-E, Ti-MIL-125, and CuMo-BPY were exploded for the same reaction achieving 90–99% conversions and 70–90% selectivity under 0.8 MPa CO_2 at 50–70 °C [671–673].

It is worth noting that rare earth elements are also used to this catalytic reaction. Three new lanthanide-based MOFs (Ln-MOFs) were synthesized under solvothermal conditions and fully characterized. MOF-590, -591, and -592 showed to be catalytically active in the oxidative carboxylation of styrene and CO_2 for a one-pot synthesis of styrene carbonate in presence of TBHP and n-Bu_4NBr under mild conditions (0.1 MPa CO_2, 80 °C, 10 h). Among the new materials, MOF-590 revealed a remarkable efficiency with conversion of 96% and selectivity of 95% [674].

2.6 Formylation of Amines with CO_2

Formamide derivatives are important solvents and key intermediates in the synthesis of adhesives, pesticides, and drugs and formulation of polymers [675, 676]. Apart from the utilization of energy-rich substrates (for example, epoxides), catalytic reductive formylation and methylation of amines represent important modern examples for cooperative effects in the activation of CO_2. The activation energy for reductive deoxygenation of CO_2 is significantly decreased by the coordination of amines to CO_2 and the formation of N–C bonds, which thus facilitates the reduction of CO_2.

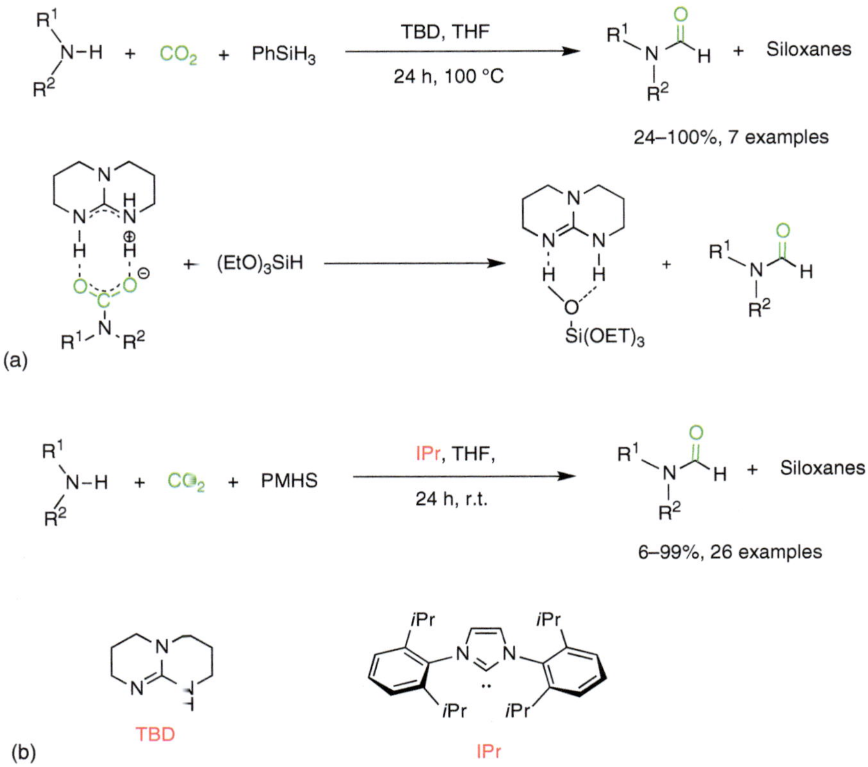

Scheme 2.84 (a, b) Organocatalytic formylation of amines with CO_2 using silanes as the reductants. Source: Modified from Jacquet et al. [677].

More specifically, Cantat and coworkers reported an organocatalytic synthesis of formamides from CO_2 up to 83% yield. Mechanistic studies uncovered the synergistic effect of the organocatalyst TBD (triazabicyclodecene) and amine substrates in the promotion of CO_2 activation (Scheme 2.84, equation a) [678]. Shortly afterward, a highly active NHC organocatalyst was designed by the same group for the formylation of N–H bonds in various amines and heterocycles under very mild conditions (0.1 MPa, 24 h and room temperature) using CO_2 and polymethylhydrosiloxane up to 90% yield (PMHS; Scheme 2.84, equation b) [677].

In 2013, Cantat et al. [679] reported zinc catalysts able to perform the methylation of amines with CO_2 and hydrosilanes at low pressure. Furthermore, the selective reduction of ureas was possible under similar reaction conditions. Although some silanes are considered to be group of Beller demonstrated the efficient N-methylation of both aromatic and aliphatic amines using CO_2/H_2 as the methylation reagent.

Applying an in situ combination of $Ru(acac)_3$, triphos and either acid additives or LiCl, the desired methylated amines were obtained with high efficiency. Notably, selective monomethylation of primary amines as well as the methylation

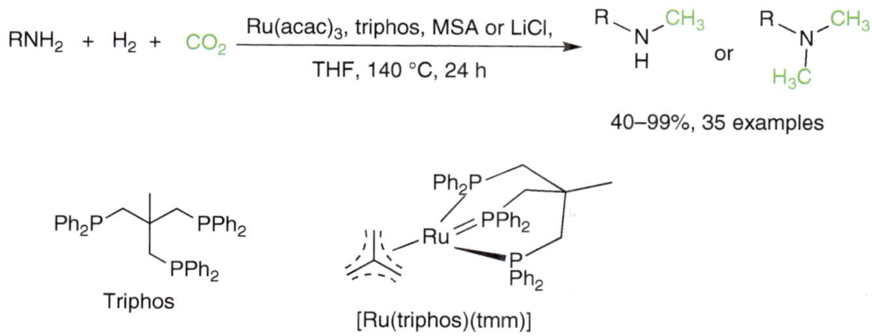

Scheme 2.85 Ru-catalyzed methylation of amines with CO_2 using H_2 as the reductant. Source: Modified from Li et al. [680].

of more challenging aliphatic amines proceeded smoothly under these conditions (Scheme 2.85) [680].

Among heterogeneous systems, the reduction of CO_2 in the presence of amines is a promising pathway for the formation of *N*-methylamines or *N*-formamides. Commonly, formamides are synthesized using CO as the carbonyl source under rigorous reaction conditions. Besides, the catalytic transformation of amines to formamides using CO_2 as feedstock is another known process (Scheme 2.86).

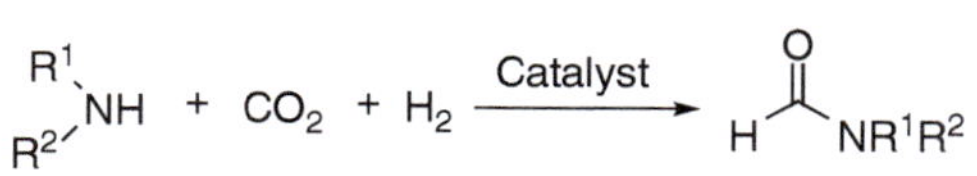

Scheme 2.86 Formylation of amines and CO_2.

In 1935, Farlow and Adkins [681] firstly reported Raney nickel as a heterogeneous catalyst for formylation of amylamine under 6 MPa CO_2 and 10–20 MPa H_2 at 150 °C for 9 h with up to 76% yield.

$RuCl_2[PMe_2(CH_2)_2Si(OEt)_3]_3$ as a catalyst [682] were prepared by incorporation of specially tailored complexes of Ru with bifunctional silylether phosphine ligands within a porous silica network by sol–gel method. It exhibited highest activities at 100% selectivity (25–94% yield) for the formylation of Me_2NH at 100 °C, 13 MPa CO_2 and 8.5 MPa H_2 for 15 h, utilizing supercritical CO_2 as both solvent and reactant.

A mesoporous ruthenium silica hybrid aerogel bidentate $RuCl_2[Ph_2P(CH_2)_3PPh_2]_2$ complexes were synthesized applying a sol–gel method, afforded highest TOF up to 18 400 h^{-1} using 75 ppm mol catalyst and 100% selectivity from CO_2, H_2, and diethylamine with 18 MPa total pressure at 110 °C for the formation of *N*,*N*-diethylformamide [683].

Ru/Al_2O_3, Cu/ZnO, $Ir/HSA\text{-}TiO_2$, and other catalysts were reported in following years for the formylation under 3–6 MPa, 140 °C for 6–16 h with up to 97% yield [684–686].

Pd/Al_2O_3–NR–RD catalyst [687] prepared by the reductive deposition (RD) of nano-Pd particles on Al_2O_3–NRs and showed high activity in the catalytic formylation of amines by CO_2–H_2 under medium conditions (1 MPa CO_2, 2 MPa H_2, 130 °C, 24 h) with up to 98% yield. However, there has been a lack of heterogeneous catalyst systems for N-formylation of primary amines with CO_2 and H_2 with sufficiently active and selective.

Based on that hydroxyl group-functionalized carbon might promote the adsorption and activation of CO_2, Shi and coworkers [688] prepared hydroxyl group-regulated active nano-Pd/C catalyst via in situ reduction of $Pd(NH_3)_xCl_y/C$ for N-formylation of amines with CO_2/H_2. The $Pd(NH_3)_xCl_y/C$ catalysts were prepared with a precipitation–deposition method, in which functionalized carbon material with a hydroxyl group was fabricated via oxidation of nitric acid. The $Pd(NH_3)_xCl_y/C$ catalysts showed excellent catalytic performance (up to 99% yield) under mild condition (1 MPa CO_2, 3 MPa H_2, 105 °C, 5 h).

After this, expect aromatic primary amine, most of amines could formylate to form corresponding formamide using CO_2 and H_2. Recently, gold nanoparticle catalyst-supported TiO_2 (Au/TiO_2) was found to allow the N-formylation of various amines, including normally unreactive anilines, and other reducible functional groups were completely retained during these formylation reactions. The Au/TiO_2 displayed yield of up to 99% under 2 MPa CO_2 and 3 MPa H_2 at 105 °C for 10 h [689].

Compared with the high pressure and temperature of formylation reactions using CO_2 and H_2, formylation reactions could be achieved in mild condition with silane and borane.

Novel metallosalen-based ionic porous organic polymers (POPs), i.e. DVB@ISZ, were synthesized for the first time using a simple unique strategy from the free-radical copolymerization reaction [690]. The catalyst of DVB@ISZ was reported to show high catalytic performance for the formylation of CO_2 with silane, giving a yield of 88–99% under the optimized reaction conditions (40 °C, 1 MPa, 20 h). The excellent catalytic performance was due to the cooperative effect between metal active center and halogen anion because it was crucial for the activation of CO_2.

Then, Gao and coworkers [691] designed $[Et_4NBr]_{50\%}$–Py–COF that the ILs were immobilizated on the channel walls of COFs using a post-synthetic strategy. The $[Et_4NBr]_{50\%}$–Py–COF afforded a high CO_2 adsorption was tested for the formylation of CO_2 and amines with $PhSiH_3$ under room temperature and atmospheric pressure for 24 h with 88–94% yield. Other catalysts, such as $[BE]_{X\%}$–TD–COFs [691], $CO_2(CO)_8$/POP [692], FIP-Im/QA [616] et al., also were reported for the same reaction with 55–99% yield under 0.4–1 MPa CO_2 for 14–24 h at 25–35 °C using $PhSiH_3$ as reducing agent [693].

Other catalytic systems, like Al_2O_3@MCM-41 [694] as a catalyst using dimethylamine-borane (DMAB) as a green reducing source, the $Co^0/ZnCl_2$-catalyzed direct N-formylation of IQs with H_2 to produce FTHIQs [695], and metal-free catalytic system [696] also have been reported in recent years for 23–99% conversion and up to 99% selectivity.

2.7 Reactions of Propargyl Alcohols/Propargyl Amines with CO_2

From the point view of green chemistry, protocols such as the coupling of CO_2 with aziridines, amino alcohols, amines, and propargylic alcohols which directly utilize CO_2 as C1 resource, are extremely attractive due to their high atom economy.

The synthesis of cyclic carbonate through coupling reaction between propargylic alcohols and CO_2 is one of the most important methods for the utilization of CO_2 (Scheme 2.87). In the past few years, many metal-free catalytic systems and transition-metal catalysts have been developed for this reaction.

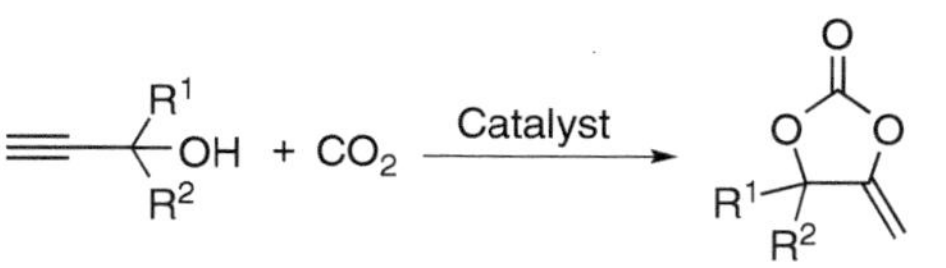

Scheme 2.87 Reaction of propargyl alcohols and CO_2 to form cyclic carbonate.

In 2008, Jiang et al. [697] presented the 87–97% yield α-alkylidene cyclic carbonates by the first heterogeneous catalyst, (dimethylamino) methyl-polystyrene-supported Cu(I) iodide (DMAM–PS–CuI) using propargyl alcohols with $scCO_2$ (14 MPa) at 40 °C for 24 h. Because of high CO_2 pressure, polystyrene-supported NHC–silver or copper complexes (PS–NHC–Ag/Cu(I)) [698], F-MOP-3–Ag [699], Ag_2WO_4/Ph_3P [700], poly(PPh_3)-azo-Ag [701], Ag–TCPE [702], TMOF-Ag [703] and et al. have been studied to reduce CO_2 pressure. It is worth noting that Ag_2WO_4/Ph_3P [700] first got product of α-alkylidene cyclic carbonates for 35–98% yield under atmospheric pressure CO_2 for 12 h at 50 °C and F-MOP-3-Ag catalyst [699] got the similar yield at room temperature with 1 MPa CO_2 for 10 h. In 2015, Han and Qian [704] investigated that AgNP catalyst could firstly achieve 86–96% yield α-alkylidene cyclic carbonates under both room temperature and atmospheric pressure CO_2 for 10 h. Then, AgBr or AgI/carbon was reported by Wang and coworkers [705] only for 4 h with 45–99% yield. Recently, the performance of the catalyst Zn/Fe_3O_4/ECS [706] was also studied to show a high yield of 90–93% under the optimized reaction conditions (30 °C, 1 MPa CO_2, 12–20 h).

2-Oxazolidinones as an important heterocyclic chemical play an important role as chemical intermediates and chiral auxiliaries in organic synthesis, and as antibacterial drugs in pharmaceutical chemistry [421, 423, 707].

Yu and coworkers [708] explored a magnetic nanocatalyst FeDOPACu (Scheme 2.88) for the construction of 2-oxazolidinones via amine, aromatic aldehydes, aromatic terminal alkynes, and CO_2 at 100 °C for 5 h with the addition of 10% KI and achieved excellent yields up to 99% (Scheme 2.89).

Through the carboxylative cyclization of propargylic amines with CO_2 to provide 2-oxazolidinones, which stands for an important clean and atom economic reaction (Scheme 2.90).

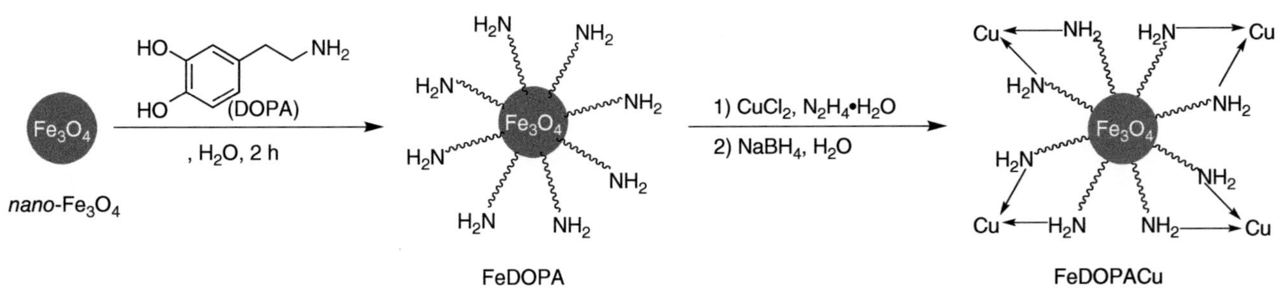

Scheme 2.88 The synthetic route of the catalyst FeDOPACu.

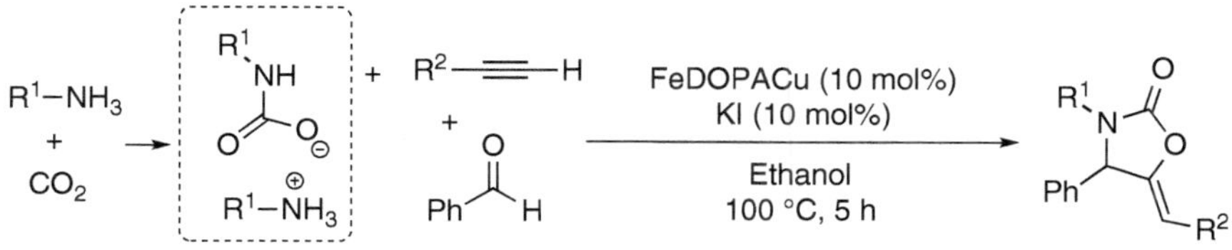

Scheme 2.89 Coupling reaction of amine, aromatic aldehydes, aromatic terminal alkynes, and CO_2 to form 2-oxazolidinones.

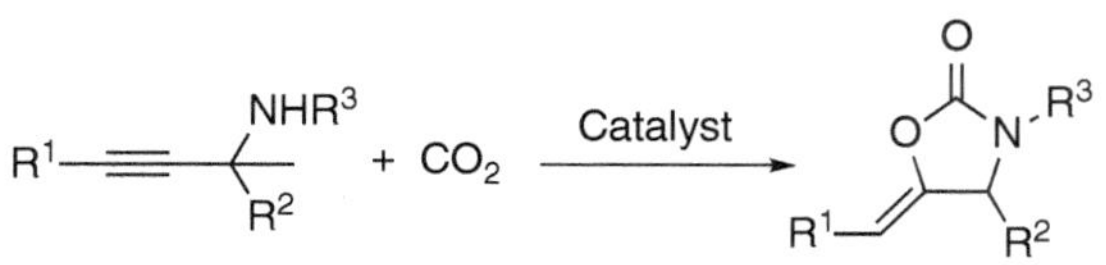

Scheme 2.90 Reaction of propargyl amines and CO_2 to form 2-oxazolidinones.

Sadeghzadeh [709] synthesized phosphine-functionalized nanoparticles (HPG@KCC-1/PPh_2) through the ring-opening polymerization of glycidol from KCC-1 and functionalized using chlorodiphenylphosphine. Next, HPG@KCC-1/PPh_2/Au was prepared by impregnation method and then used for the cyclization of propargylic amines with 0.5 MPa CO_2 to provide 2-oxazolidinones with 81–92% yields at room temperature after 20 h.

Similarly, Sadeghzadeh et al. [710] prepared KCC-1/IL/Ni@Pd composed of a fibrous nanosilica-supported nano-Ni@Pd-based IL to provide 2-oxazolidinones for the cyclization of propargylic amines with 1 MPa CO_2 under visible light irradiation at room temperature after only 3 h for 88–96% yields.

Lately, organometallic ionic complexes (OICs) owing to the cooperative effects between a Lewis acidic metal center and a nucleophilic halogen anion portion (X) of bifunctional groups have emerged as impressive catalysts for the coupling reaction with CO_2 and other organic compounds. Saadati and coworker [711] reported the design and synthesis of a novel bifunctional complex (Scheme 2.91) composed of the cross-linked KCC-1 as a substrate with the *N*,*N*-bis[(4-dimethylamino)salicylidene] ethylenediamino ruthenium(II) [Salen-Ru(II)] schiff base outer shell to serve as a catalyst. KCC-1/Salen/Ru(II) NPs nanocatalyst having core–shell structure for the synthesis of 2-oxazolidinones with propargylic amine derivatives was synthesized and got 93–98% yields under 1 MPa CO_2 at 100 °C for 1 h.

2.8 Other Reactions

2.8.1 Reactions of Aromatic Halides with CO_2

Aromatic aldehydes are important building blocks used as reactive intermediates in C–C, C–N, and C–S coupling reactions, which have found wide applications in the synthesis of materials, pharmaceuticals, pesticides, and agricultural chemicals [712]. Although several methods including reduction of carboxylic acids or esters,

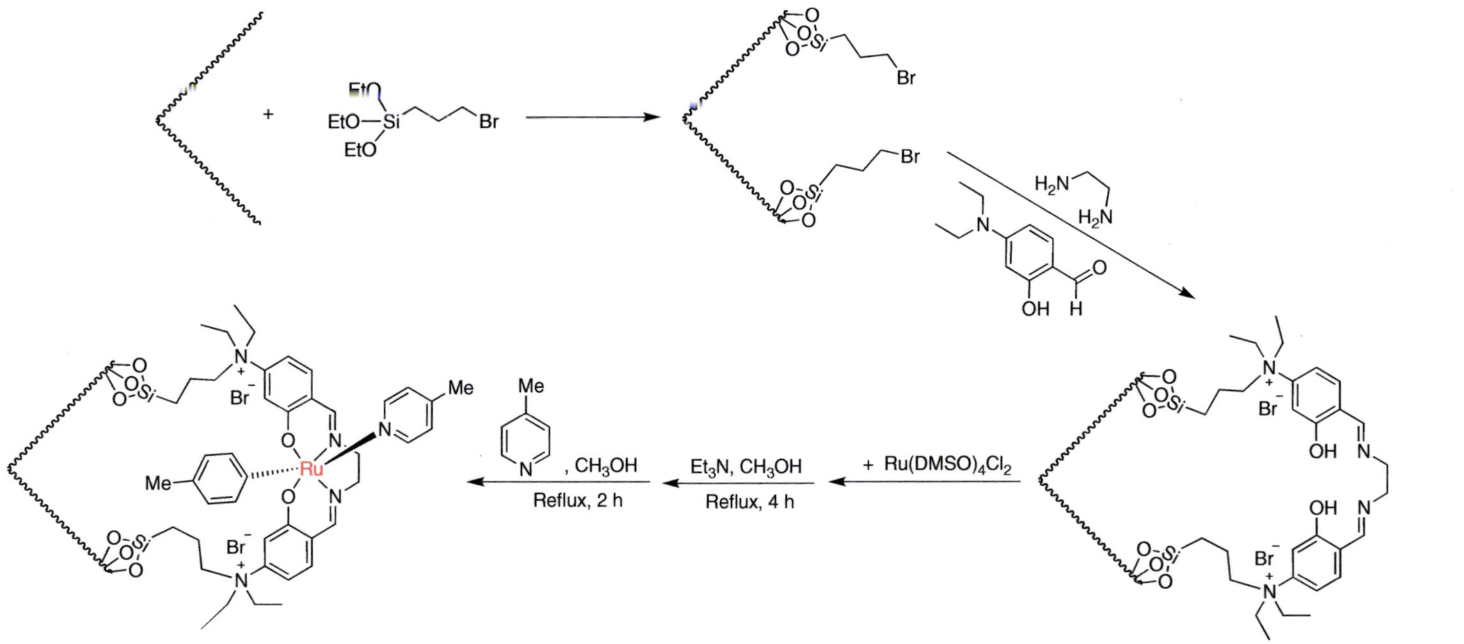

Scheme 2.91 Schematic illustration of the synthesis for KCC-1/Salen/Ru(II)NPs.

Reamer Tiemann, Duff, and Vilsmeier reactions have been developed for this transformation, there are considerable drawbacks associated with these methods such as very low yields, poor selectivity, and harsh reaction conditions. The direct formylation of aromatic halides into aromatic aldehydes using CO_2 as a C1 building block has been explored.

In 2014, commercial Pd/C as catalyst for direct formylation giving up to 81% yields aldehydes from aromatic iodides, aryl aldehydes, and 1 MPa CO_2 as a C1 resource was realized for the first time in the presence of hydrosilanes and base DBU under 80 °C for 20 h (Scheme 2.92) [713].

$$\text{Ar–I} + CO_2 + \text{PMHS} \xrightarrow{\text{Pd/C}} \text{Ar–CHO}$$

Scheme 2.92 Reactions of aromatic halides and CO_2 to aromatic aldehydes. Source: Modified from Yu et al. [713].

Amino-functionalized nanostarch (SNP-NH_2) was prepared by stirring with APTMS for 24 h in 110 °C using toluene as solution. Next, highly dispersed palladium nanoparticles were grafted onto SNP-NH_2 with a suspension of SNP-NH_2 and Li_2PdCl_4 aqueous solution kept at pH = 8; finally, SNP-NH_2/Pd(0) was obtained and used to the direct formylation of various aromatic iodides to aldehydes in the presence of poly(methyl-hydrosiloxane) as silane achieving 94% yields without using any base under mild condition (3 MPa, 90 °C, 8 h) (Scheme 2.93) [714].

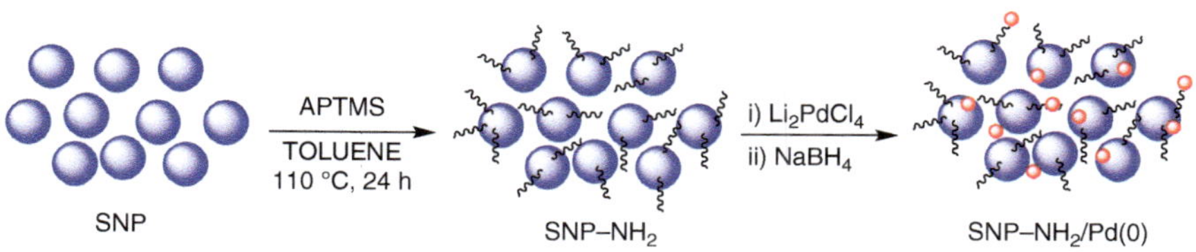

Scheme 2.93 Synthesis of SNP–NH_2/Pd(0). Source: Kumar et al. [714].

To date, conversion of carbon dioxide to alcohol synthesis has been limited to methanol only. To our knowledge, there is few reports on direct transformation of aryl halides to benzyl alcohols via a CO_2 fixation reaction. Synthesis of higher carbon chain alcohols via C–C coupling between a substrate and CO_2 is highly desirable (Scheme 2.94).

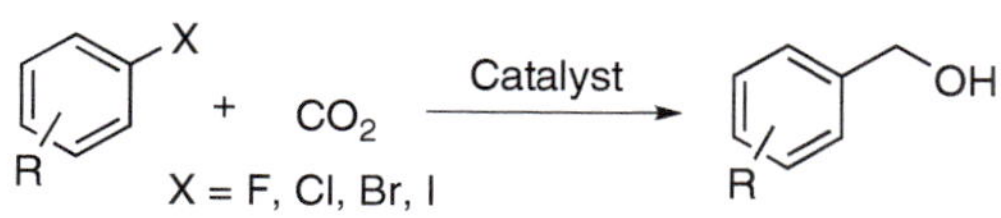

Scheme 2.94 Reactions of aromatic halides and CO_2 to benzyl alcohols.

A nitrogen-doped mesoporous carbon material containing high surface area and porous channels was synthesized from glucose and melamine under hydrothermal treatment. Then, Brij-35 as a soft template and ethylene glycol as a reducing agent were mixed with $PdCl_2$ at 170 °C for 6 h under stirring conditions and denoted as Pd@N-GMC [715]. The palladium nanoparticle embedded porous nitrogen-doped carbon material (Pd@N-GMC) was used for the synthesis of benzyl alcohols with up to 92% yield from aryl iodides and 1 MPa CO_2 with 80 °C for 9 h.

Over the past decades of years, three main kind of ligands, namely, P-containing ligands, N-containing ligands, and NHC ligands, have been successfully used in the transition metal-catalyzed activation and conversion of CO_2. Benzyl chlorides or 2-(chloromethyl)heteroarenes can be used as a carbon-based ligand for palladium-catalyzed chemical fixation of CO_2, namely, the palladium-catalyzed carboxylative Stille coupling reaction of benzyl chlorides or 2-(chloromethyl)heteroarenes with allyltributylstannane (Scheme 2.95) [716, 717].

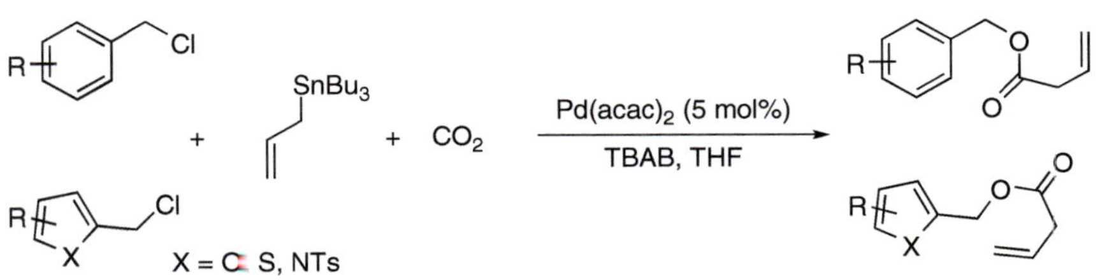

Scheme 2.95 Carboxylative Stille coupling reaction with three components. Source: Feng et al. [716]; Sun et al. [717].

The precatalyst $Pd(acac)_2$ was reduced by allyltributylstannane in the presence of stabilizer TBAB to generate palladium nanoparticles (PdNPs). Palladium-catalyzed carboxylative coupling of benzyl chlorides with allyltributylstannane was conducted to produce benzyl but-3-enoates with 64–86% yields under mild conditions (2.0 MPa, 70 °C, 24 h) using TBAB as a stabilizer [716]. The reaction process is illustrated in Scheme 2.96 [716].

Following research found that five-membered 2-(chloromethyl)heteroarenes can also react with palladium(0) to form *p*-allylpalladium(II) chloride intermediate [718]. Based on this discovery, the *p*-allyl involving heterocycles were used as a carbon-based ligand for palladium-catalyzed chemical fixation of CO_2.

PdNPs reduced by allyltributylstannane form $Pd(acac)_2$ which catalyzed three-component carboxylative Stille coupling reaction of five-membered (chloromethyl) heteroarenes, allyltributylstannane, and CO_2 through the formation of *p*-allylpalladium chloride intermediates was found to exhibit produce β,γ-unsaturated esters under mild conditions (2.0 MPa, 70 °C, 24 h) in 63–89% yields [717].

2.8.2 Reactions of 2-Aminobenzonitriles with CO_2

Quinazoline-2,4(1*H*,3*H*)-diones are important heterocyclic compounds used in the pharmaceutical and biotechnology industries [719]. The conventional synthesis of these compounds involves the reaction of anthranilic acid derivatives with urea,

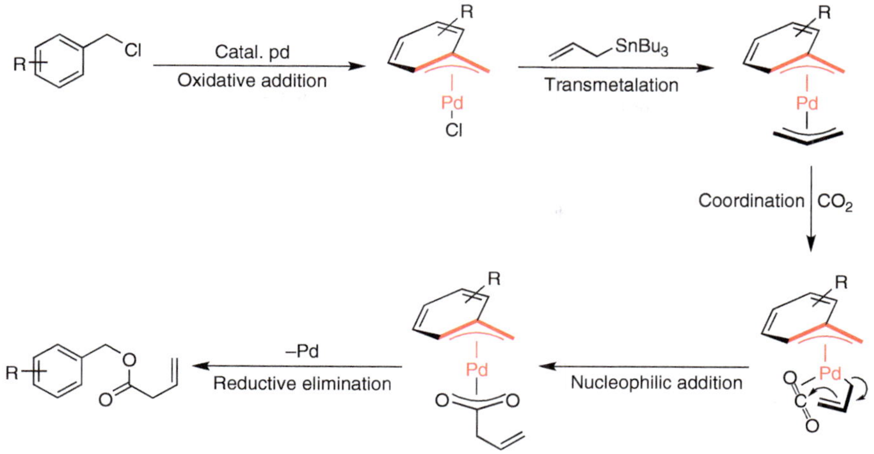

Scheme 2.96 Illustration of carboxylative Stille coupling reaction. Source: Feng et al. [716].

anthranilamides with phosgene, anthranilic acid with potassium cyanate, or chlorosulfonyl isocyanate. However, these conventional methodologies are associated with the obvious drawbacks of using toxic reagents like phosgene.

Notably, quinazoline-2,4(1*H*,3*H*)-diones can also be obtained from the reaction of 2-aminobenzonitriles and CO_2, and various kinds of catalysts have been used to promote the reactions (Scheme 2.97).

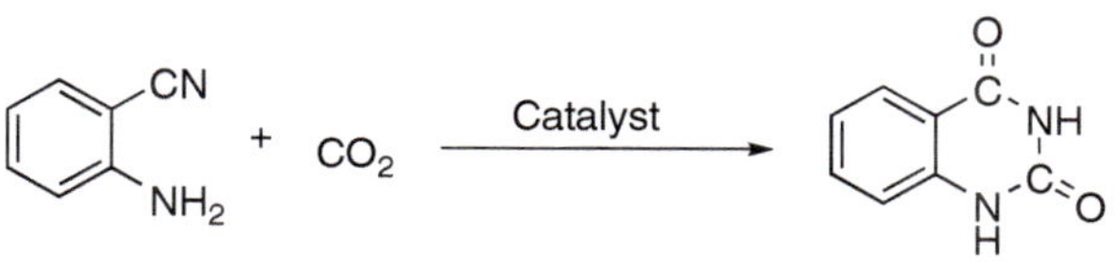

Scheme 2.97 Reactions of 2-aminobenzonitriles and CO_2 to form quinazoline-2,4(1*H*,3*H*)-diones.

In 2009, utilizing CO_2 and 2-aminobenzonitriles to synthesize quinazoline-2,4 (1*H*,3*H*)-diones with MgO/ZrO_2 as the first efficient heterogeneous base catalyst was reported by Bhanage group [720]. The MgO/ZrO_2 catalyst showed a remarkable activity with 70–92% yields under mild condition (130 °C, 3.7 MPa CO_2, 12 h). After this, Han et al. [721] reported the same reaction without any catalyst in water system under 14 MPa CO_2 and 160 °C for 19–39 h and got 80–93% yields.

Sooner after, a variety of catalyst, such as covalently linked amine-functionalized MCM-41 compound [722], 1,5,7-triazabicyclo[4.4.0]dec-5-ene (TBD)-functionalized Fe_3O_4 (TBD/Fe_3O_4 [723]), 1-hexyl-3-methyl imidazolium hydroxide IL supported on silica ([Hmim]OH/SiO_2 [724]), carbon nanofibers (CNFs) functionalized with 4-amino-2,6-dimethylpyridine (CNFs-ADMP [725]), have been studied for the synthesis of quinazoline-2,4(1*H*,3*H*)-diones (yields: 40–93%) with 2–4 MPa CO_2 and 2-aminobenzonitriles, under 100–130 °C for 10–18 h.

Subsequently, fibrous nano-silica (KCC-1)-supported IL (KCC-1/IL NPs [726]), fibrous nanosilica (KCC-1) functionalized with multi-carboxylic hyperbranched polyglycerol groups (HPG)-based nanometals (KCC-1/HPG/X [727], X = Au, Pd, and Cu) have been shown to an active catalyst for the same reaction for 87–98% yields with mild conditions (100–120 °C, 0.8–1.5 MPa, 40–60 min).

In the preparation of KCC-1/HPG/X, HPG was directly grafted on KCC-1 by surface-initiated ring-opening polymerization of glycidol. Noble metal nanopractices could be directly grown on the surface of KCC-1 firmly by impregnation method with monodisperse particles, uniform distribution, and high loading capacity, due to the unique structure of KCC-1, numerous functional groups, and amplification effect of HPG.

Srivastava and coworker [728] in 2017 explored graphitic carbon nitride (g-C_3N_4) using urea, thiourea, and a mixture of urea thiourea and then successfully introduced bifunctional acidic (–SO_3H) and basic (–NH_2) sites reacting with 60 wt% aqueous H_2SO_4. S-CN(UTU)-60 catalysts gave quinazoline-2,4(1*H*,3*H*)-dione yield up to 93% (DMSO as solvent) with 2.5 MPa CO_2 and 130 °C for 12 h.

3

Other C_1 Carbonyl Molecules

Although CO gas has been extensively applied in the bulk chemical industry, its intrinsic properties such as high toxicity, flammability, and special equipment requirement for handling limit its utilization in organic synthesis, fine chemical industry, and academia. Recently, considerable effort has been devoted to the development of CO surrogates to avoid the direct use of CO gas.

Among the various CO surrogates, formic acid and formaldehyde have a broad range of applications in organic synthesis. The direct carbonylation with formic acid (HCOOH) and formaldehyde (HCHO) represents one of the most atom-economical substitutes owing to their high weight percentage of CO. In this section, the potential roles of both formic acid and formaldehyde in transition metal-catalyzed carbonylation reactions are discussed. The potential of formic acid or formaldehyde as a renewable C1 feedstock for both bulk and fine chemical syntheses is also outlined with examples.

3.1 Formaldehyde (HCHO)

The carbonylation reaction nowadays is an extremely powerful tool for the synthesis of many carbonyl-containing compounds. Although CO is a cheap and readily available CO source, the manipulation of this toxic gas and the needed high-pressure equipment are major handicaps for its applications in many cases. Therefore, the search for CO surrogates has become an intriguing topic for chemists [175, 729–732].

Formaldehyde has been used as one-carbon sources because of its abundant, easy to handle, and inexpensive properties. Formaldehyde including its aqueous solution (formalin) and solid form (paraformaldehyde) has been applied both in the chemical industry and organic chemistry research as methylene blocks [83, 733–735], hydroxymethylation reagents [736–738], and hydrogen donor or acceptor [739, 740].

From the point view of atom efficiency, paraformaldehyde is the most ideal CO surrogates compared to other aldehyde because it contains about 93 wt% of CO. Therefore, it has garnered special attention for applications in metal-catalyzed carbonylation reactions.

Besides, these were also applied as CO surrogates in various transition metal-catalyzed carbonylative reaction (Scheme 3.1).

Carbonyl Compounds: Reactants, Catalysts and Products, First Edition.
Feng Shi, Hongli Wang and Xingchao Dai.

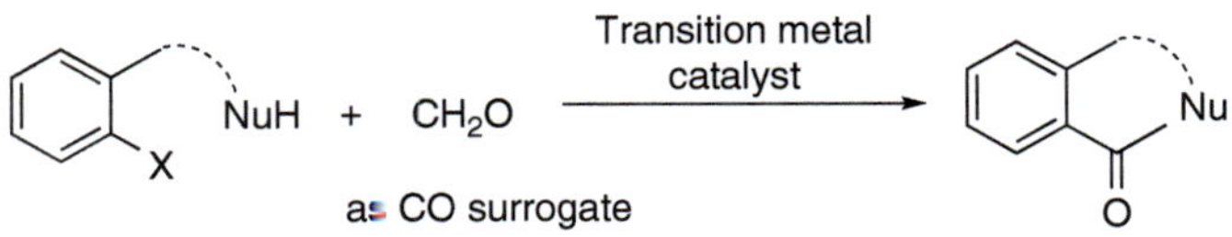

Scheme 3.1 Rh-catalyzed carbonylative cyclization of aryl halides with aldehyde.

3.1.1 Carbonylation of Halides with HCHO

In carbonylation of aryl halides, formaldehyde is an ideal CO surrogate affording the corresponding carbonyl compounds under mild conditions. For example, in 2007, Kakiuchi and coworkers reported a carbonylative cyclization of *N*-tosyl(2-bromobenzylamine) to form 2-tosylisoindolin-1-one by using 10 equiv of paraformaldehyde using paraformaldehyde as a surrogate for CO catalyzed by $[RhCl(cod)]_2$ in 8 h for 69% yield (Scheme 3.2) [741].

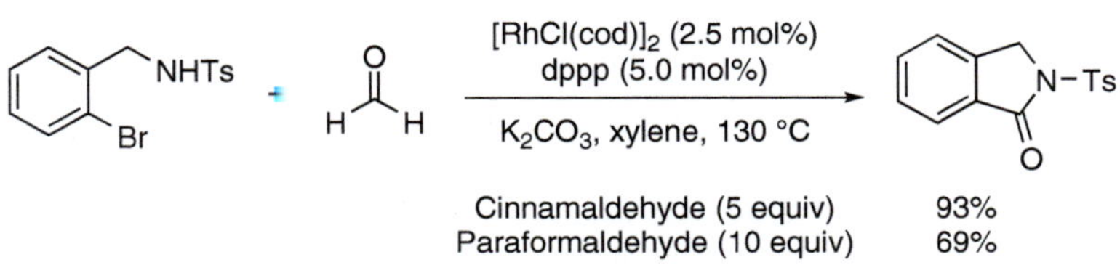

Scheme 3.2 Rh-catalyzed carbonylative cyclization of aryl halides with aldehyde. Source: Morimoto et al. [741].

Later, Kakiuchi and coworkers extended this tactic for asymmetric synthesis of 3-substituted isoindolinones, and paraformaldehyde was selected as CO surrogates with 51–73% yield (Scheme 3.3) [742].

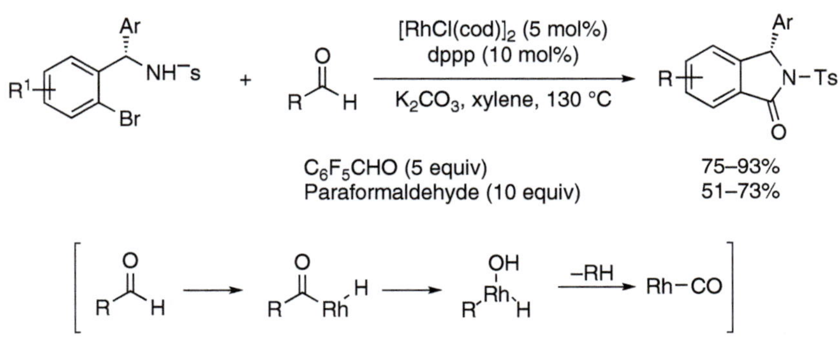

Scheme 3.3 Rh-catalyzed asymmetric carbonylative cyclization of aryl halides with aldehyde. Source: Modified from Fujioka et al. [742].

In addition, paraformaldehyde was also been utilized in palladium-catalyzed reductive carbonylations and alkoxycarbonylations of aryl bromides.

For instance, Beller et al. developed novel protocols in which aromatic aldehydes and esters were synthesized via palladium-catalyzed reductive carbonylation and

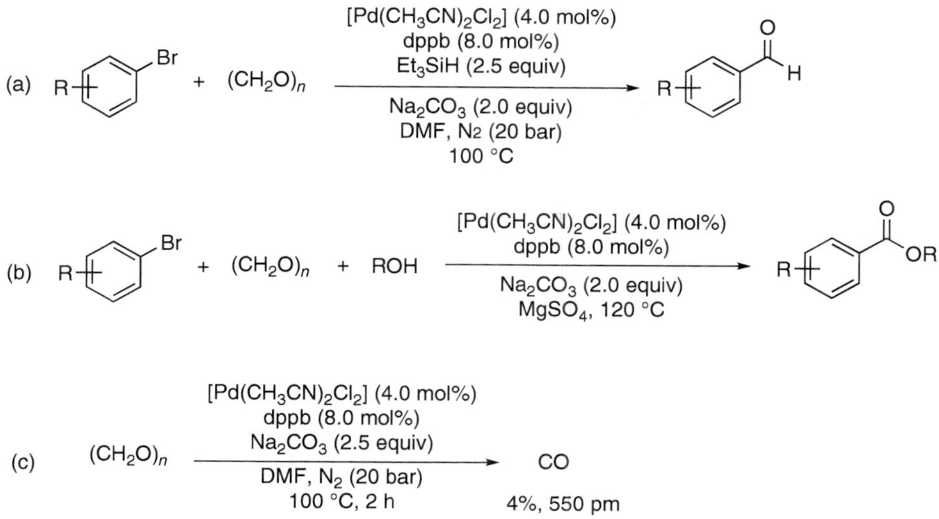

Scheme 3.4 Pd-catalyzed reductive carbonylations and alkoxycarbonylations of aryl bromides.

alkoxycarbonylation of aryl bromides using paraformaldehyde as an external CO source [743]. With Et_3SiH as hydride source, aryl bromides provided aromatic aldehydes in the presence of the $PdCl_2(CH_3CN)_2$-dppb catalyst system (Scheme 3.4, a). Alternatively, alkoxycarbonylations took place to form esters using alcohols as nucleophiles and solvents (Scheme 3.4, b). Labeling experiments using ^{13}C-labeled paraformaldehyde led to the ^{13}C-carbonyl labeled aldehyde and ester predominantly, thus confirming that paraformaldehyde was the CO source. In addition, the slow release of CO gas from paraformaldehyde was proved by detection of CO in the presence of the $PdCl_2(MeCN)_2$-dppb catalyst under basic conditions (Scheme 3.4, c).

Wu and coworker constructed various carbocyclic and heterocyclic systems by several groups using similar strategy. A palladium-catalyzed carbonylation for the synthesis of 36–86% yield substituted benzoxazinones from *N*-(*o*-bromoaryl)amides was reported (Scheme 3.5) [744]. Potassium acetate was the unique base for this reaction because other bases caused either incomplete conversion or higher amounts of debromination products.

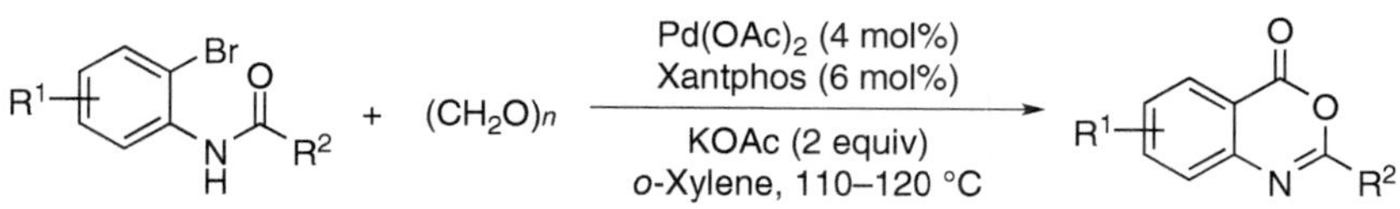

Scheme 3.5 Pd-catalyzed carbonylation for the synthesis of benzoxazinones. Source: Modified from Li and Wu [744].

In 2014, various substituted phthalazinones were selectively obtained in 25–88% yields with $Pd(OCOCF_3)$/XantPhos as the catalyst system using paraformaldehyde as the cheap carbon source by Deng and coworkers (Scheme 3.6) and the reaction gave various substituted phthalazinones at high temperature [745]. Control experiments indicated that the Pd catalyst probably worked both as a transition metal to initiate carbonylation and as a Lewis acid to promote the condensation step.

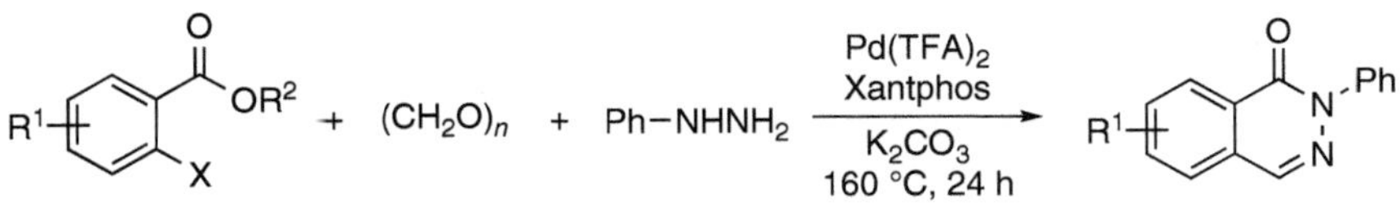

Scheme 3.6 Reactions of phenylhydrazine with various esters.

Later, Deng and coworkers revealed an efficient procedure for 2-aroylbenzofuran preparation from 2-bromophenols, phenacyl bromides, and paraformaldehyde (Scheme 3.7) [746].

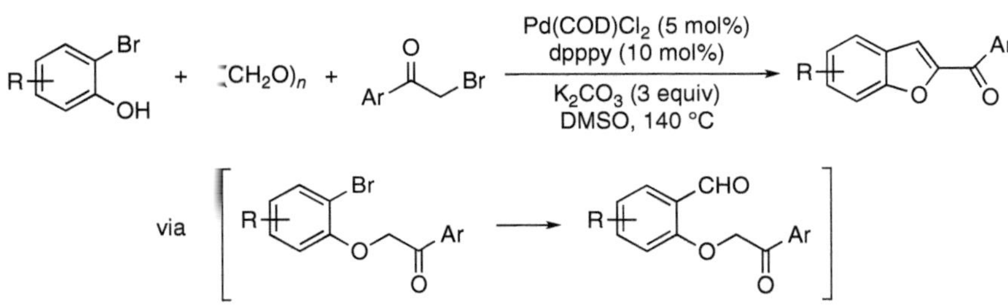

Scheme 3.7 Pd-catalyzed carbonylation for the synthesis of 2-aroylbenzofurans. Source: Modified from Cheng et al. [746].

Besides, carbocyclic compounds were also constructed. For example, Morimoto et al. developed a Pd-catalyzed cyclocarbonylation of 2-bromobiphenyls with paraformaldehyde as a carbonyl source for the synthesis of fluoren-9-one derivatives in 2016 (Scheme 3.8) [747]. The aryl group served as intermolecular nucleophile to form a C–C bond through the cleavage of a C–H bond to afford good yields catalyzed by the $Pd(OAc)_2/PCy_3$ system.

3.1.2 Carbonylation of Olefins with HCHO

Generally, carbonylation of alkenes with formaldehyde as CO surrogates has been developed in two manners: hydroformylation and alkoxycarbonylation, providing aldehyde and ester, respectively (Scheme 3.9).

Using paraformaldehyde as syngas substitution has drawn a lot of interest for many years owing to its pressure-free conditions and sometimes complementary regioselectivity to the gaseous hydroformylation reactions. In 1982, Okano, Kiji, and coworkers reported the first rhodium-catalyzed hydroformylation of alkenes using paraformaldehyde as the syngas substitution [748]. In the presence of 0.5 mol%

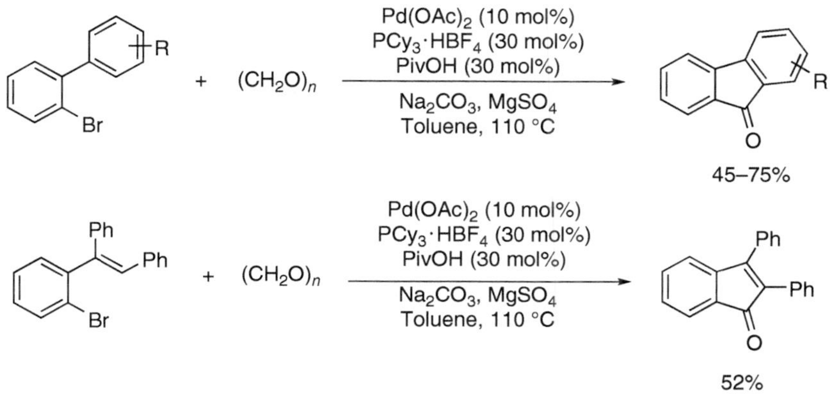

Scheme 3.8 Pd-catalyzed carbonylation for the synthesis of fluoren-9-ones and indenones. Source: Modified from Furusawa et al. [747].

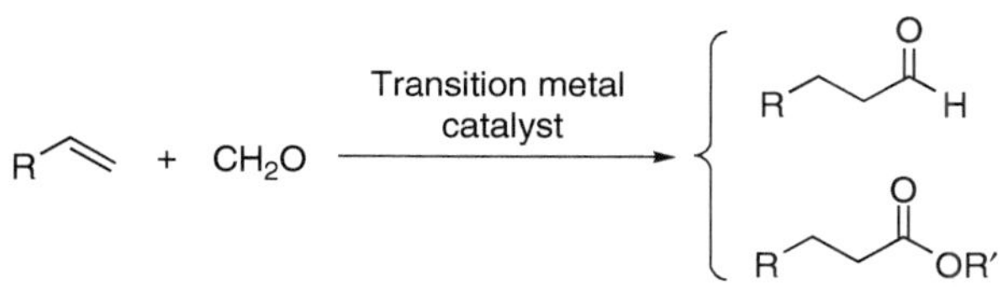

Scheme 3.9 Carbonylation of alkenes with formaldehyde as CO surrogates.

$RhH_2(O_2COH)[P(iPr)_3]_2$, six different substituted alkenes were hydroformylated to aldehydes in low to moderate yields (Scheme 3.10).

After that, a range of Rh catalysts have been studied for the synthesis of aldehydes from corresponding olefins [749–751].

Seok and coworkers studied the Rh-catalyzed hydroformylations of olefins (such as allyl alcohol, propylene, 1-hexene, styrene, methyl acrylate, and acrolein) with paraformaldehyde to form corresponding aldehydes. 4-Hydroxybutanal was formed as a major product from the reaction of allyl alcohol with formaldehyde under nitrogen atmosphere at 100 °C and 5 h for 26% yield (*n*-/*iso*- = 21). According to the experimental result, they found that allyl alcohol and methyl acrylate which have oxygen at the β-position to double bonds showed higher reactivities and *n*/*iso* ratio of the products than non-oxygenated alkenes (Scheme 3.11) [749].

In 2005, the hydroformylations of 1-hexane and allyl alcohol with paraformaldehyde were studied by Rosales, Sanchez-Delgado, and coworkers using $[Rh(CO)_2(acac)]$/2dppe as the catalytic system [751]. Under their conditions, a modest *n*/*iso* ratio was obtained for 1-hexane, while a high *iso*/*n* ratio was seen for allyl alcohol (Scheme 3.12). Unlike the hydroformylation with CO/H_2, both electronic and steric effects of the ligands affected the reaction activity of paraformaldehyde.

Later on, they performed a comparative study of the hydroformylation of 1-hexene under syngas conditions and with paraformaldehyde catalyzed by $Rh(acac)(CO)_2$ and different diphosphine ligands [750].

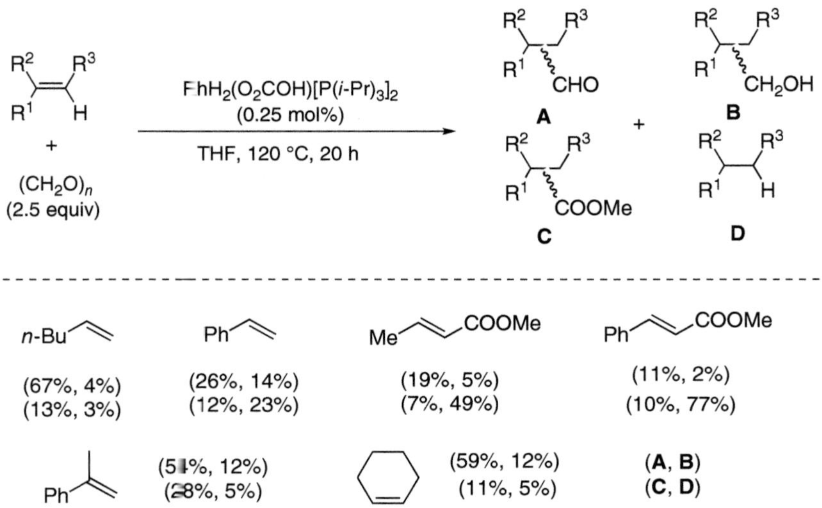

Scheme 3.10 Rh-catalyzed hydroformylation of alkenens using PFA as syngas equivalent.

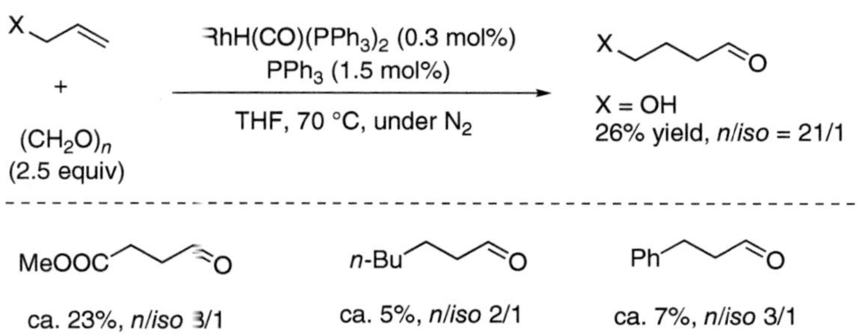

Scheme 3.11 Rh-catalyzed hydroformylations of olefins with paraformaldehyde. Source: Modified from Ahn et al. [749].

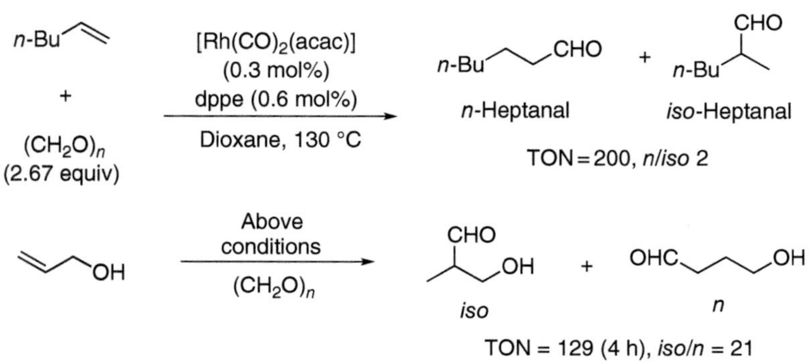

Scheme 3.12 Rh-catalyzed hydroformylation of 1-hexenes and allyl alcohols using PFA as CO source.

In 2009, Morimoto and coworkers described a highly linear-selective hydroformylation of 1-alkenes with the direct use of formaldehyde instead of syngas (Scheme 3.13) [752, 753]. In this case, two rhodium(I)-phosphine species associated with each phosphane separately derived from BIPHEP and Nixantphos proved to be a catalyst system in the hydroformylation of 1-alkenes using formaldehyde to give aldehydes with up to 95% yield and ratio of regioselectivities between linear and branched up to 98 : 2.

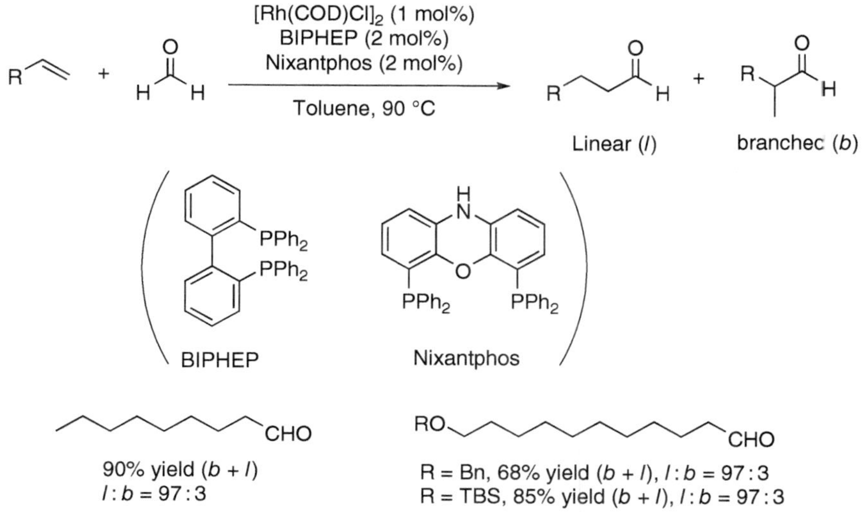

Scheme 3.13 Rh-catalyzed highly linear-selective hydroformylation of alkenes with formaldehyde. Source: Morimoto et al. [752]; Makado et al. [753].

Later, Taddei and coworkers described a microwave-assisted domino hydroformylation of β,γ-unsaturated amides on the basis of Morimoto's catalytic system (Scheme 3.14) [754].

In 2013, Boerner and coworkers revealed a Rh–dppp-catalyzed hydroformylation of alkenes with formalin as CO source [755]. They observed that addition of 1 MPa of hydrogen significantly improved the regioselectivity and accelerated the reaction for 85% yield (Scheme 3.15).

The reactivity was generally higher for the substituted styrenes. However, 3,3-dimethyl-1-butene was hydroformylated to the linear product in a very high regioselectivity and efficiency. This procedure is useful for the production of aldehydes on a small scale.

In 2015, the groups of Clark and Morimoto independently reported the asymmetric hydroformylation of alkenes using formaldehyde (Scheme 3.16) [756, 757]. Using (*S*,*S*)-Ph-BPE as the chiral phosphine ligand, a series of substituted styrenes were hydroformylated to branched aldehydes in good enantio- and regioselectivity.

The mechanistic investigations supported that the enantioselective protocol also relied on effective transfer of the CO and H_2 from formaldehyde to classic

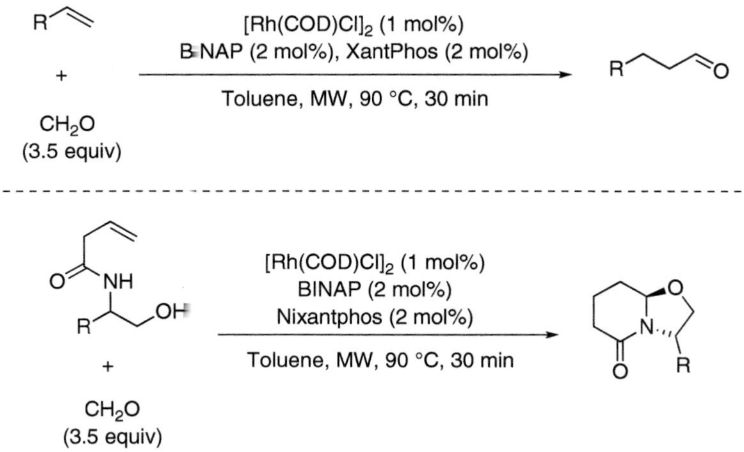

Scheme 3.14 Rh-catalyzed hydroformylation using PFA under microwave irradiation. Source: Modified from Cini et al. [754].

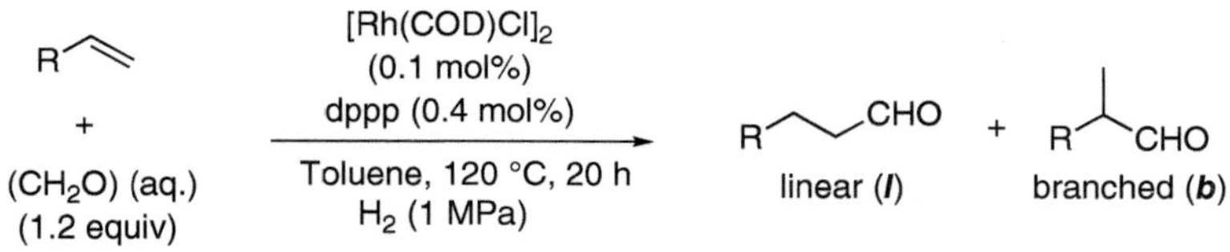

Scheme 3.15 Rh-catalyzed hydroformylation using PFA under hydrogen pressure. Source: Modified from Uhlemann et al. [755].

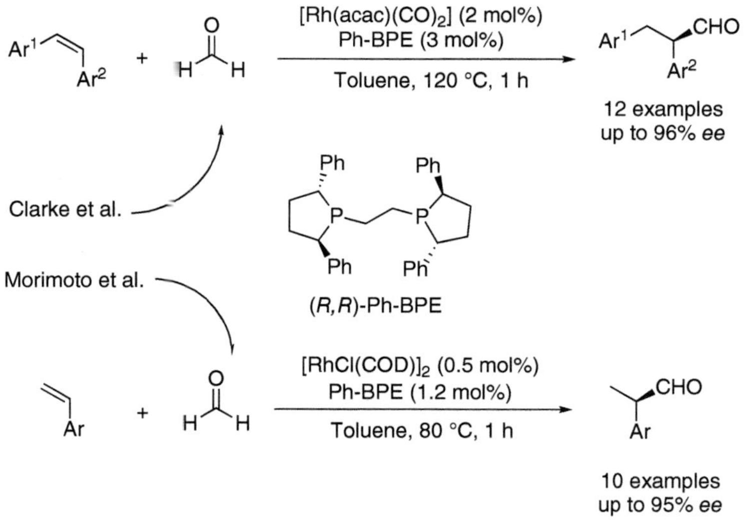

Scheme 3.16 Enantioselective Rh(I)-catalyzed HCHO-involved hydroformylation of alkenes with the aid of chiral ligand. Source: Morimoto et al. [756]; Fuentes et al. [757].

Rh(I)-catalyzed hydroformylation. The current process in terms of enantioselective transformations provides a potentially useful method for the synthesis of substituted aldehydes and their derivatives.

Alkoxycarbonylation with paraformaldehyde as a CO substitute was also achieved by Beller group. The first efficient use of paraformaldehyde in the alkoxycarbonylation of alkenes is reported by Beller in 2014 (Scheme 3.17). Using a combination of readily available $Ru_3(CO)_{12}$ and PCy_3 (Cy = cyclohexyl) as catalyst, industrially important esters are produced with up to 90% yield at 130 °C.

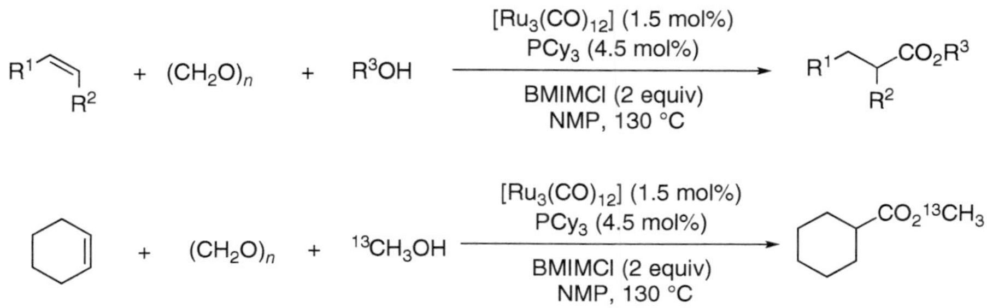

Scheme 3.17 Ru-catalyzed alkoxylcarbonylation of alkenes and alcohols.

This new method is applicable to a series of alkenes and alcohols. Moreover, the use of formaldehyde as the carbonyl source provides the potential to avoid both reliance on CO in carbonylation reactions and the requirement for special high-pressure reactors [758].

Later, Beller group developed a Pd-catalytic approach for regioselective methoxycarbonylation of alkenes with paraformaldehyde and methanol as CO substitutes. A combination of $Pd(OAc)_2$, dtbpx (1,1′-bis(di-*tert*-butylphosphino)-*o*-xylene), and PTSA (*p*-toluenesulfonic acid) was applied as the catalytic system (Scheme 3.18) [759].

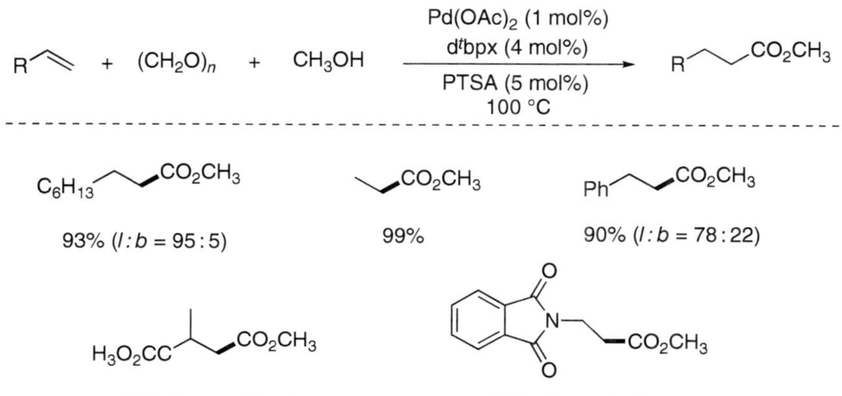

Scheme 3.18 Ru-catalyzed alkoxylcarbonylation of alkenes and alcohols. Source: Modified from Liu et al. [759].

Notably, both the concentration of the ligand dtbpx and the co-catalyst PTSA have a significant influence on the yield. Under the optimized reaction conditions (1 mol% of Pd $(OAc)_2$, 4 mol% dtbpx, and 5 mol% PTSA), a wide range of alkenes including aliphatic and aromatic alkenes with various functional groups smoothly underwent the methoxycarbonylation to afford industrially important methyl esters in good yield with excellent regioselectivity. Interestingly, ^{13}C isotope labeling experiment showed that both paraformaldehyde and methanol served as CO surrogates (Scheme 3.19).

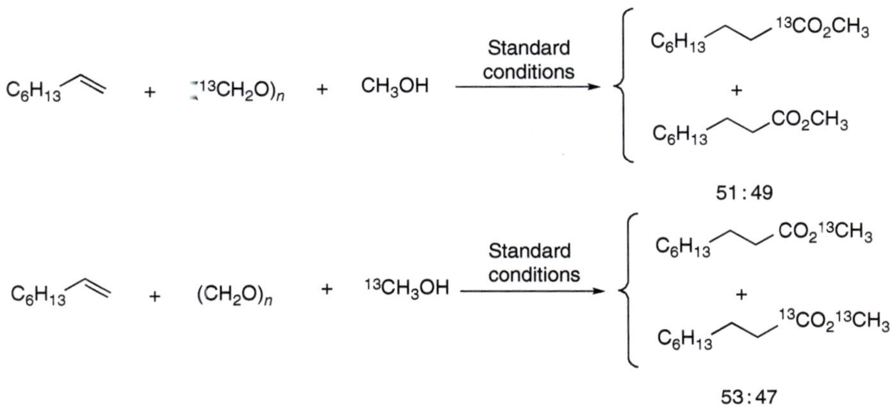

Scheme 3.19 Ru catalyzed alkoxylcarbonylation of alkenes and alcohols.

3.1.3 Carbonylation of Alkynes with HCHO

Carbonylation of alkynes using formaldehyde as CO surrogate is still limited; there are a few examples on the rhodium(I)-catalyzed carbonylation of alkynes with HCHO (Scheme 3.20) [752, 760, 761].

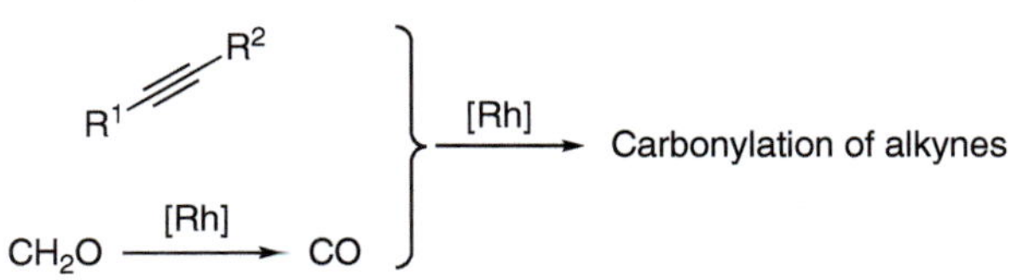

Scheme 3.20 Rh-catalyzed carbonylation of alkynes. Source: Morimoto et al. [752]; Fuji et al. [760]; Wang et al. [761].

Since 2005, there has been increased interest on the use of HCHO as C1 source for carbonylation of alkynes. For example, Morimoto group reported a rhodium(I)-catalyzed cyclohydrocarbonylation reaction of diphenylacetylene and formal for the synthesis of α,β-butenolides reacted at two sets of reaction conditions for different yields in Scheme 3.21 [760].

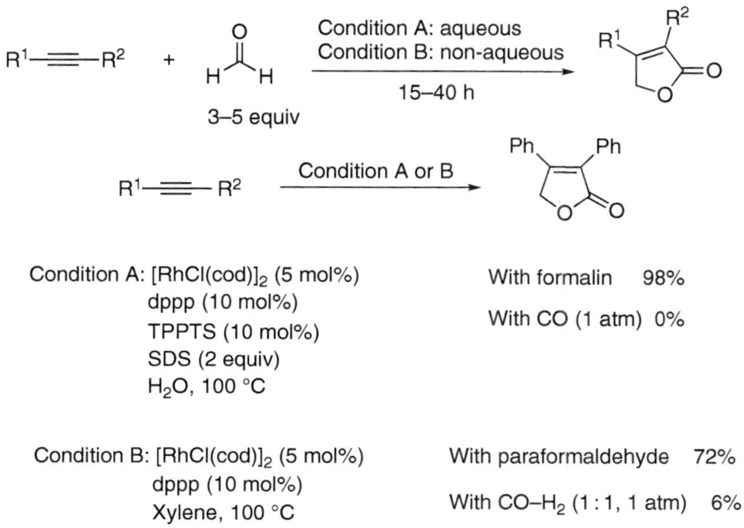

Scheme 3.21 Rh-catalyzed cyclohydrocarbonylation reaction of alkynes with formaldehyde. Source: Modified from Fuji et al. [760].

Besides, Morimoto, Chatani, and coworkers revealed a rhodium(I)-catalyzed reaction of alkynes with 2-bromophenylboronic acids using paraformaldehyde as CO source (Scheme 3.22) [752].

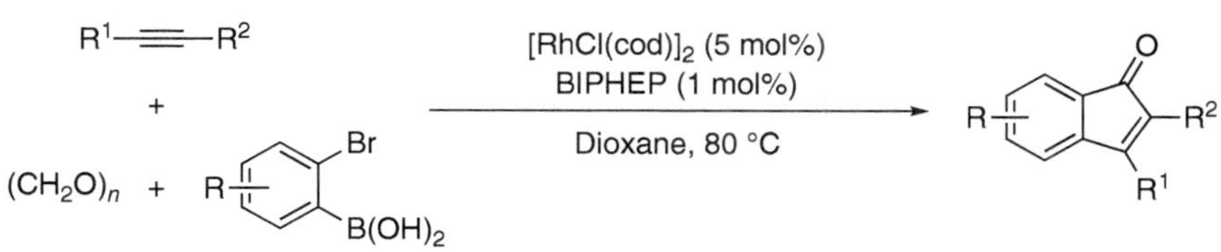

Scheme 3.22 Rh-catalyzed preparation of indenone derivatives. Source: Modified from Morimoto et al. [752].

In addition, Morimoto group described a rhodium(I)-catalyzed carbonylative arylation of alkynes and boronic acid using paraformaldehyde as CO surrogate (Scheme 3.23) [761]. Morimoto stated that phosphine-ligated complex $[RhCl(BIPHEP)]_2$ was responsible for decarbonylation of paraformaldehyde and $[Rh(COD)Cl]_2$ catalyzed the carbonylative arylation of the alkynes to deliver the product a,b-unsaturated ketones.

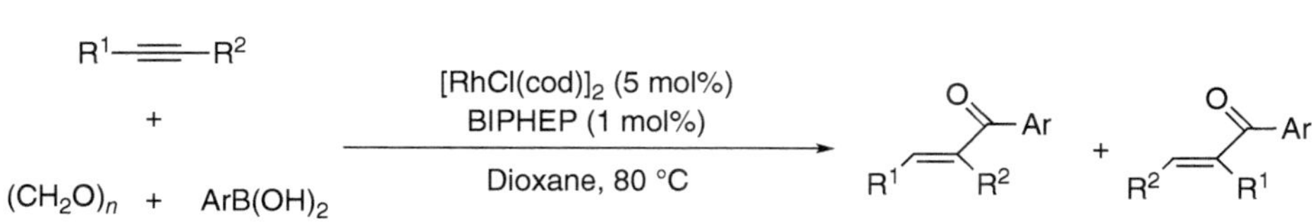

Scheme 3.23 Rh-catalyzed carbonylative arylation of alkynes and boronic acid. Source: Modified from Wang et al. [761].

3.2 Formic Acid (HCOOH)

3.2.1 Hydroxycarbonylation of Arenes with Formic Acid

In past two decades, transition metals catalyzed the cleavage and functionalization of an unreactive C–H bond was one of the hottest topics in organometallic chemistry [471, 762–765].

In 2004, Nozaki and coworkers reported an unexpected finding of a simultaneous hydroxylation–carboxylation of biphenyl to provide 4-hydroxy-4-biphenylcarboxylic acid applied as a polyester monomer in the presence of formic acid (Scheme 3.24) [766]. Besides, the reaction only produced hydroxycarbonylation product when using CO instead of formic acid. But the mechanism of neither the hydroxylation nor the carboxylation was clear.

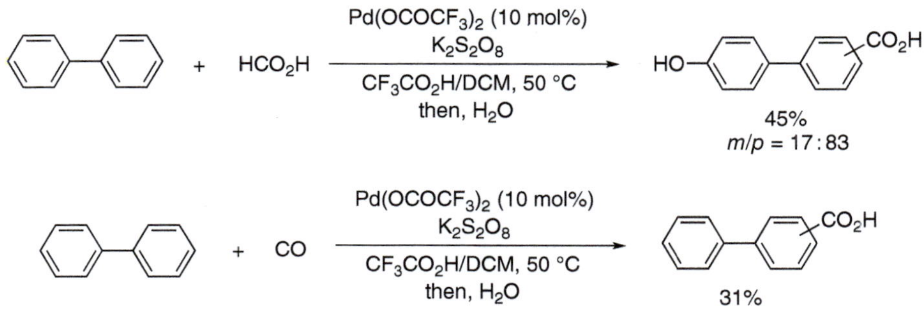

Scheme 3.24 Pd-catalyzed hydroxylation–carboxylation of biphenyl with formic acid. Source: Modified from Shibahara et al. [766].

In a later report, Nozaki revealed that addition of phosphenium significantly increased the yield (Scheme 3.25) [767].

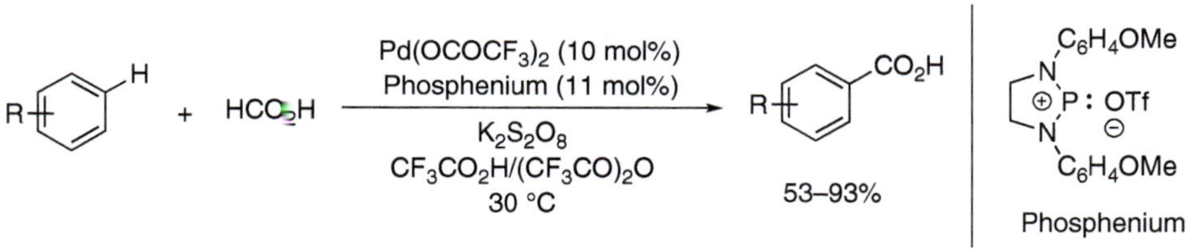

Scheme 3.25 Pd-catalyzed hydroxycarbonylation of arenes with formic acid. Source: Modified from Sakakibara et al. [767].

3.2.2 Carbonylation of Alkenes with Formic Acid

In the past years, the carbonylated addition of alkenes without CO has attracted much more attention in synthetic organic chemistry. The transition metal-catalyzed carbonylation of alkenes with formic acid as CO surrogates has proved to be an effective procedure for the synthesis of aldehydes, esters, or acids.

However, developing a general method for the direct use of HCOOH or DMF as additives remains challenging. Because formic acid readily decomposes to water and CO at elevated temperature under acidic reaction conditions, the presence of formate or its analogues is an effective method for the synthesis of carbonyl compounds.

During the past years, three paradigms were established to provide carboxylic acid, ester, and aldehyde (Scheme 3.26).

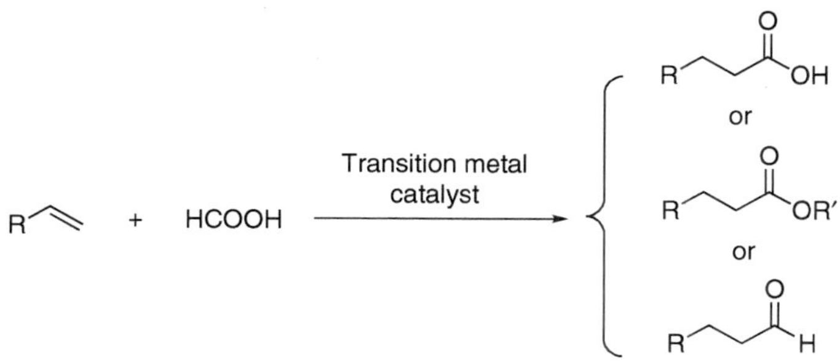

Scheme 3.26 Carbonylation of alkenes with formic acid as CO surrogates.

Transition metal-catalyzed carbonylative esterification reactions are of current interest in organic synthesis. In this context, a cooperative catalytic system of $Ru_3(CO)_{12}$ and 2-pyridinemethanol catalyzed three-component hydroesterification of olefins, alcohols using sodium formate as CO surrogate was reported first in 2014 by Jun and coworkers with up to 97% yield [768]. But, the reaction did not occur without the chelation auxiliary of 2-pyridinemethanol (Scheme 3.27).

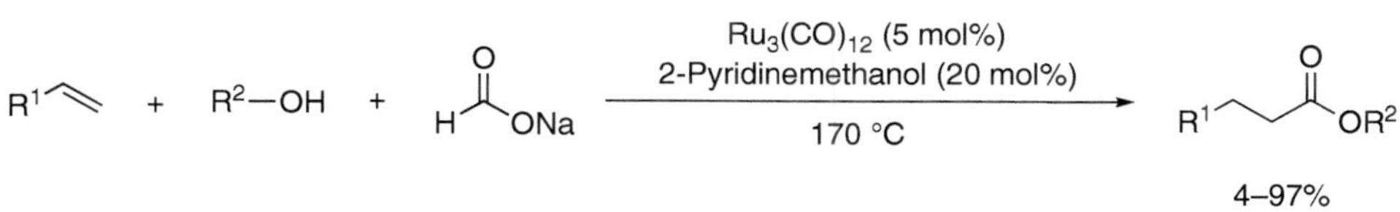

Scheme 3.27 Ru-catalyzed hydroesterification of olefins, alcohols with sodium formate.

Sterically hindered alkenes such as a methylstyrene regioselectively produced ester products which arose by the addition of terminal alkene carbons, while monosubstituted alkenes including 1-hexene and styrene formed a mixture of linear ester and branched ester. In respect of alcohols, simple primary alcohols performed well and secondary, tertiary alcohols and phenol resulted in poor yield.

Explore the direct use of HCOOH as non-gaseous C1 source in practical synthesis featured with atom-economic, operationally simple, and amenable to industrial scale, thus exploring of asymmetric carbonylation process with the direct use of HCOOH as a CO surrogate could be an interesting topic in this field.

Then, an alternative palladium-catalyzed hydrocarboxylation of olefins with HCOOH and HCOOPh was reported by Shi and coworkers (Scheme 3.28, a) [758].

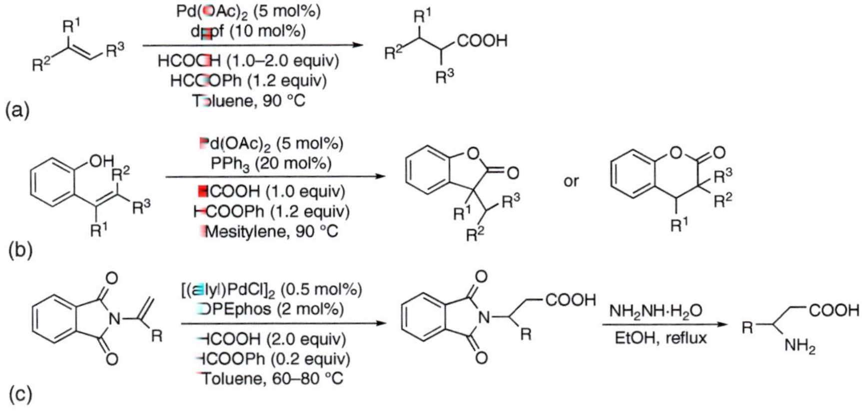

Scheme 3.28 Pd-catalyzed hydrocarboxylation of olefins with HCOOH and HCOOPh. Source: Modified from Wang et al. [758]; Modified from Wu et al. [732]; Modified from Dai et al. [769].

The hydrocarboxylation with HCOOH can be carried out in the presence of 20 mol% HCOOPh. Using this strategy, Shi and coworkers achieved the preparation of lactones (Scheme 3.28, b) [758] and β-amino acid derivatives via regioselective hydroesterification of alkenylphenols and enimides, respectively (Scheme 3.28, c) [769].

Later, Shi group discovered that the hydrocarboxylation and hydroformylation of olefins could be realized using HCOOH in the presence of a catalytic amount of acetic anhydride as a co-catalyst under mild reaction conditions (Scheme 3.29, a and b) [770, 771].

Direct reductive elimination of Pd complex A would lead to the formation of carboxylic acids (Scheme 3.29, path a), while the extrusion of CO_2 followed by reductive elimination would provide corresponding aldehydes (Scheme 3.29, path b). Notably, the ligand played an important role on the reaction pathway.

In 2016, Shi and coworkers described Pd-catalyzed regioselective hydroformylation of olefins with dppp HCOOH/Ac_2O under 80 °C for 24 h (Scheme 3.30). It was found that ligands played a crucial role in directing the reaction pathway and linear aldehydes can be obtained in up to 93% yield with more than 95% regioselectivity using 1,3-bis(diphenylphosphino) propane (dppp) as the ligand [770].

In 2018, bromoformates and iodoformates were synthesized up to 91% yield by ZnAl–BrO_3^-–LDHs with regioselectivity and stereoselectivity in the presence of HCOOH and olefin compound (Scheme 3.31). In this reaction, formic acid has function of solvent, nucleophilic reagent, and acidic medium [772].

Wu and coworkers developed a procedure for Pd/Ir-catalyzed selective synthesis of 4-arylbutanoic acids, 2-arylbutanoic acids, and 4-arylbutanals from allylbenzenes in 90–100 °C using HCOOH or TFBen as the CO surrogate in 2019 (Scheme 3.32) [773].

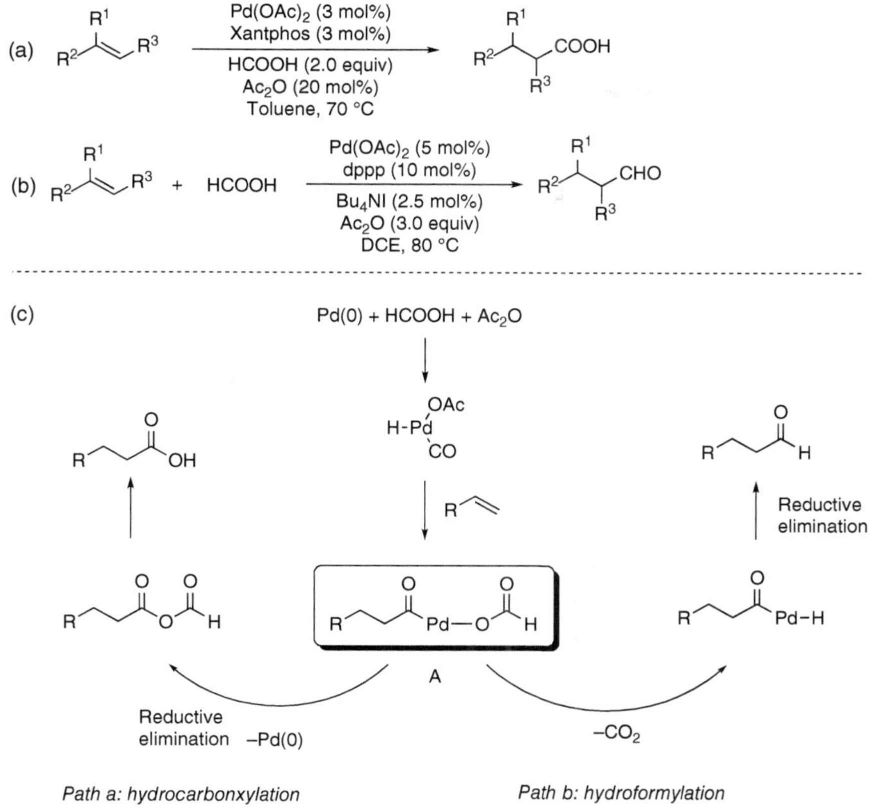

Scheme 3.29 Pd-catalyzed hydrocarboxylation and hydroformylation of olefins with HCOOH and Ac_2O. Source: Modified from Ren et al. [770]; Modified from Wang et al. [771].

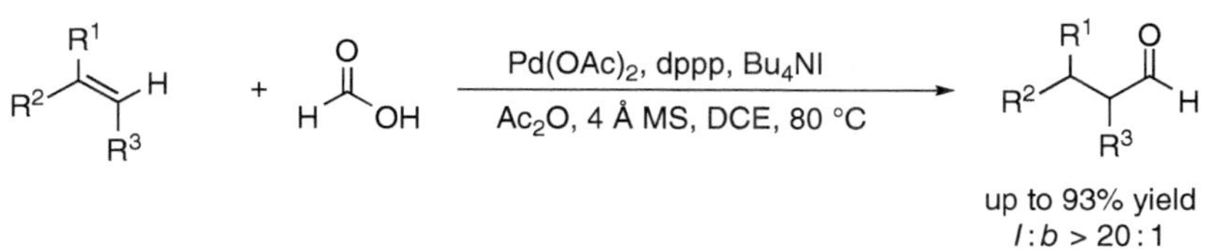

Scheme 3.30 Pd-catalyzed hydroformylation of olefins.

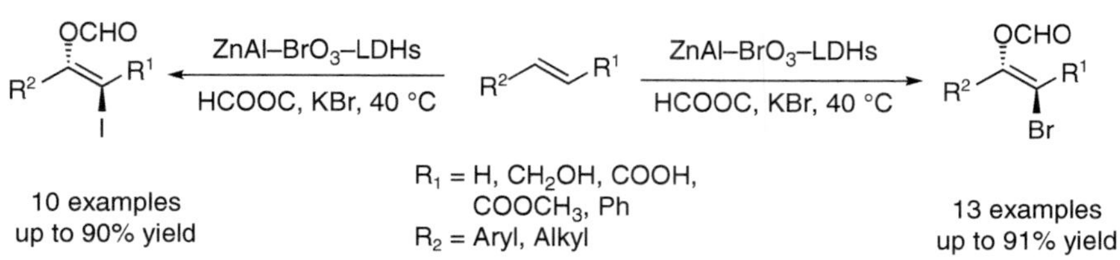

Scheme 3.31 Bromoformates and iodoformates were synthesized by $ZnAl–BrO_3–LDHs$.

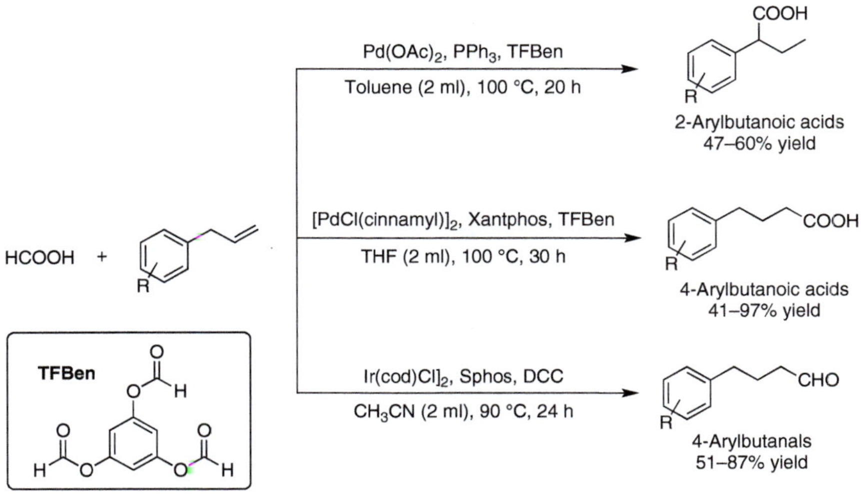

Scheme 3.32 Pd/Ir-catalyzed selective transformation of allylbenzenes. Source: Modified from Wu et al. [773].

3.2.3 Carbonylation of Alkynes with Formic Acid

As we all know, hydrocarboxylation of acetylene with CO was an atom-economical approach to acrylic acid. However, toxic and high-pressure CO was required in this process.

In 1991, the regioselective addition was catalyzed by [Rub-cymene)(PPh_3)Cl_2] from HCOOH and terminal alkynes to the enol formats with 45–95% yield at 80–100 °C for 15 h (Scheme 3.33) [774].

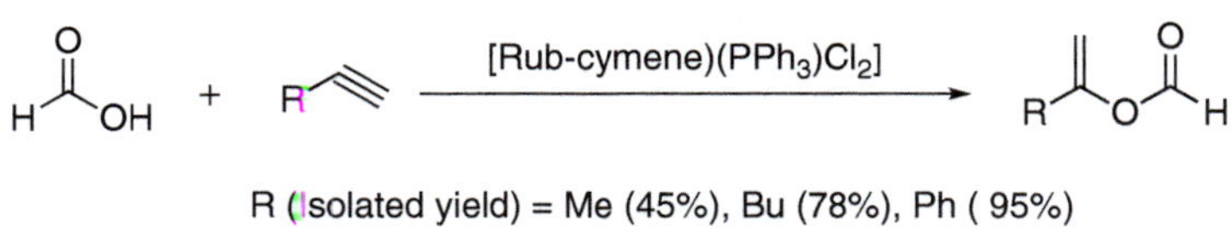

Scheme 3.33 Hydroformylation of terminal alkynes catalyzed by [Rub-cymene)(PPh_3)Cl_2]. Source: Modified from Neveux et al. [774].

In 2000, Leadbeater group reported that a range of terminal alkynes with HCOOH led to the formation of the corresponding enol formats in 15 h for 73–90% yield in the presence of Ru catalyst (Scheme 3.34) [775].

Zhou and coworkers established a palladium-catalyzed hydrocarboxylation reaction of alkynes and formic acid with Xantphos as ligand in 2015 (Scheme 3.35) [776]. Interestingly, compared with the previous methods on the carbonylation of alkenes, carbonylation of alkynes using formic acid as a CO surrogate has only been demonstrated in the hydrocarboxylation manner.

Inspired by the report on hydrocarboxylation of alkynes reported by Zhou and coworkers, Fu et al. further promoted the progress of atom-economic hydrocarboxylation of various alkynes with HCOOH, in which the combinational

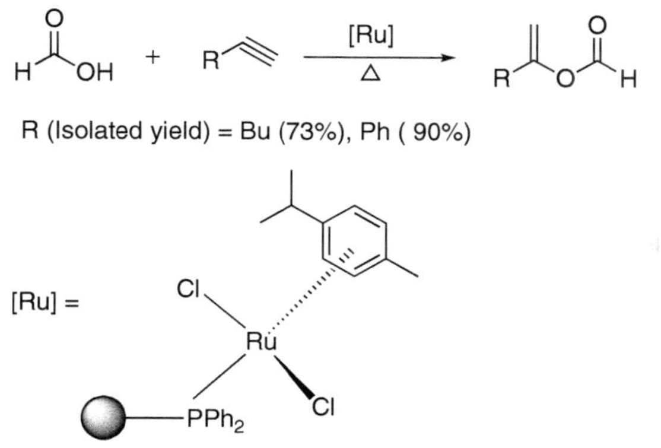

Scheme 3.34 Ru-catalyzed hydroformylation of terminal alkynes. Source: Modified from Leadbeater et al. [775].

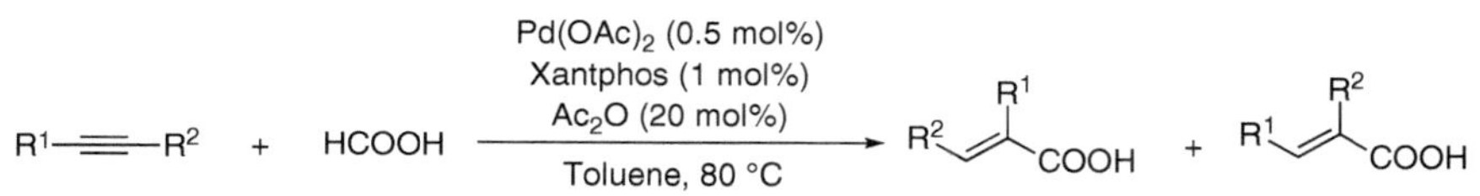

Scheme 3.35 Pd-catalyzed hydrocarboxylation of olefins with HCOOH and Ac_2O. Source: Modified from Hou et al. [776].

use of $Ni(acac)_2$ and ligand 1,2-bis(diphenylphosphino)-benzene (dppbz) or *cis*-1,2-bis(diphenylphosphino)ethylene (dppen) as well as catalytic amount of acid anhydride could catalyze this reaction with good yield and remarkable functional group compatibility (Scheme 3.36) [777]. Subsequently, another protocol for nickel-catalyzed hydrocarboxylation of alkynes with formic acid was reported by Zhou and coworkers very recently (TON up to 7700) [778]. The successful use of inexpensive nickel catalyst makes this reaction an operation-simple and user-friendly method for the preparation of structural diverse α,β-unsaturated carboxyl acids.

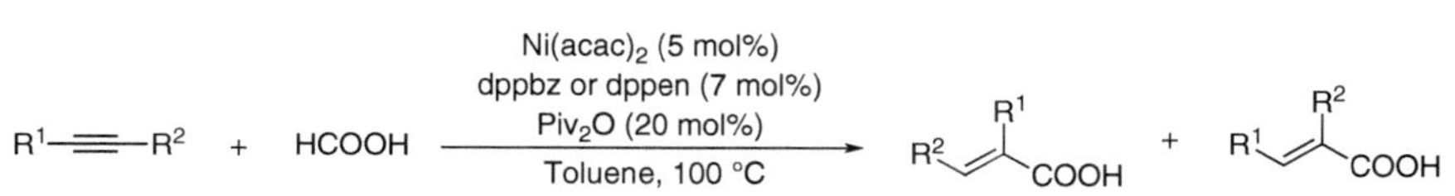

Scheme 3.36 Ni-catalyzed hydrocarboxylation of olefins with HCOOH and Piv_2O. Source: Modified from Fu et al. [777].

Although transition metal-catalyzed carbonylation reactions have been widely explored in organic chemistry, the development of highly efficient carbonylation process with formic acid is not an easy task because of its decomposition at high temperature or in the presence of transition metal-based catalyst systems.

Notably, the carbonylation chemistry with HCOOH is also related to the decarbonylated decomposition to CO that can be transferred into carbonylated metal complex as key intermediate in such carbonylation reaction.

3.2.4 N-Formylation Reactions with Formic Acid

One of the strategies for the synthesis of formamides was N-formylation of amines or nitroaromatics. Owing to their substantial range of applications in diverse areas of sciences and industries, formamides have been attracted extensive interest [779]. Formamides are an important class of compounds that appear as intermediates in fungicide and pharmaceutical syntheses, and isocyanate, formamidine, and nitrile formation [780–782].

In past studies, many approaches to synthesize formamides have been developed and formylation can be achieved by dehydration from formic acid and amines (Scheme 3.37).

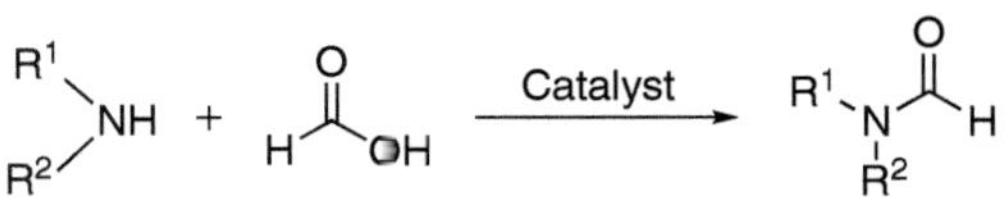

Scheme 3.37 Formylation from formic acid and amines.

3.2.4.1 Metal Oxides Catalysts

Metals and metal oxides have been paid increasing attention owing to their broader application in heterogeneous catalysis, ion exchange, and also in chemical separation processes, such as Mg, In, Zn, ZnO, and Al_2O_3 [783–787].

Hosseini-Sarvari and coworker reported that N-formylation of primary and secondary amines with HCOOH using ZnO as an acidic catalyst at 70 °C for 10–180 min with 65–99% yield (Scheme 3.38) [783]. Around 90%, 86%, and 84% yields were delivered, respectively, after three cycles of N-formylation of aniline. Subsequently, the catalytic performance of ZnO NPs and trivalent dysprosium ion-doped ZnO NPs ($ZnO : Dy^{3+}$) was studied with good yield [788, 789].

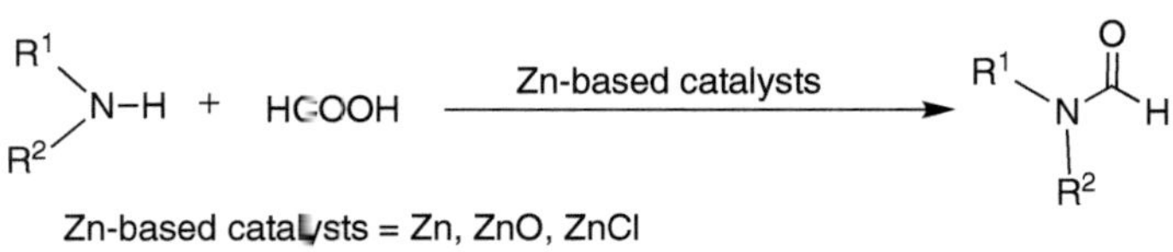

Scheme 3.38 N-Formylation of amines catalyzed by Zn-based catalysts. Source: Modified from Hosseini-Sarvari and Sharghi [783].

Then, other LAs have been applied in N-formylation of amines with HCOOH, like $ZnCl_2$, $AlCl_3$, $NiCl_2$, and $FeCl_3$, and $ZnCl_2$ catalyst provided the good results in formylation of aryl, heteroaryl, and alkyl amines to afford 60–98% yield formamides at 70 °C with 10–900 min [790]. Structurally diverse substituents were tolerated, like nitro, ester, ketone, and alkyl groups and halogens.

Besides, metal-catalyzed formylation reactions have also been developed. N-formylation of amines and HCOOH utilizing the zinc metal as a catalyst at 70 °C with 0.5–12 h for 68–96% yield [784]. Compared to other metals of Mg, In, Sn and so on, Zn appears to higher selectivity and activity [784–787].

A series of macroporous metal oxide catalysts were synthesized and investigated their catalytic activities by Thakuria et al. such as NiO, CoO, CuO, Cr_2O_3, and Mn_2O_3 [791]. Various metal oxide-based catalysts could be reused three times with no noticeable loss in activity.

Then, Reddy et al. reported that nano-MgO was developed by a solution combustion technique and used the nano-MgO catalyst in microwave-assisted N-formylation of aromatic and alkyl amines with HCOOH for 90–98% yield (Scheme 3.39) and reused for five runs without losing the catalytic activity [792].

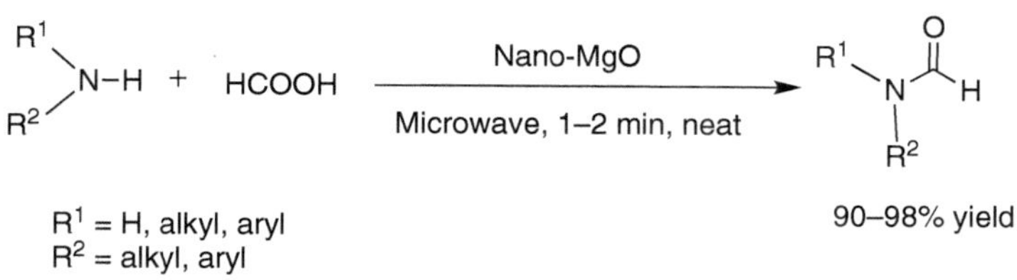

Scheme 3.39 N-Formylation of amines using nano-MgO.

Besides, other metal oxide catalysts, such as rod-shaped Al_2O_3 [796], nano-sulfated titania [794], CeO_2 NPs [795], were studied for N-formylation of different amines for mild to good yield.

3.2.4.2 Brønsted Acidic as Catalyst

Silica-functionalized perchloric acid ($HClO_4$–SiO_2) was prepared for N-formylation of substituted primary aromatic and cyclic secondary amines and HCOOH for 74–96% yield at room temperature in 15–90 min (Scheme 3.40) [796]. Among substrates bearing both the amino and hydroxyl groups, formylation occurred selectively at the amino position. Besides, other acids, such as fluoroboric (HBF_4), TFA, and sulfuric (H_2SO_4) acids afforded comparatively lower yields of N-formamides.

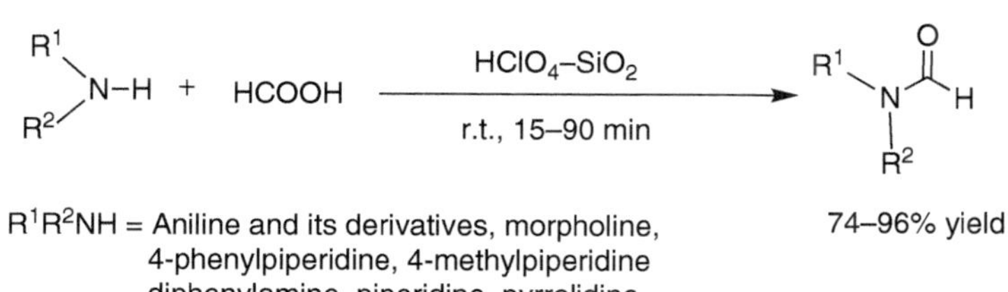

Scheme 3.40 $HClO_4$–SiO_2-catalyzed N-formylation of amines. Source: Modified from Ansari et al. [796].

Recently, sulfated tungstate, melamine trisulfonic acid (MTSA), silica sulfuric acid (SSA), FSG–Hf(NPf_2)$_4$, SBA-15/PrEn–$NHSO_3H$, and sulfonic acid-functionalized mesoporous silica (SBSA) have been studied for N-formylation of amines and HCOOH at 50–70 °C with up to 98% yield [797–801].

In 2019, a solid acidic catalyst, O-sulfonated poly(vinylpyrrolidonium) hydrogen sulfate, [PVP–SO_3H] HSO_4, has been developed for N-formylation of amines under solvent-free conditions (Scheme 3.41) [802]. The catalyst retained its stability and reusability for the formylation of 4-chloroaniline with a small decrease of activity.

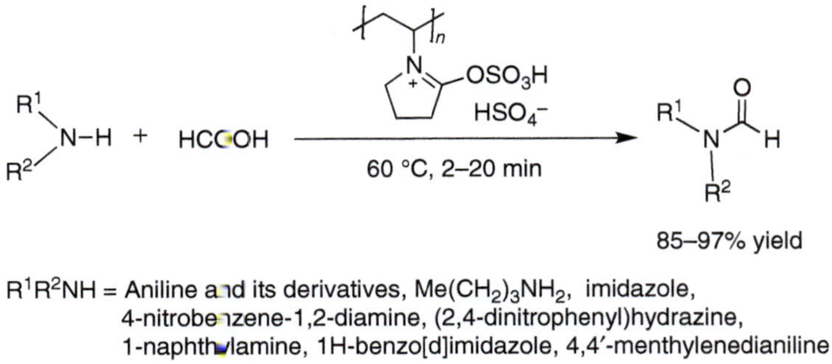

Scheme 3.41 N-Formylation of amines catalyzed by [PVP-SO_3H] HSO_4. Source: Modified from Pakpour-Roudsari et al. [802].

3.2.4.3 Amberlite IR-120 Resins as Catalysts

Amberlite IR-120 [H^+] resins as a solid acidic catalysts have been attracted more attention in organic reactions, due to their several advantages including non-toxicity, ready availability at low cost, operational simplicity, reusability, and ease of separation [803, 804].

Amberlite IR-120 [H^+] resin catalyzed N-formylation of amines for 90–97% yield under microwave irradiation (Scheme 3.42) [805]. Then, the resin catalyst was separated by simple filtration for reusable test, and its catalytic activity was reserved after five times.

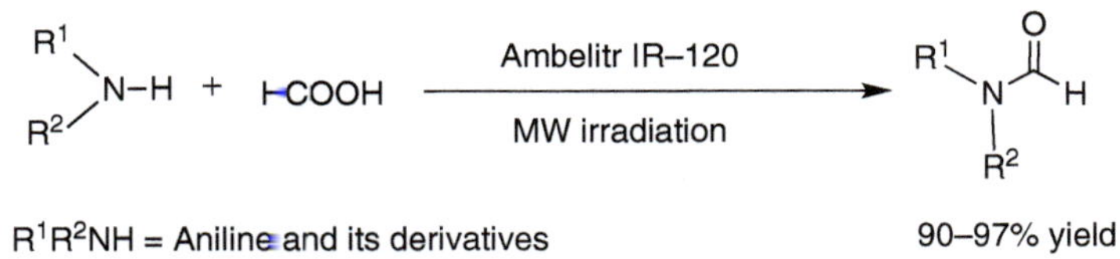

Scheme 3.42 N-Formylation of amines catalyzed by Amberlite IR-120. Source: Modified from Bhojegowd et al. [805].

3.2.4.4 Magnetic Catalysts

Magnetic catalysts have the beneficial attributes of elimination of toxic organic solvents, mild reaction conditions, and recycling possibility.

In 2010, hydroxyapatite (HAP)-encapsulated-γ-Fe_2O_3 NP-supported sulfonic acid [γ-Fe_2O_3@HAP–SO_3H] catalyzed formylation of aromatic, primary, and secondary amines with formic acid at room temperature for 90–97% yield with 15–60 min [806]. After that, magnetic $NiFe_2O_4$@SiO_2–PPA, $CoFe_2O_4$@SiO_2–PTA, and Fe_3O_4@Phendiol@Co nanocomposite were developed in similarly method (Scheme 3.43) [807–809]. In 2017, Co_3O_4 NPs with a large surface area were prepared by Marjani and colleagues to afford formamides with different functional groups [810].

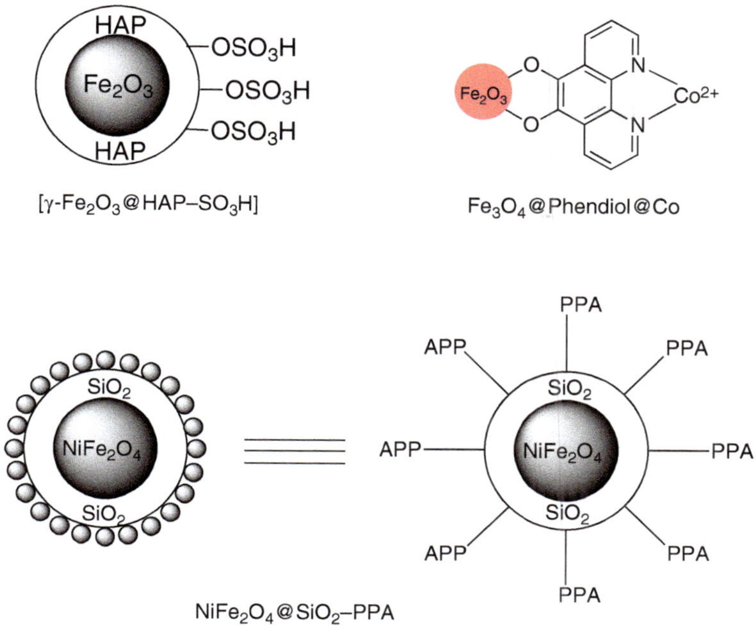

Scheme 3.43 Formylation of amines by formic acid with magnetic catalysts. Source: Khojastehnezhad et al. [807]; Kooti and Nasiri [808]; Habibi et al. [809].

3.2.4.5 Zeolite as Catalyst

Zeolite as a porous material catalyst has been used for the synthesis of formamides from various amines and HCOOH.

Zeidali and coworkers described N-formylation of various primary and secondary amines using HEU zeolite as a catalyst for 20–70 min at room temperature with 73–93% yield in 2012 and zeolite-based catalyst can remain the activity of catalyst after reused five times [811].

Next, a Lewis acid-catalyzed approach for N-formylation has been reported by Nasrollahzadeh and coworkers [812] using the natrolite zeolite to afford various formamides between 10 and 120 min (Scheme 3.44). And CuO NPs supported on the HZSM-5 nanosized zeolite (CuO/HZSM-5) were also developed in 2016 [813].

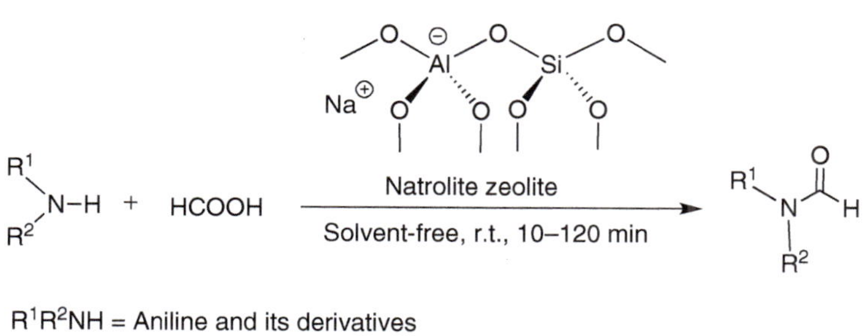

Scheme 3.44 N-Formylation of amines catalyzed by natrolite zeolite.

3.2.4.6 Ionic Liquids (ILs) as Catalyst

In 2013, N-substituted imidazolium trifluoroacetate-based IL was researched by Majumdar et al. for N-formylation at 70 °C with 1–3 h for 80–98% yield from aliphatic, aromatic heterocyclic and hindered amines, and second amines with formic acid (Scheme 3.45) [814].

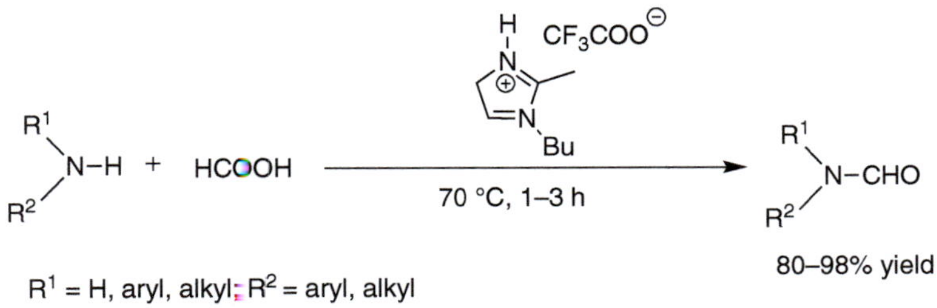

Scheme 3.45 N-Formylation catalyzed by N-substituted imidazolium trifluoroacetate-based ionic liquid. Source: Based on Majumdar et al. [814].

Baghbanian and coworker used TBD (1,5,7-triazabicyclo[4.4.0]dec-5-ene)-based protic IL, trifluoroacetate guanidinium salt [TBD][TFA], as a reusable catalyst for the synthesis of formamides and derivatives from primary and secondary amines with HCOOH at room temperature for 75–98% yield (Scheme 3.46) [815].

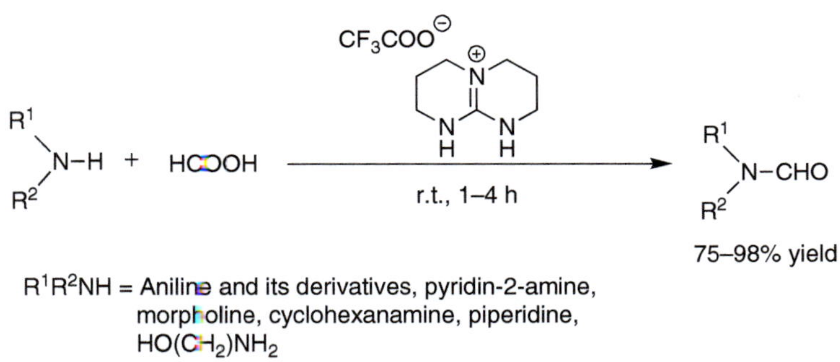

Scheme 3.46 [TBD][TFA] as a catalyst for N-formylation of amines. Source: Based on Baghbanian and Farhang [815].

In 2016, Na^+–MMT–pmim]HSO_4 was prepared by Seddighi from a Brønsted acidic IL (1-methyl-3-(trimethoxysilyl-propyl)-imidazolium hydrogen sulfate) supported on sodium montmorillonite clay (Na^+–MMT) as a nanoporous support (Scheme 3.47) [816].

Then in 2017, 1-methyl-3-(3-triethoxysilylpropyl) imidazolium chloride supported on the silica-coated ferrite NP surface (Fe_3O_4@SiO_2-IL) was introduced as a heterogeneous catalyst for N-formylation of amines and HCOOH at room temperature for 82–99% yield within 5–12 min (Scheme 3.48) [817].

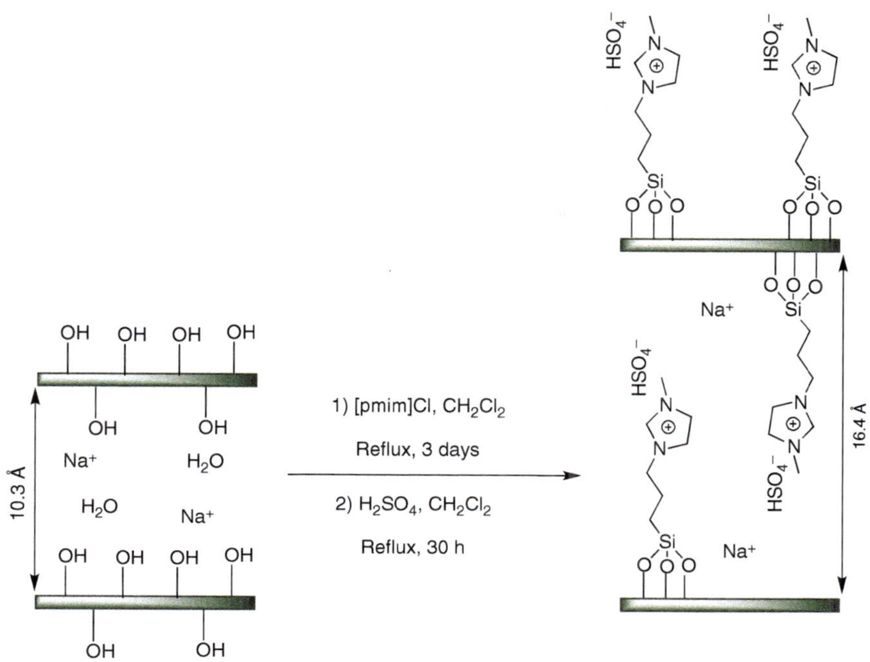

Scheme 3.47 Preparation of the Na^+–MMT–[pmim]HSO_4. Source: Shirini et al. [816].
© 2016, Springer Nature.

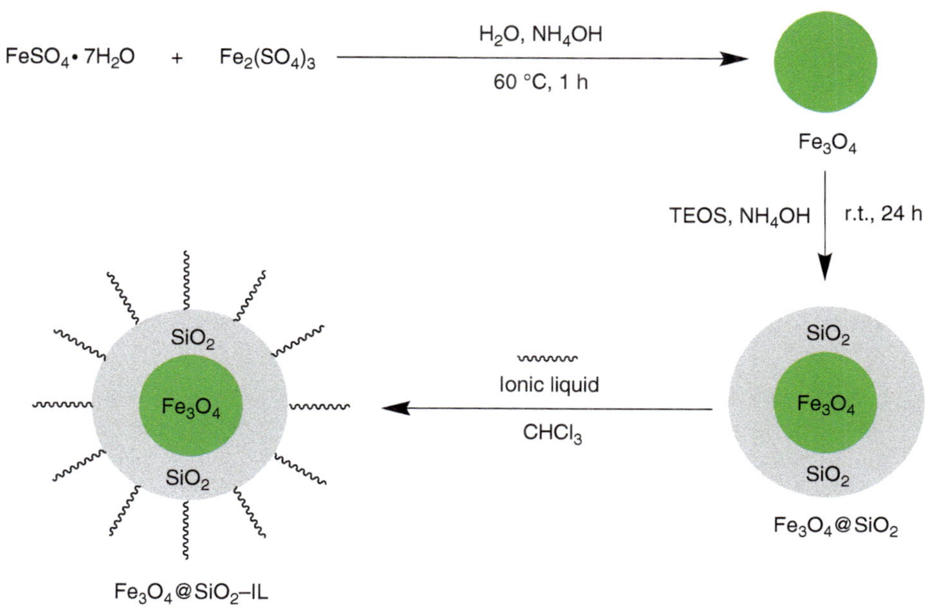

Scheme 3.48 Preparation of the Fe_3O_4@SiO_2-IL. Source: Modified from Garkoti et al. [817].

3.2.4.7 Other Catalysts

Molecular iodine (I_2) has been found to be a catalyst in N-formylation of different amines and HCOOH at 70 °C for 2–8 h with 69–96% by Jang and coworker in 2010 (Scheme 3.49) [818]. N-Formylation of amino and α-amino acid ester groups can be performed without epimerization. In this approach, iodine presumably reacts with HCOOH to generate in situ HI, which can be a real active species in the reaction.

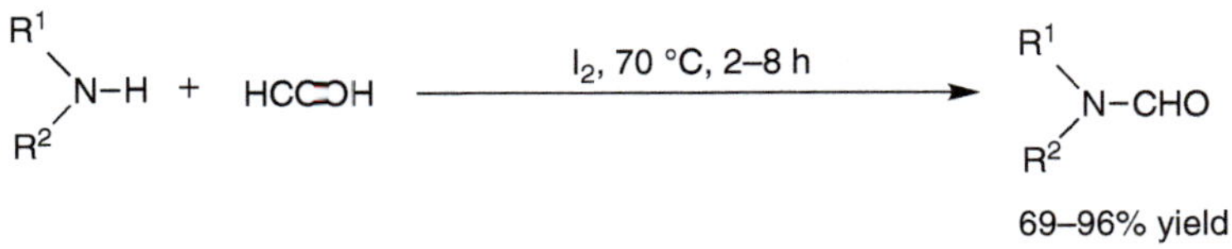

Scheme 3.49 I_2-catalyzed N-formylation of amines. Source: Modified from Kim and Jang [818].

Rice husk ash (RHA), as a natural source of amorphous silica, was used as a support for the immobilization of 1-methyl-3-(trimethoxysilylpropyl)-imidazolium hydrogen sulfate (RHA–[pmim]HSO_4) by Shirini et al. in 2014 [819]. The RHA–[pmim]HSO_4 catalyst (Scheme 3.50) was used for the N-formylation of amines or alcohols for 85–96% yield at 60 °C, and the heterogeneous catalyst can be recycled and reused 10 times without substantial loss in catalytic properties.

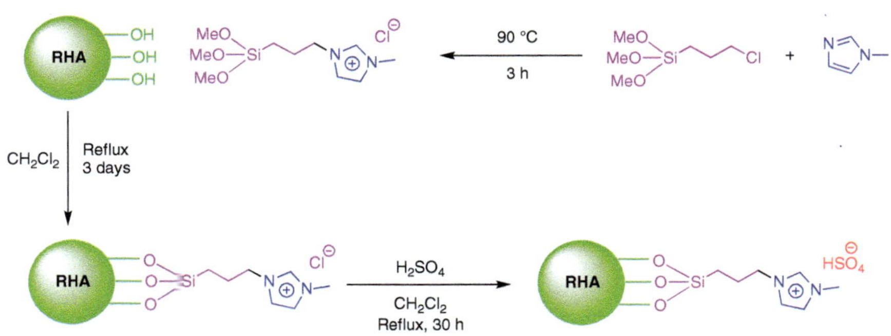

Scheme 3.50 Preparation of the acidic ionic liquid supported on RHA(RHA-[pmim]HSO_4).

Then in 2015 sulfonated rice husk ash has been used for the catalytic promotion of *N,N′*-diarylformamidines and chem-selective preparation of formamides from aryl amines (Scheme 3.51) [820].

In 2015, Rostamnia and Karimi reported the preparation and application of MOFs of amine/MIL-53 (Al) as a catalyst for N-formylation to prepare formamides interested in biology and industry from multiple amines and HCOOH (Scheme 3.52) [821]. The catalyst can be isolated from the reaction mixture and reused at least six times.

In 2017, sulfated polyborate with the benefits of easy preparation, mild acidity, and reusability catalyzed the N-formylation of amines including primary and secondary

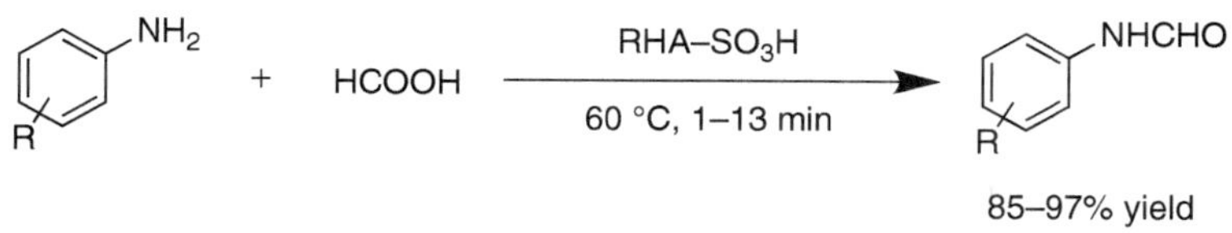

Scheme 3.51 N-Formylation of amines using RHA–SO_3H. Source: Modified from Seddighi et al. [820].

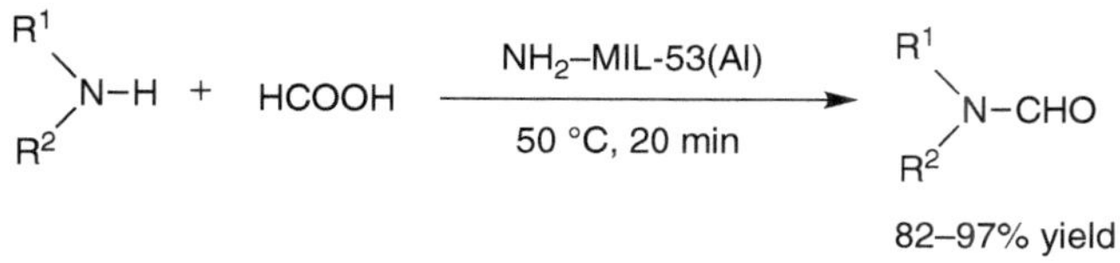

Scheme 3.52 N-Formylation of amines in the presence of NH_2-MIL-53(Al). Source: Modified from Rostamnia and Karimi [821].

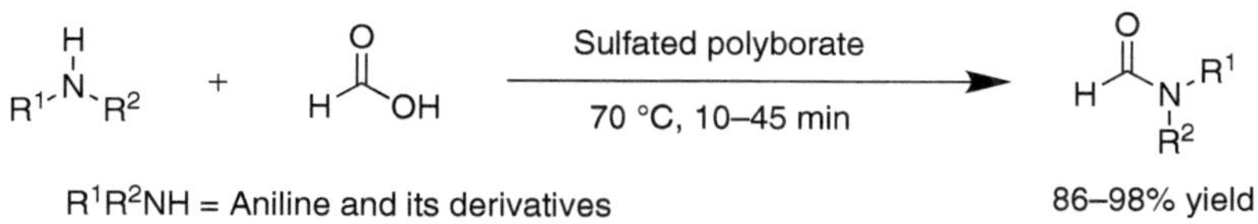

Scheme 3.53 N-Formylation of amines in the presence of sulfated polyborate. Source: Modified from Simonato et al. [822].

aromatic, aliphatic, and heterocyclic amines at 70 °C with 10–45 min for 86–98% yield (Scheme 3.53) [823].

In 2018, Azizi and coworkers described that DES/SBA-15 was utilized as a Lewis/Brønsted acid for N-formylation to provide the corresponding products in 72–98% yield at room temperature for 40–120 min and reused for four times without losing activity of catalyst [824].

3.2.5 Carbonylation of C–X with Formic Acid

As we all know, formic acid readily decomposes to CO and water under acidic conditions and at elevated temperatures. With appropriate activators, formic acid could be utilized in transition metal-catalyzed carbonylative reactions, in which CO would be released in situ under mild conditions. Early research revealed that formic acid could be utilized as a source of CO in rhodium- and iridium-catalyzed hydroxycarbonylation reactions [822, 825].

Among the many reactions involved CO, such as hydroxycarbonylation, formylation, and cross-coupling of aryl and vinyl halides or triflates [826], the use of formic acid as CO surrogate is attractive in the carbonylation of aryl halides. Then, corresponding transformations were demonstrated with formic acid (HCOOH) as CO surrogate, and many researches have been made to elaborate some organic transformations of HCOOH (Scheme 3.54).

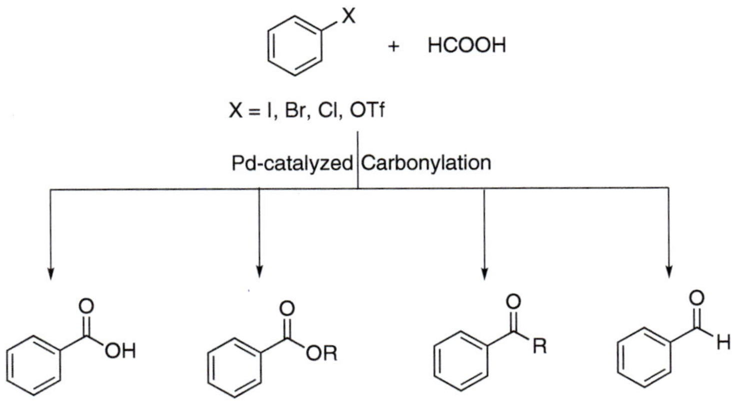

Scheme 3.54 Pd-catalyzed transfer carbonylation of aryl halides or triflates with HCOOH.

The $Pd_2(dba)_3$-catalyzed reaction of aryl and vinyl halides or triflates in the presence of acetic anhydride and lithium formate as CO source provides an route to the synthesis of the corresponding carboxylic acids for 43–92% yield. The reaction proceeds very smoothly under mild conditions and tolerates a wide range of functional groups, including ether, ketone, ester, and nitro groups (Scheme 3.55) [827].

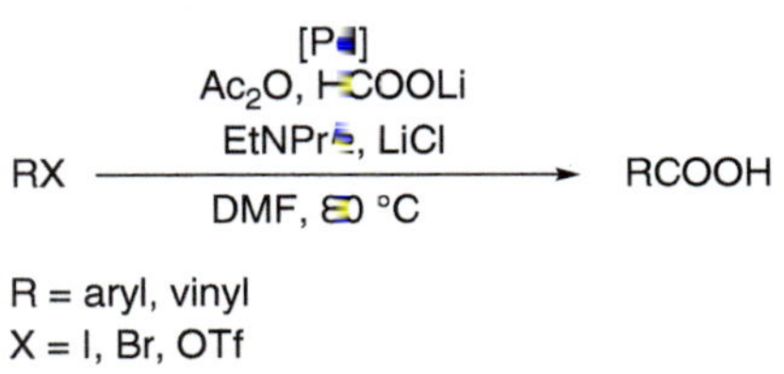

Scheme 3.55 Synthesis of carboxylic acids from halides and lithium formate. Source: Cacchi et al. [827].

Following that report, the hydroxycarbonylation of aryl and vinyl bromides with the combination of palladium(II) acetate with 1,10-bis(diphenylphosphino)ferrocene (dppf) as the catalytic system was improved by Bessmernykh and Caille for 61–89% yield at 120 °C (Scheme 3.56) [828].

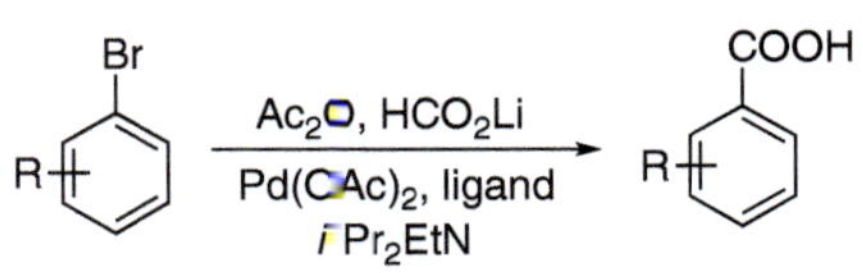

Scheme 3.56 Hydroxycarbonylation of aryl and vinyl bromides. Source: Berger et al. [828].

According to the method of Cacchi, stoichiometric acetic anhydride was indispensable for activating formate to generate CO. An efficient catalytic system using

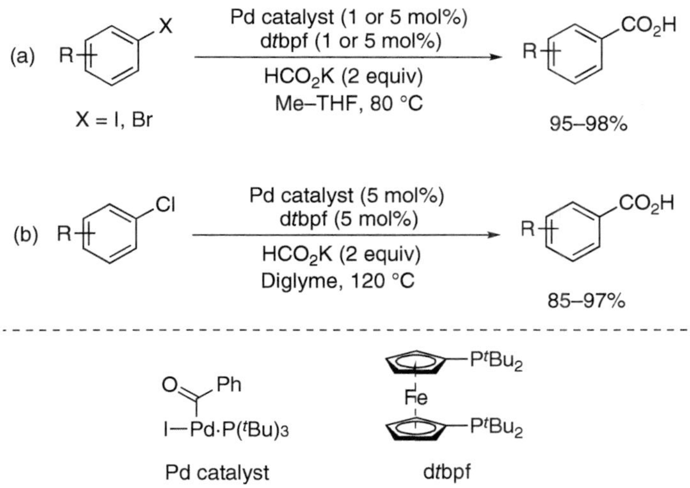

Scheme 3.57 Pd-catalyzed hydroxycarbonylation of aryl halides using potassium formate. Source: Modified from Korsager et al. [829].

potassium formate as CO source without any activator was developed by Skrydstrup and coworkers for good yield in 18 h (Scheme 3.57) [829].

The carbonylative cross-coupling reaction is a carbon–carbon bond formation process leading to useful ketones or esters. Recently, a series of carbonylative cross-coupling reactions of aryl halides employing formic acid as CO surrogates have been reported by Wu and coworkers. Using the combination of formic acid and acetic anhydride to release CO (Scheme 3.58, a), palladium-catalyzed carbonylative coupling of aryl iodides and terminal alkynes provided a variety of alkynones in good yield in a one-pot manner (Scheme 3.58, b) [830, 831]. Moreover, an array of diarylketones were produced via palladium-catalyzed carbonylative Suzuki coupling of aryl halides and arylboronic acids (Scheme 3.58, c) [832].

Besides ketones, benzoates were also prepared with phenols as coupling partner via palladium-catalyzed alkoxycarbonylation of aryl iodides or bromides (Scheme 3.58, d) [705, 833].

Recently, palladium-catalyzed reductive carbonylation of aryl iodides for 40–98% aromatic aldehydes at 80 °C using formic acid as the formyl source was reported by Wu and coworkers (Scheme 3.59) [834].

Later, Wu and coworkers also reported the hydroxycarbonylation of aryl halides under modified reaction conditions, and a variety of benzoic acid derivatives were produced in 32–88% yields from the corresponding aryl halides (Scheme 3.60) [835].

Wu and coworkers developed a palladium-catalyzed procedure for 53–88% yield of aryl formates under room temperature for 12 h with formic acid as the formyl source (Scheme 3.61) [836]. Similarly, the synthesis of benzofuran-2(3*H*)-ones from phenols and aldehydes was achieved for moderate yield (Scheme 3.62) [837].

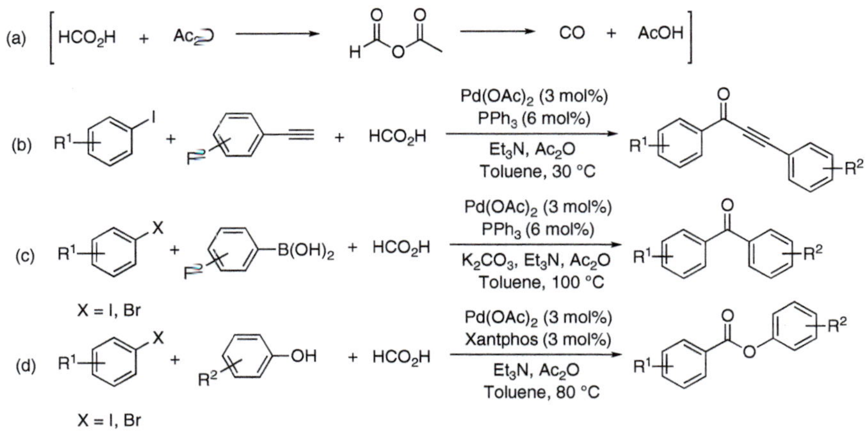

Scheme 3.58 Pd-catalyzed cross-coupling of aryl halides using formic acid. Source: Qi et al. [830]; Qi et al. [831]; modified from Qi et al. [832]; Qiu et al. [705]; Jiang et al. [833].

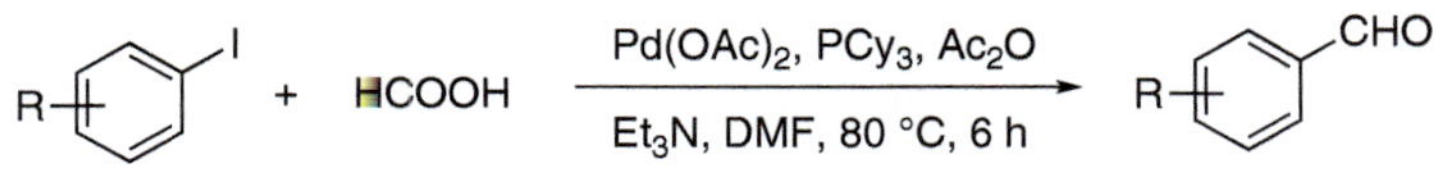

Scheme 3.59 Palladium-catalyzed reductive carbonylation of aryl iodides. Source: Qi et al. [834].

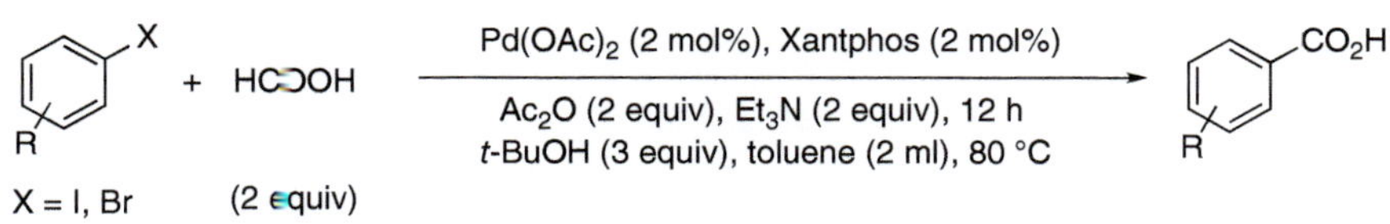

Scheme 3.60 Pd-catalyzed hydroxycarbonylation of aryl halides.

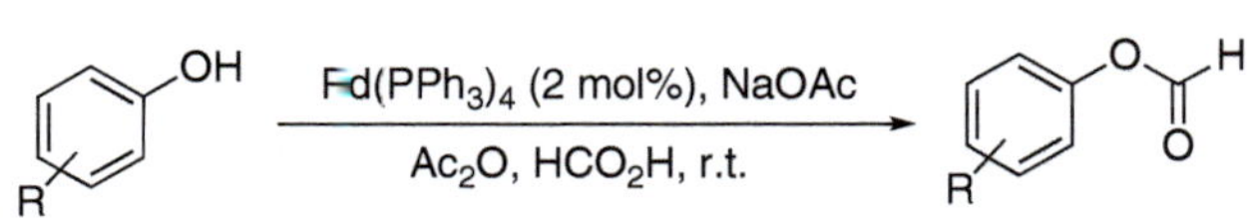

Scheme 3.61 Substrate testing for the Pd-catalyzed synthesis of aryl formates. Source: Jiang et al. [836].

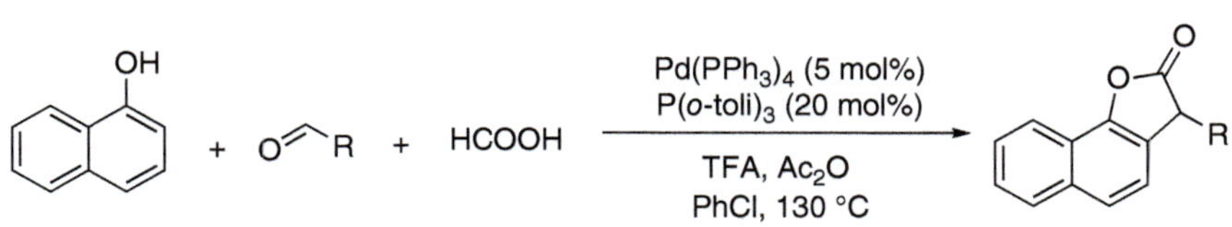

Scheme 3.62 Synthesis of benzofuran-2(3*H*)-ones from 1-naphthol and aldehydes.

In past two decades, transition metals that catalyzed the cleavage and functionalization of an unreactive C–H bond was one of the hottest topics in organometallic chemistry [471, 762, 763].

In 2004, Nozaki and coworkers reported an unexpected finding of a simultaneous hydroxylation–carboxylation of biphenyl to provide 4-hydroxy-4-biphenylcarboxylic acid applied as a polyester monomer in the presence of formic acid (Scheme 3.63) [766]. Surprisingly, the reaction only produced hydroxycarbonylation product when using CO instead of formic acid.

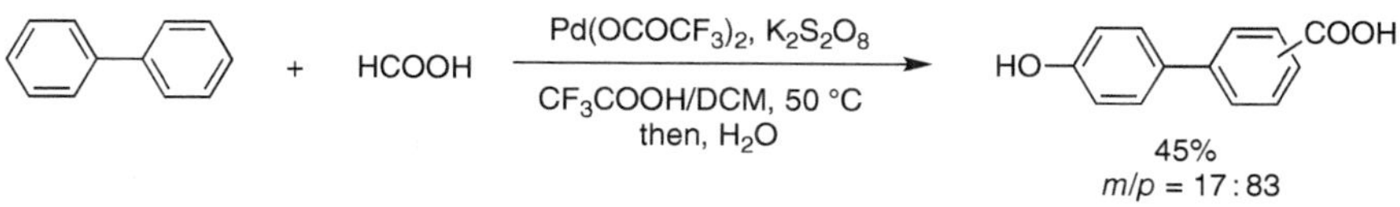

Scheme 3.63 Pd-catalyzed hydroxylation–carboxylation of biphenyl with HCOOH. Source: Shibahara et al. [766].

In a later report, Nozaki and coworkers revealed that addition of phosphenium significantly increased the yield (Scheme 3.64) [767].

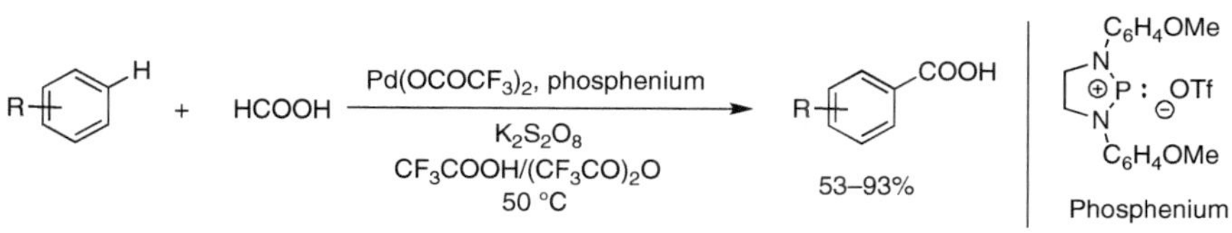

Scheme 3.64 Pd-catalyzed hydroxycarbonylation of arenes with HCOOH. Source: Modified from Sakakibara et al. [767].

3.2.6 Other Reactions

Rhodium or ruthenium-catalyzed Pauson–Khand-type reaction of 1,6-enynes is an efficient method for the synthesis of cyclopentenone derivatives. Utilization of formaldehyde as CO surrogate in rhodium(I)-catalyzed Pauson–Khand-type reaction of 1,6-enynes were demonstrated by Morimoto, Kakiuchi, and coworkers (Scheme 3.65) [129].

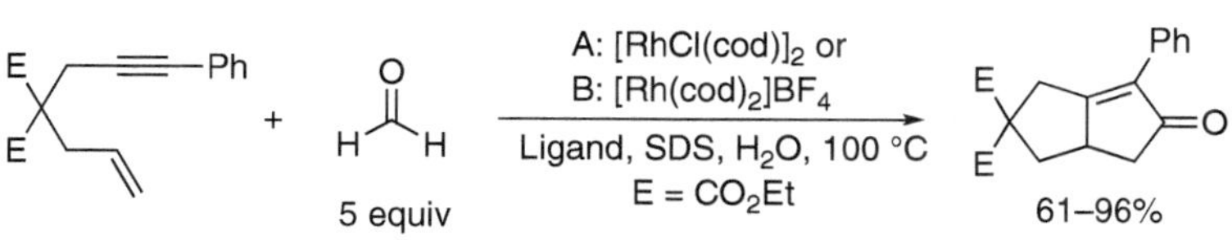

Scheme 3.65 Catalytic Pauson–Khand-type reaction of enyne in the presence of formaldehyde in an aqueous medium. Source: Fuji et al. [129].

Apart from the catalytic hydrogenolysis of C–C or C–O bond linkages discussed above, there are also a few reports dealing with the reductive cleavage of C–N bonds.

On investigating the catalytic hydrogenolysis of tertiary allylic amines, Heck's group reported that the Pd/C–FA–Et_3N system was useful for selectively removing morpholino and piperidino groups from the tertiary amines. And the yields for the mixture of two isomeric olefins was about 1 : 1 (Scheme 3.66) [838].

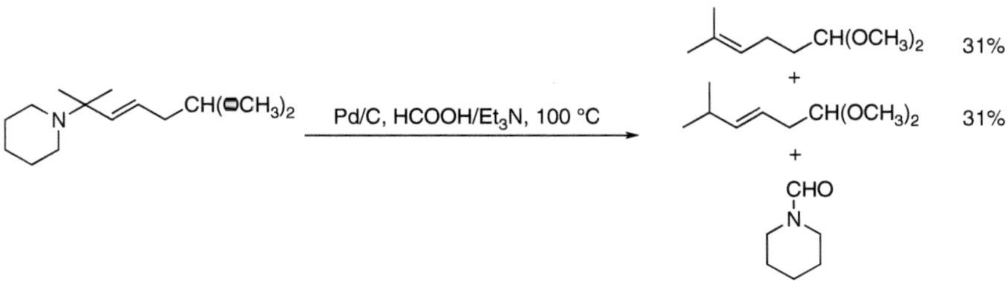

Scheme 3.66 Hydrogenolysis of tertiary allylicamines with FA. Source: Modified from Weir et al. [838].

Another interesting FA-mediated hydrogenolysis is deprotection, which is widely utilized for removing a protecting moiety introduced in order to protect a reactive functional group during a series of reactions [757, 839].

Recently, a catalytic system capable of the one-pot conversion of quinoline compounds to FTHQs has been devised utilizing FA as a source of both hydrogen and formyl groups via a reduction–formylation domino reaction sequence in the presence of gold NPs supported by TiO_2 (Scheme 3.67) [840]. Compared to traditional platinum-group-metal-based catalytic systems, the gold NPs exhibited superior catalytic activity and selectivity.

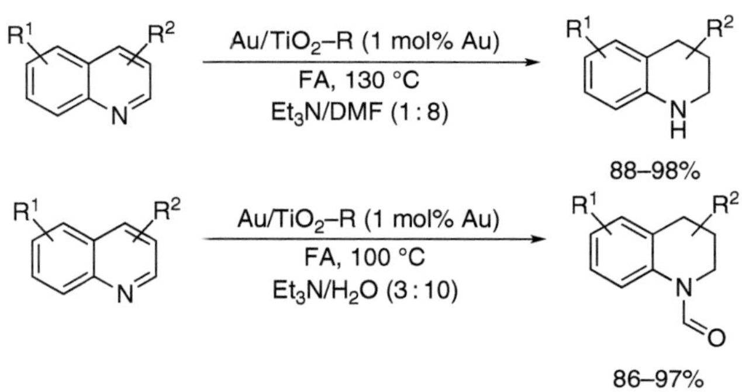

Scheme 3.67 Au–TiO_2–R-catalyzed reduction or reductive formylation of quinolines. Source: Modified from Li and Wu [840].

4

CO Surrogates

4.1 Carbonyl Metal

Developing carbonylative activation procedures with CO surrogates is under the current interesting. $M(CO)_x$ has been studied extensively and proven its worth as a CO surrogate [841, 842]. Among the possible candidates, $Mo(CO)_6$ as an air-stable solid complex was explored and applied in carbonylative transformations [843–845].

In 2002, Larhed and coworkers reported a method utilizing $Mo(CO)_6$ as a condensed source of CO for performing carbonylation reactions [841]. When a mixture of aryl bromides or aryl iodides, amines, and $Mo(CO)_6$ was heated by microwave assist for 15 min at 150 °C under an atmosphere of air using Pd/C or $Pd(OAc)_2$ as catalyst, then amides were formed in 65–83% yield (Scheme 4.1).

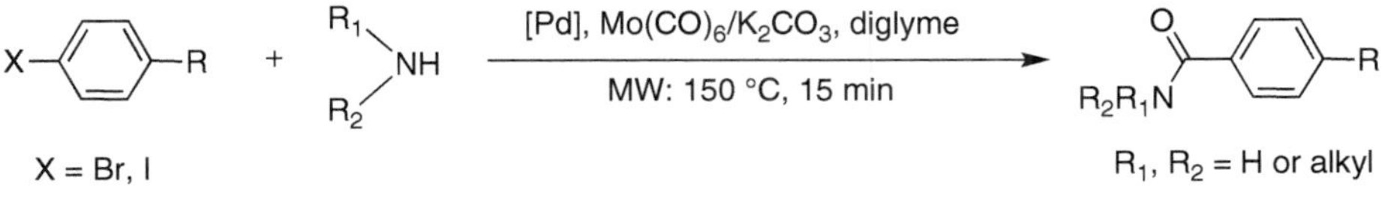

Scheme 4.1 Carbonylation utilizing $Mo(CO)_6$.

Larhed reported palladium(0)-catalyzed amino carbonylation reaction in a bridged two-vial system where $Mo(CO)_6$ functions as a suit solid CO releasing solid reagent with 71–97% yield (Scheme 4.2). In the bridged two-vial system, CO gas was first released in one of the bridged reaction vials by chemical liberation with DBU and then CO gas diffused to another vial where it could participate in the aminocarbonylation reaction from nitro group-containing aryl/heteroaryl iodides and bromides to benzamides. The above-mentioned bridged two-compartment protocol furnished good results with both primary amines and secondary amines and sluggish aniline nucleophiles at 65–85 °C reaction temperatures [846].

The synthesis of quinazolinones from 2′-bromoformanilides and nitro compounds has been developed by Wu and coworkers with 41–97% yields at 140 °C for 16 h with 26 examples of the desired products in the presence of a palladium catalyst and $Mo(CO)_6$ as a multiple promoter (Scheme 4.3) [847]. Notably, $Mo(CO)_6$ played the roles of CO source, nitro compound reducing reagent, and cyclization promoter

Carbonyl Compounds: Reactants, Catalysts and Products, First Edition.
Feng Shi, Hongli Wang and Xingchao Dai.

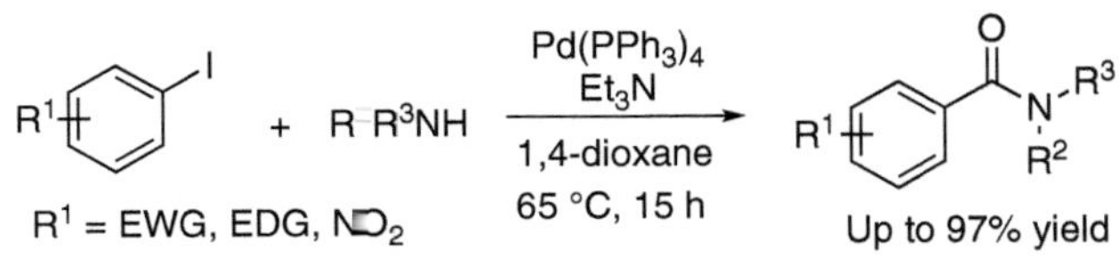

Scheme 4.2 Palladium(0)-catalyzed amino carbonylation reaction in a bridged two-vial system with $Mo(CO)_6$.

here. Not only aromatic nitros but also aliphatic nitros are suitable substrates for this novel transformation.

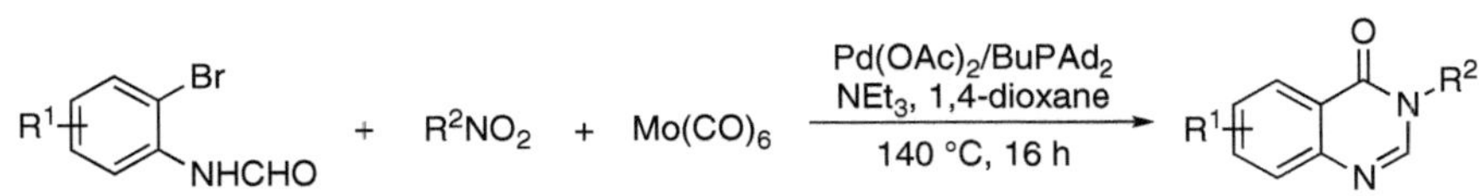

Scheme 4.3 Pd-catalyzed, $Mo(CO)_6$-mediated carbonylative coupling of 2′-bromoformanilide with nitro compounds. Source: Modified from He et al. [847].

Later, they also studied the transformation on palladium-catalyzed carbonylative C–H activation of arenes with alkene as the coupling partner via [3 + 2 + 1] reaction manner (Scheme 4.4) [848]. By applying $Mo(CO)_6$ or paraformaldehyde as the solid CO sources, various 5-(pyridin-2-yl)-hexahydro-7,10-methanophenanthridin-6 (5*H*)-ones were produced in 18–74% yields. Interestingly, the product was over oxidized to the corresponding 5-(pyridin-2-yl)-tetrahydro-7,10-methanophenanthridin-6(5*H*)-one when DDQ was applied the oxidant.

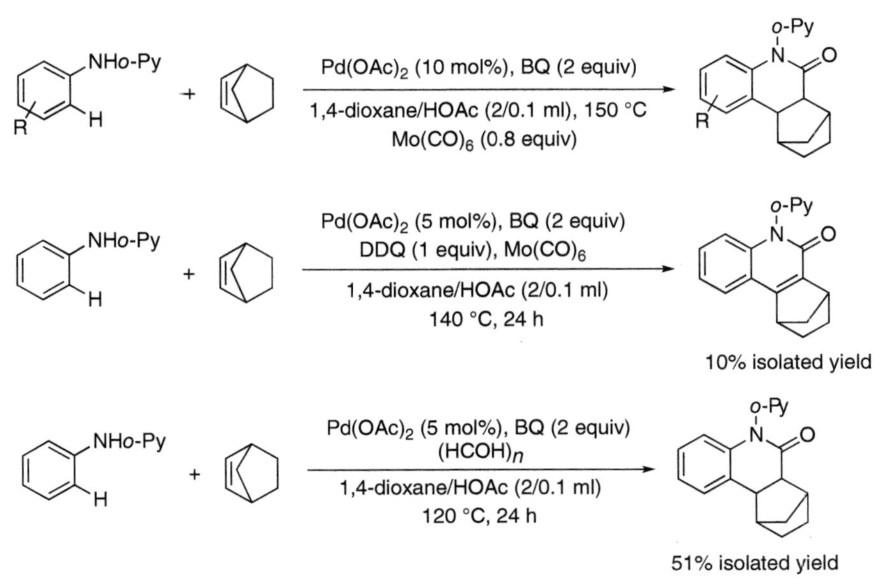

Scheme 4.4 Pd-catalyzed oxidative carbonylative C–H activation. Source: Modified from Chen et al. [848].

4.2 Formates

Alkyl formates, especially methyl formate, have been widely utilized as a C1 building block in organic synthesis. They are known to be easily decarbonylated by various transition metal catalysts to yield CO and the corresponding alcohols. This decarbonylation process is attractive as it affords a means of producing highly pure CO, thus avoiding storage and compression of the gas.

In 1983, Sneeden and coworkers discovered the Ru-catalyzed addition reaction of methyl formate to ethylene (Scheme 4.5) [849] and demonstrated that it represented an alternative to the original hydroesterification with CO [16, 53]. This reaction involved the catalytic decarbonylation process mentioned above, and ethylene, present in the reaction system, was carbonylated by the released CO and methanol to give methyl propionate. Since their discovery, numerous modifications of this reaction have been developed [41, 42, 252, 253, 264, 266, 850, 851]. In 2002, Chang and coworkers. reported on the hydroesterification of alkenes by using a unique formate, which contains a pyridyl group, as a substitute for CO and an alcohol [852]. The method can be applied to a wide variety of alkynes and dienes as well as alkenes [853].

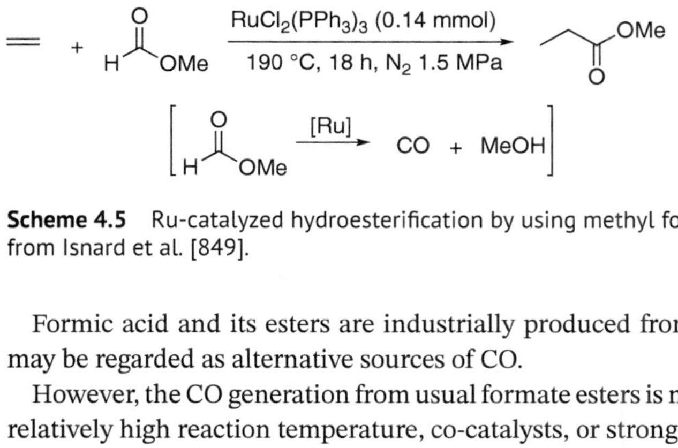

Scheme 4.5 Ru-catalyzed hydroesterification by using methyl formate. Source: Modified from Isnard et al. [849].

Formic acid and its esters are industrially produced from CO, so formate esters may be regarded as alternative sources of CO.

However, the CO generation from usual formate esters is not efficient and required relatively high reaction temperature, co-catalysts, or strong bases.

Recently, Skrydstrup et al. reported ex situ generation of CO utilizing several CO sources by the aid of palladium catalysts [854, 855]. Although efficient CO generation and the carbonylation of aryl halides proceeded, a two-chamber system must be used to separate CO generation and carbonylation processes.

In the course of studies, Chang and coworkers found the first example of a chelation-assisted hydroesterification of olefins using 2-pyridylmethyl formate instead of CO with $Ru_3(CO)_{12}$ as catalyst at 135 °C for 3–12 h [856]. In this system, one-carbon elongated esters starting from alkenes were synthesized in 65–98% total yield and 65–100% selectivity (Scheme 4.6).

This result may pave more opportunities for searches of efficient and selective new C–C forming catalytic reactions by means of the chelation strategies.

Ten years later, Manabe and coworkers researched that imidazole derivatives were revealed to be effective ligands in the Ru-catalyzed hydroesterification of alkenes

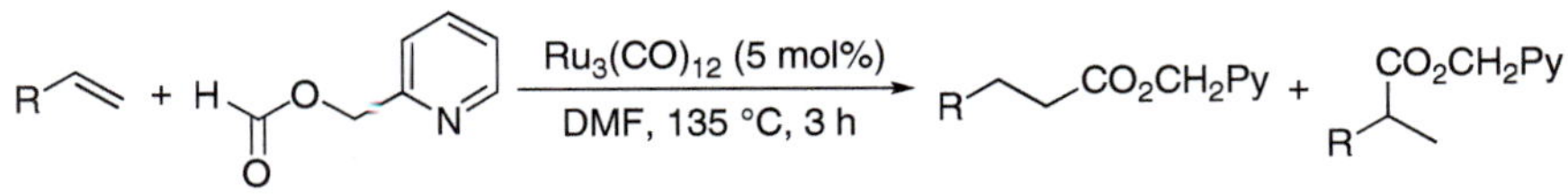

Scheme 4.6 Hydroesterification of various alkenes with 2-pyridyl-containing formates.

using formates, and intramolecular hydroesterification was performed to give lactones for the first time (Scheme 4.7) [857].

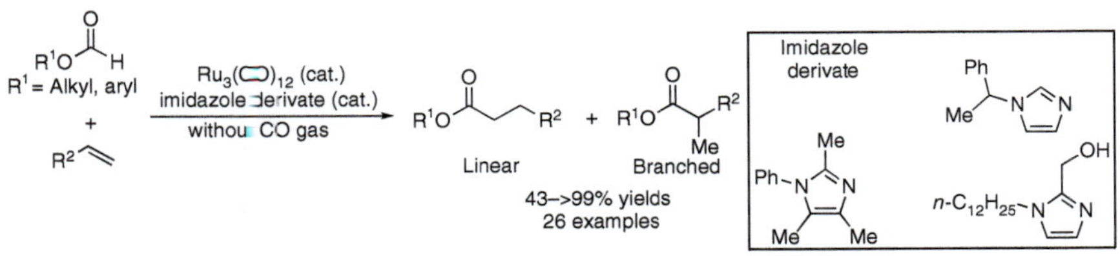

Scheme 4.7 Ru-catalyzed hydroesterification of alkenes and formates using imidazole derivatives. Source: Konishi et al. [857].

Similarly, the hydroesterification of olefins with formates was reported by the team of Profir, Bellerand, and Fleischer in the presence of $Ru_3(CO)_{12}$ (Scheme 4.8), and a range of formates can be used for selective alkoxycarbonylation of aromatic and aliphatic olefins [858].

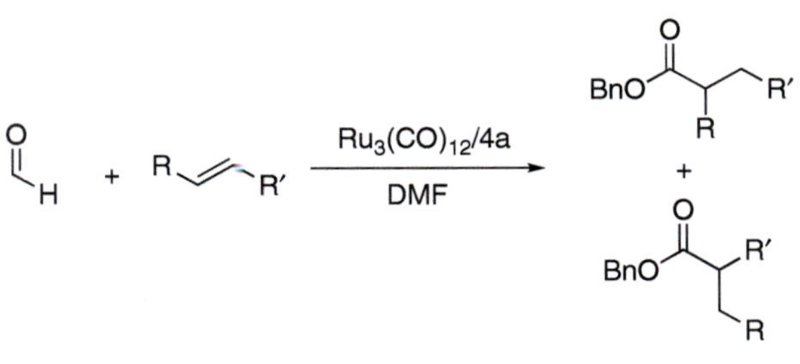

Scheme 4.8 Hydroesterification of olefins and formats in present of $Ru_3(CO)_{12}$.

In 2012, Tsuji and coworkers demonstrated that the carbonylation of aryl halides using aryl formates as CO source occurred at environmental pressure and 60 °C in the presence of a $PdCl_2(phCN)_2$ to afford aryl esters in 72–97% yields for 20 h (Scheme 4.9) [859]. During the reaction, aryl formates were converted to CO and the corresponding phenols by the aid of a base.

The Pd-catalyzed carbonylation of aryl/alkenyl halides and triflates proceeded at room temperature using 2,4,6-trichlorophenyl formate as CO surrogate, and the obtained trichlorophenyl esters can be converted to a variety of carboxylic acid derivatives in 74–99% yields (Scheme 4.10) [860].

A one-pot tandem reaction involving olefin isomerization and hydroesterification has been developed by Carreira and coworkers in 2014, which enables the

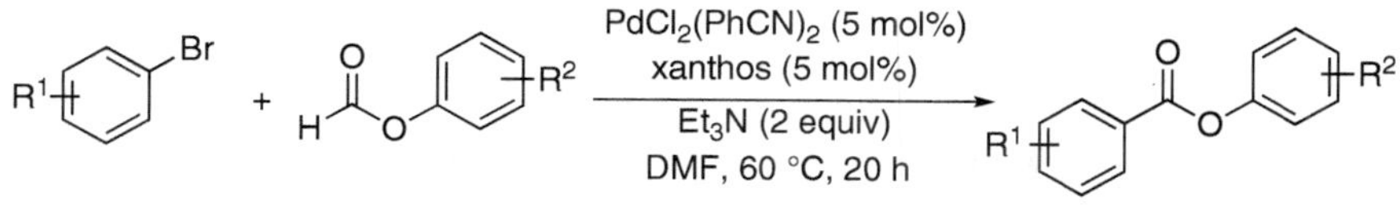

Scheme 4.9 Carbonylation of various substrates under the optimum reaction conditions. Source: Fujihara et al. [859].

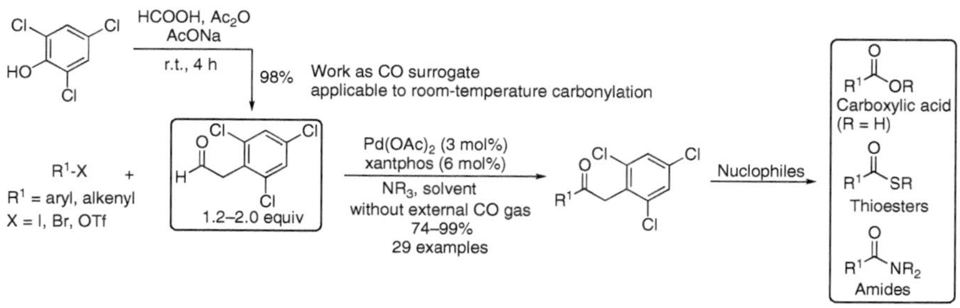

Scheme 4.10 Pd-catalyzed carbonylation using 2,4,6-trichlorophenyl formate as CO surrogate. Source: Ueda et al. [860].

incorporation of a C1-unit at the remote terminal position of allylic amides at 135 °C for 18 h with 53–85% yield (Scheme 4.11) [861].

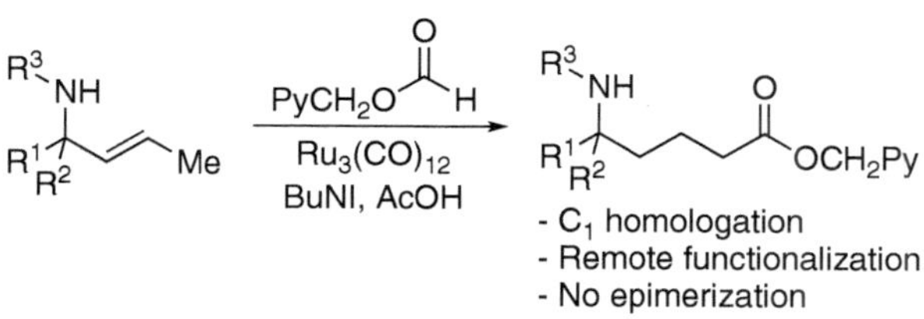

Scheme 4.11 A one-pot tandem reaction involving olefin isomerization and hydroesterification. Source: Armanino et al. [861].

After decades of development, carbonylation reactions have become one of the most powerful tools in modern organic synthesis. Carbonylative coupling reactions have been developed by Beller, Wu, and coworkers using aryl formates as the CO source and pseudohalide precursors [862]. This protocol has a high functional group tolerance and could be applied in carbonylations with C, N, and O nucleophiles. The corresponding amides, alkynones, furanones, and phenyl benzoates were synthesized upon reaction with amines, alkynes, and phenols (Scheme 4.12).

Alkyl formates readily undergo decarbonylative decomposition in the presence of strong bases to generate CO, along with the corresponding alcohols. Mortreux et al. successfully applied this decarbonylation of methyl formate to the Heck-type esterification of aryl and vinyl halides. Thus, the reaction of aryl and vinyl halides with alkyl formates in the presence of sodium alkoxide and a catalytic amount of a palladium complex results in the alkoxycarbonylation of aryl and vinyl halides to

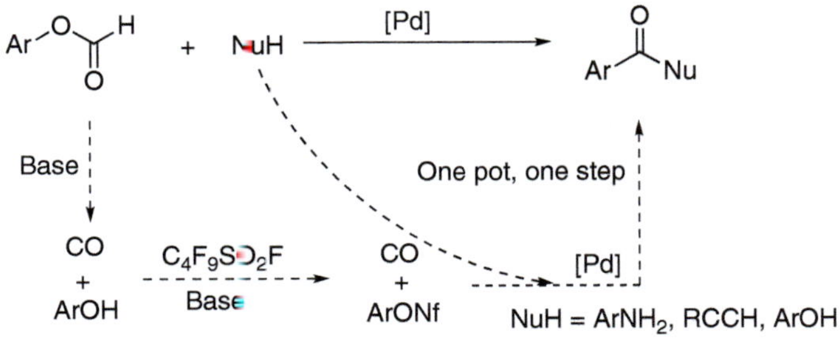

Scheme 4.12 Carbonylative coupling reactions using aryl formates.

give aromatic and vinyl esters (Scheme 4.13) [863, 864]. When a combination of ethyl formate and sodium ethoxide was used, the reaction proceeded smoothly at room temperature (85% yield of ethyl benzoate). Potassium *t*-butoxide was unsuitable as a decarbonylating reagent because the decarbonylation of alkyl formats occurred violently, even at 0 °C.

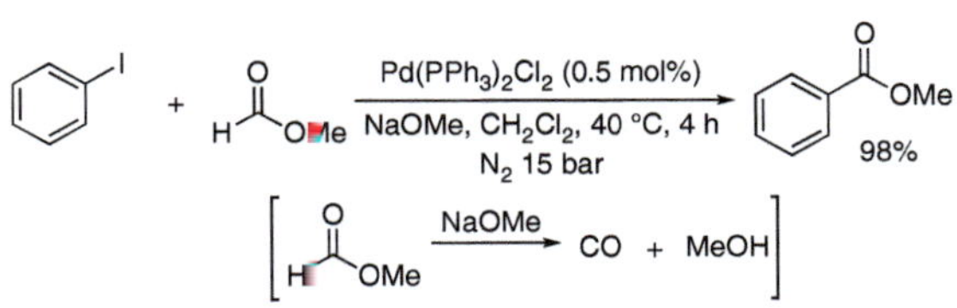

Scheme 4.13 Pd-catalyzed alkoxycarbonylation by using methyl formate. Source: Adak et al. [863]; Adamek et al. [864].

4.3 Formamides

In 2002, the first example of aminocarbonylation of aryl and alkenyl halides in the absence of CO in 10–36 h at 120 °C for 66–92% yield was reported by Nozaki and Hiyama. In the presence of $Pd_2(dba)_3$, the coupling reaction of aryl and alkenyl iodides with *N*,*N*-dimethylformamide (DMF) produces the corresponding aromatic amides in one step using phosphoryl chloride ($POCl_3$) as an additive (Scheme 4.14) [865].

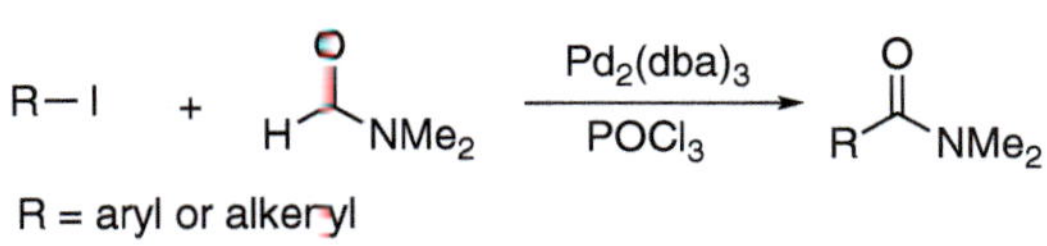

Scheme 4.14 Aminocarbonylation of aryl and alkenyl iodide with *N*,*N*-dimethylformamide. Source: Hosoi et al. [865].

Similarly, aminocarbonylation (Heck carbonylation) of *p*-tolyl bromide with DMF as an source of CO and dimethylamine was catalyzed by $Pd(OAc)_2$ to provide

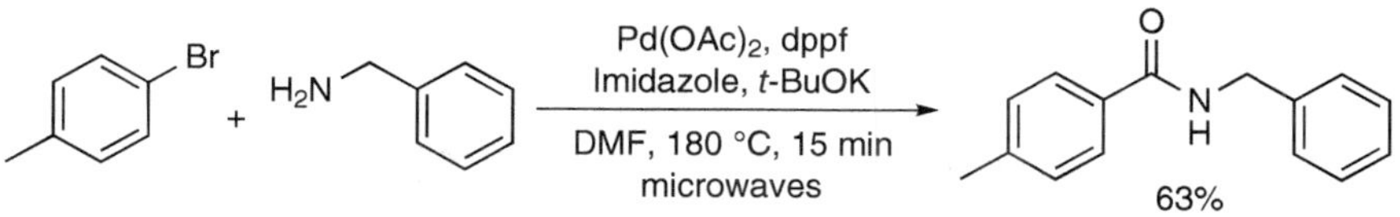

Scheme 4.15 Aminocarbonylation with DMF as a source of CO and dimethylamine. Source: Based on Hosoi et al. [865].

corresponding aryl amides in 59–94% yields for 15 min with microwave heating to 180–190 °C (Scheme 4.15) [865].

The reaction proceeded smoothly with bromobenzene and more electron-rich aryl bromides, but electron-deficient aryl bromides fail to undergo aminocarbonylation. The carbonylation procedure relied on the in situ generation of CO, was appropriate for small-scale reactions where short reaction times were desired and the direct use of CO gas was impractical.

Quinazolinones were omnipresent motif in many pharmaceuticals and agrochemicals. In 2012, a direct carbonylation of C–H bonds of aromatic or aliphatic amides was developed by Ge group through Ni/Cu synergistic catalysis under atmospheric oxygen at 160 °C in 24 h for middle yield of quinazolinones using DMF as the CO source. What is more, the sp^2 C–H bond functionalization process was featured with high regioselectivity and a good compatibility with a broad range of functional groups (Scheme 4.16) [866].

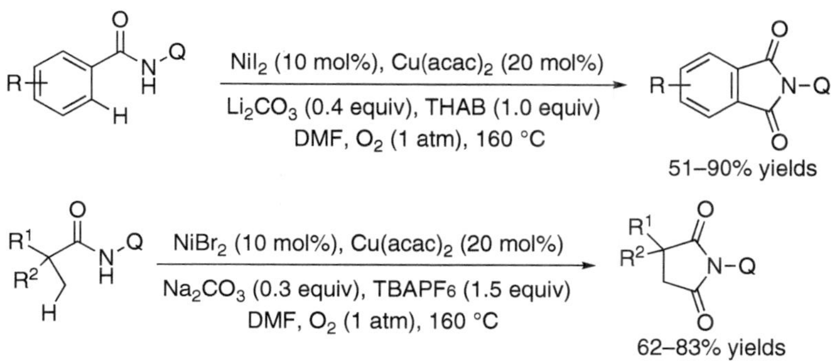

Scheme 4.16 Synthesis of quinazolinones using DMF. Source: Wu et al. [866].

Based on continual interest on carbonylation reactions, Wu group investigated the Pd-catalyzed carbonylation of *N*-phenylpyridin-2-amine with DMF in the presence of TFA as the co-solvent and using $K_2S_2O_8$ (potassium persulfate) as oxidant in the presence of TFA as the co-solvent for 79% yield at 140 °C (Scheme 4.17) [867].

4.4 Formic Anhydride

In 2003, Cacchi et al. demonstrated that the palladium-catalyzed hydroxycarbonylation of aryl and vinyl halides or triflates by using formate salts in the presence of

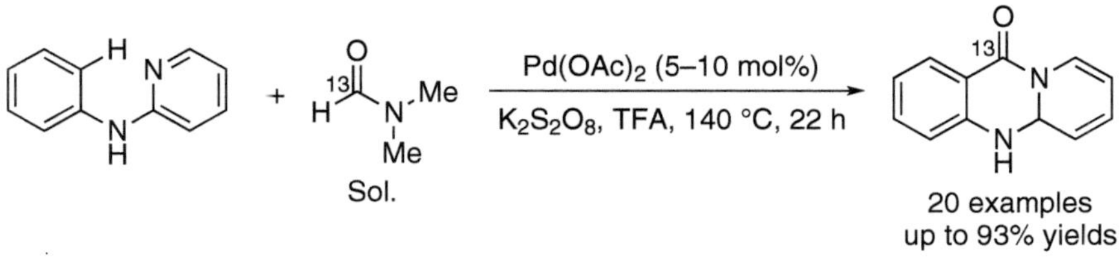

Scheme 4.17 DMF as CO source in Pd-catalyzed carbonylation. Source: Chen et al. [867].

acetic anhydride provided a straightforward route to carboxylic acids (Scheme 4.18) [827]. The reaction involves the formation of formic acetic anhydride as a condensed source of CO, which was generated in situ by the reaction of a formate salt with acetic anhydride, followed by thermal decomposition to give CO and AA.

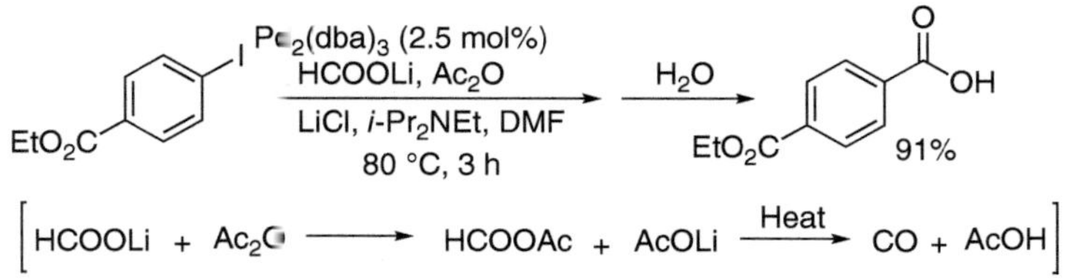

Scheme 4.18 Pd-catalyzed hydrocarbonylation using a mixed anhydride as the surrogate. Source: Modified from Profir et al. [858].

4.5 Silacarboxylic Acid

Silacarboxylic acids have been presented as versatile, benchtop-stable, easy-to-handle, metal-free CO-releasing molecules. Different silacarboxylic acids were synthesized by Skrydstrup and coworkers from the corresponding chlorosilanes and carbon dioxide. The $MePh_2SiCO_2H$ was applied in palladium-catalyzed carbonylative couplings because of its efficient decarbonylation in different substrates for 62–91% (Schemes 4.19 and 4.20) [854].

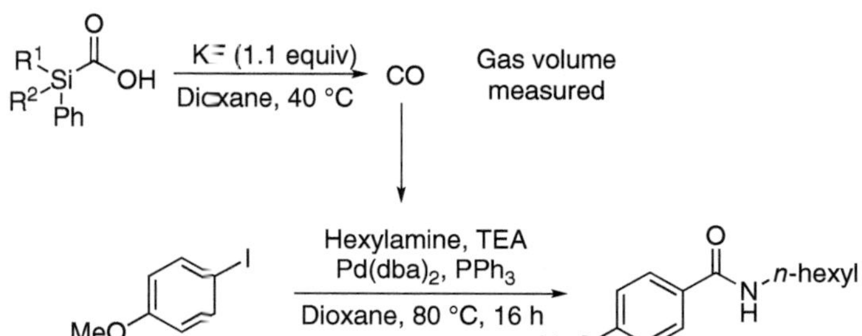

Scheme 4.19 Decarbonylation of silacarboxylic acids using potassium fluoride. Source: Friis et al. [854].

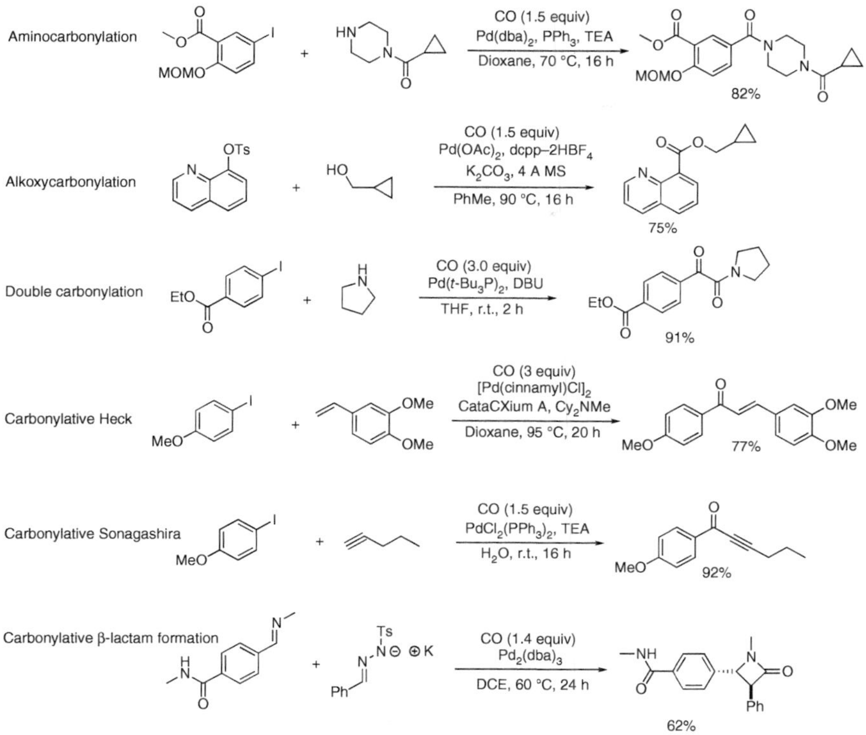

Scheme 4.20 Carbonylative coupling using $MePh_2SiCO_2H$ as the CO. Source: Friis et al. [854].

The activation of formamides by transition-metal complexes can also donate a carbonyl group in carbonylation reactions. In 1987, it was reported that the reaction of alkenes with formamides in the presence of a $[Ru_3(CO)_{12}]$ catalyst afforded the hydroamidation products [868]. Although this is the first use of formamides as a carbonyl source in carbonylation reactions that can substitute the reaction of alkenes with CO and amines [869], external CO was essential for maintaining catalytic activity in this reaction. Kondo et al. subsequently found that the use of the [PPN] $[Ru_3H(CO)_{11}]/PCy_3$ catalyst system (PPN = bis(triphenylphosphane) iminium) in the reaction with formamides proceeds without the need for external CO (Scheme 4.21) [870]. In 2003, Chang and coworkers reported that the use of formamide, which has a coordinating pyridyl group, led to a similar hydroamidation under milder reaction conditions (Scheme 4.22) [871].

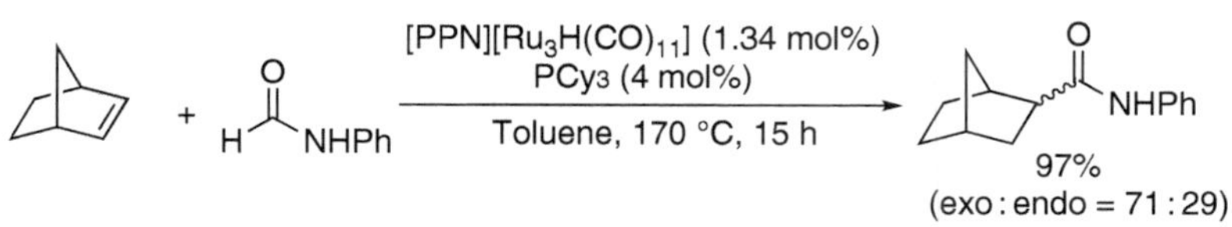

Scheme 4.21 Ru-catalyzed hydroamidation by using formanilide. Source: Modified from Kondo et al. [870].

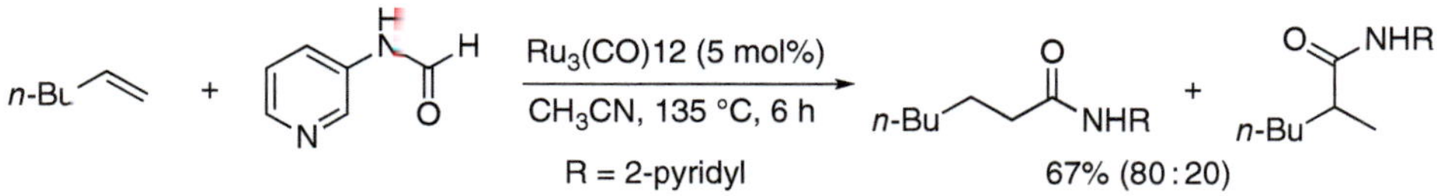

Scheme 4.22 Ru-catalyzed hydroamidation by using formamide. Source: Modified from Ko and Chang [871].

4.6 N-Formylsaccharin

In 2013, Palladium-catalyzed reductive carbonylation of aryl bromides was applied by Manabe and coworkers in a wide range of aryl bromides at 65–90 °C with N-formylsaccharin as CO source for 45–93% yield (Scheme 4.23) [872]. Compared to other existing CO sources for palladium catalyzed reactions, N-formylsaccharin has a great number of advantages, including low cost, good availability, ease of handling, stability, and high reactivity as a CO source.

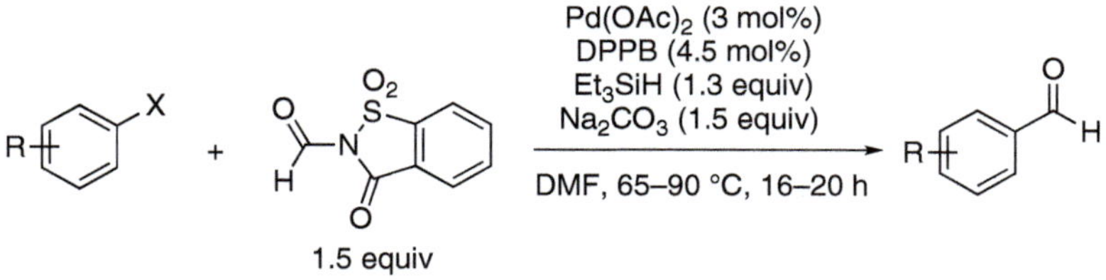

Scheme 4.23 Pd-catalyzed reductive carbonylation of aryl bromides, iodides, and triflate. Source: Ueda et al. [872].

Further investigations concerning the application of this methodology to other CO-free, palladium-catalyzed reactions were still under studied.

In 2015, Fleischer et al. report a Pd-catalyzed alkoxycarbonylation of alkenes based on N-formylsaccharin as a CO surrogate at room temperature (Scheme 4.24) [873]. The carbonylation proceeds with a complementary regioselective yielding the desired branched products from styrene derivatives and valuable linear esters from alkyl-substituted alkenes up to 97% total yield.

4.7 Acyl Chloride

A new method for the ex situ generation of CO and its incorporation in palladium-catalyzed carbonylation reactions was achieved using a simple sealed two-chamber system. The ex-situ generation of CO was derived by a palladium-catalyzed decarbonylation of tertiary acid chlorides using $Pd(dba)_2$ and $P(tBu)_3$ as catalysts [855]. Releasing CO ex situ from pivaloyl chloride was a new Pd-catalyzed decarbonylative protocol. This method was applied for the synthesis of heteroaromatic amides at 80 °C using 2-pyridyl tosylates in palladium-catalyzed aminocarbonylations for 87% yield (Scheme 4.25).

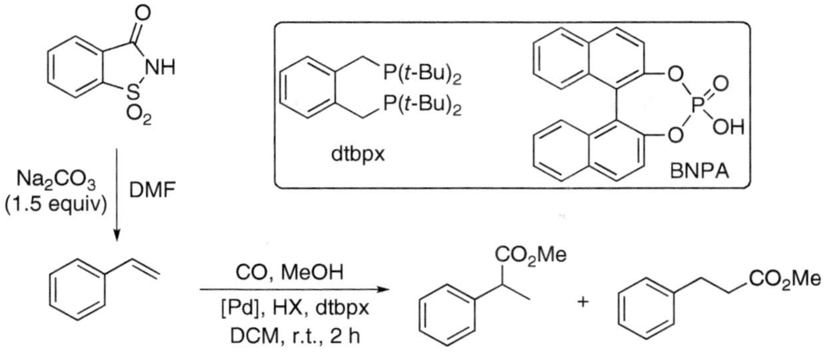

Scheme 4.24 Optimization of the methoxycarbonylation of styrene. Source: Modified from Gehrtz et al. [873].

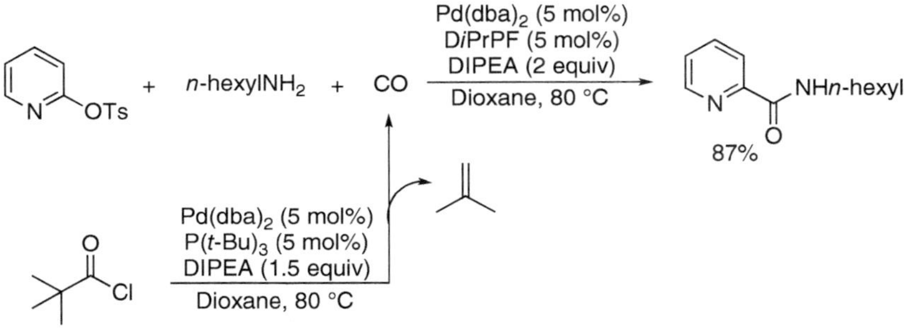

Scheme 4.25 Crossover of CO in sealed two-chamber system.

Further development of this class of CO-equivalents led to a stable and solid acid chloride derivative capable of releasing CO at room temperature mediated by a palladium-catalyzed decarbonylation (Scheme 4.26) [855].

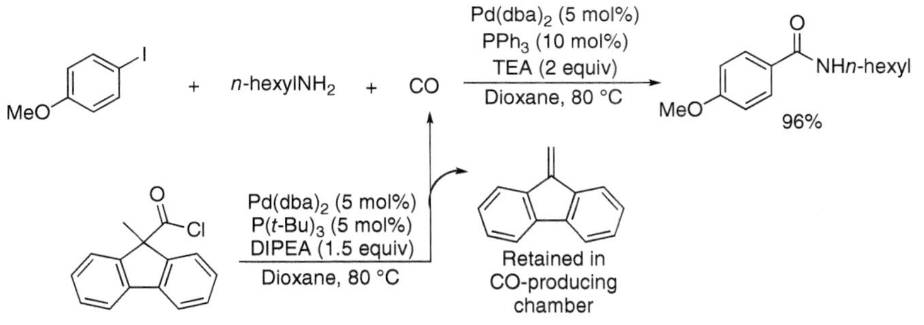

Scheme 4.26 Palladium-catalyzed decarbonylation with a stable and solid acid chloride. Source: Modified from Hermange et al. [855].

Later, Gracza reported a protocol for the generation of CO by Zn-mediated reduction of oxalyl chloride in a simple two-chamber system [874]. Oxalyl chloride

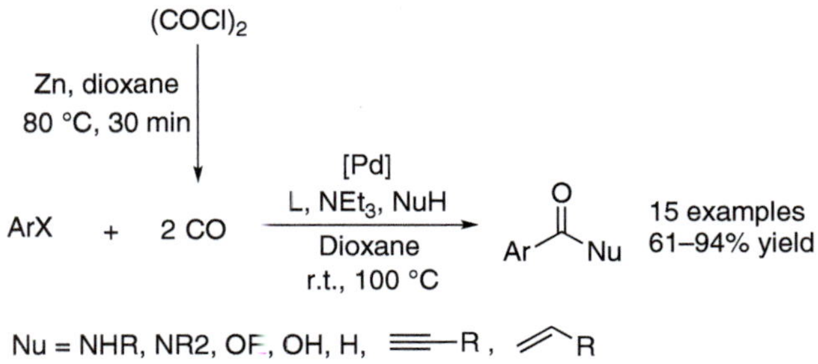

Scheme 4.27 Oxalyl chloride as substitute of CO in the palladium-catalyzed carbonylation.

was applied as a substitute for CO in the palladium-catalyzed various carbonylation reactions processes providing corresponding products in 61–94% yields at room temperature to 100 °C, such as esters, amides, aldehydes, and carboxylic acids (Scheme 4.27).

4.8 In Situ Generated Carbonyl Source

4.8.1 Methanol

In recent decades, considerable effort has focused on the development of novel carbonylative transformations using CO surrogates. Consequently, toxic CO gas can be replaced by more convenient inorganic or organic carbonyl compounds [731, 732, 875]. Methanol is an abundant and potentially renewable chemical that is produced on the order of 35 million metric tons every year; however, the use of methanol as a carbonyl source is underdeveloped (Scheme 4.28). This can be explained by the relatively high energetic demand of methanol dehydrogenation (DH = +84 kJ mol^{-1}).

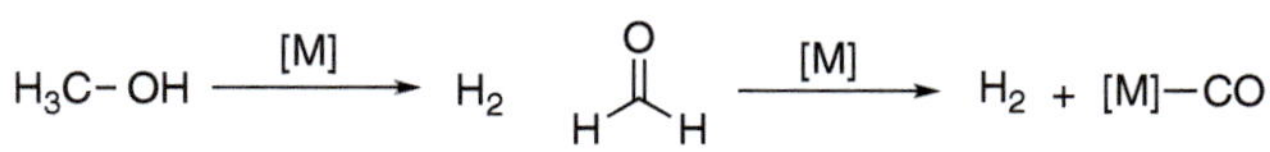

Scheme 4.28 Methanol decarbonylation.

In 1986, Keim and coworkers reported the use of methanol as a CO and H_2 source in ruthenium-catalyzed hydroesterfication reactions (Scheme 4.29) [876]. Due to the high energetic demand of the dehydrogenation of methanol, the reaction was temperature-dependent. Thus, at temperatures of 230–250 °C, methyl butanoate was obtained from propylene and methanol in 65% yield. In addition, high pressure (15–53 MPa propylene with nitrogen or carbon dioxide to ensure high pressure) was required.

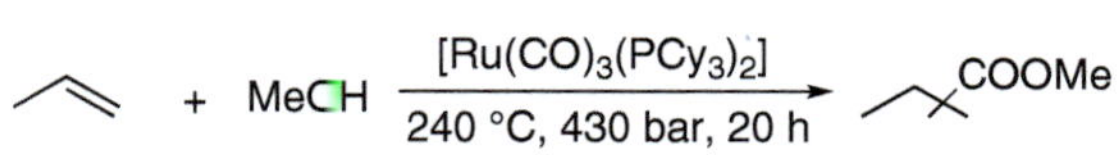

Scheme 4.29 Alkene hydroesterification reactions using methanol as CO and H_2.

In 2008, Jun and coworkers reported the synthesis of dialkyl ketones from methanol and alkenes (Scheme 4.30) [877]. Though the yields were not very high, the temperature (150 °C) was much lower than that reported in Keim's work (230–250 °C). Their catalytic system consisted of Rh(I), 2-amino-4-picoline, and benzoic acid.

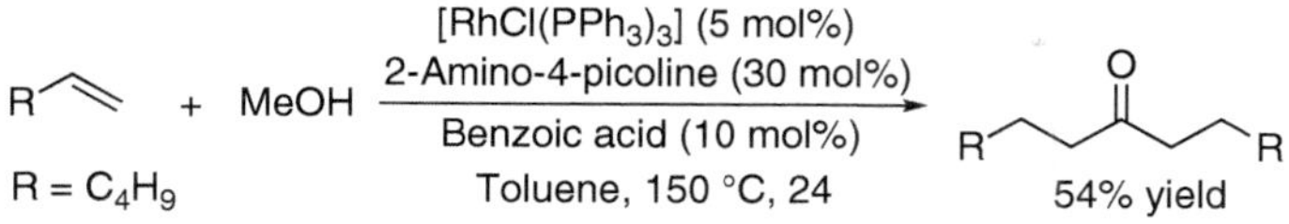

Scheme 4.30 Rh-catalyzed synthesis of dialkyl ketones from methanol and alkenes. Source: Modified from Jo et al. [877].

In the course of their development of hydrogen-mediated C–C coupling of alcohols with unsaturated reactants, Krische's group established an iridium-catalyzed reaction of allenes with methanol to furnish higher alcohols with quaternary centers under milder reaction conditions in 2011 (Scheme 4.31) [878].

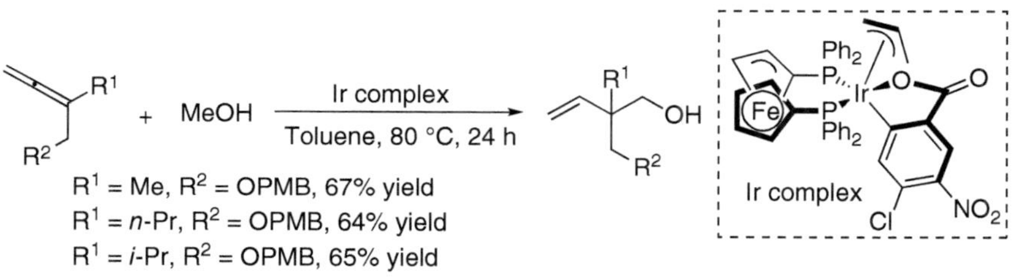

Scheme 4.31 Ir-catalyzed reactions of alkene with methanol. Source: Modified from Moran et al. [878].

In 2015, Beller and coworkers reported for the first time highly linear selective methoxycarbonylation reactions of alkenes by using paraformaldehyde and methanol as inexpensive and atom-efficient CO surrogates simultaneously (Scheme 4.32) [879]. Interestingly, this reaction proceeds through synergetic multiple catalytic cycles. This new procedure is applicable to a series of alkenes in the presence of a palladium catalyst under relatively mild conditions and is highly atom efficient and offers a convenient synthesis of various methyl esters.

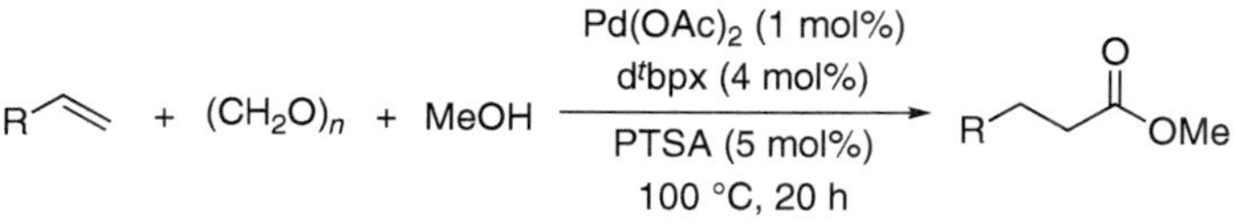

Scheme 4.32 Pd-catalyzed methoxycarbonylation of alkenes with paraformaldehyde and methanol. Source: Qiang Liu et al. [879].

4.8.2 Glycerol

Glycerol is formed as the major by-product (10 wt%) of biodiesel industry from the basic hydrolysis of triglycerides, which corresponds to the annual production of millions of tons of this polyol. Glycerol and other alcohols have been used for the generation of syngas (CO/H_2), which could be utilized in ensuing reductive formylation or hydroformylation reactions. Therefore, much effort has been devoted to investigating the use of this compound as a feedstock to other valuable compounds such as lactic acid [880, 881], acrolein [882, 883], dihydroxyacetone [884, 885], and others [886–888].

In 2017, Nielsen et al. reported on Ir-catalyzed decarbonylation of glycerol, which could be coupled to Pd-catalyzed alkoxycarbonylation of styrenes. A range of styrenes were converted to the corresponding methyl ester under the optimized conditions in good regioselectivity and up to 99% yield (Scheme 4.33) [889].

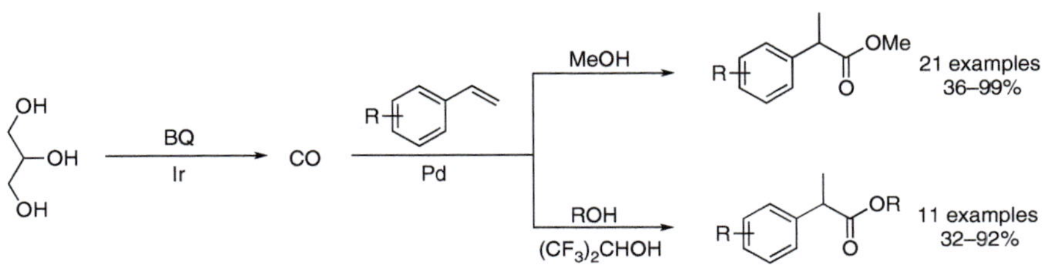

Scheme 4.33 Ir-catalyzed decarbonylation of glycerol. Source: Based on Nielsen et al. [889].

The formation of hydrogen could be avoided by employing BQ as an external oxidant to intercept the intermediary iridium hydride complex. Fortunately, the suitability of this methodology for the preparation of three nonsteroidal anti-inflammatory drugs (NSAIDs) (Scheme 4.34).

The conversion of biomass to value-added renewable products is a highly topical and promising research area. In this respect, polyols constitute interesting platform chemicals, which are readily available. Well-developed methods for transforming biomass to syngas include gasification at higher temperatures (>700 °C) and aqueous-phase reforming processes (Scheme 4.35) [890].

Still, the use of polyols as a syngas source in carbonylation reactions was not known until Andersson's group reported ex situ produced CO and hydrogen in hydroformylation using a dual-reactor setup in 2013 [891]. More specifically, iridium-catalyzed dehydrogenation–decarbonylation of polyols took place in reactor A and the rhodium-catalyzed hydroformylation of styrene occurred in reactor B. Relatively inexpensive C3–C6 polyols were used as the syngas sources. Yields from 46% to 83% were achieved with glycerol as the best syngas source (Scheme 4.36).

Glycerol is one of the main biomass-based alcohols available in industry, which can be applied to the synthesis of multifunctional carbonyl compounds as a CO surrogate. In 2016, Shi group reported the generation of highly valuable prochiral aminoketones and *N*-formamide via catalytically activated glycerol and primary and

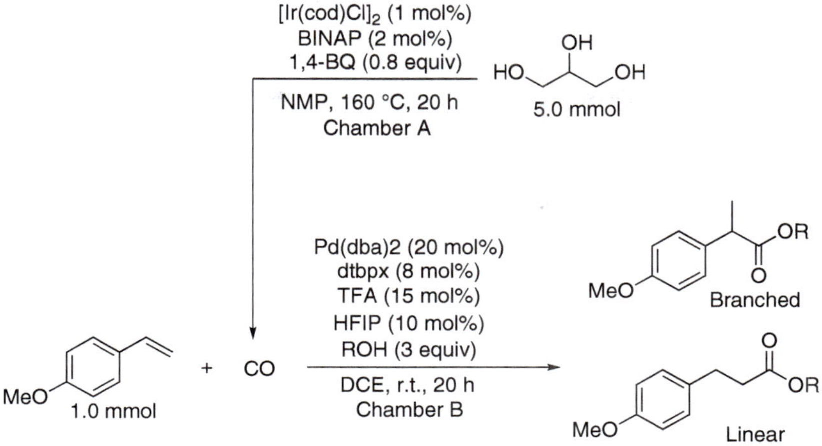

Scheme 4.34 Employing different alcohols as nucleophiles.

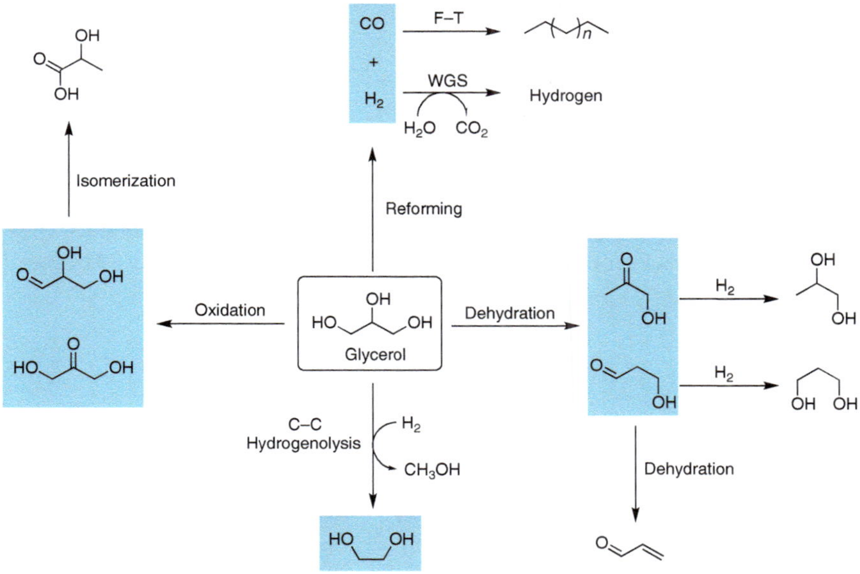

Scheme 4.35 Main routes for the aqueous-phase transformation of glycerol into fuels and chemicals. Source: Serrano-Ruiz et al. [890].

secondary amines over CuNiAlOx catalyst (Scheme 4.37). The incorporation of Cu and Ni into one catalyst was essential to realize these transformations [892].

Soon after, Shi group for the first time reported that the biomass-based 1, 3-dihydroxyacetone as a source of carbonyl group could be catalytically activated and directly reacted with amines and alcohols for the co-synthesis of glycolic acid, formamides, and formates in excellent yields at room temperature by Cu/Al_2O_3 catalyst (Scheme 4.38) [893]. The reactive copper is present by means of single active Cu^{II}

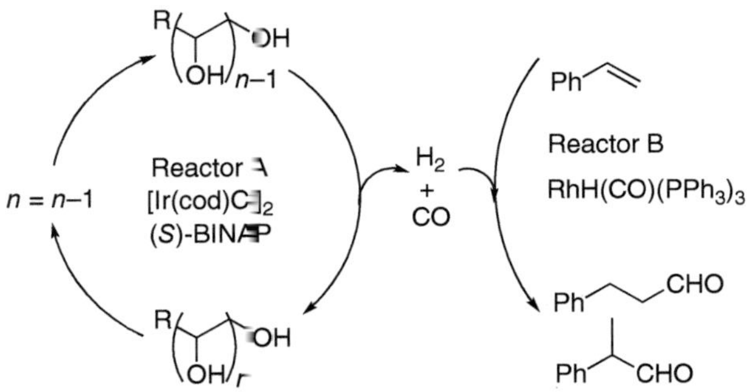

Scheme 4.36 Transfer of H_2 and CO from polyols to alkenes.

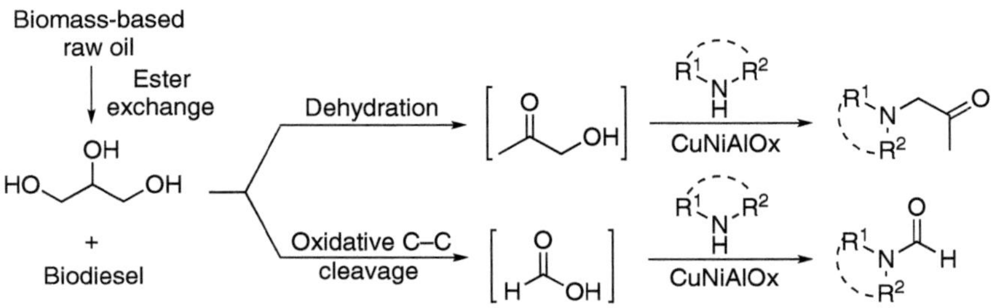

Scheme 4.37 One-pot synthesis of prochiral aminoketones and *N*-formamide with amine and glycerol.

species. By combining EPR spin-trapping and operando ATR-IR experiments, they thought •OH radicals formed from H_2O_2 via a Fenton-like reaction play a key role.

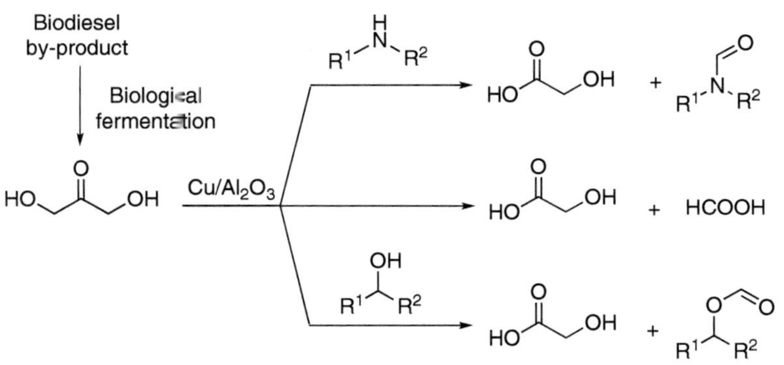

Scheme 4.38 One-pot synthesis of glycolic acid, formamides, and formats with 1,3-dihydroxyacetone. Source: Dai et al. [893].

4.8.3 Aldoses

Aldoses are widespread in nature, in the form of oligosaccharides, polysaccharides, glycoproteins, lipopolysaccharides, nucleotidyl sugars, and nucleic acids, and they

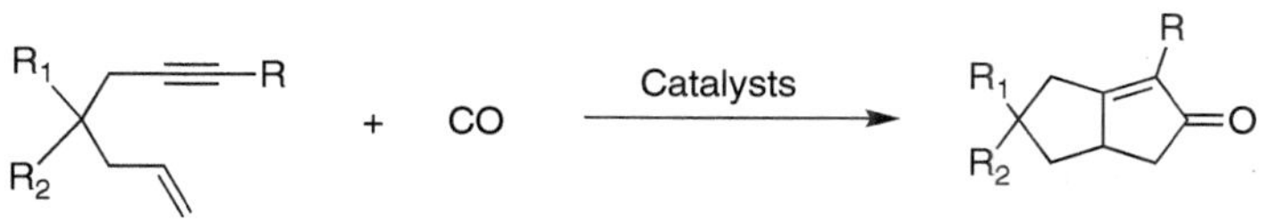

Scheme 4.39 Cyclocarbonylation of enynes with carbon monoxide. Source: Gibson and Stevenazzi [894]; Shibata [895].

are one of the most reliable and sustainable carbon resources. Therefore, aldoses could be regarded as one of the most favorable carbonyl sources.

The cyclocarbonylation of enynes with CO, so-called the PKR, is a powerful tool for the direct one-step synthesis of bicyclic cyclopentenones (Scheme 4.39) [894, 895].

Similarly, Morimoto group reported a new method for the cyclocarbonylation of variety of enynes in 22–67% yields using aldose derivatives in the presence of a rhodium(I) catalyst at 130 °C for 40 h [896]. In rhodium catalysis of this reaction, aldoses serve as CO source by donating their carbonyl moieties to enynes proved by ^{13}C labeling experiment (Scheme 4.40).

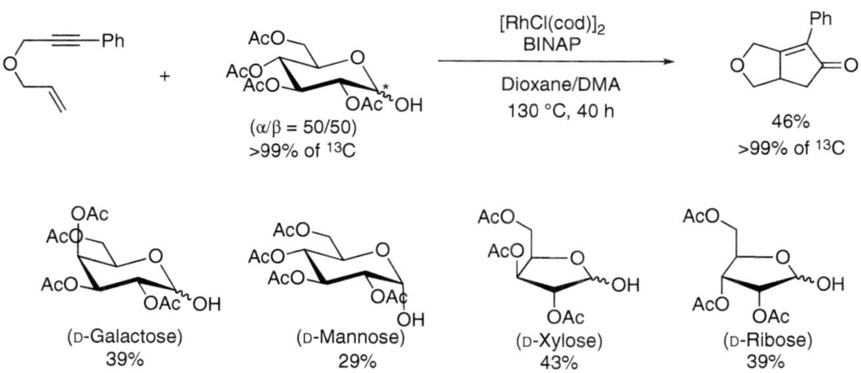

Scheme 4.40 Cyclocarbonylation of variety of enynes using aldose derivatives.

According to the results of studies, variety of aldoses, including D-glucose, D-mannose, D-galactose, D-xylose, and D-ribose, could be used as a carbonyl source and asymmetric variant also proceeded with moderate enantioselectivity.

Later, Chung et al. developed a cyclocarbonylation reaction of an enyne using just glucose as a carbonyl source, which is based on the same strategy as mentioned above.

4.8.4 Epoxide

Jun and coworkers reported that carbonylative esterification reaction with aryl bromides, oxiranes, and alcohols was catalyzed by Pd/C at 150 °C for 6 h in 35–88% yield [897]. In this process, oxiranes serve as carbonyl source by their conversion to corresponding aldehydes through a palladium-promoted Meinwald rearrangement (Scheme 4.41).

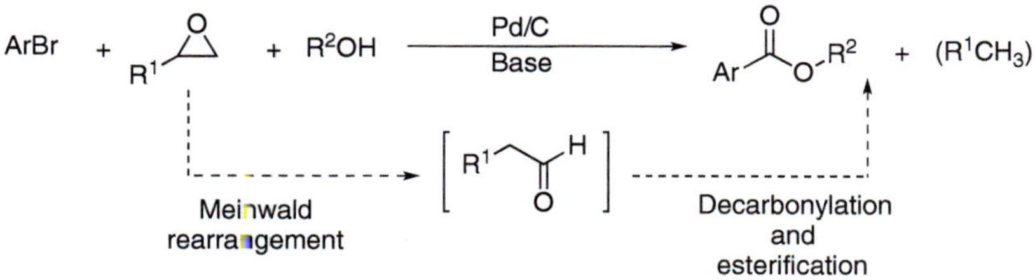

Scheme 4.41 Carbonylative esterification strategy employing Meinwald rearrangements of oxiranes.

Intramolecular versions of this process serve as methods for the synthesis of lactones and phthalimides. For example, the reaction of 2-bromobenzyl alcohol with 2-phenyloxirane in the presence of Pd/C and NaF takes place to generate isobenzofuran-1(3*H*)-one in 94% yield for 24 h (Scheme 4.42).

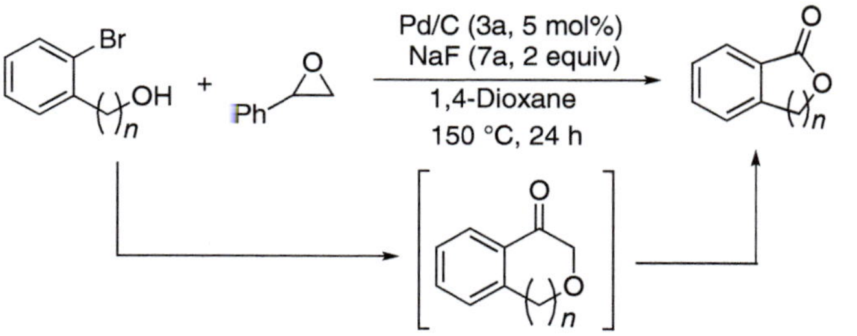

Scheme 4.42 Lactonization of various 2-bromoaryl alcohol.

Lactonization using styrene and *meta*-chloroperbenzoic acid (*m*CPBA) as a CO source was also conducted in 94% yield. During the reaction, 2-phenyloxirane might be generated in situ by the reaction of styrene with *m*CPBA (Scheme 4.43).

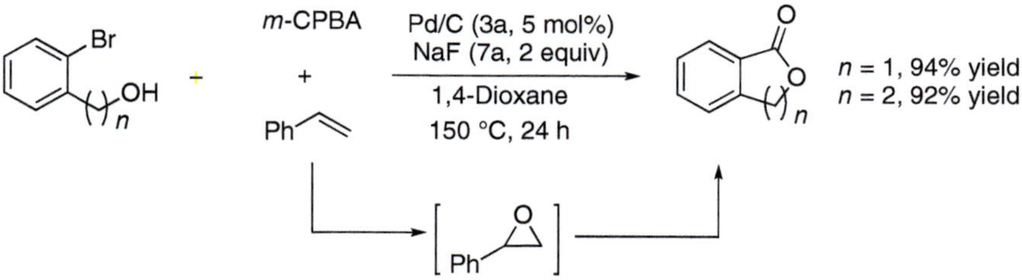

Scheme 4.43 Lactonization using styrene and *meta*-chloroperbenzoic acid (*m*CPBA).

Next, the intramolecular cyclization of 2-bromobenzamide with oxirane was observed to generate phthalimide in 68% isolated yield (Scheme 4.44).

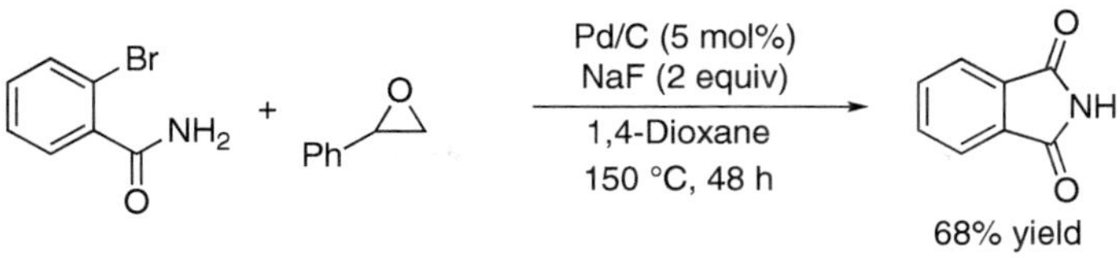

Scheme 4.44 Intramolecular cyclization of 2-bromobenzamide with oxirane.

4.8.5 Chloroform

Besides being a solvent, $CHCl_3$ is a widely available chemical in organic synthesis that undergoes hydrolysis under strongly basic aqueous hydroxide conditions to generate CO gas in situ [898]. The hydrolysis of chloroform is well known; however, the reaction is not considered to be economical as the yield of the released CO is poor and involves slower reaction rates due to the side reactions of the intermediate dichlorocarbene, such as disproportionation or cycloadditions to an unsaturated system, thus competing with the hydrolysis reaction [899–901]. But recent research revealed that the combination of chloroform surrogate with transition metal catalysts allows the easy installation of a carbonyl group into an organic molecule under mild conditions with high efficacy. Also, commercially available $^{13}CHCl_3$ and $^{14}CHCl_3$ can be effectively used for incorporating ^{13}CO and ^{14}CO in several organic moieties for isotopic labeling of the products and thus making chloroform. Our discussion is organized according to the different transition metals used in carbonylation reactions.

A general iron-catalyzed carbonylative Suzuki–Miyaura coupling of aryl halides or arylborons was reported by Han group utilizing stoichiometric $CHCl_3$ as the CO source to obtain a range of useful biaryl ketones in 65–91% yields at 120 °C with high selectivities (Scheme 4.45) [902].

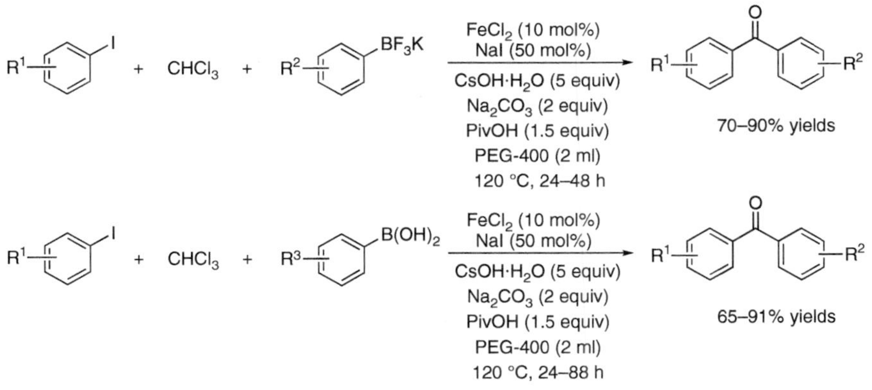

Scheme 4.45 Iron-catalyzed carbonylative Suzuki–Miyaura coupling of aryl halides or arylborons. Source: Zhao et al. [902].

The efficiency, economy, and operational simplicity of this catalytic system linked with the substrate generality make this method particularly attractive in organic synthesis. Mechanistic studies and the extension to other carbonylation reactions need further study.

4.8.5.1 Pd-catalyzed Carbonylation Reactions

The first carbonylation reaction using chloroform as the CO surrogate was realized by Alper and Grushin in 1993 [903]. Their reaction involved the carbonylation of aromatic, vinylic, and benzylic halides using tertiary phosphine–palladium complexes (1–5 mol%) as the catalytic system at ambient temperature under nitrogen atmosphere leading to the formation of the corresponding carboxylic acids up to 92% isolated yield (Scheme 4.46). The CO generated in situ from chloroform and aqueous alkali was unambiguously confirmed by isotopic labeling with $^{13}CHCl_3$ which proved that the source of the C1 unit in the carbonyl comes from chloroform. They successfully carbonylated several iodoarenes containing various substituents at the ortho, meta, and para positions of the ring with chloroform under biphasic conditions.

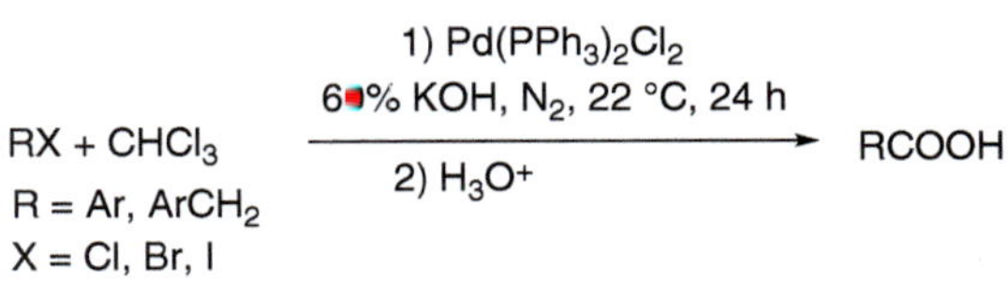

Scheme 4.46 Pd-catalyzed carbonylation of organic halides using $CHCl_3$ as CO source.

Later in 1999, Chen and Xia reported a simple and efficient method for the carbonylation of diaryliodonium salts using CO generated in situ from chloroform and aqueous alkali in the presence of $[(Ph_3P)_2PdCl_2]$ in good yields under the nitrogen atmosphere (Scheme 4.47) [904]. They productively utilized the high reactivity of organoiodine(III) compounds and the reaction succeeded smoothly within one hour. Several diaryliodonium salts bearing various substituents such as methyl, chloro, bromo, methoxy, and nitro groups were successfully carbonylated with chloroform. The reaction also worked well with unsymmetric iodonium salts which produced single products with the elimination of iodobenzene.

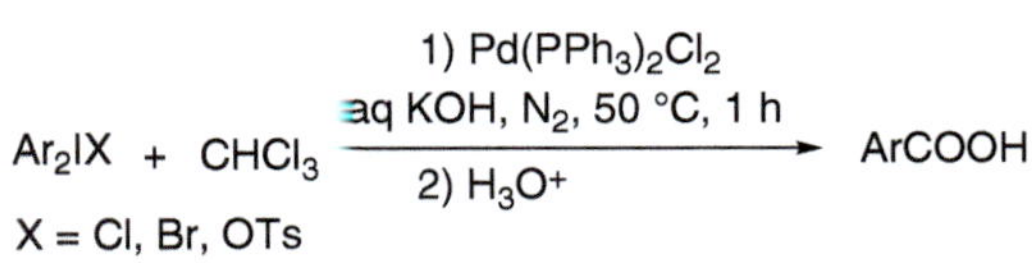

Scheme 4.47 Pd-catalyzed carbonylation of diaryliodonium salts by chloroform and alkali. Source: Modified from Xia and Chen [214].

In 2015, Gu and coworker employed chloroform for the in situ generation of CO to perform a Pd-catalyzed domino cyclization and carboxylation of neopentylic alkenes using "borrowed hydrogen" from Hecktype β-hydride elimination. Chloroform, when used as a CO source, provided better linear or branched selectivity as compared to CO gas (Scheme 4.48) [905]. This reaction shows chemoselectivity toward sterically hindered neopentylic C=C, whereas saturated C–C bonds and other less sterically hindered bonds remain unaffected without any additional reducing agent or organometallic reagent.

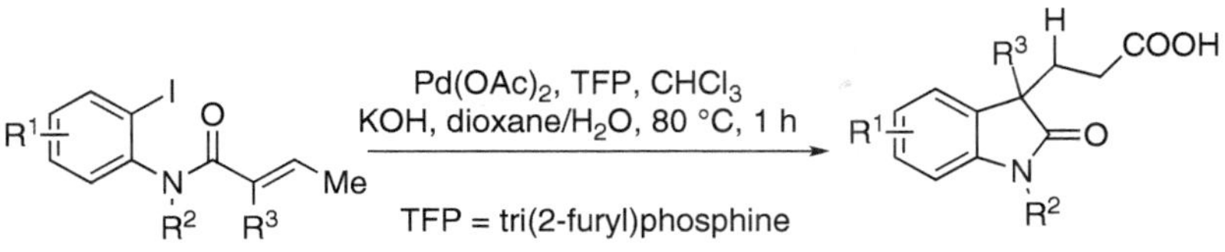

Scheme 4.48 Pd-catalyzed hydrocarboxylation reaction using chloroform as CO surrogate. Source: Modified from Liu and Gu [905].

In the same year, Hull and coworker for the first time developed an efficient methodology for the aminocarbonylation of aryl, vinyl, and benzyl halides using chloroform as the carbonyl source [906]. They effectively utilized a heterogeneous base system, $CsOH{\cdot}H_2O$, for the in situ generation of CO, and outcompeted the noncarbonylative couplings with excellent functional group tolerance and amine scope (Scheme 4.49). Further, they extended the space of the reaction to the synthesis of several pharmaceuticals along with an isotopically ^{13}C-labeled amide.

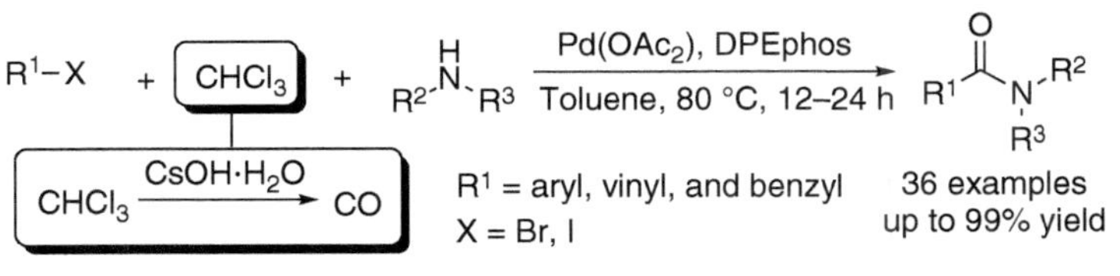

Scheme 4.49 Aminocarbonylation of amines with iodobenzene.

The practice of carbonylation reaction with chloroform as CO alternate was further utilized by Hu and coworkers in 2016 for the generation of alkynones from aryl iodides and terminal alkyne via carbonylative Sonogashira coupling [907]. The alkynones, an integral intermediate in many heterocycles synthesis, were synthesized under mild reaction conditions with a short reaction time, high yields and were compatible with various functional groups (Scheme 4.50). In addition, they explored the generality of the method by modifying natural products and the corresponding carbonylated products were obtained in good yields compared to the existing methods.

In 2017, Yuan and coworkers described the first palladium-catalyzed aroylation of NH–sulfoximes for the efficient synthesis of biologically important *N*-aroylsulfoximes from aryl halides and chloroform [908]. This transformation involved mild reaction conditions and employed KOH as the base to generate CO from chloroform instead of expensive and hygroscopic $CsOH{\cdot}H_2O$. In contrast to 1,1′-binaphthyl-2.2′-diphenyl phosphine and other phosphine ligands, **1**,

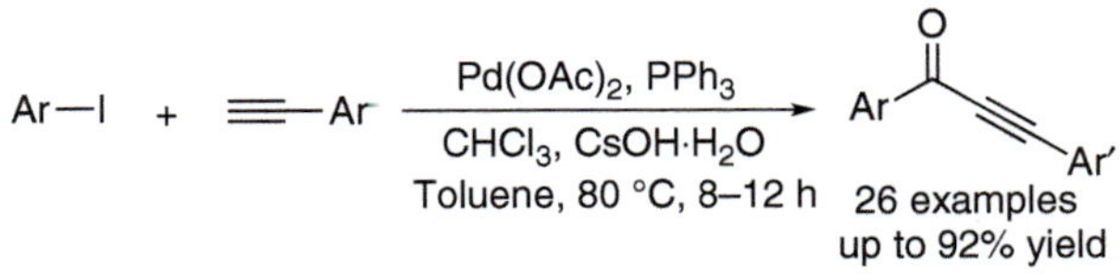

Scheme 4.50 Reaction of aryliodides with ethynyl benzene.

8-diazabicyclo[5.4.0]undec-7-ene and 4-dimethylaminopyridine (DMAP) formed the desired products in good yields. A wide range of sulfoximes underwent reaction with various aryl halides to give the corresponding products in good yields (Scheme 4.51).

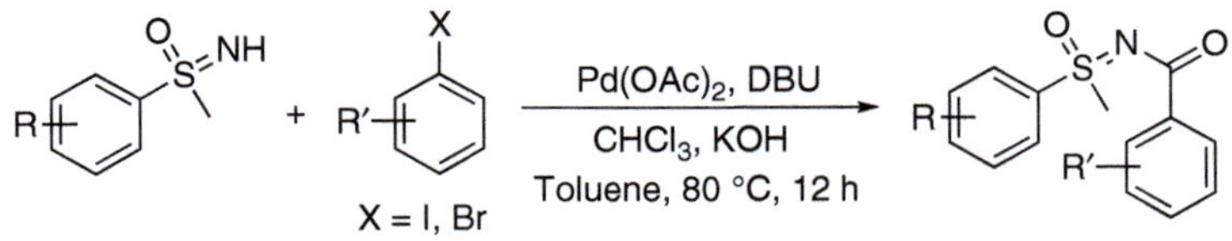

Scheme 4.51 Pd-catalyzed N-arylation of sulfoximines with aryl halides.

Meanwhile, Das and coworkers developed a direct aminocarbonylation reaction of 7-azaindoles because of the interesting biological property exhibited by this class of compounds (Scheme 4.52) [909]. They utilized chloroform as the carbonyl surrogate for introducing amide group over halo-substituted 7-azaindole and thereafter extended the reaction with other heteroarenes, such as pyrazolopyridines and indazoles. Also, through controlled experiments, they showed that phosphine ligand is beneficial for the smooth reaction and the yield of the targeted product reduces drastically in the absence of phosphine ligands. Through this reaction, they have realized a new route to introduce amide group on many pharmaceutically important heterocycles to produce medicinally important scaffolds in good yields.

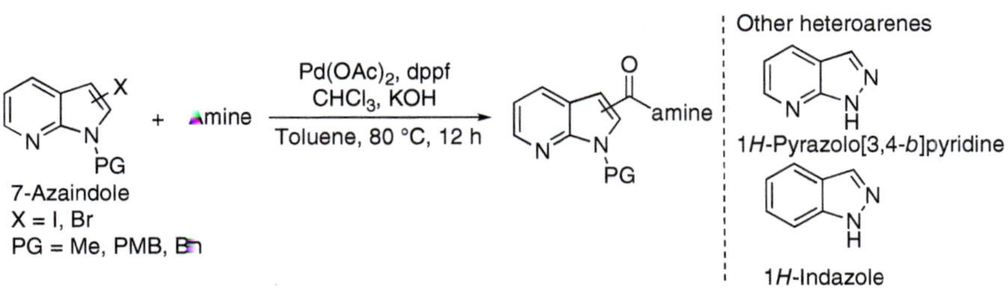

Scheme 4.52 C-3 aminocarbonylation of halo-substituted 7-azaindoles using chloroform as CO source. Source: Modified from Kannaboina et al. [909].

In 2018, Das and coworkers once again exploited chloroform as a carbonyl source for the palladium-catalyzed carbonylative preparation of biologically relevant 2-amidoimidazopyridine scaffolds in good yields from different amines and substituted imidazopyridines (Scheme 4.53) [910]. This amino-carbonylation reaction was found to be general with a wide range of amines and substituted imidazopyridines.

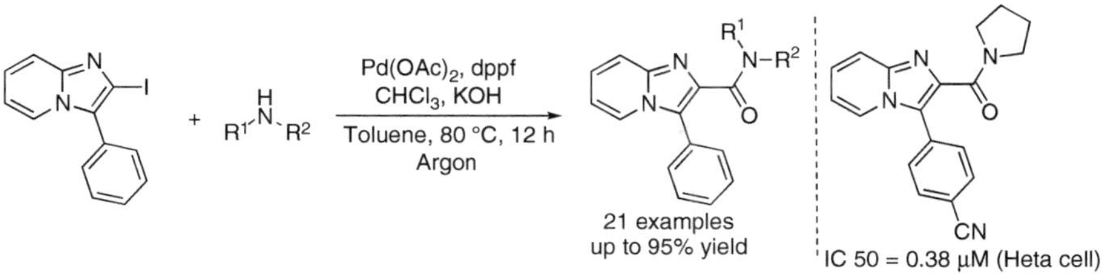

Scheme 4.53 Carbonylation of imidazopyridines. Source: Modified from Nitha et al. [910].

Recently, Wang and coworkers utilized chloroform as CO surrogate in water for Pd/C-catalyzed domino-type reaction to synthesize urea derivatives. Both symmetrical and unsymmetrical urea were prepared in moderate to good yields from aryl iodides, sodium azide, amines and in situ generated CO (Scheme 4.54) [368]. Pd/C catalyst proved to be environment-friendly because of its safe handling, easy removal, reuse, and recovery even after six consecutive runs with little decrease in catalytic activity. Chloroform was employed as a green source of CO for the in situ CO generation in the presence of KOH as compared to CsOH·H_2O because of its low cost. Tocofersolan (TGPS)/H_2O was used as solvent as it can be easily recycled and reused, thus eliminating the need of and organic co-solvent.

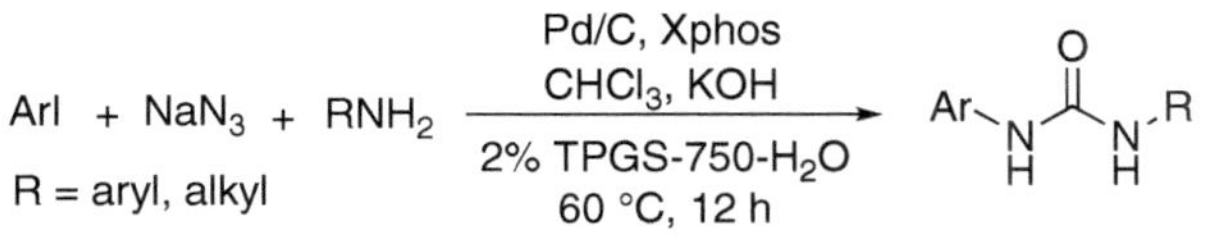

Scheme 4.54 Pd/C-catalyzed domino-type reaction for synthesis of urea. Source: Modified from Wang et al. [368].

4.8.5.2 Fe-Catalyzed Carbonylation Reactions

Investigations on first-row transition metals reveal that iron can be used as a suitable candidate for the aforesaid catalysis because it is a widely available, eco-friendly and the second most abundant element in the earth crust [863, 911–913]. Despite these advantages, the iron-catalyzed carbonylative coupling of aryl halides using CO surrogate remained undiscovered until the report of Han and coworkers in 2016 [902]. In their work, they developed an iron-catalyzed carbonylation of aryl halides with arylborons via Suzuki–Miyaura coupling using stoichiometric ratio of chloroform as CO surrogate. From their previous experiences, they commenced with a novel and efficient $FeCl_2$-catalyzed catalytic system using chloroform with cesium hydroxide as CO source, Na_2CO_3 as base, and PivOH as an additive in PEG-400. They effectively employed this general methodology for the synthesis of biaryl ketones with high selectivity (Scheme 4.55).

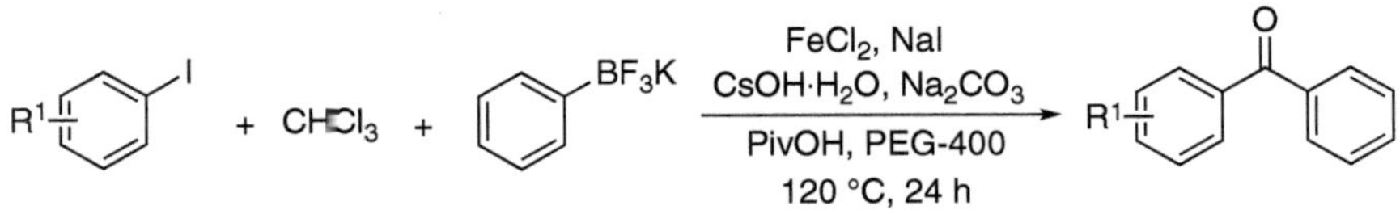

Scheme 4.55 Fe-catalyzed carbonylative Suzuki–Miyaura reaction of aryl halides with arylborons.

4.8.5.3 Zn-Catalyzed Carbonylation Reactions

In 2018, Ueda and coworkers reported a novel Friedel–Crafts type acylative coupling of amines with indoles using chloroform as a phosgene precursor. The dimethylzinc–O_2 mediated three-component coupling of indole, amine, and in situ generated phosgene from chloroform resulted in the formation of indolecarboxamides, whereas the intramolecular version of the same led to the formation of dihydroisoquinolinones [914]. Though phosgene is toxic and inconvenient to handle, it was generated in situ and undergoes immediate reaction with nucleophiles thus maintaining low stationary concentration, which provides a safer pathway. They established the feasibility of the reaction with various amine substrates. Cyclic amines such as tetrahydroisoquinoline and piperidine furnished the corresponding products in good yields, while a lower yield was obtained with morpholine. Furthermore, the primary amines did not yield the desired product while simple dialkylamines such as diisopropylamine, diethylamine, etc. abided the desired product in good yields (Scheme 4.56).

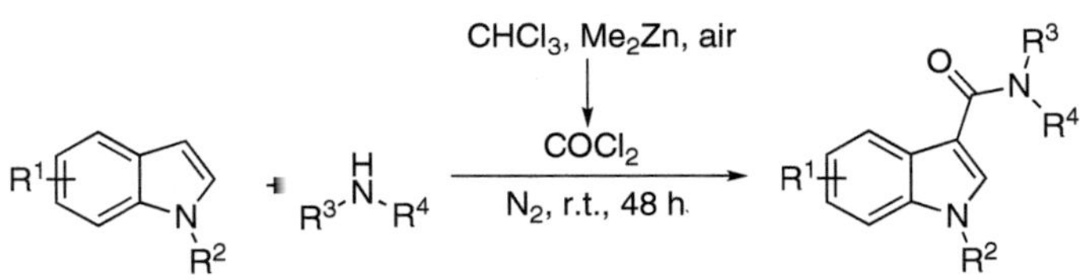

Scheme 4.56 Reaction of various amines with Me_2Zn and $CHCl_3$.

Part I References

1 Franke, R., Selent, D., and Boerner, A. (2012). *Chem. Rev.* 112: 5675–5732.
2 Roelen, O. (1938). DE Patent 849,548.
3 Roelen, O. (1943). US Patent 2,327,066.
4 Li, S., Li, Z., You, C. et al. (2019). *Chin. J. Org. Chem.* 39: 1568–1582.
5 Heck, R.F. and Breslow, D.S. (1961). *J. Am. Chem. Soc.* 83: 4023–4027.
6 Beller, M. (ed.) (2006). *Catalytic Carbonylation Reactions*, vol. 18. Springer.
7 Roelen, O. (1948). *Angew. Chem.* 60: 62.
8 Aldridge, C.L., Fasce, E.V., and Jonassen, H.B. (1958). *J. Phys. Chem.* 62: 869–870.
9 Cornils, B., Herrmann, W.A., and Rasch, M. (1994). *Angew. Chem., Int. Ed. Engl.* 33: 2144–2163.
10 Tucci, E.R. (1968). *Ind. Eng. Chem. Prod. Res. Dev.* 7: 32–38.
11 Hood, D.M., Johnson, R.A., Carpenter, A.E. et al. (2020). *Science* 367: 542–548.
12 Hartwig, J.F. (2010). *Organotransition Metal Chemistry: from Bonding to Catalysis*. Univ Science Books.
13 Davis, M.E., Rode, E., Taylor, D., and Hanson, B.E. (1984). *J. Catal.* 86: 67–74.
14 Cornils, B. and Herrmann, W.A. (2004). *Aqueous-phase Organometallic Catalysis: Concepts and Applications*. Wiley.
15 Herrmann, W.A., Kohlpaintner, C.W., Bahrmann, H., and Konkol, W. (1992). *J. Mol. Catal.* 73: 191–201.
16 Ajjou, A.N. and Alper, H. (1998). *J. Am. Chem. Soc.* 120: 1466–1468.
17 Arhancet, J.P., Davis, M.E., Merola, J.S., and Hanson, B.E. (1989). *Nature* 339: 454–455.
18 Arya, P., Rao, N.V., Singkhonrat, J. et al. (2000). *J. Org. Chem.* 65: 1881–1885.
19 Mukhopadhyay, K., Mandale, A.B., and Chaudhari, R.V. (2003). *Chem. Mater.* 15: 1766–1777.
20 Liu, S., Dai, X., Wang, H. et al. (2020). *Chin. J. Chem.* 38: 139–143.
21 Fang, W. and Breit, B. (2018). *Angew. Chem., Int. Ed.* 57: 14817–14821.
22 Liu, X., Haruta, M., and Tokunaga, M. (2008). *Chem. Lett.* 37: 1290–1291.
23 Han, D., Li, X., Zhang, H. et al. (2006). *J. Catal.* 243: 318–328.
24 Damoense, L., Datt, M., Green, M., and Steenkamp, C. (2004). *Coord. Chem. Rev.* 248: 2393–2407.

Carbonyl Compounds: Reactants, Catalysts and Products, First Edition.
Feng Shi, Hongli Wang and Xingchao Dai.

25 Kamer, P.C.J. van Rooy, A., Schoemaker, G.C., and van Leeuwen, P. (2004). *Coord. Chem. Rev.* 248: 2409–2424.

26 Sudheesh, N. Chaturvedi, A.K., and Shukla, R.S. (2011). *Appl. Catal. A: Chem.* 409: 99–105.

27 Gao, L., Meng, K., Jiang, W. et al. (2018). *Chem. Ind. Eng. Prog.* 37: 1433–1441.

28 Wang, T., Wang, W., Lyu, Y. et al. (2017). *Chin. J. Catal.* 38: 691–698.

29 Tan, M., Ding, W., Ai, P. et al. (2016). *Appl. Catal. A: Chem.* 527: 53–59.

30 Sun, Q., Jiang, M., Shen, Z. et al. (2014). *Chem. Commun.* 50: 11844–11847.

31 Sun, Q., Dai, Z., Liu, X. et al. (2015). *J. Am. Chem. Soc.* 137: 5204–5209.

32 Wang, Y., Yan, L., Li, C. et al. (2018). *Appl. Catal. A: Chem.* 551: 98–105.

33 Li, C., Yan, L., Lu, L. et al. (2016). *Green Chem.* 18: 2995–3005.

34 Tan, M., Ishikuro, Y., Hosoi, Y. et al. (2017). *Chem. Eng. J.* 330: 863–869.

35 Tao, L., Zhong, M., Chen, J. et al. (2018). *Green Chem.* 20: 188–196.

36 Liang, Z., Chen, J., Chen, X. et al. (2019). *Chem. Commun.* 55: 13721–13724.

37 Lang, R., Li T., Matsumura, D. et al. (2016). *Angew. Chem., Int. Ed.* 55: 16054–16058.

38 Wang, L., Zhang, W., Wang, S. et al. (2016). *Nat. Commun.* 7: 14036.

39 Von Kutepow, N.; Bittler, K.; Neubauer, D. (1969). US Patent 3437676.

40 Fenton, D.M. (1973). *J. Org. Chem.* 38: 3192–3198.

41 Alper, H., Woell, J.B., Despeyroux, B., and Smith, D.J.H. (1983). *J. Chem. Soc., Chem. Commun.*: 1270–1271.

42 Alper, H. and Hamel, N. (1990). *J. Am. Chem. Soc.* 112: 2803–2804.

43 Stahly, G. and Starrett, R. (1997). *Chirality in Industry II*. New York: Wiley.

44 Papadogianakis, G., Verspui, G., Maat, L., and Sheldon, R.A. (1997). *Catal. Lett.* 47: 43–46.

45 Verspui, G., Feiken, J., Papadogianakis, G., and Sheldon, R.A. (1999). *J. Mol. Catal. A: Chem.* 146: 299–307.

46 Monflier, E., Tilloy, S., Bertoux, F. et al. (1997). *New J. Chem.* 21: 857–859.

47 Bertoux, F., Monflier, E., Castanet, Y., and Mortreux, A. (1999). *J. Mol. Catal. A: Chem.* 143: 11–22.

48 Goedheijt, M.S., Reek, J.N.H., Kamer, P.C.J., and van Leeuwen, P. (1998). *Chem. Commun.*: 2431–2432.

49 Elali, B. and Alper, H. (1992). *J. Mol. Catal.* 77: 7–13.

50 Sang, R., Kucmierczyk, P., Duehren, R. et al. (2019). *Angew. Chem. Int. Ed.* 58: 14365–14373.

51 Cornils, B. and Herrmann, W.A. (2002). *Applied Homogeneous Catalysis with Organometallic Compounds: Applications; Volume 2. Developments; Volume 3. Developments*. Wiley-VCH.

52 Weissermel, K. and Arpe, H.J. (2008). *Industrial Organic Chemistry*, 4e. Wiley.

53 Amer, I. and Alper, H. (1990). *J. Organomet. Chem.* 383: 573–577.

54 Lee, J.T. and Alper, H. (1991). *Tetrahedron Lett.* 32: 1769–1770.

55 Satyanarayana, N., Alper, H., and Amer, I. (1990). *Organometallics* 9: 284–286.

56 Satyanarayana, N. and Alper, H. (1991). *Organometallics* 10: 804–807.

57 Gabriele, B., Salerno, G., Costa, M., and Chiusoli, G.P. (1999). *Chem. Commun.*: 1381–1382.

58 Gabriele, B., Salerno, G., Costa, M., and Chiusoli, G.P. (1999). *Tetrahedron Lett.* 40: 989–990.

59 Yang, D., Liu, H., Liu, L. et al. (2019). *Green Chem.* 21: 5336–5344.

60 Heck, R.F. (1969). *J. Am. Chem. Soc.* 91: 6707–6714.

61 James, D.E. and Stille, J.K. (1976). *J. Am. Chem. Soc.* 98: 1810–1823.

62 Knifton, J.F. (1976). *J. Org. Chem.* 41: 2885–2890.

63 Knifton, J.F. (1978). *J. Am. Oil Chem. Soc.* 55: 496–500.

64 Semmelhack, M.F. and Bodurow, C. (1984). *J. Am. Chem. Soc.* 106: 1496–1498.

65 Orejon, A. and Alper, H. (1999). *J. Mol. Catal. A: Chem.* 143: 137–142.

66 Maffei, A., Mele, G., Ciccarella, G. et al. (2002). *Appl. Organomet. Chem.* 16: 543–546.

67 Touzani, R. and Alper, H. (2005). *J. Mol. Catal. A: Chem.* 227: 197–207.

68 Dong, K., Fang, X., Gülak, S. et al. (2017). *Nat. Commun.* 8: 1–7.

69 Tsuji, J., Takahashi, M., and Takahashi, T. (1980). *Tetrahedron Lett.* 21: 849–850.

70 Vasilevsky, S.F., Trofimov, B.A., Malkina, A.G., and Brandsma, L. (1994). *Synth. Commun.* 24: 85–88.

71 Okuro, K. and Alper, H. (1997). *J. Org. Chem.* 62: 1566–1567.

72 Kadnikov, D.V. and Larock, R.C. (2000). *Org. Lett.* 2: 3643–3646.

73 Kadnikov, D.V. and Larock, R.C. (2003). *J. Org. Chem.* 68: 9423–9432.

74 Kadnikov, D.V. and Larock, R.C. (2003). *J. Organomet. Chem.* 687: 425–435.

75 Bonardi, A., Costa, M., Gabriele, B. et al. (1994). *J. Chem. Soc., Chem. Commun.*: 2429–2430.

76 Gabriele, B., Salerno, G., DePascali, F. et al. (1997). *J. Chem. Soc., Perkin Trans.* 1: 147.

77 Tamaru, Y., Hojo, M., and Yoshida, Z. (1991). *J. Org. Chem.* 56: 1099–1105.

78 Yang, D., Liu, L., Wang, D.-L. et al. (2019). *J. Catal.* 371: 236–244.

79 Shinohara, T., Arai, M.A., Wakita, K. et al. (2003). *Tetrahedron Lett.* 44: 711–714.

80 Fang, X., Jackstell, R., and Beller, M. (2013). *Angew. Chem., Int. Ed.* 52: 14089–14093.

81 Liu, H., Yan, N., and Dyson, P.J. (2014). *Chem. Commun.* 50: 7848–7851.

82 Cheng, J., Qi, X., Li, M. et al. (2015). *J. Am. Chem. Soc.* 137: 2480–2483.

83 Zhang, G., Gao, B., and Huang, H. (2015). *Angew. Chem., Int. Ed.* 54: 7657–7661.

84 Liu, S., Wang, H., Dai, X., and Shi, F. (2018). *Green Chem.* 20: 3457–3462.

85 Reppe, W. (1953). *Liebigs Ann. Chem.* 582: 1–37.

86 Li, Y., Alper, H., and Yu, Z. (2006). *Org. Lett.* 8: 5199–5201.

87 Suleiman, R., Tijani, J., and El Ali, B. (2010). *Appl. Organomet. Chem.* 24: 38–46.

88 Suleiman, R., Ibdah, A., and El Ali, B. (2011). *J. Organomet. Chem.* 696: 2355–2363.

89 Driller, K.M., Prateeptongkum, S., Jackstell, R., and Beller, M. (2011). *Angew. Chem., Int. Ed.* 50: 537–541.

90 Sha, F. and Alper, H. (2017). *ACS Catal.* 7: 2220–2229.

91 Huang, Z., Dong, Y., Li, Y. et al. (2019). *Chemcatchem* 11: 5236–5240.

92 Schore, N.E. (1988). *Chem. Rev.* 88: 1081–1119.

93 Geis, O. and Schmalz, H.G. (1998). *Angew. Chem., Int. Ed.* 37: 911–914.

94 Khand, I.U., Knox, G.R., Pauson, P.L. et al. (1973). *J. Chem. Soc., Perkin Trans. 1*: 977–981.

95 Rautenstrauch, V., Megard, P., Conesa, J., and Kuster, W. (1990). *Angew. Chem., Int. Ed.* 29: 1413–1416.

96 Jeong, N., Hwang, S.H., Lee, Y.S., and Chung, Y.K. (1994). *J. Am. Chem. Soc.* 116: 3159–3160.

97 Sugihara, T. and Yamaguchi, M. (1998). *J. Am. Chem. Soc.* 120: 10782–10783.

98 Lee, N.Y. and Chung, Y.K. (1996). *Tetrahedron Lett.* 37: 3145–3148.

99 Rajesh, T. and Periasamy, M. (1999). *Tetrahedron Lett.* 40: 817–818.

100 Jeong, N., Hwang, S.H., Lee, Y.W., and Lim, J.S. (1997). *J. Am. Chem. Soc.* 119: 10549–10550.

101 Hayashi, M., Hashimoto, Y., Yamamoto, Y. et al. (2000). *Angew. Chem., Int. Ed.* 39: 631–633.

102 Son, S.U., Lee, S.I., and Chung, Y.K. (2000). *Angew. Chem., Int. Ed.* 39: 4158–4160.

103 Hicks, F.A., Kablaoui, N.M., and Buchwald, S.L. (1996). *J. Am. Chem. Soc.* 118: 9450–9451.

104 Kondo, T., Suzuki, N., Okada, T., and Mitsudo, T. (1997). *J. Am. Chem. Soc.* 119: 6187–6188.

105 Morimoto, T., Chatani, N., Fukumoto, Y., and Murai, S. (1997). *J. Org. Chem.* 62: 3762–3765.

106 Koga, Y., Kobayashi, T., and Narasaka, K. (1998). *Chem. Lett.*: 249–250.

107 Jeong, N., Lee, S., and Sung, B.K. (1998). *Organometallics* 17: 3642–3644.

108 Krafft, M.E. (1988). *J. Am. Chem. Soc.* 110: 968–970.

109 Krafft, M.E., Juliano, C.A., Scott, I.L. et al. (1991). *J. Am. Chem. Soc.* 113: 1693–1703.

110 Krafft, M.E. and Juliano, C.A. (1992). *J. Org. Chem.* 57: 5106–5115.

111 Itami, K., Mitsudo, K., and Yoshida, J. (2002). *Angew. Chem. Int. Ed.* 41: 3481–3484

112 Itami, K., Mitsudo, K., Fujita, K. et al. (2004). *J. Am. Chem. Soc.* 126: 11058–11066.

113 Castro, J., Moyano, A., Pericas, M.A. et al. (1996). *J. Org. Chem.* 61: 9016–9020.

114 Jeong, N., Yoo, S.E., Lee, S.J. et al. (1991). *Tetrahedron Lett.* 32: 2137–2140.

115 Castro, J., Sorensen, H., Riera, A. et al. (1990). *J. Am. Chem. Soc.* 112: 9388–9389.

116 Jamison, T.F., Shambayati, S., Crowe, W.E., and Schreiber, S.L. (1994). *J. Am. Chem. Soc.* 116: 5505–5506.

117 Kerr, W.J., Lindsay, D.M., Rankin, E.M. et al. (2000). *Tetrahedron Lett.* 41: 3229–3233.

118 Derdau, V., Laschat, S., and Jones, P.G. (1998). *Heterocycles* 48: 1445–1453.

119 Tormo, J., Moyano, A., Pericas, M.A., and Riera, A. (1997). *J. Org. Chem.* 62: 4851–4856.

120 Hicks, F.A. and Buchwald, S.L. (1996). *J. Am. Chem. Soc.* 118: 11688–11689.
121 Hicks, F.A. and Buchwald, S.L. (1999). *J. Am. Chem. Soc.* 121: 7026–7033.
122 Jeong, N., Sung, B.K., and Choi, Y.K. (2000). *J. Am. Chem. Soc.* 122: 6771–6772.
123 Shibata, T. and Takagi, K. (2000). *J. Am. Chem. Soc.* 122: 9852–9853.
124 Sturla, S.J. and Buchwald, S.L. (2002). *J. Org. Chem.* 67: 3398–3403.
125 Lledo, A., Fuster, A., Reves, M. et al. (2013). *Chem. Commun.* 49: 3055–3057.
126 Mottalib, M.A., Haukka, M., and Nordlander, E. (2016). *Polyhedron* 103: 275–282.
127 Morimoto, T., Fuji, K., Tsutsumi, K., and Kakiuchi, K. (2002). *J. Am. Chem. Soc.* 124: 3806–3807.
128 Shibata, T., Toshida, N., and Takagi, K. (2002). *J. Org. Chem.* 67: 7446–7450.
129 Fuji, K., Morimoto, T., Tsutsumi, K., and Kakiuchi, K. (2003). *Angew. Chem., Int. Ed.* 42: 2409–2411.
130 Lang, X.-D., You, F., He, X. et al. (2019). *Green Chem.* 21: 509–514.
131 Yoneda, N., Kusano, S., Yasui, M. et al. (2001). *Appl. Catal., A: Chem.* 221: 253–265.
132 Von Kutepow, N., Himmele, W., and Hohenschutz, H. (1965). *Chem. Ing. Tech.* 37: 383–388.
133 Hohenschurz, H. and N, v. K.; Himmele, W. (1966). *Hydrocarbon Process* 45: 141.
134 Paulik, F. and Roth, J. (1968). *Chem. Commun.*: 1578.
135 Howard, M., Jones, M., Roberts, M., and Taylor, S. (1993). *Catal. Today* 18: 325–354.
136 Schultz, R.G. (1969). *J. Catal.* 13: 105–106.
137 Maneck, H.-E., Gutschick, D., Burkhardt, I. et al. (1988). *Catal. Today* 3: 421–429.
138 Nam, J.S., Kim, A.R., Kim, D.M. et al. (2017). *Catal. Commun.* 99: 141–145.
139 Qi, J. and Christopher, P. (2019). *Ind. Eng. Chem. Res.* 58: 12632–12641.
140 Ren, Z., Liu, Y., Lyu, Y. et al. (2019). *J. Catal.* 369: 249–256.
141 Forster, D. (1976). *J. Am. Chem. Soc.* 98: 846–848.
142 Forster, D. (1979). *Adv. Organomet. Chem.* 17: 255–267.
143 Dekleva, T.W. and Forster, D. (1986). *Adv. Catal.* 34: 81–130.
144 Torrence, P. (2002). Carbonylations: synthesis of acetic acid and acetic acid anhydride from methanol. In: *Applied Homogeneous Catalysis with Organometallic Compounds*, 2e, vol. 1 (eds. B. Cornils and W.A. Herrmann), 104. Weinheim: Wiley.
145 Thomas, C.M. and Süss-Fink, G. (2003). *Coord. Chem. Rev.* 243: 125–142.
146 Forster, D. (1977). *Ann. N.Y. Acad. Sci.* 295: 79–82.
147 Fulford, A., Hickey, C.E., and Maitlis, P.M. (1990). *J. Organomet. Chem.* 398: 311–323.
148 Coux, O., Tanaka, K., and Goldberg, A.L. (1996). *Annu. Rev. Biochem* 65: 801–847.
149 Jones, J.H. (2000). *Platinum Met. Rev.* 44: 94–105.
150 Howard, M.J., Sunley, G.J., Poole, A.D. et al. (1999). *Stud. Surf. Sci. Catal.* 121: 61–68.

151 Gautron, S., Lassauque, N., Le Berre, C. et al. (2006). *Top. Catal.* 40: 83–90.
152 Feng, S., Lin, X., Song, X. et al. (2019). *J. Catal.* 377: 400–408.
153 Forster, D. (1979). *J. Chem. Soc., Dalton Trans.*: 1639–1645.
154 Matsumoto, T., Mizoroki, T., and Ozaki, A. (1978). *J. Catal.* 51: 96–100.
155 Haynes, A., Maitlis, P.M., Morris, G.E. et al. (2004). *J. Am. Chem. Soc.* 126: 2847–2861.
156 Ellis, P.R., Pearson, J.M., Haynes, A. et al. (1994). *Organometallics* 13: 3215–3226.
157 Bassetti, M., Monti, D., Haynes, A. et al. (1992). *Gazz. Chim. Ital.* 122: 391–393.
158 Brennführer, A., Neumann, H., and Beller, M. (2009). *Angew. Chem., Int. Ed.* 48: 4114–4133.
159 Beller, M. and Wu, X.-F. (2013). *Transition Metal Catalyzed Carbonylation Reactions*. Springer.
160 Bumagin, N., Nikitin, K., and Beletskaya, I. (1988). *J. Organomet. Chem.* 358: 563–565.
161 Uozumi, Y. and Watanabe, T. (1999). *J. Org. Chem.* 64: 6921–6923.
162 Wittmann, S., Schätz, A., Grass, R.N. et al. (2010). *Angew. Chem., Int. Ed.* 49: 1867–1870.
163 Schoenbe.A; Bartolet.I; Heck, R. F. (1974). *J. Org. Chem.* 39: 3318–3326.
164 Magerlein, W., Beller, M., and Indolese, A.F. (2000). *J. Mol. Catal. A: Chem.* 156: 213–221.
165 Wu, X.-F., Neumann, H., and Beller, M. (2010). *Chemcatchem* 2: 509–513.
166 Wu, X.-F., Neumann, H., and Beller, M. (2012). *Chem. Eur. J.* 18: 3831–3834.
167 Kulkarni, S.M., Kelkar, A.A., and Chaudhari, R.V. (2001). *Chem. Commun.*: 1276–1277.
168 Uozumi, Y., Arii, T., and Watanabe, T. (2001). *J. Org. Chem.* 66: 5272–5274.
169 Schnyder, A. and Indolese, A.F. (2002). *J. Org. Chem.* 67: 594–597.
170 Wu, X.-F., Neumann, H., and Beller, M. (2010). *Chem. Eur. J.* 16: 9750–9753.
171 Wu, X.-F., Neumann, H., and Beller, M. (2010). *Chem. Asian J.* 5: 2168–2172.
172 Wu, X.-F., Schranck, J., Neumann, H., and Beller, M. (2011). *Tetrahedron Lett.* 52: 3702–3704.
173 Wu, X.-F., Schranck, J., Neumann, H., and Beller, M. (2012). *Chemcatchem* 4: 69–71.
174 Appukkuttan, P., Axelsson, L., Van der Eycken, E., and Larhed, M. (2008). *Tetrahedron Lett.* 49: 5625–5628.
175 Odell, L.R., Russo, F., and Larhed, M. (2012). *Synlett* 23: 685–698.
176 Schoenberg, A. and Heck, R.F. (1974). *J. Am. Chem. Soc.* 96: 7761–7764.
177 Baillargeon, V.P. and Stille, J.K. (1986). *J. Am. Chem. Soc.* 108: 452–461.
178 Baillargeon, V.P. and Stille, J.K. (1983). *J. Am. Chem. Soc.* 105: 7175–7176.
179 Kotsuki, H., Datta, P.K., and Suenaga, H. (1996). *Synthesis*: 470–472.
180 Kikukawa, K., Totoki, T., Wada, F., and Matsuda, T. (1984). *J. Organomet. Chem.* 270: 283–287.
181 Pribar, I. and Buchman, O. (1984). *J. Org. Chem.* 49: 4009–4011.
182 Ashfield, L. and Barnard, C.F.J. (2007). *Org. Process Res. Dev.* 11: 39–43.
183 Okano, T., Harada, N., and Kiji, J. (1994). *Bull. Chem. Soc. Jpn.* 67: 2329–2332.

184 Pribar, I. and Buchman, O. (1988). *J. Org. Chem.* 53: 624–626.
185 Cai, M.Z., Zhao, H., Zhou, J., and Song, C.S. (2002). *Synth. Commun.* 32: 923–926.
186 Bendavid, Y., Portnoy, M., and Milstein, D. (1989). *J. Chem. Soc., Chem. Commun.*: 1816–1817.
187 Klaus, S., Neumann, H., Zapf, A. et al. (2006). *Angew. Chem.* 118: 161–165.
188 Klaus, S., Neumann, H., Zapf, A. et al. (2006). *Angew. Chem. Int. Ed.* 45: 154–158.
189 Sergeev, A.G., Zapf, A., Spannenberg, A., and Beller, M. (2008). *Organometallics* 27: 297–300.
190 Brennfuhrer, A., Neumann, H., Klaus, S. et al. (2007). *Tetrahedron* 63: 6252–6258.
191 Brenrifuehrer, A., Neumann, H., and Beller, M. (2007). *Synlett*: 2537–2540.
192 Neumann, H., Kadyrov, R., Wu, X.-F., and Beller, M. (2012). *Chem. Asian J.* 7: 2213–2216.
193 Hone, C.A., Lopatka, P., Munday, R. et al. (2019). *ChemSusChem* 12: 326–337.
194 Wakita, Y., Yasunaga, T., Akita, M., and Kojima, M. (1986). *J. Organomet. Chem.* 301: C17–C20.
195 Ishiyama, T., Kizaki, H., Miyaura, N., and Suzuki, A. (1993). *Tetrahedron Lett.* 34: 7595–7598.
196 Prediger, P., Moro, A.V., Nogueira, C.W. et al. (2006). *J. Org. Chem.* 71: 3786–3792.
197 Skoda-Foldes, R., Szekvolgyi, Z., Kollar, L. et al. (2000). *Tetrahedron* 56: 3415–3418.
198 Couve-Bonnaire, S., Carpentier, J.F., Mortreux, A., and Castanet, Y. (2001). *Tetrahedron Lett.* 42: 3689–3691.
199 Maerten, E., Hassouna, F., Couve-Bonnaire, S. et al. (2003). *Synlett*: 1874–1876.
200 Couve-Bonnaire, S., Carpentier, J.F., Mortreux, A., and Castanet, Y. (2003). *Tetrahedron* 59: 2793–2799.
201 Maerten, E., Sauthier, M., Mortreux, A., and Castanet, Y. (2007). *Tetrahedron* 63: 682–689.
202 Neumann, H., Brennfuehrer, A., and Beller, M. (2008). *Chem. Eur. J.* 14: 3645–3652.
203 Gotov, B., Kaufmann, J., Schumann, H., and Schmalz, H.G. (2002). *Synlett*: 1161–1163.
204 Wu, X.-F., Neumann, H., and Beller, M. (2010). *Tetrahedron Lett.* 51: 6146–6149.
205 Wu, X.-F., Neumann, H., and Beller, M. (2011). *Adv. Synth. Catal.* 353: 788–792.
206 Miloserdov, F.M. and Grushin, V.V. (2012). *Angew. Chem., Int. Ed.* 51: 3668–3672.
207 Ishiyama, T., Miyaura, N., and Suzuki, A. (1991). *Bull. Chem. Soc. Jpn.* 64: 1999–2001.
208 Jackson, R.F., Turner, D., and Block, M.H. (1997). *J. Chem. Soc., Perkin Trans. 1*: 865–870.

209 O'Keefe, B.M. Simmons, N., and Martin, S.F. (2008). *Org. Lett.* 10: 5301–5304.
210 Hatanaka, Y. and Hiyama, T. (1989). *Chem. Lett.* 18: 2049–2052.
211 Hatanaka, Y., Fukushima, S., and Hiyama, T. (1992). *Tetrahedron* 48: 2113–2126.
212 Wu, X.-F., Neumann, H., and Beller, M. (2012). *Tetrahedron Lett.* 53: 582–584.
213 Kang, S.K., Lim, K.H., Ho, P.S. et al. (1998). *Synth. Commun.* 28: 1481–1489.
214 Xia, M. and Chen, Z.C. (1999). *J. Chem. Res.* 23: 400–401.
215 Saeki, T., Son E.C., and Tamao, K. (2004). *Org. Lett.* 6: 617–619.
216 Sonogashira, K., Tohda, Y., and Hagihara, N. (1975). *Tetrahedron Lett.*: 4467–4470.
217 Fawcett, C.H., Firn, R.D., and Spencer, D.M. (1971). *Physiol. Plant Pathol.* 1: 163–166.
218 Quesnelle, C.A., Gill, P., Dodier, M. et al. (2003). *Bioorg. Med. Chem. Lett.* 13: 519–524.
219 Kobayashi, T. and Tanaka, M. (1981). *J. Chem. Soc., Chem. Commun.*: 333–334.
220 Huang, Y.J. and Alper, H. (1991). *J. Org. Chem.* 56: 4534–4536.
221 Ciattini, P.G. Morera, E., and Ortar, G. (1991). *Tetrahedron Lett.* 32: 6449–6452.
222 Delaude, L., Masdeu, A.M., and Alper, H. (1994). *Synthesis*: 1149–1151.
223 Kang, S.K., Lim, K.H., Ho, P.S., and Kim, W.Y. (1997). *Synthesis*: 874–876.
224 Luo, S.J., Liang, Y.M., Liu, C.M., and Ma, Y.X. (2001). *Synth. Commun.* 31: 343–347.
225 Ahmed, M.S.M. and Mori, A. (2003). *Org. Lett.* 5: 3057–3060.
226 Ahmed, M.S.M., Sekiguchi, A., Masui, K., and Mori, A. (2005). *Bull. Chem. Soc. Jpn.* 78: 160–168.
227 Bishop, B.C. Brands, K.M.J., Gibb, A.D., and Kennedy, D.J. (2004). *Synthesis*: 43–52.
228 Ma, W., Li, X., Yang, J. et al. (2006). *Synthesis*: 2489–2492.
229 Liu, J., Chen, J., and Xia, C. (2008). *J. Catal.* 253: 50–56.
230 Wu, X.-F., Neumann, H., and Beller, M. (2010). *Chem. Eur. J.* 16: 12104–12107.
231 Xiang, K., Tong, P., Yan, B. et al. (2019). *Org. Lett.* 21: 412–416.
232 Kobayashi, T. and Tanaka, M. (1986). *Tetrahedron Lett.* 27: 4745–4748.
233 Negishi, E., Zhang, Y.T., Shimoyama, I., and Wu, G.Z. (1989). *J. Am. Chem. Soc.* 111: 8018–8020.
234 Negishi, E., Makabe, H., Shimoyama, I. et al. (1998). *Tetrahedron* 54: 1095–1106.
235 Campo, M.A. and Larock, R.C. (2002). *J. Org. Chem.* 67: 5616–5620.
236 Campo, M.A. and Larock, R.C. (2000). *Org. Lett.* 2: 3675–3677.
237 Peng, C., Cheng, J., and Wang, J. (2007). *J. Am. Chem. Soc.* 129: 8708–8709.
238 Wu, X.-F., Anbarasan, P., Neumann, H., and Beller, M. (2010). *Angew. Chem., Int. Ed.* 49: 7316–7319.
239 Sen, A. and Lai, T.W. (1982). *J. Am. Chem. Soc.* 104: 3520–3522.
240 Negishi, E. and Miller, J.A. (1983). *J. Am. Chem. Soc.* 105: 6761–6763.
241 Negishi, E. and Tour, J.M. (1986). *Tetrahedron Lett.* 27: 4869–4872.
242 Negishi, E., Ma, S.M., Amanfu, J. et al. (1996). *J. Am. Chem. Soc.* 118: 5919–5931.

243 Negishi, E., Coperet, C., Ma, S.M. et al. (1996). *J. Am. Chem. Soc.* 118: 5904–5918.
244 Coperet, C., Ma, S.M., and Negishi, E.I. (1996). *Angew. Chem., Int. Ed.* 35: 2125–2126.
245 Torii, S., Okumoto, H., and Xu, L.H. (1990). *Tetrahedron Lett.* 31: 7175–7178.
246 Satoh, T., Itaya, T., Okuro, K. et al. (1995). *J. Org. Chem.* 60: 7267–7271.
247 Wu, X.-F., Neumann, H., and Beller, M. (2010). *Angew. Chem., Int. Ed.* 49: 5284–5288.
248 Wu, X.-F., Neumann, H., Spannenberg, A. et al. (2010). *J. Am. Chem. Soc.* 132: 14596–14602.
249 Aumann, R. and Ring, H. (1977). *Angew. Chem., Int. Ed.* 16: 50–50.
250 Aumann, R., Ring, H., Krüger, C., and Goddard, R. (1979). *Chem. Ber.* 112: 3644–3671.
251 Lee, J.T., Thomas, P.J., and Alper, H. (2001). *J. Org. Chem.* 66: 5424–5426.
252 Alper, H., Arzoumanian, H., Petrignani, J.F., and Saldanamaldonado, M. (1985). *J. Chem. Soc., Chem. Commun.*: 340–341.
253 Alper, H., Eisenstat, A., and Satyanarayana, N. (1990). *J. Am. Chem. Soc.* 112: 7060–7061.
254 Ganji, P. and Ibrahim, H. (2012). *Chem. Commun.* 48: 10138–10140.
255 Yokokawa, C., Watanabe, Y., and Takegami, Y. (1964). *Bull. Chem. Soc. Jpn.* 37: 677–680.
256 Roos, L., Goetz, R.W., and Orchin, M. (1965). *J. Org. Chem.* 30: 3023–3027.
257 Weber, R., Englert, U., Ganter, B. et al. (2000). *Chem. Commun.*: 1419–1420.
258 Nakano, K., Katayama, M., Ishihara, S. et al. (2004). *Synlett*: 1367–1370.
259 Seki, Y., Murai, S., Yamamoto, I., and Sonoda, N. (1977). *Angew. Chem., Int. Ed.* 16: 789–789.
260 Murai, S., Kato, T., Sonoda, N. et al. (1979). *Angew. Chem., Int. Ed.* 18: 393–394.
261 Fukumoto, Y., Chatani, N., and Murai, S. (1993). *J. Org. Chem.* 58: 4187–4188.
262 Chatani, N. and Murai, S. (1996). *Synlett*: 414–424.
263 Furukawa, J., Iseda, Y., Saegusa, T., and Fujii, H. (1965). *Makromol. Chem.* 89: 263–268.
264 Allmendinger, M., Eberhardt, R., Luinstra, G., and Rieger, B. (2002). *J. Am. Chem. Soc.* 124: 5646–5647.
265 Allmendinger, M., Eberhardt, R., Luinstra, G.A., and Rieger, B. (2003). *Macromol. Chem. Phys.* 204: 564–569.
266 Allmendinger, M., Zintl, M., Eberhardt, R. et al. (2004). *J. Organomet. Chem.* 689: 971–979.
267 Takeuchi, D., Sakaguchi, Y., and Osakada, K. (2002). *J. Polym. Sci. Pol. Chem.* 40: 4530–4537.
268 Drent, E. and Kragtwijk, E. (1994). *Chem. Abstr.* 120: 191517c.
269 Lee, J.T. and Alper, H. (2004). *Macromolecules* 37: 2417–2421.
270 Eisenmann, J., Yamartino, R.L., and Howard, J.F. (1961). *J. Org. Chem.* 26: 2102–2104.

271 Kawabata, Y., Tanaka, M., Hayashi, T., and Ogata, I. (1979). *Nippon Kagaku Kaishi*: 635–640.

272 Hinterding, K. and Jacobsen, E.N. (1999). *J. Org. Chem.* 64: 2164–2165.

273 Zhang, W., Han, F., Tong, J. et al. (2017). *Chin. J. Catal.* 38: 805–812.

274 Tsuji, Y., Kobayashi, M., Okuda, F., and Watanabe, Y. (1989). *J. Chem. Soc., Chem. Commun.*: 1253–1254.

275 Goodman, S.N. and Jacobsen, E.N. (2002). *Angew. Chem., Int. Ed.* 41: 4703–4705.

276 Wakamatsu, H., Uda, J., and Yamakami, N. (1971). *J. Chem. Soc. D*: 1540.

277 Wakamatsu, H.; Uda, J.; Yamakami, N. J. K. (1989), 44, 448.

278 Lin, J.J. and Knifton, J.F. (1991). *J. Organomet. Chem.* 417: 99–110.

279 Knifton, J.F., Lin, J.J., Storm, D.A., and Wong, S.F. (1993). *Catal. Today* 18: 355–384.

280 Drent, E. and Kragtwijk, E. (1991). p GB Patent 2,252,770.

281 Ojima, I., Hirai, K., Fujita, M., and Fuchikami, T. (1985). *J. Organomet. Chem.* 279: 203–214.

282 Hirai, K., Takahashi, Y., and Ojima, I. (1982). *Tetrahedron Lett.* 23: 2491–2494.

283 Beller, M., Eckert, M., Vollmuller, F. et al. (1997). *Angew. Chem., Int. Ed.* 36: 1494–1496.

284 Beller, M., Eckert, M.K., and Vollmuller, F. (1998). *J. Mol. Catal. A: Chem.* 135: 23–33.

285 Beller, M., Eckert, M., Moradi, W.A., and Neumann, H. (1999). *Angew. Chem., Int. Ed.* 38: 1454–1457.

286 Becke, F. and Gnad, J. (1968). *Liebigs Ann. Chem.* 713: 212–214.

287 Becke, F., Fleig, H., and Päßler, P. (1971). *Liebigs Ann. Chem.* 749: 198–201.

288 Beller, M., Eckert, M., and Holla, E.W. (1998). *J. Org. Chem.* 63: 5658–5661.

289 Chan, A. (1983). *J. Mol. Catal.* 19: 377.

290 Okano, T., Makino, M., Konishi, H., and Kiji, J. (1985). *Chem. Lett.*: 1793–1796.

291 Murai, S., Kato, T., Sonoda, N. et al. (1979). *Angew. Chem.* 91: 421–422.

292 Wright, M.E. and Cochran, B.B. (1993). *J. Am. Chem. Soc.* 115: 2059–2060.

293 Kablaoui, N.M., Hicks, F.A., and Buchwald, S.L. (1996). *J. Am. Chem. Soc.* 118: 5818–5819.

294 Crowe, W.E. and Vu, A.T. (1996). *J. Am. Chem. Soc.* 118: 1557–1558.

295 Kablaoui, N.M., Hicks, F.A., and Buchwald, S.L. (1997). *J. Am. Chem. Soc.* 119: 4424–4431.

296 Chatani, N., Morimoto, T., Fukumoto, Y., and Murai, S. (1998). *J. Am. Chem. Soc.* 120: 5335–5336.

297 Kang, S.K., Kim, K.J., and Hong, Y.T. (2002). *Angew. Chem., Int. Ed.* 41: 1584–1586.

298 Seyferth, D., Weinstein, R.M., Hui, R.C. et al. (1992). *J. Org. Chem.* 57: 5620–5629.

299 Seyferth, D., Weinstein, R.M., Wang, W.L., and Hui, R.C. (1983). *Tetrahedron Lett.* 24: 4907–4910.

300 Harada, S., Taguchi, T., Tabuchi, N. et al. (1998). *Angew. Chem.* 110: 1796–1798.

301 Satoh, T., Tsuda, T., Kushino, Y. et al. (1996). *J. Org. Chem.* 61: 6476–6477.

302 Satoh, T., Tsuda, T., Terao, Y. et al. (1999). *J. Mol. Catal. A: Chem.* 143: 203–210.

303 Woo, E.P. and Cheng, F.C.W. (1986). *J. Org. Chem.* 51: 3706–3707.

304 Cavinato, G. and Toniolo, L. (1991). *J. Mol. Catal.* 69: 283–297.

305 Gabriele, B., Mancuso, R., and Salerno, G. (2012). *Eur. J. Org. Chem.*: 6825–6839.

306 Wu, X.-F., Neumann, H., and Beller, M. (2013). *ChemSusChem* 6: 229–241.

307 Liu, Q., Zhang, H., and Lei, A. (2011). *Angew. Chem., Int. Ed.* 50: 10788–10799.

308 Tsuji, J., Morikawa, M., and Kiji, J. (1963). *Tetrahedron Lett.*: 1061–1064.

309 Tsuji, J., Morikawa, M., and Kiji, J. (1964). *J. Am. Chem. Soc.* 86: 4851–4853.

310 Yukawa, T. and Tsutsumi, S. (1969). *J. Org. Chem.* 34: 738–740.

311 Cometti, G. and Chiusoli, G.P. (1979). *J. Organomet. Chem.* 181: C14–C16.

312 Inomata, K., Toda, S., and Kinoshita, H. (1990). *Chem. Lett.*: 1567–1570.

313 Takeuchi, S., Ukaji, Y., and Inomata, K. (2001). *Bull. Chem. Soc. Jpn.* 74: 955–958.

314 Hayashi, M., Takezaki, H., Hashimoto, Y. et al. (1998). *Tetrahedron Lett.* 39: 7529–7532.

315 Fergusson, S.B. and Alper, H. (1984). *J. Chem. Soc., Chem. Commun.*: 1349–1351.

316 Tsujihara, T., Shinohara, T., Takenaka, K. et al. (2009). *J. Org. Chem.* 74: 9274–9279.

317 Li, Z., Gao, Y., Tang, Y. et al. (2008). *Org. Lett.* 10: 3017–3020.

318 Hayes, P.Y. and Kitching, W. (2002). *J. Am. Chem. Soc.* 124: 9718–9719.

319 John, J.P., Jost, J., and Novikov, A.V. (2009). *J. Org. Chem.* 74: 6083–6091.

320 Hegedus, L.S., Allen, G.F., and Olsen, D.J. (1980). *J. Am. Chem. Soc.* 102: 3583–3587.

321 Tsuji, J., Morikawa, M., and Iwamoto, N. (1964). *J. Am. Chem. Soc.* 86: 2095–2095.

322 Tsuji, J. and Nogi, T. (1966). *J. Am. Chem. Soc.* 88: 1289–1292.

323 Li, J.H., Jiang, H.F., Feng, A.Q., and Jia, L.Q. (1999). *J. Org. Chem.* 64: 5984–5987.

324 Okumoto, H., Nishihara, S., Nakagawa, H., and Suzuki, A. (2000). *Synlett*: 217–218.

325 Ma, S.M., Wu, B., and Zhao, S.M. (2003). *Org. Lett.* 5: 4429–4432.

326 Kato, K., Nishimura, A., Yamamoto, Y., and Akita, H. (2001). *Tetrahedron Lett.* 42: 4203–4205.

327 Kato, K., Motodate, S., Mochida, T. et al. (2009). *Angew. Chem., Int. Ed.* 48: 3326–3328.

328 Motodate, S., Kobayashi, T., Fujii, M. et al. (2010). *Chem. Asian J.* 5: 2221–2230.

329 Gabriele, B., Salerno, G., Veltri, L., and Costa, M. (2001). *J. Organomet. Chem.* 622: 84–88.

330 Gadge, S.T., Khedkar, M.V., Lanke, S.R., and Bhanage, B.M. (2012). *Adv. Synth. Catal.* 354: 2049–2056.

331 Zhang, C., Liu, J., and Xia, C. (2015). *Catal. Sci. Technol.* 5: 4750–4754.

332 Stille, J.K. and Wong, P.K. (1975). *J. Org. Chem.* 40: 335–339.

333 Larock, R.C. (1975). *J. Org. Chem.* 40: 3237–3242.

334 Tamao, K., Kakui, T., and Kumada, M. (1979). *Tetrahedron Lett.*: 619–622.

335 Zhao, Y., Jin, L., Li, P., and Lei, A. (2008). *J. Am. Chem. Soc.* 130: 9429–9433.

336 Devasagayaraj, A. and Knochel, P. (1995). *Tetrahedron Lett.* 36: 8411–8414.

337 Kang, S.-K., Ryu, H.-C., and Choi, S.-C. (2001). *Synth. Commun.* 31: 1035–1039.

338 Miyaura, N. and Suzuki, A. (1981). *Chem. Lett.*: 879–882.

339 Yamashina, N., Hyuga, S., Hara, S., and Suzuki, A. (1989). *Tetrahedron Lett.* 30: 6555–6558.

340 Liu, Q., Li, G., He, J. et al. (2010). *Angew. Chem., Int. Ed.* 49: 3371–3374.

341 Fujiwara, Y., Kawauchi, T., and Taniguchi, H. (1980). *J. Chem. Soc., Chem. Commun.* 220-221.

342 Lu, W., Yamaoka, Y., Taniguchi, Y. et al. (1999). *J. Organomet. Chem.* 580: 290–294.

343 Taniguchi, Y., Yamaoka, Y., Nakata, K. et al. (1995). *Chem. Lett.* 24: 345–346.

344 Liu, C. and Widenhoefer, R.A. (2004). *J. Am. Chem. Soc.* 126: 10250–10251.

345 Liu, C. and Widenhoefer, R.A. (2006). *Chem. Eur. J.* 12: 2371–2382.

346 Giri, R. and Yu, J.-Q. (2008). *J. Am. Chem. Soc.* 130: 14082–14083.

347 Houlden, C.E., Hutchby, M., Bailey, C.D. et al. (2009). *Angew. Chem., Int. Ed.* 48: 1830–1833.

348 Tsuji, J. and Iwamoto, N. (1966). *Chem. Commun.*: 380–380.

349 Shi, F. and Deng, Y. (2001). *Chem. Commun.*: 443–444.

350 Shi, F. and Deng, Y. (2002). *J. Catal.* 211: 548–551.

351 Shi, F., Deng, Y., SiMa, T., and Yang, H. (2001). *J. Catal.* 203: 525–528.

352 Shi, F., Peng, J., and Deng, Y. (2003). *J. Catal.* 219: 372–375.

353 Shi, F., Zhang, Q., Gu, Y., and Deng, Y. (2005). *Adv. Synth. Catal.* 347: 225–230.

354 Zheng, S., Peng, X., Liu, J. et al. (2007). *Helv. Chim. Acta* 90: 1471–1476.

355 Zheng, S.Z., Peng, X.G., Liu, J.M. et al. (2007). *Chin. J. Chem.* 25: 1065–1068.

356 Didgikar, M.R., Roy, D., Gupte, S.P. et al. (2010). *Ind. Eng. Chem. Res.* 49: 1027–1032.

357 Didgikar, M.R., Joshi, S.S., Gupte, S.P. et al. (2011). *J. Mol. Catal. A: Chem.* 334: 20–28.

358 Giannoccaro, P., Ferragina, C., Gargano, M., and Quaranta, E. (2010). *Appl. Catal., A: Chem.* 375: 78–84.

359 Gabriele, B., Salerno, G., Brindisi, D. et al. (2000). *Org. Lett.* 2: 625–627.

360 Liu, S., Dai, X., Wang, H., and Shi, F. (2019). *Green Chem.* 21: 4040–4045.

361 Chen, B., Peng, J.B., Ying, J. et al. (2018). *Adv. Synth. Catal.* 360: 2820–2824.

362 Mancuso, R., Raut, D.S., Della Ca', N. et al. (2015). *ChemSusChem* 8: 2204–2211.

363 Casiello, M., Iannone, F., Cotugno, P. et al. (2015). *J. Mol. Catal. A: Chem.* 407: 8–14.

364 Saliu, F., Putomatti, B., and Rindone, B. (2012). *Tetrahedron Lett.* 53: 3590–3593.

365 Zhu, B. and Angelici, R.J. (2006). *J. Am. Chem. Soc.* 128: 14460–14461.

366 Kim, H.S., Kim, Y.J., Lee, H. et al. (1999). *J. Catal.* 184: 526–534.

367 Kanagasabapathy, S., Gupte, S.P., and Chaudhari, R.V. (1994). *Ind. Eng. Chem. Res.* 33: 1–6.

368 Wang, L., Wang, H., Li, G. et al. (2018). *Adv. Synth. Catal.* 360: 4585–4593.

369 Li, J. and Tu, D.-h.; Li, Y.; Wang, W.; Yu, Q.; Yang, J.; Lu, J. (2018). *Appl. Catal. A: Chem.* 549: 112–116.

370 Gadge, S.T., Kusumawati, E.N., Harada, K. et al. (2015). *J. Mol. Catal. A: Chem.* 400: 170–178.

371 Kollár, L. (2008). *Modern Carbonylation Methods*. Wiley.

372 Tidwell, T.T. (2006). *Ketenes II*. Wiley Online Library.

373 Lyashchuk, S.N. and Skrypnik, Y.G. (1994). *Tetrahedron Lett.* 35: 5271–5274.

374 Lavallo, V., Canac, Y., Donnadieu, B. et al. (2006). *Angew. Chem., Int. Ed.* 45: 3488–3491.

375 Ozawa, F., Park, J.W., Mackenzie, P.B. et al. (1989). *J. Am. Chem. Soc.* 111: 1319–1327.

376 Kron, J. and Schubert, U. (1989). *J. Organomet. Chem.* 373: 203–219.

377 Morrison, E.D., Steinmetz, G.R., Geoffroy, G.L. et al. (1983). *J. Am. Chem. Soc.* 105: 4104–4105.

378 Morrison, E.D., Steinmetz, G.R., Geoffroy, G.L. et al. (1984). *J. Am. Chem. Soc.* 106: 4783–4789.

379 Werner, H., Schwab, P., Bleuel, E. et al. (2000). *Chem. Eur. J.* 6: 4461–4470.

380 Miyashita, A., Shitara, H., and Nohira, H. (1985). *Organometallics* 4: 1463–1464.

381 Tuba, R. and Ungváry, F. (2003). *J. Mol. Catal. A: Chem.* 203: 59–67.

382 Kegl, T. and Ungváry, F. (2010). *Lett. Org. Chem.* 7: 634–644.

383 Zhang, Z., Liu, Y., Ling, L. et al. (2011). *J. Am. Chem. Soc.* 133: 4330–4341.

384 Paul, N.D., Chirila, A., Lu, H. et al. (2013). *Chem. Eur. J.* 19: 12953–12958.

385 Ramakrishna, K. and Sivasankar, C. (2017). *Eur. J. Org. Chem.* 2017: 4035–4043.

386 Paul, F. (2000). *Coord. Chem. Rev.* 203: 269–323.

387 Tafesh, A.M. and Weiguny, J. (1996). *Chem. Rev.* 96: 2035–2052.

388 Ragaini, F. and Cenini, S. (1996). *J. Mol. Catal. A: Chem.* 109: 1–25.

389 Leconte, P., Metz, F., Mortreux, A. et al. (1990). *J. Chem. Soc., Chem. Commun.*: 1616–1617.

390 Paul, F., Fischer, J., Ochsenbein, P., and Osborn, J.A. (1998). *Organometallics* 17: 2199–2206.

391 Wang, X.F., Lu, S.W., and Yu, Z.K. (2004). *Adv. Synth. Catal.* 346: 929–932.

392 Chen, J.Z., Ling, G., Yu, Z. et al. (2004). *Adv. Synth. Catal.* 346: 1267–1270.

393 Cenini, S. and Ragaini, F. (2013). *Catalytic Reductive Carbonylation of Organic Nitro Compounds*, vol. 20. Springer Science & Business Media.

394 Sakakura, T., Choi, J.-C., and Yasuda, H. (2007). *Chem. Rev.* 107: 2365–2387.

395 Fujita, S. and Arai, M. (2005). *J. Jpn. Petrol Inst.* 48: 67–75.

396 Pagani, G. (1982). *Hydrocarbon Process (United States)* 61: 11.

397 Hayashi, T. and Kuyama, M. (1951). *J. Nat. Sci. Ochanomizu Univ.* 2: 79–86.

398 Yamazaki, N., Higashi, F., and Iguchi, T. (1974). *Tetrahedron Lett.* 15: 1191–1194.

399 Nomura, R., Hasegawa, Y., Ishimoto, M. et al. (1992). *J. Org. Chem.* 57: 7339–7342.

400 Inoue, S. and Yamazaki, N. (1982). *Organic and Bio-organic Chemistry of Carbon dioxide*. Halsted Press.

401 Tai, C.C., Huck, M.J., McKoon, E.P. et al. (2002). *J. Org. Chem.* 67: 9070–9072.

402 Shi, F., Deng, Y.Q., SiMa, T.L. et al. (2003). *Angew. Chem., Int. Ed.* 42: 3257–3260.

403 Kong, D.-L., He, L.-N., and Wang, J.-Q. (2010). *Synlett*: 1276–1280.

404 Primo, A., Aguado, E., and Garcia, H. (2013). *ChemCatChem* 5: 1020–1023.

405 Tamura, M., Noro, K., Honda, M. et al. (2013). *Green Chem.* 15: 1567–1577.

406 Tamura, M., Ito, K., Nakagawa, Y., and Tomishige, K. (2016). *J. Catal.* 343: 75–85.

407 Dell'Amico, D.B., Calderazzo, F., Labella, L. et al. (2003). *Chem. Rev.* 103: 3857–3897.

408 Chaturvedi, D. and Ray, S. (2006). *Monatsh. Chem. Chem. Monthly* 137: 127–145.

409 Aresta, M., Ballivet-Tkatchenko, D., Dell'Amico, D.B. et al. (2000). *Chem. Commun.*: 1099–1100.

410 Dibenedetto, A., Aresta, M., Fragale, C., and Narracci, M. (2002). *Green Chem.* 4: 439–443.

411 Kayaki, Y., Suzuki, T., and Ikariya, T. (2008). *Chem. Asian J.* 3: 1865–1870.

412 Aresta, M. and Quaranta, E. (1992). *Tetrahedron* 48: 1515–1530.

413 Waldman, T.E. and McGhee, W.D. (1994). *J. Chem. Soc., Chem. Commun.*: 957–958.

414 Yoshida, M., Hara, N., and Okuyama, S. (2000). *Chem. Commun.*: 151–152.

415 Abla, M., Choi, J.C., and Sakakura, T. (2001). *Chem. Commun.*: 2238–2239.

416 Abla, M., Choi, J.C., and Sakakura, T. (2004). *Green Chem.* 6: 524–525.

417 Bhanage, B.M., Fujita, S., Ikushima, Y., and Arai, M. (2003). *Green Chem.* 5: 340–342.

418 Srivastava, R., Manju, M.D., Srinivas, D., and Ratnasamy, P. (2004). *Catal. Lett.* 97: 41–47.

419 Krawczyk, T., Jasiak, K., Kokolus, A., and Baj, S. (2016). *Catal. Lett.* 146: 1163–1168.

420 McGhee, W.D. and Riley, D.P. (1992). *Organometallics* 11: 900–907.

421 Mukhtar, T.A. and Wright, G.D. (2005). *Chem. Rev.* 105: 529–542.

422 Barbachyn, M.R. and Ford, C.W. (2003). *Angew. Chem., Int. Ed. Engl.* 42: 2010–2023.

423 Aurelio, L., Brownlee, R.T.C., and Hughes, A.B. (2004). *Chem. Rev.* 104: 5823–5846.

424 Kawanami, H., Matsumoto, H., and Ikushima, Y. (2005). *Chem. Lett.* 34: 60–61.

425 Miller, A.W. and Nguyen, S.T. (2004). *Org. Lett.* 6: 2301–2304.

426 Sudo, A., Morioka, Y., Sanda, F., and Endo, T. (2004). *Tetrahedron Lett.* 45: 1363–1365.

427 Sudo, A., Morioka, Y., Koizumi, E. et al. (2003). *Tetrahedron Lett.* 44: 7889–7891.

428 Shi, M., Jiang, J.K., Shen, Y.M. et al. (2000). *J. Org. Chem.* 65: 3443–3448.

429 Ihata, O., Kayaki, Y., and Ikariya, T. (2005). *Chem. Commun.*: 2268–2270.

430 Ochiai, B., Inoue, S., and Endo, T. (2005). *J. Polym. Sci. Pol. Chem.* 43: 6613–6618.

431 Ihata, O., Kayaki, Y., and Ikariya, T. (2004). *Angew. Chem., Int. Ed.* 43: 717–719.

432 Nomura, R., Nakano, T., Nishio, Y. et al. (1989). *Chem. Ber.* 122: 2407–2409.

433 Tascedda, P. and Duñach, E. (2000). *Chem. Commun.*: 449–450.

434 Kawanami, H. and Ikushima, Y. (2002). *Tetrahedron Lett.* 43: 3841–3844.

435 Hancock, M.T. and Pinhas, A.R. (2003). *Tetrahedron Lett.* 44: 5457–5460.

436 Ihata, O., Kayaki, Y., and Ikariya, T. (2005). *Kobunshi Ronbunshu* 62: 196–199.

437 Ihata, O., Kayaki, Y., and Ikariya, T. (2005). *Macromolecules* 38: 6429–6434.

438 Gu, Y.L., Zhang, Q.H., Duan, Z.Y. et al. (2005). *J. Org. Chem.* 70: 7376–7380.

439 Xu, H., Liu, X.F., Cao, C.S. et al. (2016). *Adv. Sci.* 3: 1–7.

440 Liu, A.H., Dang, Y.L., Zhou, H. et al. (2018). *ChemCatChem* 10: 2686–2692.

441 Connolly, T.J., McGarry, P., and Sukhtankar, S. (2005). *Green Chem.* 7: 586–589.

442 Mizuno, T. and Ishino, Y. (2002). *Tetrahedron* 58: 3155–3158.

443 Mizuno, T., Iwai, T., and Ishino, Y. (2004). *Tetrahedron Lett.* 45: 7073–7075.

444 Manjolinho, F., Arndt, M., Gooßen, K., and Gooßen, L.J. (2012). *ACS Catal.* 2: 2014–2021.

445 Kitamura, T. (2009). *Eur. J. Org. Chem.* 2009: 1111–1125.

446 Correa, A. and Martin, R. (2009). *Angew. Chem., Int. Ed. Engl.* 48: 6201–6204.

447 Chiba, K., Tagaya, H., Miura, S., and Karasu, M. (1992). *Chem. Lett.*: 923–926.

448 Mankad, N.P., Gray, T.G., Laitar, D.S., and Sadighi, J.P. (2004). *Organometallics* 23: 1191–1193.

449 (a) Abstracts of Papers, t. A. N. M., New Orleans, LA, United States, March 23–27, 2003; (b) Abe, H. and Inoue, S. (1994). *J. Chem. Soc., Chem. Commun.*: 1197–1198.

450 Quirk, R.P., Yin, J., Fetters, L.J., and Kastrup, R.V. (1992). *Macromolecules* 25: 2262–2267.

451 Shi, M. and Nicholas, K.M. (1997). *J. Am. Chem. Soc.* 119: 5057–5058.

452 Ukai, K., Aoki, M., Takaya, J., and Iwasawa, N. (2006). *J. Am. Chem. Soc.* 128: 8706–8707.

453 Galindo, A., Pastor, A., Perez, P.J., and Carmona, E. (1993). *Organometallics* 12: 4443–4451.

454 Schubert, G. and Papai, I. (2003). *J. Am. Chem. Soc.* 125: 14847–14858.

455 Papai, I., Schubert, G., Mayer, I. et al. (2004). *Organometallics* 23: 5252–5259.

456 Saito, S., Nakagawa, S., Koizumi, T. et al. (1999). *J. Org. Chem.* 64: 3975–3978.

457 Takimoto, M. and Mori, M. (2001). *J. Am. Chem. Soc.* 123: 2895–2896.

458 Six, Y. (2003). *Eur. J. Org. Chem.*: 1157–1171.

459 Takimoto, M., Kawamura, M., and Mori, M. (2003). *Org. Lett.* 5: 2599–2601.

460 Aoki, M., Kaneko, M., Izumi, S. et al. (2004). *Chem. Commun.*: 2568–2569.

461 Perez, E.R., Santos, R.H.A., Gambardella, M.T.P. et al. (2004). *J. Org. Chem.* 69: 8005–8011.

462 Heldebrant, D.J., Jessop, P.G., Thomas, C.A. et al. (2005). *J. Org. Chem.* 70: 5335–5338.

463 Takimoto, M. and Mori, M. (2002). *J. Am. Chem. Soc.* 124: 10008–10009.

464 Takimoto, M., Nakamura, Y., Kimura, K., and Mori, M. (2004). *J. Am. Chem. Soc.* 126: 5956–5957.

465 Kosugi, Y., Imaoka, Y., Gotoh, F. et al. (2003). *Org. Biomol. Chem.* 1: 817–821.

466 Takimoto, M. Kawamura, M., Mori, M., and Sato, Y. (2005). *Synlett*: 2019–2022.

467 Rahim, M.A., Matsui, Y., Matsuyama, T., and Kosugi, Y. (2003). *Bull. Chem. Soc. Jpn.* 76: 2191–2195.

468 Sclafani, A., Palmisano, L., and Farneti, G. (1997). *Chem. Commun.*: 529–530.

469 Zerella, M., Mukhopadhyay, S., and Bell, A.T. (2003). *Org. Lett.* 5: 3193–3196.

470 Fujiwara, Y., Taniguchi, Y., Takaki, K. et al. (1997). *Natural Gas Conversion IV*, vol. 107 (eds. M. dePontes, R.L. Espinoza, C.P. Nicolaides, et al.), 275–278. Elsevier.

471 Jia, C.G., Kitamura, T., and Fujiwara, Y. (2001). *Acc. Chem. Res.* 34: 633–639.

472 Olah, G.A., Torok, A., Joschek, J.P. et al. (2002). *J. Am. Chem. Soc.* 124: 11379–11391.

473 Hattori, T., Suzuki, Y., and Miyano, S. (2003). *Chem. Lett.* 32: 454–455.

474 Suzuki, Y., Hattori, T., Okuzawa, T., and Miyano, S. (2002). *Chem. Lett.*: 102–103.

475 Fukuoka, A., Gotoh, N., Kobayashi, N. et al. (1995). *Chem. Lett.* 24: 567–568.

476 Osakada, K. Sato, R., and Yamamoto, T. (1994). *Organometallics* 13: 4645–4647.

477 Liu, X.H., Ma, J.G., Niu, Z. et al. (2015). *Angew. Chem.* 54: 1002–1005.

478 Trivedi, M., Bhaskaran, Kumar, A. et al. (2016). *New J. Chem.*: 3109–3118.

479 Molla, R.A., Ghosh, K., Banerjee, B. et al. (2016). *J. Colloid Interface Sci.* 477: 220–229.

480 Wu, Z.L., Liu, Q.G., Yang, X.F. et al. (2017). *ACS Sustainable Chem. Eng.* 5: 9634–9639.

481 Das, S., Mondal, P., Ghosh, S. et al. (2018). *New J. Chem.* 42: 7314–7325.

482 Zhu, N.N., Liu, X.H., Li, T. et al. (2017). *Inorg. Chem.* 56: 3414–3420.

483 Wu, Z.L., Sun, L., Liu, Q.Q. et al. (2017). *Green Chem.* 19: 2080–2085.

484 Sun, L.L., Yun, Y.P., Sheng, H.T. et al. (2018). *J. Mater. Chem. A* 6: 15371–15376.

485 Decortes, A., Castilla, A.M., and Kleij, A.W. (2010). *Angew. Chem., Int. Ed. Engl.* 49: 9822–9837.

486 Sakakura, T. and Kohno, K. (2009). *Chem. Commun. (Camb.)*: 1312–1330.

487 North, M., Pasquale, R., and Young, C. (2010). *Green Chem.* 12: 1514.

488 Darensbourg, D.J. and Holtcamp, M.W. (1996). *Coord. Chem. Rev.* 153: 155–174.

489 Zhao, T.S., Han, Y.Z., and Sun, Y.H. (1999). *PCCP* 1: 3047–3051.

490 Sun, Y.H. (2004). *Carbon Dioxide Utilization for Global Sustainability*, vol. 153 (eds. S.E. Park, J.S. Chang and K.W. Lee), 9–16. International Atomic Energy Agency.
491 Wang, J.Q., Kong, D.L., Chen, J.Y. et al. (2006). *J. Mol. Catal. A: Chem.* 249: 143–148.
492 Srivastava, R., Srinivas, D., and Ratnasamy, P. (2005). *Appl. Catal. A: Chem.* 289: 128–134.
493 Lu, X.B., Zhang, Y.J., Jin, K. et al. (2004). *J. Catal.* 227: 537–541.
494 Lu, X.B., Zhang, Y.J., Liang, B. et al. (2004). *J. Mol. Catal. A: Chem.* 210: 31–34.
495 Lu, X.B., Feng, X.J., and He, R. (2002). *Appl. Catal. A: Chem.* 234: 25–33.
496 Paddock, R.L. and Nguyen, S.T. (2001). *J. Am. Chem. Soc.* 123: 11498–11499.
497 Darensbourg, D.J., Fang, C.C., and Rodgers, J.L. (2004). *Organometallics* 23: 924–927.
498 Shen, Y.M., Duan, W.L., and Shi, M. (2003). *J. Org. Chem.* 68: 1559–1562.
499 Lu, X.B., Liang, B., Zhang, Y.J. et al. (2004). *J. Am. Chem. Soc.* 126: 3732–3733.
500 Jing, H.W., Edulji, S.K., Gibbs, J.M. et al. (2004). *Inorg. Chem.* 43: 4315–4327.
501 Kasuga, K., Kato, T., Kabata, N., and Handa, M. (1996). *Bull. Chem. Soc. Jpn.* 69: 2885–2888.
502 Sugimoto, H. and Inoue, S. (1998). *Pure Appl. Chem.* 70: 2365–2369.
503 Kasuga, K., Kabata, N., Kato, T. et al. (1998). *Inorg. Chim. Acta* 278: 223–225.
504 Ji, D.F., Lu, X.B., and He, R. (2000). *Appl. Catal. A: Chem.* 203: 329–333.
505 Sun, J.M., Fujita, S., and Arai, M. (2005). *J. Organomet. Chem.* 690: 3490–3497.
506 Peng, J.J. and Deng, Y.Q. (2001). *New J. Chem.* 25: 639–641.
507 Kim, Y.J. and Cheong, M. (2002). *Bull. Korean Chem. Soc.* 23: 1027–1028.
508 Kim, H.S., Kim, J.J., Kim, H., and Jang, H.G. (2003). *J. Catal.* 220: 44–46.
509 Palgunadi, J., Kwon, O.S., Lee, H. et al. (2004). *Catal. Today* 98: 511–514.
510 Sun, J.M., Fujita, S., Zhao, F.Y., and Arai, M. (2004). *Green Chem.* 6: 613–616.
511 Calo, V., Nacci, A., Monopoli, A., and Fanizzi, A. (2002). *Org. Lett.* 4: 2561–2563.
512 Li, F.W., Xia, C.G., Xu, L.W. et al. (2003). *Chem. Commun.*: 2042–2043.
513 Man, M.L., Lam, K.C., Sit, W.N. et al. (2006). *Chem. Eur. J.* 12: 1004–1015.
514 Kim, H.S., Bae, J.Y., Lee, J.S. et al. (2005). *J. Catal.* 232: 80–84.
515 Srivastava, R., Bennur, T.H., and Srinivas, D. (2005). *J. Mol. Catal. A: Chem.* 226: 199–205.
516 Darensbourg, D.J., Ganguly, P., and Billodeaux, D.R. (2004). *Organometallics* 23: 6025–6030.
517 Kasuga, K. and Kabata, N. (1997). *Inorg. Chim. Acta* 257: 277–278.
518 Huang, J.W. and Shi, M. (2003). *J. Org. Chem.* 68: 6705–6709.
519 Kossev, K., Koseva, N., and Troev, K. (2003). *J. Mol. Catal. A: Chem.* 194: 29–37.
520 Sun, J.M., Fujita, S.I., Zhao, F.Y., and Arai, M. (2005). *Appl. Catal. A: Chem.* 287: 221–226.
521 Yasuda, H., He, L.N., Sakakura, T., and Hu, C.W. (2005). *J. Catal.* 233: 119–122.
522 Shen, Y.M., Duan, W.L., and Shi, M. (2004). *Eur. J. Org. Chem.*: 3080–3089.
523 Shen, Y.M., Duan, W.L., and Shi, M. (2003). *Adv. Synth. Catal.* 345: 337–340.

524 Baba, A., Kashiwagi, H., and Matsuda, H. (1985). *Tetrahedron Lett.* 26: 1323–1324.

525 Du, Y., Cai, F., Kong, D.L., and He, L.N. (2005). *Green Chem.* 7: 518–523.

526 Shi, F., Zhang, Q.H., Ma, Y.B. et al. (2005). *J. Am. Chem. Soc.* 127: 4182–4183.

527 Alvaro, M., Baleizao, C., Carbonell, E. et al. (2005). *Tetrahedron* 61: 12131–12139.

528 Park, D.W., Yu, B.S., Jeong, E.S. et al. (2004). *Catal. Today* 98: 499–504.

529 Kim, H.S., Kim, J.J., Kwon, H.N. et al. (2002). *J. Catal.* 205: 226–229.

530 Nishikubo, T., Kameyama, A., Yamashita, J. et al. (1993). *J. Polym. Sci. Pol. Chem.* 31: 939–947.

531 Nishikubo, T., Kameyama, A., Yamashita, J. et al. (1995). *J. Polym. Sci. Pol. Chem.* 33: 1011–1017.

532 Ramin, M., Grunwaldt, J.D., and Baiker, A. (2005). *J. Catal.* 234: 256–267.

533 Barbarini, A., Maggi, R., Mazzacani, A. et al. (2003). *Tetrahedron Lett.* 44: 2931–2934.

534 Alvaro, M., Baleizao, C., Das, D. et al. (2004). *J. Catal.* 228: 254–258.

535 Takahashi, T., Watahiki, T., Kitazume, S. et al. (2006). *Chem. Commun.*: 1664–1666.

536 Coates, G.W. and Moore, D.R. (2004). *Angew. Chem., Int. Ed.* 43: 6618–6639.

537 Aresta, M. and Dibenedetto, A. (2002). *J. Mol. Catal. A: Chem.* 182: 399–409.

538 Tomishige, K., Yasuda, H., Yoshida, Y. et al. (2004). *Green Chem.* 6: 206–214.

539 Du, Y., Kong, D.L., Wang, H.Y. et al. (2005). *J. Mol. Catal. A: Chem.* 241: 233–237.

540 Aresta, M., Dibenedetto, A., Dileo, C. et al. (2003). *J. Supercrit. Fluids* 25: 177–182.

541 Yoshida, M., Fujita, M., Ishii, T., and Ihara, M. (2003). *J. Am. Chem. Soc.* 125: 4874–4881.

542 Yoshida, M. and Ihara, M. (2004). *Chem. Eur. J.* 10: 2886–2893.

543 Tomishige, K., Sakaihori, T., Ikeda, Y., and Fujimoto, K. (1999). *Catal. Lett.* 58: 225–229.

544 Tomishige, K., Ikeda, Y., Sakaihori, T., and Fujimoto, K. (2000). *J. Catal.* 192: 355–362.

545 Jiang, C.J., Guo, Y.H., Wang, C.G. et al. (2003). *Appl. Catal. A: Chem.* 256: 203–212.

546 Wu, X.L., Xiao, M., Meng, Y.Z., and Lu, Y.X. (2005). *J. Mol. Catal. A: Chem.* 238: 158–162.

547 Wu, X.L., Meng, Y.Z., Xiao, M., and Lu, Y.X. (2006). *J. Mol. Catal. A: Chem.* 249: 93–97.

548 Zevenhoven, R., Eloneva, S., and Teir, S. (2006). *Catal. Today* 115: 73–79.

549 Ballivet-Tkatchenko, D., Burgat, R., Chambrey, S. et al. (2006). *J. Organomet. Chem.* 691: 1498–1504.

550 Jessop, P.G., Joo, F., and Tai, C.C. (2004). *Coord. Chem. Rev.* 248: 2425–2442.

551 Sakakura, T. and Kohno, K. (2009). *Chem. Commun.*: 1312–1330.

552 Yokoyama, C., Kawase, Y., Shibasaki-Kitakawa, N., and Smith, R.L. (2003). *J. Appl. Polym. Sci.* 89: 3167–3174.

553 Soga, K., Hosoda, S., Tazuke, Y., and Ikeda, S. (1975). *J. Polym. Sci. C Polym. Lett.* 13: 265–268.

554 Soga, K., Sato, M., Hosoda, S., and Ikeda, S. (1975). *J. Polym. Sci. C Polym. Lett.* 13: 543–548.

555 Saegusa, T., Kobayashi, S., and Kimura, Y. (1977). *Macromolecules* 10: 68–72.

556 Behr, A. and Juszak, K.D. (1983). *J. Organomet. Chem.* 255: 263–268.

557 Behr, A. and Heite, M. (2000). *Chem. Ing. Tech.* 72: 58–61.

558 Tsuda, T., Yamamoto, T., and Saegusa, T. (1992). *J. Organomet. Chem.* 429: C46–C48.

559 Hoberg, H., Schaefer, D., Burkhart, G. et al. (1984). *J. Organomet. Chem.* 266: 203–224.

560 Tsuda, T. and Maruta, K. (1992). *Macromolecules* 25: 6102–6105.

561 Oi, S.C., Fukue, Y., Nemoto, K., and Inoue, Y. (1996). *Macromolecules* 29: 2694–2695.

562 Yoshida, H., Fukushima, H., Ohshita, J., and Kunai, A. (2006). *J. Am. Chem. Soc.* 128: 11040–11041.

563 Louie, J., Gibby, J.E., Farnworth, M.V., and Tekavec, T.N. (2002). *J. Am. Chem. Soc.* 124: 15188–15189.

564 Yamashita, M., Goto, K., and Kawashima, T. (2005). *J. Am. Chem. Soc.* 127: 7294–7295.

565 Aresta, M., Dibenedetto, A., and Tommasi, I. (2001). *Eur. J. Inorg. Chem.* 1801-1806.

566 Franks, R.J. and Nicholas, K.M. (2000). *Organometallics* 19: 1458–1460.

567 Yano, T., Matsui, H., Koike, T. et al. (1997). *Chem. Commun.*: 1129–1130.

568 Yamaguchi, K., Ebitani, K., Yoshida, T. et al. (1999). *J. Am. Chem. Soc.* 121: 4526–4527.

569 An, Q., Li, Z.F., Graff, R. et al. (2015). *ACS Appl. Mater. Interf.* 7: 4969–4978.

570 Ravi, S., Puthiaraj, P., Park, D.W., and Ahn, W.S. (2018). *New J. Chem.* 42: 12429–12436.

571 Doskocil, E.J., Bordawekar, S.V., Kaye, B.G., and Davis, R.J. (1999). *J. Phys. Chem. B* 103: 6277–6282.

572 Srivastava, R., Srinivas, D., and Ratnasamy, P. (2005). *J. Catal.* 233: 1–15.

573 Shiels, R.A. and Jones, C.W. (2007). *J. Mol. Catal. A: Chem.* 261: 160–166.

574 Ravi, S., Roshan, R., Tharun, J. et al. (2015). *J. Co2 Utiliz.* 10: 88–94.

575 Khatun, R., Bhanja, P., Molla, R.A. et al. (2017). *Mol. Catal.* 434: 25–31.

576 Dai, Z.F., Sun, Q., Liu, X.L. et al. (2017). *ChemSusChem* 10: 1186–1192.

577 Sharma, R.K., Gaur, R., Yadav, M. et al. (2018). *Scientific Rep.* 8: 1–12.

578 D'Elia, V., Dong, H.L., Rossini, A.J. et al. (2015). *J. Am. Chem. Soc.* 137: 7728–7739.

579 Kelly, M.J., Barthel, A., Maheu, C. et al. (2017). *J. Co2 Utiliz.* 20: 243–252.

580 Khatun, R., Bhanja, P., Mondal, P. et al. (2017). *New J. Chem.* 41: 12937–12946.

581 Ji, G., Yang, Z., Zhang, H. et al. (2016). *Angew. Chem., Int. Ed. Engl.* 55: 9685–9689.

582 Biswas, S., Khatun, R., Sengupta, M., and Islam, S.M. (2018). *Mol. Catal.* 452: 129–137.

583 Ghosh, S., Mondal, P., Das, D. et al. (2018). *J. Organomet. Chem.* 866: 1–12.
584 Ansari, M.B., Min, B.-H., Mo, Y.-H., and Park, S.-E. (2011). *Green Chem.*: 13.
585 Biswas, T. and Mahalingam, V. (2017). *New J. Chem.* 41: 14839–14842.
586 Zhang, S., Zhang, H., Cao, F.X. et al. (2018). *ACS Sustainable Chem. Eng.* 6: 4204–4211.
587 Lu, X.B. and Darensbourg, D.J. (2012). *Chem. Soc. Rev.* 41: 1462–1484.
588 Sorokin, A.B. (2013). *Chem. Rev.* 113: 8152–8191.
589 Martín, C., Fiorani, G., and Kleij, A.W. (2015). *ACS Catal.* 5: 1353–1370.
590 Lu, X.B., Wang, H., and He, R. (2002). *J. Mol. Catal. A: Chem.* 186: 33–42.
591 Xie, Y., Wang, T.T., Liu, X.H. et al. (2013). *Nat. Commun.* 4: 1–7.
592 Kim, M.H., Song, T., Seo, U.R. et al. (2017). *J. Mater. Chem. A* 5: 23612–23619.
593 Dai, Z.F., Sun, Q., Liu, X.L. et al. (2016). *J. Catal.* 338: 202–209.
594 Wang, S.L., Song, K.P., Zhang, C.X. et al. (2017). *J. Mater. Chem. A* 5: 1509–1515.
595 Parvulescu, V.I. and Hardacre, C. (2007). *Chem. Rev.* 107: 2615–2665.
596 Xu, B.-H., Wang, J.-Q., Sun, J. et al. (2015). *Green Chem.* 17: 108–122.
597 Zeng, S., Zhang, X., Bai, L. et al. (2017). *Chem. Rev.* 117: 9625–9673.
598 Xiao, L.-F., Li, F.-W., Peng, J.-J., and Xia, C.-G. (2006). *J. Mol. Catal. A: Chem.* 253: 265–269.
599 Zhang, X.L., Wang, D.F., Zhao, N. et al. (2009). *Catal. Commun.* 11: 43–46.
600 Xie, Y., Ding, K.L., Liu, Z.M. et al. (2010). *Chem. Eur. J.* 16: 6687–6692.
601 Besse, V., Illy, N., David, G. et al. (2016). *Chemsuschem* 9: 2167–2173.
602 Jose, T., Cañellas, S., Pericàs, M.A., and Kleij, A.W. (2017). *Green Chem.* 19: 5488–5493.
603 Shi, L.J., Xu, S.B., Zhang, Q.R. et al. (2018). *Ind. Eng. Chem. Res.* 57: 15319–15323.
604 Han, L., Li, H.Q., Choi, S.J. et al. (2012). *Appl. Catal. A: Gen.* 429: 67–72.
605 Calabrese, C., Liotta, L.F., Carbonell, E. et al. (2017). *Chemsuschem* 10: 1202–1209.
606 Liu, M.S., Lan, J.W., Liang, L. et al. (2017). *J. Catal.* 347: 138–147.
607 Xie, Y., Zhang, Z., Jiang, T. et al. (2007). *Angew. Chem. Int. Ed. Engl.* 46: 7255–7258.
608 Qiao, K., Ono, F., Bao, Q.X. et al. (2009). *J. Mol. Catal. A: Chem.* 303: 30–34.
609 Wang, W.L., Li, C.Y., Yan, L. et al. (2016). *ACS Catal.* 6: 6091–6100.
610 Mavlyankariev, S.A., Ahlers, S.J., Kondratenko, V.A. et al. (2016). *ACS Catal.*: 3317–3325.
611 Guo, Z.J., Jiang, Q.W., Shi, Y.M. et al. (2017). *ACS Catal.* 7: 6770–6780.
612 Sun, Q., Aguila, B., Perman, J. et al. (2016). *J. Am. Chem. Soc.* 138: 15790–15796.
613 Cho, H.C., Lee, H.S., Chun, J. et al. (2011). *Chem. Commun.* 47: 917–919.
614 Wang, J.Q., Leong, J.Y., and Zhang, Y.G. (2014). *Green Chem.* 16: 4515–4519.
615 Zhong, W., Bobbink, F.D., Fei, Z.F., and Dyson, P.J. (2017). *ChemSusChem* 10: 2728–2732.
616 Chen, Y.J., Luo, R.C., Bao, J.H. et al. (2018). *J. Mater. Chem. A* 6: 9172–9182.

617 Wang, J.-Q., Yue, X.-D., Cai, F., and He, L.-N. (2007). *Catal. Commun.* 8: 167–172.
618 Martínez-Ferraté, O., Chacón, G., Bernardi, F. et al. (2018). *Catal. Sci. Technol.* 8: 3081–3089.
619 Kim, M.I., Kim, D.K., Bineesh, K. et al. (2013). *Catal. Today* 200: 24–29.
620 Wu, X.H., Wang, M.P., Xie, Y.Z. et al. (2016). *Appl. Catal. A-Gen.* 519: 146–154.
621 Guo, L.Y., Deng, L.L., Jin, X.C. et al. (2017). *Catal. Lett.* 147: 2290–2297.
622 Zhou, H.C., Long, J.R., and Yaghi, O.M. (2012). *Chem. Rev.* 112: 673–674.
623 Lee, J., Farha, O.K., Roberts, J. et al. (2009). *Chem Soc Rev* 38: 1450–1459.
624 Zhou, H.C. and Kitagawa, S. (2014). *Chem. Soc. Rev.* 43: 5415–5418.
625 Chughtai, A.H., Ahmad, N., Younus, H.A. et al. (2015). *Chem. Soc. Rev.* 44: 6804–6849.
626 He, H., Perman, J.A., Zhu, G., and Ma, S. (2016). *Small* 12: 6309–6324.
627 Song, J., Zhang, Z., Hu, S. et al. (2009). *Green Chem.* 11: 1031.
628 Miralda, C.M., Macias, E.E., Zhu, M. et al. (2011). *ACS Catal.* 2: 180–183.
629 Zalomaeva, O.V., Chibiryaev, A.M., Kovalenko, K.A. et al. (2013). *J. Catal.* 298: 179–185.
630 Gao, W.Y., Chen, Y., Niu, Y. et al. (2014). *Angew. Chem., Int. Ed. Engl.* 53: 2615–2619.
631 Beyzavi, M.H., Klet, R.C., Tussupbayev, S. et al. (2014). *J. Am. Chem. Soc.* 136: 15861–15864.
632 Zhou, Z., He, C., Xiu, J.H. et al. (2015). *J. Am. Chem. Soc.* 137: 15066–15069.
633 Zhang, G.Y., Wei, G.F., Liu, Z.P. et al. (2016). *Chem. Mater.* 28: 6276–6281.
634 Wang, H.H., Hou, L., Li, Y.Z. et al. (2017). *ACS Appl. Mater. Interf.* 9: 17969–17976.
635 He, H.M., Sun, Q., Gao, W.Y. et al. (2018). *Angew. Chem., Int. Ed.* 57: 4657–4662.
636 Patel, P., Parmar, B., Kureshy, R.I. et al. (2018). *ChemCatChem* 10: 2401–2408.
637 Han, Y.H., Zhou, Z.Y., Tian, C.B., and Du, S.W. (2016). *Green Chem.* 18: 4086–4091.
638 Kathalikkattil, A.C., Roshan, R., Tharun, J. et al. (2016). *Chem. Commun.* 52: 280–283.
639 Lu, B.B., Jiang, W., Yang, J. et al. (2017). *ACS Appl. Mater. Interf.* 9: 39441–39449.
640 Nguyen, P.T.K., Nguyen, H.T.D., Nguyen, H.N. et al. (2018). *ACS Appl. Mater. Interf.* 10: 733–744.
641 Leng, Y., Lu, D., Zhang, C.J. et al. (2016). *Chem. Eur. J.* 22: 8368–8375.
642 Liu, T.T., Liang, J., Huang, Y.B., and Cao, R. (2016). *Chem. Commun.* 52: 13288–13291.
643 Chen, Y.J., Luo, R.C., Xu, Q.H. et al. (2017). *ChemSusChem* 10: 2534–2541.
644 Wang, W.L., Wang, Y.Q., Li, C.Y. et al. (2017). *ACS Sustainable Chem. Eng.* 5: 4523–4528.
645 Chen, Y.J., Luo, R.C., Xu, Q.H. et al. (2018). *ACS Sustainable Chem. Eng.* 6: 1074–1082.

646 Feng, D.W., Chung, W.C., Wei, Z.W. et al. (2013). *J. Am. Chem. Soc.* 135: 17105–17110.

647 Zadehahmadi, F., Ahmadi, F., Tangestaninejad, S. et al. (2015). *J. Mol. Catal. A: Chem.* 398: 1–10.

648 Liu, C.J., Liu X.H., Li, B. et al. (2017). *J. Energy Chem.* 26: 821–824.

649 Dutta, G., Jana, A.K., and Natarajan, S. (2018). *Chem. Asian J.* 13: 66–72.

650 Ding, M.L. and Jiang, H.L. (2018). *ACS Catal.* 8: 3194–3201.

651 Sun, Y.X., Huang, H.L., Vardhan, H. et al. (2018). *ACS Appl. Mater. Interfaces* 10: 27124–27130.

652 Talapaneni, S.N., Buyukcakir, O., Je, S.H. et al. (2015). *Chem. Mater.* 27: 6818–6826.

653 Zhi, Y.F., Shao, P.P., Feng, X. et al. (2018). *J. Mater. Chem. A* 6: 374–382.

654 Sun, J., Liang, L., Sun, J. et al. (2010). *Catal. Surv. Asia* 15: 49–54.

655 Shaikh, A.-A.G. and Sivaram, S. (1996). *Chem. Rev.* 96: 951–976.

656 Clements, J.H. (2003). *Ind. Eng. Chem. Res.* 42: 663–674.

657 Aresta, M. (2017). *Coord. Chem. Rev.* 334: 150–183.

658 Sonnati, M.O., Amigoni, S., Taffin de Givenchy, E.P. et al. (2013). *Green Chem.* 15: 283–306

659 Ma, J., Sun, N., Zhang, X. et al. (2009). *Catal. Today* 148: 221–231.

660 Dai, W.-L., Luo, S.-L., Yin, S.-F., and Au, C.-T. (2009). *Appl. Catal. A: Gen.* 366: 2–12.

661 Ozorio, L.P. and Mota, C.J.A. (2017). *ChemPhysChem* 18: 3260–3265.

662 Aresta, M., Dibenedetto, A., and Tommasi, I. (2000). *Appl. Organomet. Chem.* 14: 799–802.

663 Jasiak, K., Krawczyk, T., Pawlyta, M. et al. (2016). *Catal. Lett.* 146: 893–901.

664 Dias, L.D., Carrilho, R.M.B., Henriques, C.A. et al. (2018). *ChemCatChem* 10: 2792–2803.

665 Sun, J., Fujita, S., Zhao, F. et al. (2005). *J. Catal.* 230: 398–405.

666 Wang, Y.L. Sun, J.H., Xiang, D. et al. (2009). *Catal. Lett.* 129: 437–443.

667 Xiang, D., Liu, X.F., Sun, J.S. et al. (2009). *Catal. Today* 148: 383–388.

668 Zalomaeva O.V., Maksimchuk, N.V., Chibiryaev, A.M. et al. (2013). *J. Energy Chem.* 22: 130–135.

669 Han, Q., Qi, B., Ren, W. et al. (2015). *Nat. Commun.* 6: 10007.

670 Sharma, N., Dhankhar, S.S., Kumar, S. et al. (2018). *Chem. Eur. J.* 24: 16662–16669.

671 Maksimchuk, N.V., Ivanchikova, I.D., Ayupov, A.B., and Kholdeeva, O.A. (2016). *Appl. Catal. B: Environ.* 181: 363–370.

672 Evangelisti, C., Guidotti, M., Tiozzo, C. et al. (2018). *Inorg. Chim. Acta* 470: 393–401.

673 Shi, Z.L., Niu, G.Q., Han, Q.X. et al. (2018). *Mol. Catal.* 461: 10–18.

674 Nguyen, H.T.D., Tran, Y.B.N., Nguyen, H.N. et al. (2018). *Inorg. Chem.* 57: 13772–13782.

675 Kennedy, G.L. (1986). *CRC Crit. Rev. Toxicol.* 17: 129–182.

676 Tlili, A., Blondiaux, E., Frogneux, X., and Cantat, T. (2015). *Green Chem.* 17: 157–168.

677 Jacquet, O., Gomes, C.D.N., Ephritikhine, M., and Cantat, T. (2012). *J. Am. Chem. Soc.* 134: 2934–2937.

678 Gomes, C.D.N., Jacquet, O., Villiers, C. et al. (2012). *Angew. Chem., Int. Ed.* 51: 187–190.

679 Jacquet, O., Frogneux, X., Gomes, C.D.N., and Cantat, T. (2013). *Chem. Sci.* 4: 2127–2131.

680 Li, Y., Sorribes, I., Yan, T. et al. (2013). *Angew. Chem., Int. Ed.* 52: 12156–12160.

681 Farlow, M.W. and Adkins, H. (1935). *J. Am. Chem. Soc.* 57: 2222–2223.

682 Kröcher, O., Köppel, R.A., Fröba, M., and Baiker, A. (1998). *J. Catal.* 178: 284–298.

683 Schmid, L., Rohr, M., and Baiker, A. (1999). *Chem. Commun.*: 2303–2304.

684 Rohr, M., Grunwaldt, J.D., and Baiker, A. (2005). *Journal of Catalysis* 229: 144–153.

685 Liu, J.L., Guo, C.K., Zhang, Z.F. et al. (2010). *Chem. Commun.* 46: 5770–5772.

686 Bi, Q.Y., Lin, J.D., Liu, Y.M. et al. (2014). *Chem. Commun.* 50: 9138–9140.

687 Cui, X.J., Zhang, Y., Deng, Y.Q., and Shi, F. (2014). *Chem. Commun.* 50: 189–191.

688 Zhang, Y.J., Wang, H.L., Yuan, H.K., and Shi, F. (2017). *ACS Sustainable Chem. Eng.* 5: 5758–5765.

689 Mitsudome, T., Urayama, T., Fujita, S. et al. (2017). *ChemCatChem* 9: 3632–3636.

690 Luo, R.C., Chen, Y.J., He, Q. et al. (2016). *ChemSusChem*: 1–10.

691 Dong, B., Wang, L.Y., Zhao, S. et al. (2016). *Chem. Commun.* 52: 7082–7085.

692 Lv, H., Wang, W.L., and Li, F.W. (2018). *Chem. Eur. J.* 24: 16588–16594.

693 Mu, Z.J., Ding, X., Chen, Z.Y., and Han, B.H. (2018). *ACS Appl. Mater. Interf.*: 41350–41358.

694 Nale, D.B., Rath, D., Parida, K.M. et al. (2016). *Catal. Sci. Technol.* 6: 4872–4881.

695 He, Z.H., Liu, H.Y., Liu, H.Z. et al. (2017). *Chemcatchem* 9: 1947–1952.

696 Wang, Y., Zhang, J., Liu, J. et al. (2015). *ChemSusChem* 8: 2066–2072.

697 Jiang, H.F., Wang, A.Z., Liu, H.L., and Qi, C.R. (2008). *Eur. J. Org. Chem.*: 2309–2312.

698 Tang, X.D., Qi, C.R., He, H.T. et al. (2013). *Adv. Synth. Catal.* 355: 2019–2028.

699 Yang, Z.Z., Zhao, Y.F., Zhang, H.Y. et al. (2014). *Chem. Commun.* 50: 13910–13913.

700 Song, Q.W., Yu, B., Li, X.D. et al. (2014). *Green Chem.* 16: 1633–1638.

701 Yang, Z.Z., Yu, B., Zhang, H.Y. et al. (2016). *ACS Catal.*: 1268–1273.

702 Zhou, Z., He, C., Yang, L. et al. (2017). *ACS Catal.*: 2248–2256.

703 Zhang, G.Y., Yang, H.M., and Fei, H.H. (2018). *ACS Catal.* 8: 2519–2525.

704 Cui, M., Qian, Q.L., He, Z.H. et al. (2015). *Chem. Eur. J.* 21: 15924–15928.

705 Qiu, J.K., Zhao, Y.L., Wang, H.Y. et al. (2016). *RSC Adv.*: 54020–54026.

706 Panwar, V. and Jain, S.L. (2018). *J. CO2 Utiliz.* 24: 306–314.

707 Chung, C.W.Y. and Toy, P.H. (2004). *Tetrahedron: Asymmetry* 15: 387–399.

708 Cheng, B.B., Yu, B., and Hu, C.W. (2016). *RSC Adv.*: 87179–87187.

709 Sadeghzadeh, S.M. (2016). *J. Mol. Catal. A-Chem.* 423: 216–223.

710 Sadeghzadeh, S.M., Zhiani, R., and Emrani, S. (2018). *Appl. Organomet. Chem.* 32: 1–10.
711 Saadati, S.M. and Sadeghzadeh, S.M. (2018). *Catal. Lett.* 148: 1692–1702.
712 Ferguson, L.N. (1946). *Chem. Rev.* 38: 227–254.
713 Yu, B., Zhao, Y.F., Zhang, H.Y. et al. (2014). *Chem. Commun.* 50: 2330–2333.
714 Kumar, S., Verma, S., and Jain, S.L. (2015). *Tetrahedron Lett.*: 2430–2433.
715 Molla, R.A., Iqubal, M.A., Ghosh, K., and Islam, S.M. (2016). *Green Chem.*: 4649–4656.
716 Feng, X., Sun, A., Zhang, S. et al. (2012). *Org. Lett.* 15: 108–111.
717 Sun, J., Bao, M., Feng, X.J. et al. (2015). *Tetrahedron Lett.*: 6747–6750.
718 Zhang, S., Yu, X., Feng, X. et al. (2015). *Chem. Commun. (Camb.)* 51: 3842–3845.
719 Vessally, E., Soleimani-Amiri, S., Hosseinian, A. et al. (2017). *J. CO2 Utiliz.* 21: 342–352.
720 Patil, Y.P., Tambade, P.J., Parghi, K.D. et al. (2009). *Catal. Lett.* 133: 201–208.
721 Ma, J., Han, B., Song, J. et al. (2013). *Green Chem.* 15: 1485–1489.
722 Nale, D.B., Rana, S., Parida, K., and Bhanage, B.M. (2014). *Catal. Sci. Technol.*: 1608–1614.
723 Zhao, Y.N., Yu, B., Yang, Z.Z., and He, L.N. (2014). *RSC Adv.*: 28941–28946.
724 Nale, D.B., Saigaonkar, S.D., and Bhanage, B.M. (2014). *J. CO2 Utiliz.*: 67–73.
725 Kumar, S., Verma, S., Shawat, E. et al. (2015). *RSC Adv.*: 24670–24674.
726 Sadeghzadeh, S.M. (2015). *Catal. Commun.*: 91–96.
727 Sadeghzadeh, S.M. (2016). *Catal. Sci. Technol.*: 1435–1441.
728 Samanta, S. and Srivastava, R. (2017). *Sustainable Energy Fuels*: 1390–1404.
729 Gautam, P. and Bhanage, B.M. (2015). *Catal. Sci. Technol.* 5: 4663–4702.
730 Konishi, H. and Manabe, K. (2014). *Synlett* 25: 1971–1986.
731 Morimoto, T. and Kakiuchi, K. (2004). *Angew. Chem., Int. Ed.* 43: 5580–5588.
732 Wu, L., Liu, Q., Jackstell, R., and Beller, M. (2014). *Angew. Chem., Int. Ed.* 53: 6310–6320.
733 Yu, S. and Ma, S. (2012). *Angew. Chem., Int. Ed.* 51: 3074–3112.
734 Khan, A., Zheng, R., Kan, Y. et al. (2014). *Angew. Chem., Int. Ed.* 53: 6439–6442.
735 Xie, Y., Hu, J., Wang, Y. et al. (2012). *J. Am. Chem. Soc.* 134: 20613–20616.
736 Kawamoto, T., Fukuyama, T., and Ryu, I. (2012). *J. Am. Chem. Soc.* 134: 875–877.
737 Koepfer, A., Sam, B., Breit, B., and Krische, M.J. (2013). *Chem. Sci.* 4: 1876–1880.
738 Bausch, C.C., Patman, R.L., Breit, B., and Krische, M.J. (2011). *Angew. Chem., Int. Ed.* 50: 5686–5689.
739 Suenobu, T., Isaka, Y., Shibata, S., and Fukuzumi, S. (2015). *Chem. Commun.* 51: 1670–1672.
740 Heim, L.E., Schloerer, N.E., Choi, J.-H., and Prechtl, M.H.G. (2014). *Nat. Commun.*: 5.
741 Morimoto, T., Fujioka, M., Fuji, K. et al. (2007). *J. Organomet. Chem.* 692: 625–634.

742 Fujioka, M., Morimoto, T., Tsumagari, T. et al. (2012). *J. Org. Chem.* 77: 2911–2923.
743 Natte, K., Dumrath, A., Neumann, H., and Beller, M. (2014). *Angew. Chem., Int. Ed.* 53: 10090–10094.
744 Li, W. and Wu, X.-F. (2014). *J. Org. Chem.* 79: 10410–10416.
745 Wang, H., Cai, J., Huang, H., and Deng, G.-J. (2014). *Org. Lett.* 16: 5324–5327.
746 Cheng, X., Peng, Y., Wu, J., and Deng, G.-J. (2016). *Org. Biomol. Chem.* 14: 2819–2823.
747 Furusawa, T., Morimoto, T., Oka, N. et al. (2016). *Chem. Lett.* 45: 406–408.
748 Okano, T., Kobayashi, T., Konishi, H., and Kiji, J. (1982). *Tetrahedron Lett.* 23: 4967–4968.
749 Ahn, H.S., Han, S.H., Uhm, S.J. et al. (1999). *J. Mol. Catal. A: Chem.* 144: 295–306.
750 Rosales, M., Arrieta, F., Baricelli, P. et al. (2008). *Catal. Lett.* 126: 367–370.
751 Rosales, M., Gonzalez, A., Gonzalez, B. et al. (2005). *J. Organomet. Chem.* 690: 3095–3098.
752 Morimoto, T., Yamasaki, K., Hirano, A. et al. (2009). *Org. Lett.* 11: 1777–1780.
753 Makado, G., Morimoto, T., Sugimoto, Y. et al. (2010). *Adv. Synth. Catal.* 352: 299–304.
754 Cini, E., Airiau, E., Girard, N. et al. (2011). *Synlett*: 199–202.
755 Uhlemann, M., Doerfelt, S., and Boerner, A. (2013). *Tetrahedron Lett.* 54: 2209–2211.
756 Morimoto, T., Fujii, T., Miyoshi, K. et al. (2015). *Org. Biomol. Chem.* 13: 4632–4636.
757 Fuentes, J.A., Pittaway, R., and Clarke, M.L. (2015). *Chem. Eur. J.* 21: 10645–10649.
758 Wang, Y., Ren, W., Li, J. et al. (2014). *Org. Lett.* 16 (22): 5960–5963.
759 Liu, Q., Yuan, K., Arockiam, P.-B. et al. (2015). *Angew. Chem., Int. Ed.* 54: 4493–4497.
760 Fuji, K., Morimoto, T., Tsutsumi, K., and Kakiuchi, K. (2005). *Chem. Commun.*: 3295–3297.
761 Wang, C., Morimoto, T., Kanashiro, H. et al. (2014). *Synlett* 25: 1155–1159.
762 Shilov, A.E. and Shul'pin, G.B. (1997). *Chem. Rev.* 97: 2879–2932.
763 Jun, C.H., Moon, C.W., and Lee, D.Y. (2002). *Chem. Eur. J.* 8: 2423–2428.
764 Kakiuchi, F. and Murai, S. (2002). *Acc. Chem. Res.* 35: 826–834.
765 Ritleng, V., Sirlin, C., and Pfeffer, M. (2002). *Chem. Rev.* 102: 1731–1769.
766 Shibahara, F., Kinoshita, S., and Nozaki, K. (2004). *Org. Lett.* 6: 2437–2439.
767 Sakakibara, K., Yamashita, M., and Nozaki, K. (2005). *Tetrahedron Lett.* 46: 959–962.
768 Kim, D.-S., Park, W.-J., Lee, C.-H., and Jun, C.-H. (2014). *J. Org. Chem.* 79: 12191–12196.
769 Dai, J., Ren, W., Wang, H., and Shi, Y. (2015). *Org. Biomol. Chem.* 13: 8429–8432.
770 Ren, W., Chang, W., Dai, J. et al. (2016). *J. Am. Chem. Soc.* 138: 14864–14867.
771 Wang, Y., Ren, W., and Shi, Y. (2015). *Org. Biomol. Chem.* 13: 8416–8419.

772 Wang, L., Zhang, H., Yu, Q. et al. (2018). *Synlett* 29: 1611–1616.

773 Wu, F.-P., Li, D., Peng, J.-B., and Wu, X.-F. (2019). *Org. Lett.* 21: 5699–5703.

774 Neveux, M., Bruneau, C., and Dixneuf, P.H. (1991). *J. Chem. Soc., Perkin Trans.* 1: 1197–1199.

775 Leadbeater, N.E., Scott, K.A., and Scott, L.J. (2000). *J. Org. Chem.* 65: 3231–3232.

776 Hou, J., Xie, J.-H., and Zhou, Q.-L. (2015). *Angew. Chem., Int. Ed.* 54: 6302–6305.

777 Fu, M.-C., Shang, R., Cheng, W.-M., and Fu, Y. (2016). *ACS Catal.* 6: 2501–2505.

778 Hou, J., Yuan, M.-L., Xie, J.-H., and Zhou, Q.-L. (2016). *Green Chem.* 18: 2981–2984.

779 Nasrollahzadeh, M., Motahharifar, N., Sajjadi, M. et al. (2019). *Green Chem.* 21: 5144–5167.

780 Grant, H.G. and Summers, L.A. (1980). *Aust. J. Chem.* 33: 613–617.

781 Jackson, A. and Methcohn, O. (1995). *J. Chem. Soc., Chem. Commun.*: 1319–1319.

782 Pettit, G.R., Parent, K., Kalnins, M.V. et al. (1961). *J. Org. Chem.* 26: 2563.

783 Hosseini-Sarvari, M. and Sharghi, H. (2006). *J. Org. Chem.* 71: 6652–6654.

784 Kim, J.-G. and Jang, D.O. (2010). *Bull. Korean Chem. Soc.* 31: 2989–2991.

785 Auge, J., Lubin-Germain, N., and Uziel, J. (2007). *Synthesis-Stuttgart*: 1739–1764.

786 Nair, V., Ros, S., Jayan, C.N., and Pillai, B.S. (2004). *Tetrahedron* 60: 1959–1982.

787 Kim, J.-G. and Jang, D.O. (2010). *Synlett*: 1231–1234.

788 Yadav, L.S.R., Raghavendra, M., Udayabhanu et al. (2018). *J. Mater. Sci. Mater. Electr.* 29: 8747–8759.

789 Yadav, L.S.R., Raghavendra, M., Kumar, K.H.S. et al. (2018). *Eur. Phys. J. Plus*: 133.

790 Shekhar, A.C., Kurnar, A.R., Sathaiah, G. et al. (2009). *Tetrahedron Lett.* 50: 7099–7101.

791 Thakuria, H., Borah, B.M., and Das, G. (2007). *J. Mol. Catal. A: Chem.* 274: 1–10.

792 Reddy, M.B.M., Ashoka, S., Chandrappa, G.T., and Pasha, M.A. (2010). *Catal. Lett.* 138: 82–87.

793 Das, V.K., Devi, R.R., Raul, P.K., and Thakur, A.J. (2012). *Green Chem.* 14: 847–854.

794 Hosseini-Sarvari, M., Safary, E., Jarrahpour, A., and Heiran, R. (2012). *Comp. Rendus Chim.* 15: 980–987.

795 Patil, U.B., Singh, A.S., and Nagarkar, J.M. (2013). *Chem. Lett.* 42: 524–526.

796 Ansari, M.I., Hussain, M.K., Yadav, N. et al. (2012). *Tetrahedron Lett.* 53: 2063–2066.

797 Pathare, S.P., Sawant, R.V., and Akamanchi, K.G. (2012). *Tetrahedron Lett.* 53: 3259–3263.

798 Yang, X.J. and Zhang, Y.S. (2013). *Res. Chem. Intermed.* 39: 2843–2848.

799 Habibi, D., Rahmani, P., and Akbaripanah, Z. (2013). *J. Chem.*: 972960.

800 Hong, M. and Xiao, G. (2013). *J. Fluorine Chem.* 146: 11–14.
801 Rostamnia, S. and Doustkhah, E. (2016). *J. Mol. Catal. A: Chem.* 411: 317–324.
802 Pakpour-Roudsari, F., Seddighi, M., Shirini, F., and Tajik, H. (2019). *Chemistry-select* 4: 6382–6389.
803 Bhattacharya, A.K. and Rana, K.C. (2008). *Tetrahedron Lett.* 49: 2598–2601.
804 Tewari, N., Katiyar, D., Tiwari, V.K., and Tripathi, R.P. (2003). *Tetrahedron Lett.* 44: 6639–6642.
805 Bhojegowd, M.R.M., Nizam, A., and Pasha, M.A. (2010). *Chin. J. Catal.* 31: 518–520.
806 Ma'mani, L., Sheykhan, M., Heydari, A. et al. (2010). *Appl. Catal. A: Chem.* 377: 64–69.
807 Khojastehnezhad, A., Rahimizadeh, M., Moeinpour, F. et al. (2014). *Comp. Rendus Chim.* 17: 459–464.
808 Kooti, M. and Nasiri, E. (2015). *J. Mol. Catal. A: Chem.* 406: 168–177.
809 Habibi, D., Heydari, S., and Afsharfarnia, M. (2017). *Appl. Organomet. Chem.*: 31.
810 Marjani, A.P., Hosseini, S.A., Shokri, Z., and Maleki, N. (2017). *Res. Chem. Intermed.* 43: 413–422.
811 Bahari, S., Mohammadi-Aghdam, B., Sajadi, S.M., and Zeidali, F. (2012). *Bull. Korean Chem. Soc.* 33: 2251–2254.
812 Habibi, D., Nasrollahzadeh, M., and Sahebekhtiari, H. (2013). *J. Mol. Catal. A: Chem.* 378: 148–155.
813 Tajbakhsh, M., Alinezhad, H., Nasrollahzadeh, M., and Karnali, T.A. (2016). *J. Colloid Interface Sci.* 471: 37–47.
814 Majumdar, S., De, J., Hossain, J., and Basak, A. (2013). *Tetrahedron Lett.* 54: 262–266.
815 Baghbanian, S.M. and Farhang, M. (2013). *J. Mol. Liq.* 183: 45–49.
816 Shirini, F., Mazloumi, M., and Seddighi, M. (2016). *Res. Chem. Intermed.* 42: 1759–1776.
817 Garkoti, C., Shabir, J., and Mozumdar, S. (2017). *New J. Chem.* 41: 9291–9298.
818 Kim, J.-G. and Jang, D.O. (2010). *Synlett*: 2093–2096.
819 Shirini, F., Seddighi, M., and Mamaghani, M. (2014). *RSC Adv.* 4: 50631–50638.
820 Seddighi, M., Shirini, F., and Mamaghani, M. (2015). *J. Iranian Chem. Soc.* 12: 433–439.
821 Rostamnia, S. and Karimi, Z. (2015). *Inorg. Chim. Acta* 428: 133–137.
822 Simonato, J.P., Walter, T., and Metivier, P. (2001). *J. Mol. Catal. A: Chem.* 171: 91–94.
823 Khatri, C.K. and Chaturbhuj, G.U. (2017). *J. Iranian Chem. Soc.* 14: 2513–2519.
824 Azizi, N., Edrisi, M., and Abbasi, F. (2018). *Appl. Organomet. Chem.*: 32.
825 Simonato, J.P. (2003). *J. Mol. Catal. A: Chem.* 197: 61–64.
826 Wu, X.-F., Neumann, H., and Beller, M. (2013). *Chem. Rev.* 113: 1–35.
827 Cacchi, S., Fabrizi, G., and Goggiamani, A. (2003). *Org. Lett.* 5: 4269–4272.
828 Berger, P., Bessmernykh, A., Caille, J.-C., and Mignonac, S. (2006). *Synthesis-Stuttgart*: 3106–3110.

829 Korsager, S., Taaning, R.H., and Skrydstrup, T. (2013). *J. Am. Chem. Soc.* 135: 2891–2894.

830 Qi, X., Li, R., and Wu, X.-F. (2016). *RSC Adv.* 6: 62810–62813.

831 Qi, X., Jiang, L.-B., Li, C.-L. et al. (2015). *Chem. Asian J.* 10: 1870–1873.

832 Qi, X., Jiang, L.-B., Li, H.-P., and Wu, X.-F. (2015). *Chem. Eur. J.* 21: 17650–17656.

833 Jiang, L.-B., Qi, X., and Wu, X.-F. (2016). *Tetrahedron Lett.* 57: 3368–3370.

834 Qi, X., Li, C.-L., and Wu, X.-F. (2016). *Chem. Eur. J.* 22: 5835–5838.

835 Li, C.-L., Qi, X., and Wu, X.-F. (2016). *Chemistryselect* 1: 1702–1704.

836 Jiang, L.-B., Li, R., Li, H.-P. et al. (2016). *Chemcatchem* 8: 1788–1791.

837 Qi, X., Li, H.-P., and Wu, X.-F. (2016). *Chem. Asian J.* 11: 2453–2457.

838 Weir, J.R., Patel, B.A., and Heck, R.F. (1980). *J. Org. Chem.* 45: 4926–4931.

839 Chen, J., Natte, K., Spannenberg, A. et al. (2014). *Chem. Eur. J.* 20: 14189–14193.

840 Li, W. and Wu, X.-F. (2015). *Eur. J. Org. Chem.*: 331–335.

841 Kaiser, N.F.K., Hallberg, A., and Larhed, M. (2002). *J. Comb. Chem.* 4: 109–111.

842 Wannberg, J. and Larhed, M. (2003). *J. Org. Chem.* 68: 5750–5753.

843 Wu, X.Y., Ekegren, J.K., and Larhed, M. (2006). *Organometallics* 25: 1434–1439.

844 Wu, X.Y. and Larhed, M. (2005). *Org. Lett.* 7: 3327–3329.

845 Wu, X.Y., Wannberg, J., and Larhed, M. (2006). *Tetrahedron* 62: 4665–4670.

846 Nordeman, P., Odell, L.R., and Larhed, M. (2012). *J. Org. Chem.* 77: 11393–11398.

847 He, L., Sharif, M., Neumann, H. et al. (2014). *Green Chem.* 16: 3763–3767.

848 Chen, J., Natte, K., and Wu, X.-F. (2016). *J. Organomet. Chem.* 803: 9–12.

849 Isnard, P., Denise, B., Sneeden, R.P.A. et al. (1983). *J. Organomet. Chem.* 256: 135–139.

850 Choudhary, V.R., Dumbre, D.K., and Narkhede, V.S. (2012). *J. Chem. Sci.* 124: 835–839.

851 Hajimohammadi, M., Safari, N., Mofakham, H., and Shaabani, A. (2010). *Tetrahedron Lett.* 51: 4061–4065.

852 Ko, S., Na, Y., and Chang, S. (2002). *J. Am. Chem. Soc.* 124: 750–751.

853 Na, Y., Ko, S., Hwang, L.K., and Chang, S. (2003). *Tetrahedron Lett.* 44: 4475–4478.

854 Friis, S.D., Taaning, R.H., Lindhardt, A.T., and Skrydstrup, T. (2011). *J. Am. Chem. Soc.* 133: 18114–18117.

855 Hermange, P., Lindhardt, A.T., Taaning, R.H. et al. (2011). *J. Am. Chem. Soc.* 133: 6061–6071.

856 Huber, G.W., Iborra, S., and Corma, A. (2006). *Chem. Rev.* 106: 4044–4098.

857 Konishi, H., Ueda, T., Muto, T., and Manabe, K. (2012). *Org. Lett.* 14: 4722–4725.

858 Profir, I., Beller, M., and Fleischer, I. (2014). *Org. Biomol. Chem.* 12: 6972–6976.

859 Fujihara, T., Hosoki, T., Katafuchi, Y. et al. (2012). *Chem. Commun.* 48: 8012–8014.

860 Ueda, T., Konishi, H., and Manabe, K. (2012). *Org. Lett.* 14: 5370–5373.

861 Armanino, N., Lafrance, M., and Carreira, E.M. (2014). *Org. Lett.* 16: 572–575.

862 Li, H., Neumann, H., Beller, M., and Wu, X.-F. (2014). *Angew. Chem., Int. Ed.* 53: 3183–3186.

863 Adak, L., Kawamura, S., Toma, G. et al. (2017). *J. Am. Chem. Soc.* 139: 10693–10701.

864 Adamek, J., Mazurkiewicz, R., Pazdzierniok-Holewa, A. et al. (2014). *J. Org. Chem.* 79: 2765–2770.

865 Hosoi, K., Nozaki, K., and Hiyama, T. (2002). *Org. Lett.* 4: 2849–2851.

866 Wu, X., Zhao, Y., and Ge, H. (2015). *J. Am. Chem. Soc.* 137: 4924–4927.

867 Chen, J., Feng, J.-B., Natte, K., and Wu, X.-F. (2015). *Chem. Eur. J.* 21: 16370–16373.

868 Esfandiari, H., Jameh-bozorghi, S., Esmaielzadeh, S. et al. (2013). *Res. Chem. Intermed.* 39: 3319–3325.

869 Akabori, S. and Takanohashi, Y. (1991). *J. Chem. Soc., Perkin Trans. 1*: 479–482.

870 Kondo, T., Okada, T., and Mitsudo, T.-a. (1999). *Organometallics* 18: 4123–4127.

871 Ko, S., Han, H., and Chang, S. (2003). *Org. Lett.* 5: 2687–2690.

872 Ueda, T., Konishi, H., and Manabe, K. (2013). *Angew. Chem., Int. Ed.* 52: 8611–8615.

873 Gehrtz, P.H., Hirschbeck, V., and Fleischer, I. (2015). *Chem. Commun.* 51: 12574–12577.

874 Markovic, M., Lopatka, P., Koos, P., and Gracza, T. (2015). *Org. Lett.* 17: 5618–5621.

875 Mondal, K., Halder, P., Gopalan, G. et al. (2019). *Org. Biomol. Chem.* 17: 5212–5222.

876 Behr, A., Kanne, U., and Keim, W. (1986). *J. Mol. Catal.* 35: 19–28.

877 Jo, E.-A., Lee, J.-H., and Jun, C.-H. (2008). *Chem. Commun.*: 5779–5781.

878 Moran, J., Preetz, A., Mesch, R.A., and Krische, M.J. (2011). *Nat. Chem.* 3: 287–290.

879 Qiang Liu, K.Y., Arockiam, P.-B., Franke, R. et al. (2015). *Angew. Chem., Int. Ed.* 54: 1–6.

880 Sharninghausen, L.S., Campos, J., Manas, M.G., and Crabtree, R.H. (2014). *Nat. Commun.*: 5.

881 Li, Y., Nielsen, M., Li, B. et al. (2015). *Green Chem.* 17: 193–198.

882 Haider, M.H., Dummer, N.F., Zhang, D. et al. (2012). *J. Catal.* 286: 206–213.

883 Wang, Z., Wang, L., Jiang, Y. et al. (2014). *ACS Catal.* 4: 1144–1147.

884 Painter, R.M., Pearson, D.M., and Waymouth, R.M. (2010). *Angew. Chem., Int. Ed.* 49: 9456–9459.

885 Chung, K., Banik, S.M., De Crisci, A.G. et al. (2013). *J. Am. Chem. Soc.* 135: 7593–7602.

886 Pagliaro, M., Ciriminna, R., Kimura, H. et al. (2007). *Angew. Chem., Int. Ed.* 46: 4434–4440.

887 Gu, Y., Azzouzi, A., Pouilloux, Y. et al. (2008). *Green Chem.* 10: 164–167.

888 Sun, Q., Wang, S., and Liu, H. (2017). *ACS Catal.* 7: 4265–4275.

889 Nielsen, D.B., Wahlqvist, B.A., Nielsen, D.U. et al. (2017). *ACS Catal.* 7: 6089–6093.

890 Serrano-Ruiz, J.C., Luque, R., and Sepulveda-Escribano, A. (2011). *Chem. Soc. Rev.* 40: 5266–5281.

891 Verendel, J.J., Nordlund, M., and Andersson, P.G. (2013). *ChemSusChem* 6: 426–429.

892 Dai, X., Rabeah, J., Yuan, H. et al. (2016). *ChemSusChem* 9: 3133–3138.

893 Dai, X., Adomeit, S., Rabeah, J. et al. (2019). *Angew. Chem., Int. Ed.* 58: 5251–5255.

894 Gibson, S.E. and Stevenazzi, A. (2003). *Angew. Chem., Int. Ed.* 42: 1800–1810.

895 Shibata, T. (2006). *Adv. Synth. Catal.* 348: 2328–2336.

896 Ikeda, K., Morimoto, T., and Kakiuchi, K. (2010). *J. Org. Chem.* 75: 6279–6282.

897 Min, B.-H., Kim, D.-S., Park, H.-S., and Jun, C.-H. (2016). *Chem. Eur. J.* 22: 6234–6238.

898 Geuther, A. and J. L. (1862). *Ann. Chem.* 123: 121–122.

899 Makosza, M. (1975). *Pure Appl. Chem.* 43: 439–463.

900 Meilahn, M.K., Olsen, D.K., Brittain, W.J., and Anders, R.T. (1978). *J. Org. Chem.* 43: 1346–1350.

901 Fedorynski, M. (2003). *Chem. Rev.* 103: 1099–1132.

902 Zhao, H., Du, H., Yuan, X. et al. (2016). *Green Chem.* 18: 5782–5787.

903 Khodakov, A.Y., Chu, W., and Fongarland, P. (2007). *Chem. Rev.* 107: 1692–1744.

904 Waugh, K.C. (1992). *Catal. Today* 15: 51–75.

905 Liu, X. and Gu, Z. (2015). *Org. Chem. Front.* 2: 778–782.

906 Gockel, S.N. and Hull, K.L. (2015). *Org. Lett.* 17: 3236–3239.

907 Sun, G., Lei, M., and Hu, L. (2016). *RSC Adv.* 6: 28442–28446.

908 Guo, S.-R., Santhosh Kumar, P., Yuan, Y.-Q., and Yang, M.-H. (2017). *Tetrahedron Lett.* 58: 2681–2684.

909 Kannaboina, P., Raina, G., Anil Kumar, K., and Das, P. (2017). *Chem. Commun.* 53: 9446–9449.

910 Nitha, P.R., Joseph, M.M., Gopalan, G. et al. (2018). *Org. Biomol. Chem.* 16: 6430–6437

911 Gopalaiah, K. (2013). *Chem. Rev.* 113: 3248–3296.

912 Parmar, D., Henkel, L., Dib, J., and Rueping, M. (2015). *Chem. Commun.* 51: 2111–2113.

913 Iwamoto, T., Okuzono, C., Adak, L. et al. (2019). *Chem. Commun.* 55: 1128–1131.

914 Nishida, Y., Takeda, N., Matsuno, K. et al. (2018). *Eur. J. Org. Chem.* 2018: 3928–3935.

Part II

Carbonyl Compounds as Catalysts

5

Acid-Catalyzed Reactions with $-CO_2H$

5.1 Carboxylic Acid Molecules Catalyzed Reactions

Carboxylic acids, as hydrogen bond donors and acceptors, can form relatively strong complexes with organic molecules, and this is particularly true for organic substrates functionalized with carbonyl, alcohol, and other heteroatomic groups [1]. Meanwhile, the carbonyl group in carboxylic acids is generally regarded as a nucleophilic center, that is, making the α-carbon in the carbonyl electrophilic, would be of significant synthetic utility and provide a complementary strategy to access derivatives that are otherwise difficult to prepare conventionally.

5.1.1 Hydrolysis/Aminolysis/Ethanolysis Reactions

In fact, as early as 1964, the first example of an acid-catalyzed reaction in which one amide function aids in hydrolysis of another amide group was reported by Cohen. Under the same conditions, the hydrolysis of *o*-benzamido-*N*,*N*-dicyclohexylbenzamide(I) in aqueous acetic acid containing sulfuric acid at 80 °C occurred at the *N*,*N*-dicyclohexylamide linkage and was at least 10^4 times faster than the hydrolysis of the *N*,*N*-dicyclohexylbenzamide(IV), which is less sterically hindered model compound. Several mechanisms were presented by which the *o*-benzamido group might participate in the hydrolysis by acting as a general base, a nucleophile, or, when protonated, as a general acid (Scheme 5.1) [2].

Carboxyl-catalyzed intramolecular aminolysis was also reported by Gisin and coworker in 1972 [3]. The intramolecular aminolysis of D-valyl-L-prolyl-resin of polymer-supported peptide ester could be catalyzed by carboxylic acids, but this process could be repressed if adding the carbodiimide reagent prior to the carboxyl component. The rate of the intramolecular aminolysis was greatly determined by the composition and configuration of the dipeptide.

The kinetics of carboxylic acid (benzoic acid, *n*-butyric acid, and chloroacetic acid) catalysis of the ethanolysis of *p*-chlorophenyl isocyanate in diethyl ether at room temperature has been investigated by Lammiman and Satchell in 1974 [4]. The results suggested that an intermediate of ethanol and a carboxylic acid adduct was rapidly formed, and the attack of this intermediate to the isocyanate was a rate-determining step. The intermediate was argued to have a cyclic structure.

Carbonyl Compounds: Reactants, Catalysts and Products, First Edition.
Feng Shi, Hongli Wang and Xingchao Dai.

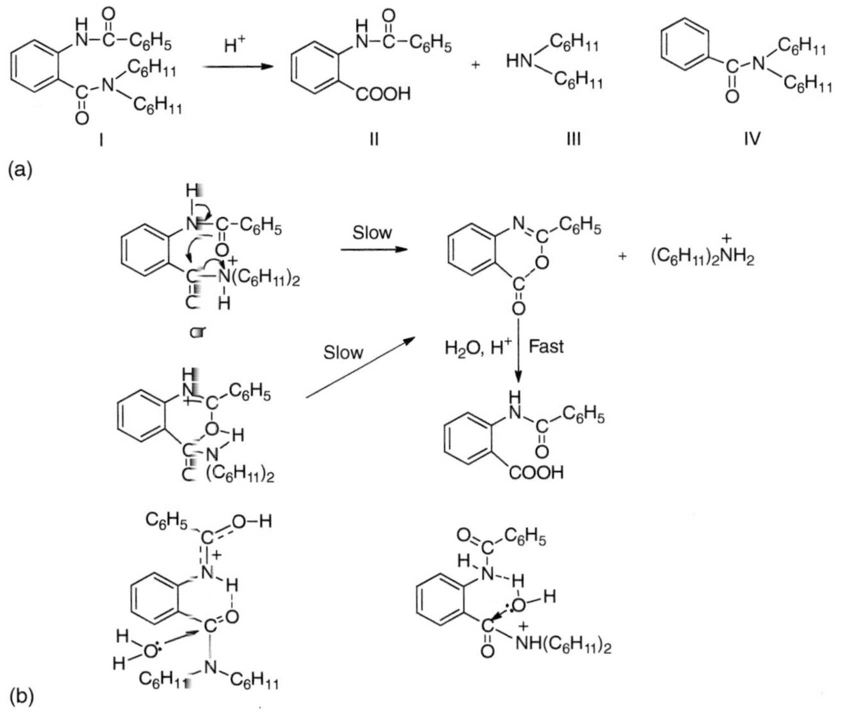

Scheme 5.1 (a) Acid-catalyzed hydrolysis of *o*-benzamido-*N*,*N*-dicyclohexylbenzamide; (b) Mechanisms of acid-catalyzed hydrolysis of *o*-benzamido-*N*,*N*-dicyclohexylbenzamide. Source: Modified from Cohen and Lipowitz [2].

In 1981, the kinetics of acetic, formic, propionic, trifluoroacetic, mono-, di-, and trichloro acid-catalyzed hydrolysis of formamide at 65 °C was explored [5]. An analysis of kinetic results showed that the acid-catalyzed hydrolysis was attributed to special hydronium ion rather than the general acid catalysis. In the same year, the reductions of $Co(NH_3)_5py^{3+}$ and $Co(en)_3{}^{3+}$ with Eu^{2+} were catalyzed by pyridinecarboxylic acids [6]. The catalytic performance depended mainly on the features of –C(=O)N– or –C(=O)O– group in the 4-position of the pyridine ring. Performance enhancements in the reduction of radical intermediate resulting from interposition of "insulating" side chains were attributed to intervention of homoallylic intermediates of types VIII and IX (Scheme 5.2), in which europium was bound to the carboxyl but interacted with the pyridine ring as well.

In 1990, synthesis of 3,6-anhydro-D-glucono-lactone and 3,6-anhydro-D-mannono-1,4-lactone, respectively, from D-glucono-l,5-lactone and D-mannono-1,4-lactone catalyzed by formic and acetic acid in anhydrous hydrogen fluoride was reported. Also, this protocol was applicable for the synthesis of 1,4 : 3,6-dianhydride and 1,4 : 3,6-dianhydrooglucitol, respectively, from D-mannitol and D-glucitol. The mechanism by ^{13}C NMR spectroscopy indicated that dioxolanylium ions might be the intermediate [7].

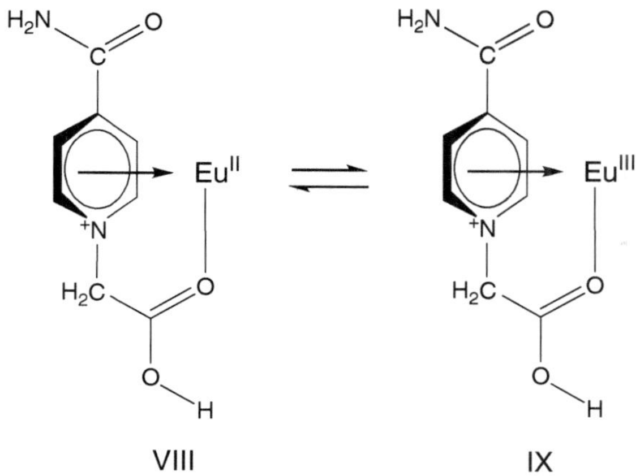

Scheme 5.2 Homoallylic intermediates of types VIII and IX.

5.1.2 Mutarotation of 2,3,4,6-Tetramethyl-d-glucose (TM-G)

After that, Kergomard and Renard showed that benzoic acid and other carboxylic acids could act as bifunctional catalysts for the mutarotation of 2,3,4,6-tetramethyl-D-glucose (TM-G) in benzene in 1968 [8]. The same year, Lillford and Satchell also reported a kinetic study of the spontaneous and the carboxylic acid-catalyzed additions of water and a series of alcohols to dimethylketen in ether at 25 °C. The results showed that addition of water was an autocatalytic process, and the catalytic effect of an acid was inversely related to its conventional strength [9].

5.1.3 Depolymerization of Polyoxymethylenes

In 1969, the depolymerization of polyoxymethylenes in $PhNO_2$ solution at 135 °C under the influence of phenylacetic acid, caproic acid, and propionic acid was investigated. It is found that phenylacetic acid does not noticeably cleave chains but only catalyzes depolymerization at the hemiacetal chain ends, and the rate of depolymerization is roughly proportional to the concern of acid [10].

5.1.4 Elimination Reactions

In 1974, Hirata and Fukuzumi found that the reaction of 1-(*p*-nitrophenyl)-2-(2-quinolyl)ethyl acetate with carboxylic acids could give quantitatively an elimination product, *trans*-2-*p*-nitrostyrylquinoline [11]. The concentration of acid had a great influence on the reaction mechanism. At low concentration of acid, the reaction proceeded by the protonation of the N atom by monomeric carboxylic acids to give an ion-pair intermediate, followed by *E*2 attack of the counter anion, carboxylate anion, on the β-H to give trans olefin. The latter step was rate-determining step. At high concentration of acid, the reaction proceeded by dimerization of carboxylic acids, the protonation of the quinolyl N atom, and attack of the carboxylate anion on the β-H atom.

Following this report, carboxylic acids and acetate anion-catalyzed elimination reactions of 1-(*p*-nitrophenyl)-2-(2-quinolyl)ethyl acetate and 1-(*p*-nitrophenyl)-2-(4-quinolyl)ethyl acetate to give *trans*-2-(*p*-nitrostyryl)quinoline and 4-(*p*-nitrostyryl) quinoline, respectively, had been studied in aprotic solvents. The order of acid-catalytic power was acetic acid > acetate anion [12].

5.1.5 Hydrogen–Deuterium Exchange Reactions

Carboxylic acid-catalyzed hydrogen–deuterium exchange reaction of 2-methylpyridine in deuterium oxide had been reported by Hirata and Fukuzumi in 1976 [13]. The reaction was carried out in a sealed pyrex tube at 120 °C, and the results showed that the exchange rate catalyzed by acids (DCl, CH_3CO_2H, $ClCH_2CO_2H$, Cl_2CHCO_2H, Cl_3CCO_2H, and CF_3CO_2H) was larger than that catalyzed by bases such as NaOD and NaOAc. The strong acids were more effective than acetic acid at lower acid concentrations, while at higher acid concentrations vice versa.

5.1.6 Reduction Reactions

In 1986, the mechanism of carboxylic acid-catalyzed reduction of substituted nitrosobenzenes with 1-benzyl-3,5-bis(1-pyrrolidinylcarbonyl)-1,4-dihydropyridine as the reducing agent had been studied in acetonitrile at 25 °C [14]. When excess dihydropyridine was used, hydroxylamine was observed as the major product, while that was azoxybenzene in the reverse case. The kinetic experiments results indicated that a general acid catalysis by carboxylic acids played an important role in the reduction of nitrosobenzene.

5.1.7 Decomposition of Diazodiphenylmethane

Acid-catalyzed decomposition of diazodiphenylmethane (DDM) had been studied in DMSO containing varying water content in 1999. The kinetics showed that the reactions of DDM with chloroacetic and trifluoromethanesulfonic acids were found to be of first order. Moreover, water concentration on kinetic hydrogen isotope effect suggested that the carboxylic acid-catalyzed reactions in near-anhydrous DMSO did not use the rate-limiting proton transfer mechanism attributed to the reaction of DDM in hydroxylic solvents. A mechanism, leading to diphenylcarbene, which would account for the observed products: benzophenone, benzhydrol, and benzhydryl esters, was suggested (Scheme 5.3) [15].

5.1.8 Amino–Imino Tautomerism Reactions

In 2000, acetic acid-catalyzed amino–imino tautomerism in 9-cyclohexylmethyladenine (9CHA) in cyclohexane solvent through the excited-state double-proton-transfer (ESDPT) mechanism was first reported (Scheme 5.4) [16]. There were two reasons for synthesizing 9CHA, on the one hand, the cyclohexylmethyl functional group acted as a substituent to simulate the linkage of the deoxyribose

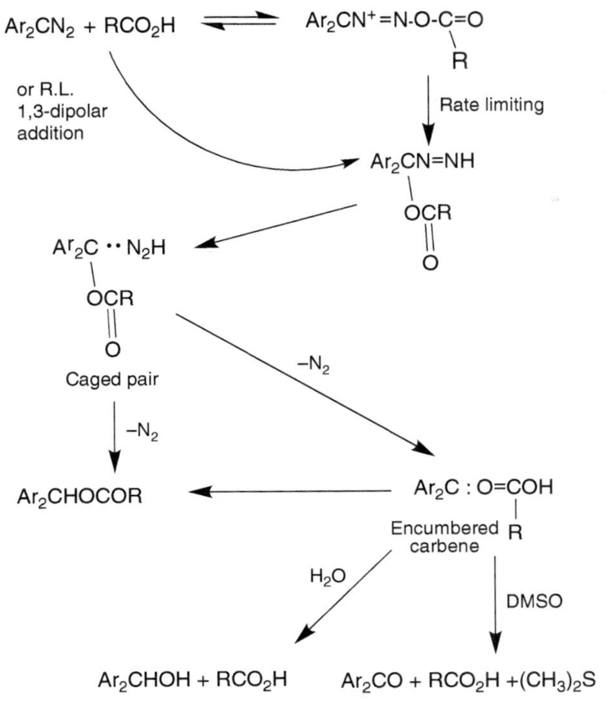

Scheme 5.3 Mechanism of acid-catalyzed decomposition of DDM. Source: Eliason [15].

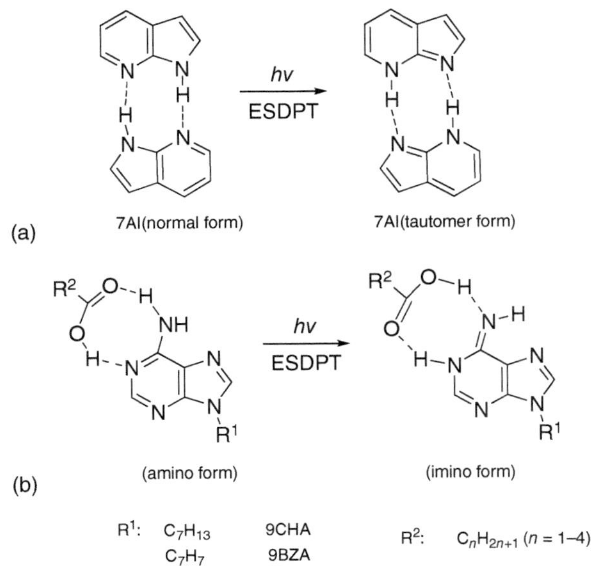

Scheme 5.4 Proposed ESDPT mechanism of (a) the 7AI dimmer and (b) adenines/carboxylic acid complexes. Source: Chou et al. [16].

site at the N(9) position, and, on the other hand, the hydrophobic moiety of the cyclohexylmethyl group increases the solubility of 9CHA in cyclohexane. Acid concentration-dependent electronic absorption spectra were observed for 9CHA in which the 258 nm peak characterized gradually disappeared upon increasing the acetic acid concentration, accompanied by a red shift of the absorption profile throughout the titration in cyclohexane.

5.1.9 Aldol Reaction

The asymmetric amino acid-catalyzed aldol reaction is plausibly an ancient transformation, which enzymes have catalyzed for billions of years, and the ability of amino acids to catalyze the asymmetric aldol reaction was discovered in the 1970s by Eguiazabal and coworkers [17] and Wang and coworkers [18]. In 2005, acyclic amino acid- and their derivatives-catalyzed asymmetric intermolecular aldol reactions with high stereoselectivity were presented. Density functional theory (DFT) calculation and experimental data were combined to understand the stereochemistry of primary amino acid-catalyzed intermolecular aldol reactions that involve enamine intermediates [19]. The results showed that only one amino acid molecule was involved in the transition state, and the carboxylic acid-catalyzed enamine mechanism was more favorable than other mechanisms because it required the lowest activation energy (Scheme 5.5). Furthermore, the calculations demonstrated that such a mechanism accurately predicted the proper stereochemistry of the reaction.

5.1.10 Friedel–Crafts Reaction

The 3-substituted indoles synthesis has attracted much attention because of the broad scope of their biological activity. In 2006, the carboxylic acid-catalyzed 3-substituted indoles synthesis by three-component aza-Friedel–Crafts (AFC) reactions of aldehydes, primary amines, and indoles in water was reported (Scheme 5.6) [20]. The AFC reactions had been known to be difficult to control, but the present reaction system enabled the desired products to be obtained in good yields in Scheme 5.7 [20]. The reaction proceeded under nonmetallic conditions in water, which offers an efficient method for the synthesis of 3-substituted indoles with diverse structures.

Friedel–Crafts alkylation of β-arylethylimines was widely used in synthetic chemistry, which was first reported by Levchik et al. over a century ago [21]. In 2017, the cooperative mechanism of benzoic acid and thiourea co-catalysis in the enantioselective protio-Pictet–Spengler reaction by computational and experimental analyses was revealed. Studies of kinetic experiments, isotope effect, structure–enantioselectivity relationship, and computational analyses indicated rearomatization via deprotonation of the pentahydro-β-carbolinium ion intermediate as both the rate- and the enantioselectivity-determining step. Pre-equilibrium substrate protonation was promoted by the thiourea coordination with acidification of the weak benzoic acid co-catalyst, and the pentahydro-β-carbolinium ion intermediate was stabilized by the interactions of the conjugate base with anion binding.

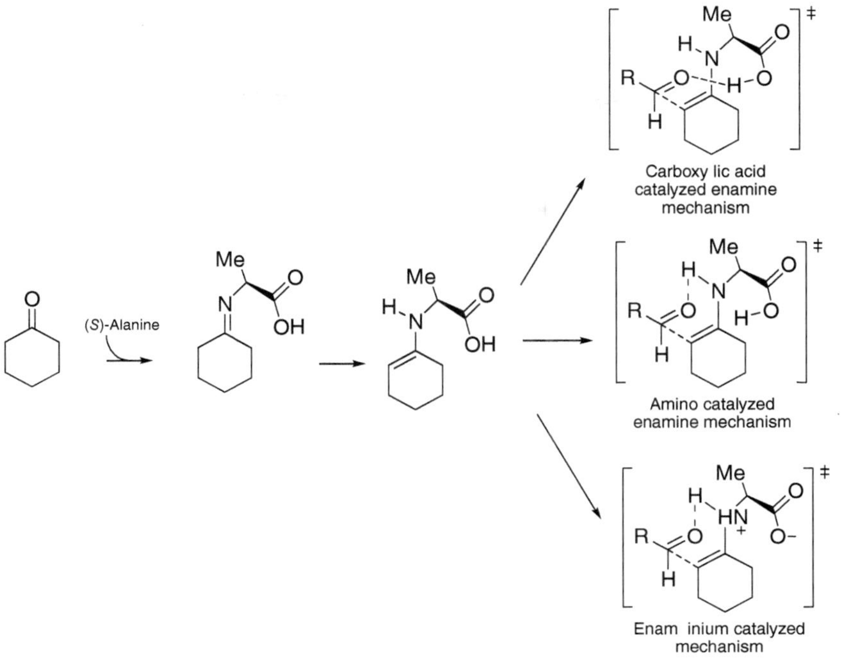

Scheme 5.5 The three possible mechanisms in the alanine-mediated aldol reaction that involve an enamine intermediate.

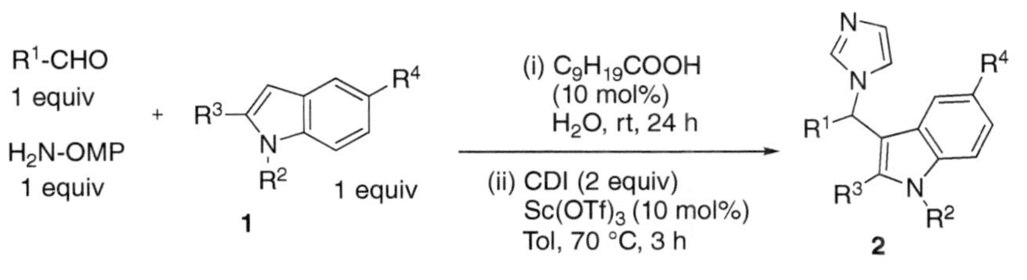

Scheme 5.6 Synthesis of aromatase inhibitor-type compounds **2** via three-component aza-Friedel–Crafts reaction in water. Source: Shirakawa and Kobayashi [20].

These interactions increased the concentrations of high-energy intermediates en route to the rate-determining step and thereby contribute to rate acceleration, whereas enantioinduction is mediated by differential $\pi\cdots\pi$ and C–H$\cdots\pi$ interactions within a scaffold organized by multiple hydrogen-bonding interactions (Scheme 5.8) [22].

5.1.11 Hydrogen Shifts Reaction

In 2010, carboxylic acid-catalyzed intramolecular hydrogen shifts in the gas phase through the initial formation of a hydrogen bonded complex followed by a double-hydrogen-shift reaction were demonstrated. Quantum chemistry and

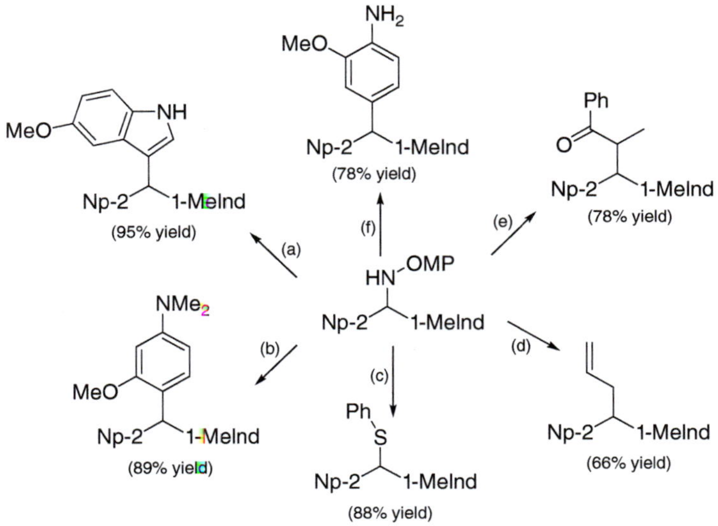

Scheme 5.7 Transformations of aza-Friedel–Crafts product. Source: Shirakawa and Kobayashi [20].

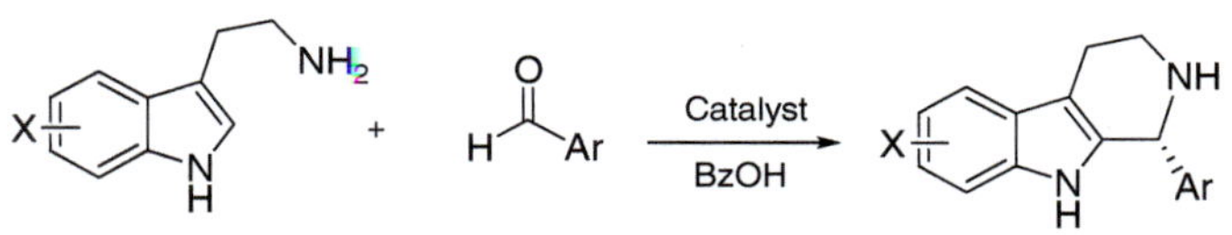

Scheme 5.8 Benzoic acid and thiourea co-catalysis in the enantioselective protio-Pictet–Spengler reaction. Source: Klausen et al. [22].

variational transition state theory for the barrierless association were used to calculate the rate constants for formic acid-catalyzed conversion of vinyl alcohol to acetaldehyde, that is, the simplest keto–enol tautomerization [23].

5.1.12 Cyclization Reaction

Cascade reactions have attracted considerable attention in organic synthesis because of the high atom efficiency. In 2011, a novel strategy for the synthesis of biologically important dihydrofuro[3,2-*c*]pyridinones and 3(2*H*)-furanones by a carboxylic acid-catalyzed bromonium-initiated cascade process was achieved by Liang's group (Scheme 5.9) [24]. Further, they demonstrated that functionalized 5-amino-3(2*H*)-furanones could also be synthesized by carboxylic acid-catalyzed reaction of 1-acetylcyclopropanecarboxamides with *N*-halosuccinimide (NXS) (Scheme 5.10) [25]. A possible mechanism, in which halonium ion was supposed to generate in situ in the presence of NXS and then directly captured by the amide oxygen, giving the oxonium intermediate, was proposed.

The catalytic asymmetric 1,3-dipolar cycloaddition (1,3-DC) is one of the most established methods for the stereoselective synthesis of five-membered

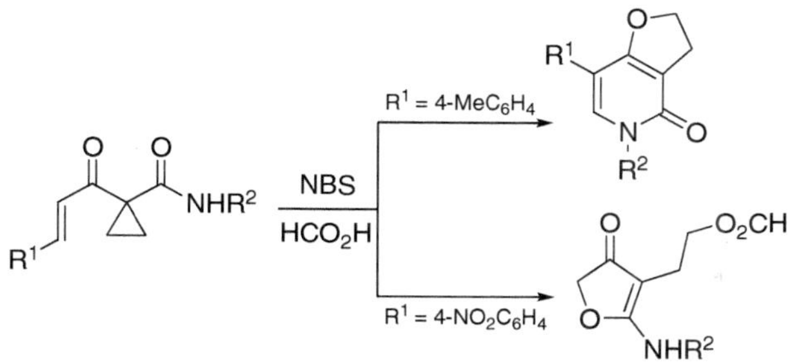

Scheme 5.9 Halonium initiated electrophilic cascades of 1-alkenoylcyclopropane carboxamides. Source: Wei et al. [24].

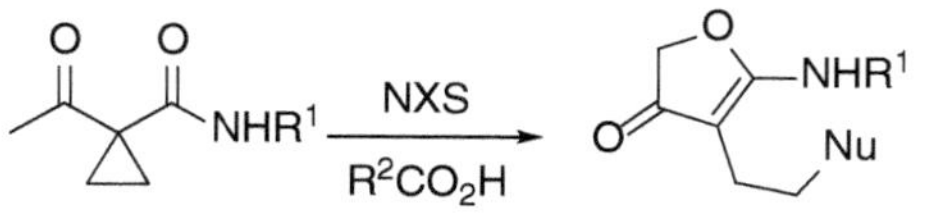

R^1 = alkyl; R^2 = H, alkyl, aryl
X = Cl, Br, I; Nu = OR, O_2CR, X

Scheme 5.10 5-Amino-3(2*H*)-furanones synthesized by carboxylic acid-catalyzed reaction of 1-acetylcyclopropanecarboxamides with *N*-halosuccinimide (NXS). Source: Wei et al. [25].

heterocycles. In 2011, an asymmetric inverse-electron-demand (IED) 1,3-DC of *C,N*-cyclic azomethine imines and vinyl ether was reported (Scheme 5.11) [26], which catalyzed by an axially chiral dicarboxylic acid having diarylmethyl groups at the 3,3′-positions. The concept can be applied for switching the regioselectivity of the cycloaddition of vinylogous aza-enamines catalyzed by titanium/binolate.

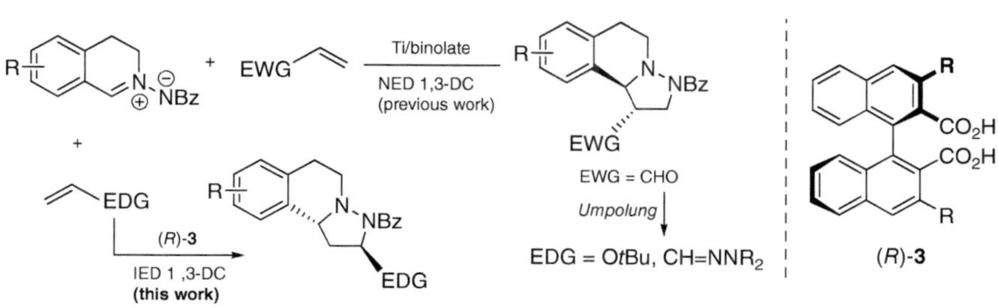

Scheme 5.11 Asymmetric inverse-electron-demand (IED) 1,3-DC of *C,N*-cyclic azomethine imines and vinyl ether. Source: Hashimoto et al. [26].

In 2013, a chiral Brønsted acid catalyst composed of two independent organic molecules, a chiral diol and 2-boronobenzoic acid, was developed to catalyze asymmetric *trans*-aziridinations of *N*-Boc and *N*-benzyl imines with *N*-phenyldiazoacetamide. A boronate ester-assisted chiral carboxylic acid was in situ generated, which acted as the active catalyst. ^1H NMR and X-ray crystallography confirmed the nature of catalyst. This binary system offers an opportunity to tune the chiral environment and acidity of the catalyst independently and avoids a tedious synthesis of a variety of different catalysts (Scheme 5.12) [27].

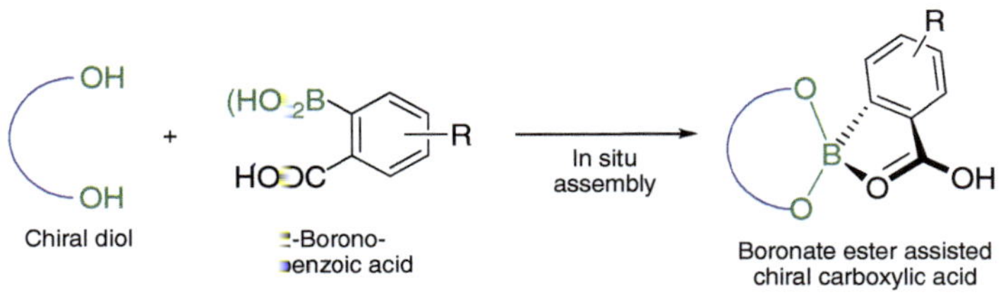

Scheme 5.12 Chiral diol and 2-boronobenzoic acid catalyze asymmetric *trans*-aziridinations of *N*-Boc and *N*-benzyl imines with *N*-phenyldiazoacetamide. Source: Hashimoto et al. [27].

In 2018, one-pot synthesis of 2,3-dihydroquinazolin-4(1*H*)-ones with 2-aminobenzamide and aldehydes catalyzed by carboxylic acids such as benzylic acid, phthalic acid, and salicylic acid was reported. This protocol has some advantages such as low cost, nontoxic solvents, and simple separation of products [28].

Cyclic aminals are common structural elements of diverse commercial drugs. In 2018, an enantioselective aminalization of 2-aminobenzenesulfonamides and aldehydes was achieved using a cyclopentadiene-based chiral carboxylic acid as a catalyst (Scheme 5.13) [29]. A series of benzothiadiazine class of chiral cyclic aminals were synthesized in good to excellent yields and with up to 98% ee.

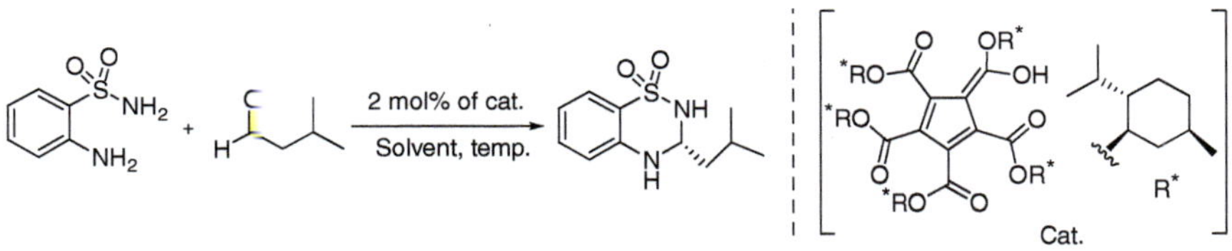

Scheme 5.13 Cyclopentadiene-based chiral carboxylic acid-catalyzed enantioselective aminalization of 2-aminobenzenesulfonamides and aldehydes. Source: Sui et al. [29].

Tetraalkylpyrazaboles are usually synthesized in the reactions of pyrazole or its derivatives with trialkylboranes or tetraalkyldiboranes. However, due to the high stability of the boron–carbon bond, these reactions require elevated temperatures, such as 130–150 °C. In 2000, the reaction of triethylborane with pyrazole leading to 4,4,8,8-tetraethylpyrazabole was reported to be catalyzed by pivalic and benzoic acids in xylene at 80 °C. Based on the isolated intermediates, which had been characterized by NMR, infrared (IR) spectroscopy, and X-ray crystallography, a reaction mechanism was proposed in Scheme 5.14 [30]. The rate of reaction was most probably limited mainly by the equilibrium between complex **1** and uncomplexed Et_3B and pyrazole, which were active reagents in the cycle.

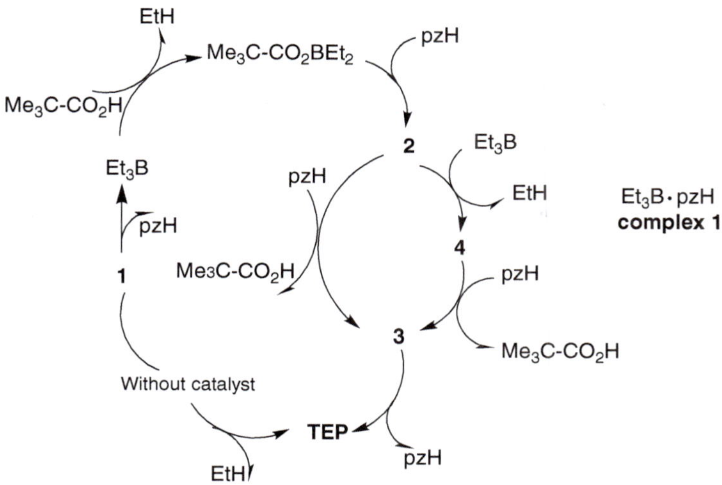

Scheme 5.14 The simplified pathway of the catalyzed reaction of triethylborane with pyrazole. The reversibility of certain stages is omitted for clarity. Source: Dabrowski et al. [30].

5.1.13 Hydroboration Reaction

The direct hydroboration of alkynes is one of the most straightforward methods to synthesize alkenylboronates, which are important synthetic intermediates in various useful transformations. As is well known, carboxylic acid is an efficient reagent for C–B bond cleavage. In 2014, carboxylic acid-catalyzed direct hydroboration of various alkynes with pinacolborane [HB(pin)] without using any metal catalysts and additives was demonstrated (Scheme 5.15) [31]. This protocol not only afforded the synthetically important alkenyl diboronates and monoboronates in good to excellent yields with exclusive regio- and stereoselectivities but also might open a new avenue for hydroboration of versatile unsaturated C–C multiple bonds using organocatalysts.

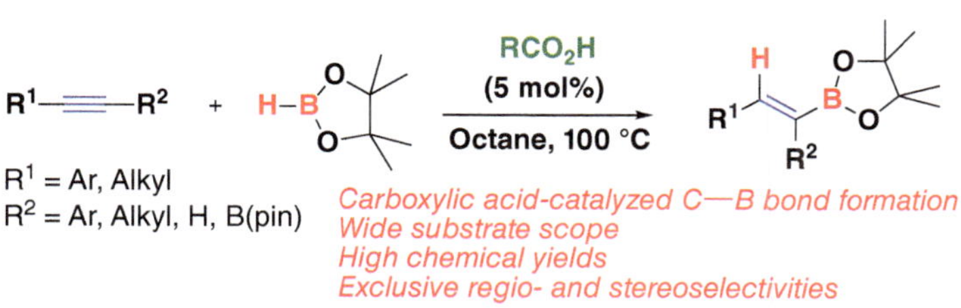

Scheme 5.15 Carboxylic acid-catalyzed direct hydroboration of alkynes with pinacolborane [HB(pin)] without using any metal catalysts and additives. Source: Ho et al. [31].

5.1.14 Trifluoromethylation Reaction

Synthesis of fluorine-containing organic molecules has attracted more notice due to their potential applications in medicinal chemistry, agrichemicals, and materials science. The trifluoromethylation is an efficient method to introduce fluorine atom. Among the existing methods, photoredox-catalyzed radical trifluoromethylations have attracted significant attention due to the clean and sustainability of energy source. In 2019, cheap and commercially available anthraquinone-2-carboxylic acid (AQN-2-CO_2H) was used as the photo-organocatalyst in the visible-light-induced trifluoromethylation of imidazo[1,2-*a*]pyridines with Langlois reagent as the trifluoromethylating reagent (Scheme 5.16) [32]. A series of 3-(trifluoromethyl)imidazo[1,2-*a*]pyridine derivatives with broad functionalities were synthesized in moderate to good yields by direct regioselective functionalization. Other imidazoheterocycles such as imidazo[2,1-*b*]thiazole and benzo[*d*]imidazo-[2,1-*b*]thiazole were also well tolerated.

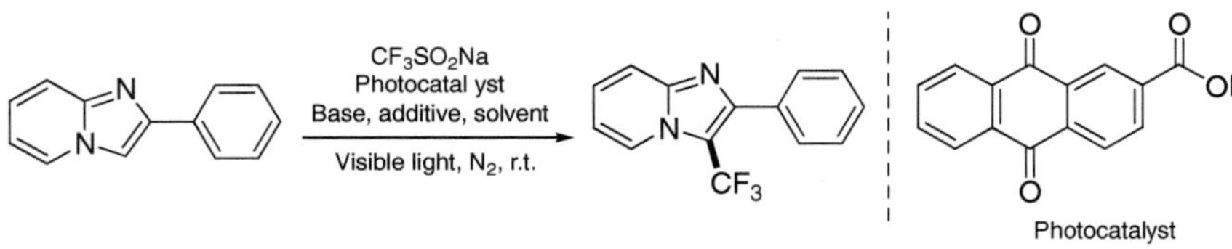

Scheme 5.16 Anthraquinone-2-carboxylic acid as the photo-organocatalyst catalyzed the visible-light-induced trifluoromethylation of imidazo[1,2-*a*]pyridines with Langlois reagent. Source: Zhou et al. [32].

5.2 Carbon Material-Catalyzed Reactions

Carbon materials were widely applied in different catalytic reactions, in which oxygen-containing functional groups on the surface play crucial roles in generating active anchor sites for further functionalization and act as acid centers for acid catalysis. Therefore, the development of novel functionalization methods that enable the introduction of acid functional groups more efficiently and controllably is desirable. Carboxyl groups may be the most significant of these oxygen-containing moieties because of the relatively high acidity and good thermal stability.

5.2.1 Reduction of Nitric Oxide

As early as 1983, the effect of the oxygen-containing functional groups on the surface of carbon on the catalytic activity was discovered. For the reduction of nitric oxide with ammonia, the original coke exhibited poor catalytic performance and only 10% yield of nitrogen was obtained after 90 minutes reaction at 150 °C. However, the yield could be increased to 70% when coke was activated with sulfuric acid and subsequent heat treatment at ~450 °C, which might be attributed to the formation of oxidized functional groups induced on the coke surface by the sulfuric acid [33].

After that, in 1993, the effects of varying the gas phase and surface oxygen concentrations on the reduction of NO to N_2 overactivated carbons were studied in detail. Surface oxygen-containing groups were introduced by treating a coconut shell-activated carbon with sulfuric acid at 100–300 °C. The type and concentration of surface oxides were investigated by IR spectroscopy and linear temperature-programmed desorption. The activity of activated carbon at reaction temperatures >150 °C was enhanced by sulfuric acid treatment, which acted as an oxidizing agent and created acidic surface functionalities including carboxyl and carbonyl groups [34].

In 2001, the activity of carbon catalysts in catalytic NO reduction with NH_3, which can be significantly affected by treatments with sulfuric and nitric acids, was examined in a fixed bed reactor at 110–200 °C. The transient behavior of NO reduction over the original and the treated carbons reflected that the adsorption of reactants, either physical or chemical, plays an important role in determining the rate of NO conversion, which varies with the surface characteristics of the carbons and the number of oxygen-containing sites. In oxygen-free systems, the catalytic activity of the nitric-acid-treated (ADN) carbon was higher than original carbon (AD), but it was lower for the sulfuric-acid-treated carbon (ADS). The activity of these carbons in the absence of oxygen was found to be an increasing function of the amount of CO_2 evolved, while in the presence of oxygen the activity had an order of ADN > ADS > AD, which was found to be identical to that for the amount of CO evolved in the desorption process. The reaction scheme of Scheme 5.17(a) could involve the interaction of NH_3 with the acidic part of carboxyl groups, i.e. COH, to form $CO^-(NH_4)^+$, then following the adsorption of NH, NO molecule would diffuse to the carboxyl group and interact with the C=O part to form C(ONO). And on the hydroxyl groups and the carbonyl sites of the pyrone-like groups are described in Scheme 5.17(b). Therefore, NO reduction with NH_3 over carbon suggested that the formation of $CO^-(NH_4)^+$ and C(ONO) on surface oxides and further interactions of these complexes were important steps (Scheme 5.17) [35].

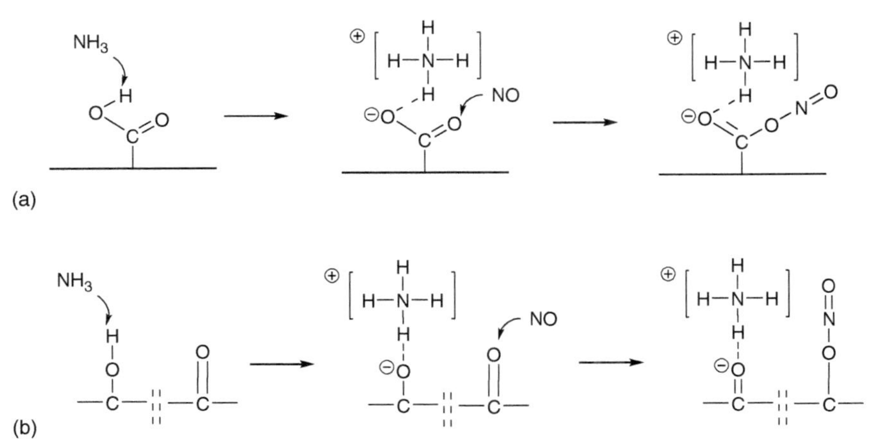

Scheme 5.17 Reaction scheme for the formation of surface complexes during NO reduction with NH_3 over carbon catalyst, (a) the interaction of NH_3 with the acidic part of carboxyl groups; (b) the interaction of NH_3 with the hydroxyl groups and the carbonyl sites. Source: Teng et al. [35].

In 2004, carbon samples with different surface chemistry were obtained from commercially activated carbon by chemical modification involving oxidation with concentration nitric acid (DOx) (i), high-temperature treatment (~1000 K) under vacuum (DHT) (ii) or in ammonia (iii) for the reduction of nitrogen oxide with ammonia [36]. Additionally, a portion of the DOx sample was promoted with iron(III) ions (DOxFe). The catalytic tests were performed in a microreactor at 413–573 K. The DHT carbon exhibited the lowest activity, and the DOx sample showed a slight enhanced catalytic activity. As expected, promotion of the DOx sample with iron(III) ions increased drastically its catalytic activity, but the selectivity slightly decreased and N_2O side product was produced. This adverse effect in selectivity can be avoided using ammonia-treated carbons. The most active and selective sample was that first oxidized with nitric acid and then heated in an ammonia stream (DOxN). Surface nitrogen species seemed to play an important role in catalytic reduction of nitrogen oxide with ammonia, possibly facilitating the formation of a reaction intermediate NO_2.

In 2008, carbon-based briquette catalyst that was prepared from a low-rank coal and a commercial tar pitch and treated by sulfuric and nitric acids was applied for NO reduction (SCR) with NH_3 in a fixed-bed reactor at low temperature (150 °C) [37]. The catalytic behavior of acid-treated briquettes was found to vary with the surface characteristics of the carbon support, which suggested that the number of oxygen-containing sites and the loading and dispersion of vanadium affect the reaction activity. In the presence of oxygen, the SCR activity was enhanced with a nitric acid treatment that led to the formation of acidic surface groups such as carboxyl and lactone. Therefore, the results suggested that the formation of acidic sites on the surface was an important step for NO reduction with NH_3 over carbon-based catalysts.

In 2017, graphene oxide (GO) was utilized to provide a detailed investigation on the role of carboxylic acid groups on carbon in reducing nitric oxide at low temperature (100 °C) [38]. As a result, GO with abundant carboxylic acid groups reduced 45% of NO at 100 °C, and GO without these oxygen-containing groups barely reduced <5% of NO under the same conditions. After reducing NO, the content of carboxylic acid groups on GO decreased, meanwhile, the anhydride group content increased. The results suggested that the carboxylic acid groups directly participated in NO reduction and were transformed into anhydride and lactone groups, and the transformation was confirmed by FTIR spectroscopy. Moreover, transient kinetics and TG-MS studies showed the transfer and conversion process of oxygen atoms on GO during reduction of NO and revealed that [1] the carboxylic acid group reacted easily with carbon to form CO_2, generating a free active site ($-C_n-H$), which had a high affinity toward NO reduction; [2] the NO reacted with the in situ generated active site to form N_2 and regenerated the carboxylic acid groups (Scheme 5.18). This mechanism explained why carbon with abundant carboxylic acid groups is usually an effective reductant and catalyst support for NO removal at low temperature.

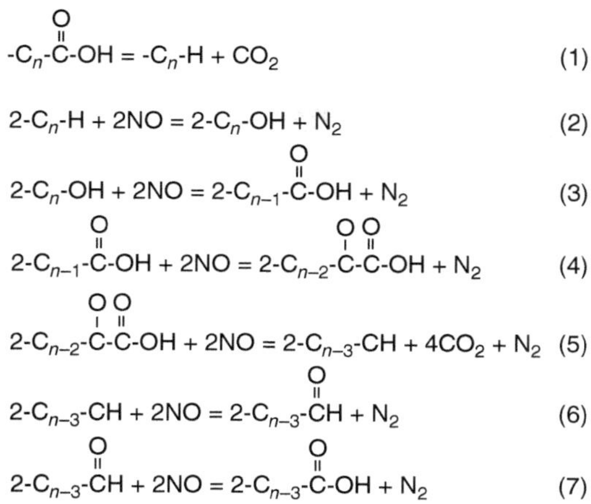

Scheme 5.18 The transfer and conversion process of oxygen atoms on graphene oxide during the reduction of NO.

5.2.2 Oxidative Coupling of Amines to Imines

GO, a two-dimensional aromatic scaffold decorated by oxygen-containing functional groups, possesses rich chemical properties and may present a green alternative to precious metal catalysts. In 2012, a simple chemical treatment that could create and enlarge the defects in GO imparted GO-enhanced catalytic activities for the oxidative coupling of amines to imines, and up to 98% yield was obtained at 5 wt% catalyst loading under solvent-free and open-air conditions. This study examined the origin of the enhanced catalytic activity, which could be linked to the synergistic effect of carboxylic acid groups and localized unpaired electrons at the edges of defects [39].

5.2.3 Depolymerization of Cellulose and Lignocellulose

The conversion of cellulose, the most abundant form of lignocellulosic biomass on earth, to glucose is generally recognized as a central cost controlling bottleneck for renewable approaches to fuels and chemicals. In 2014, depolymerization of adsorbed (1 → 4)-β-D-glucans (β-glu) derived from crystalline cellulose was catalyzed by weak-acid sites of post-synthetically surface-functionalized mesoporous carbon nanoparticle (MCN) catalysts, HT_5–HSO_3–MCN and COOH–MCN. The effect of acid-site density and β-glu molecular weight on this depolymerization was investigated (Scheme 5.19) [40]. Both HT_5–HSO_3–MCN and COOH–MCN catalyzed the depolymerization of adsorbed β-glu strands via glycosidic bond hydrolysis to afford glucose in 73% and 90% yields, respectively, at a buffered pH of 2.0, after treatment at 180 °C for three hours. The yields were significantly higher than an unfunctionalized MCN-control catalyst under identical conditions, which demonstrated the importance of post-synthetic surface functionalization for achieving weak-acid catalytic hydrolysis. All yields were generally decreased when

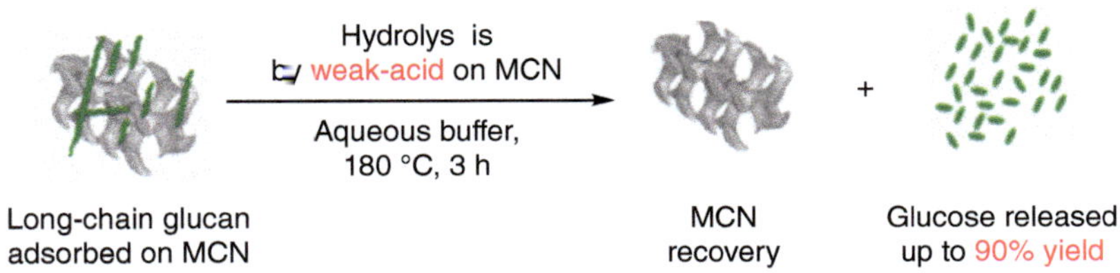

Scheme 5.19 Weak-acid sites of post-synthetically surface-functionalized mesoporous carbon nanoparticle (MCN)-catalyzed depolymerization of adsorbed (1 → 4)-β-D-glucans (β-glu) derived from crystalline cellulose. Source: Charmot et al. [40].

using a lower rather than higher molecular weight of adsorbed β-glu strands on the same catalyst, and catalytic hydrolysis rate generally increases upon decreasing buffer pH particularly for the more acidic carboxylic acid-functionalized catalyst COOH–MCN.

The direct hydrolysis of crystalline cellulose to glucose in water without prior pretreatment enables the transformation of biomass into fuels and chemicals. In 2015, the post-synthetic surface-functionalized nanoporous carbon material MSC-30 by oxidation was used to depolymerize crystalline cellulose without pretreatment in water (Scheme 5.20) [41]. The most active catalyst provided glucose in 70% yield, and no reaction was observed when using MSC-30 without oxidation under identical conditions. This comparison strongly suggested that the catalytic activity was greatly influenced by a high local density of weak-acid sites on the external surface. These weak-acid sites endowed the post-synthetically functionalized catalyst with

Scheme 5.20 Functionalization of nanoporous carbon material MSC-30-catalyzed depolymerization of crystalline cellulose without pretreatment in water. Source: To et al. [41]. © 2015, John Wiley & Sons.

a more negative zeta potential and could catalyze the depolymerization reaction by direct solid–solid contact with the crystalline cellulose surface.

In 2016, the reusable and cost-effective heterogeneous catalyst, which was prepared by air oxidation of biomass feedstock (Eucalyptus) bearing an aromatic skeleton with carboxylic groups and aliphatic moieties, for the depolymerization of lignocellulose was reported. The catalyst quickly converts lignin-containing Eucalyptus to glucose and xylose in high yields within one hour. Furthermore, after the reaction, the solid residue composed of catalysts and insoluble substrate of woody biomass was easily transformed back into fresh catalyst by the same air oxidation method (Scheme 5.21) [42]. The strategy is a self-contained system using woody biomass as both the catalyst source and the substrate for improving the efficiency of catalytic processes. At the same year, a carbon catalyst containing a high density of carboxyl groups, which was prepared by solvent-free mechanochemical oxidation of activated carbon with persulfate salts as the oxidant without introducing sulfonic acid groups, for the hydrolysis of cellulose was reported. The carbon catalyst was analyzed by X-ray photoelectron spectroscopy, which showed that the newly generated oxygenated functional groups were mainly present in the form of carboxyl groups. The oxidized carbon showed that good catalytic activity for the hydrolysis of cellulose and glucose was obtained in 85% yield from mix-milled cellulose in the presence of a trace amount of HCl after 20 minutes.

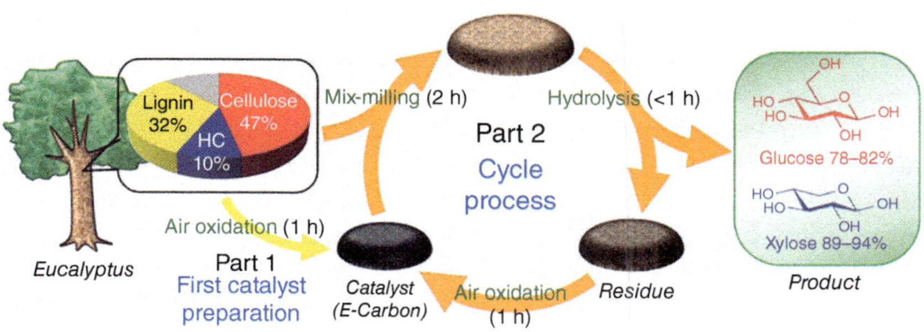

Scheme 5.21 Schematic representation of the *E-Carbon* system. HC, xylan hemicellulose. Source: Kobayashi et al. [42]. © 2016, Royal Society of Chemistry. Licensed under CCBY 3.0.

In 2019, carbon catalysts bearing only weakly acidic –COOH and phenolic –OH groups were prepared with black liquor from rice straw pretreated with potassium hydroxide (KOH) aqueous solution (Scheme 5.22) [43]. The obtained carbon catalyst exhibited excellent activity for the hydrolysis of cellulose and the alkali-pretreated rice straw in water. After mixed ball-milling pretreatment, 76.3% yield of glucose was obtained from cellulose, and yields of 52.1% glucose and 66.5% xylose were obtained from alkali-pretreated rice straw, in 0.015 wt% HCl solution at 200 °C for 60 minutes. Moreover, the weakly acidic catalyst showed good stability and recyclability in the aqueous reaction system. The work showed that weakly acidic catalysts from biomass provided a promising process for the efficient hydrolysis of lignocellulose.

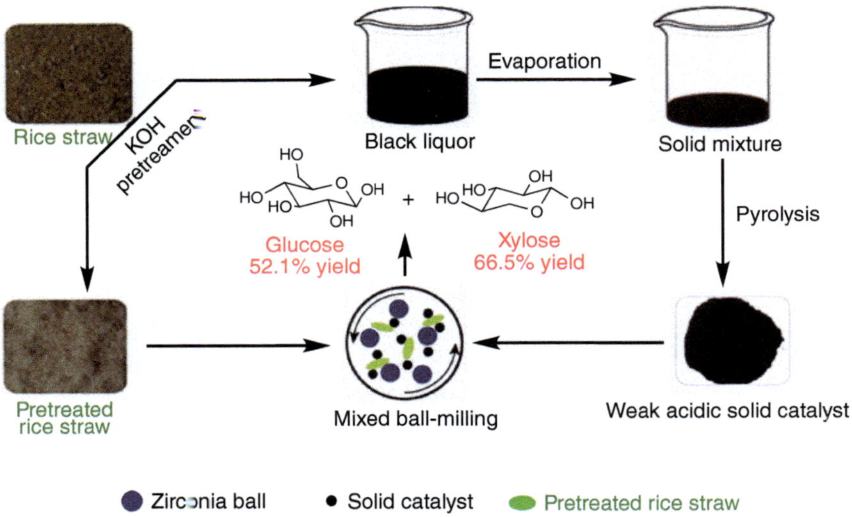

Scheme 5.22 The preparation of carbon catalysts bearing only weakly acidic –COOH and phenolic –OH groups. Source: Qi et al. [43]. © 2019, Elsevier.

5.2.4 Nitrobenzene Reduction Reaction and Beckmann Rearrangement Reaction

In 2016, hydroxy- and carboxyl-enriched hydrothermal carbon (HTC) with different morphology and size was synthesized from glucose by hydrothermal carbonization, and the number of surface carboxyl groups could be obviously increased by treatment in air in an open tube furnace. The HTC could be used in nitrobenzene reduction, and higher conversion was obtained on the sphere morphology with smaller size. Carboxyl-enriched HTC could be applied in Beckmann rearrangement of cyclohexanone oxime, and the performance was even better than conventionally used solid acids (HY and HZSM-5), which could be attributed to the significantly larger number of weak Brønsted acid sites on HTC provided. These results demonstrated that the weak Brønsted acid sites on carbon could exhibit significant advantages to other conventionally used solid acids in some weak Brønsted acid-catalyzed reactions [44].

5.2.5 Ring-Opening Reaction of Styrene Oxide

In 2017, the acid-catalyzed ring-opening reaction of styrene oxide was used as a probe reaction for evaluating the acidic properties of carboxylated carbocatalysts, in which the acidities of four carboxylated carbocatalysts were quantitatively determined by Boehm titrations [45]. The results showed that the initial reaction rates were normalized by the number of acid sites as well as their activation energies E_a and pre-exponential factor A were derived from Arrhenius analysis. Using the reported pK_a values of (substituted) benzoic acids, a linear relationship was established between pK_a and $-\log(A/\text{overall carboxyls})$, from which the fitted pK_a values of the carbocatalysts were obtained. The variation in the calculated pK_a values of the carboxyl groups was attributed to the electronic effects of the nitro groups, which was supported by the temperature-programmed desorption profiles of nitrogen monoxide ions. The kinetic profiling of the catalytic behaviors of carboxylated carbocatalysts in the styrene oxide ring-opening reaction might assist in evaluating the acidity of carbocatalysts in terms of acid strength and the number of available acid sites.

6

Reactions via Carbonyl and Hydroxyl Groups Recycling

6.1 Carbon-Catalyzed Selective Oxidation Reactions

Oxidative dehydrogenation (ODH) of alkanes is an economical alternative to nonoxidative thermal dehydrogenation for alkane-to-alkene conversion. Traditionally, supported metal oxide catalysts were used for ODH because of their high conversion capability but are limited by reduced selectivity at higher conversions due to increased combustion side reactions and structural complexity, which hinders tailoring them for selective dehydrogenation. Hence, there is a need for alternative catalysts for ODH, which can be highly selective to alkenes and are environmentally friendly and sustainable in nature. Carbon catalysts for ODH of alkanes are promising alternatives to metal oxides because of the inexpensive, abundant, sustainable, and renewable nature of carbon materials. Various oxygen-functionalized carbons were reported as active catalysts for ODH of alkane.

6.1.1 Oxidative Dehydrogenation of Ethylbenzene

As early as 1987, an active coke was reported to catalyze ODH of ethylbenzene [46]. The results of different analyzing techniques such as burn off experiments, thermodesorption analysis, ESCA, SIMS, and ESR studies showed that quinone/hydroquinone or aroxyl/phenol groups were most probably to be considered as the catalytically active centers in the reaction. Reaction mechanism study showed that the adsorbed ethylbenzene reacted with oxygen most probably via a radical mechanism to form larger molecules, and these molecules condensed by oxidation arrange to larger molecules on the basis of C_6-rings with quinoid or aroxylic functions at the edges of the layer planes.

In 1999, a detailed kinetic study of the ODH of ethylbenzene catalyzed by carbon catalysts was studied. The type and amount of functional groups on the surface of catalysts were modified by oxidative treatments in the gas or in the liquid phase [47]. The results showed that the higher catalytic performances of the gas phase treatments were associated with the increase in the amount of carbonyl/quinone groups on the surface of catalysts, which were identified as the active sites for the reaction by selectively removing the surface functional groups. Moreover, the catalytic activity of the catalyst increased with the degree of oxidation. A more

Carbonyl Compounds: Reactants, Catalysts and Products, First Edition.
Feng Shi, Hongli Wang and Xingchao Dai.

detailed kinetic study of the ODH of ethylbenzene in the presence of an activated carbon catalyst was presented in 2000 [48]. The kinetic results obtained in five hour runs showed that activated carbons exhibited high conversions and selectivities at lower temperatures compared with conventional catalysts (metal oxides). The experimental results could be well described by a kinetic model in which the main reaction occurred by a redox mechanism involving quinone/hydroquinone groups on the surface of the carbon catalyst. The activation energies determined for the main reaction and for the formation of carbon oxides were comparable to those determined on coked oxide catalysts.

In 2007, a novel mechanistic study on ODH of ethylbenzene catalyzed by nanostructured carbons such as activated carbon, nanodiamond, and carbon nanotubes (CNTs) was reported [49]. The reaction obeyed the dual-site Langmuir–Hinshelwood (L–H) mechanism, in which the cleavage of C–H bonds was kinetically relevant. IR spectroscopy of nanodiamond before and after reaction showed a strong band at 1790 cm^{-1}, which could be assigned to the stretching C=O vibration in quinine or carbonyl groups. The π-conjugation effect of the graphite structure did not seem to be essential for nanocarbon to act as a catalyst, and the mechanistic findings differed in from those for the metal or metal oxide catalysts involving lattice oxygen that had been seen as one of the most active phases. In 2010, the application of nanodiamond as a catalyst for direct dehydrogenation of ethylbenzene under steam-free conditions was reported. The unique sp^2–sp^3 hybrid structure results in superior activity and stability and freedom from coke formation (Scheme 6.1) [50].

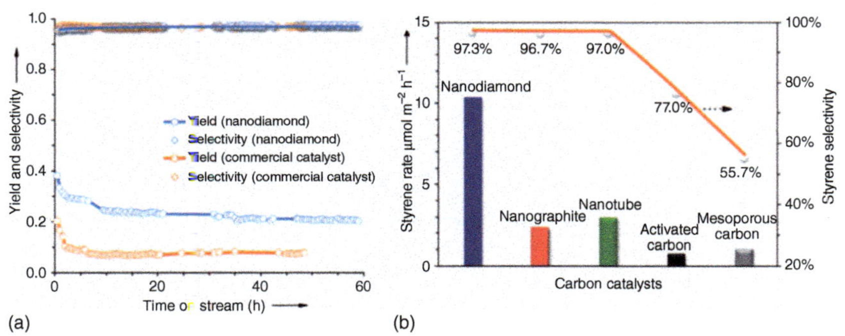

Scheme 6.1 Ethylbenzene (EB) dehydrogenation activity of nanodiamond. (a) Long-term performance with activity of commercial Fe_2O_3 catalyst for comparison. (b) Steady-state activities of various carbons. Reaction conditions: 0.05 g, 550 °C, 2.8% EB in helium, 10 ml min^{-1}. Source: Zhang et al. [50].

In 2018, a detailed catalytic reaction mechanism (elementary steps) for nanocarbon-catalyzed ODH of ethylbenzene was revealed by in situ active site titration, kinetic, isotopic, TPSR, and spectroscopic methods [51]. The chemical titration results suggested that the redox cycle of ketonic carbonyl–hydroxyl pairs on nanocarbon served as the active sites (Scheme 6.2), and the turnover frequency (TOF) of carbon catalysts could be calculated through chemical titration process.

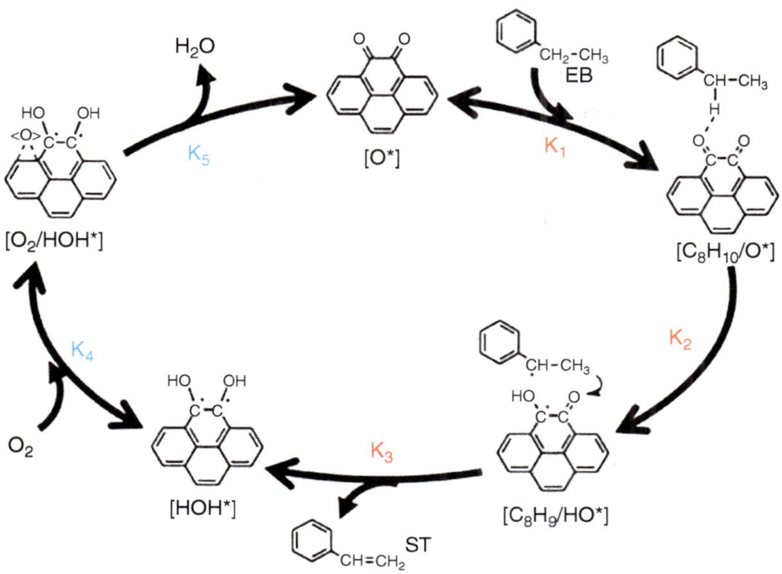

Scheme 6.2 Proposed reaction pathway for ethylbenzene (EB) ODH on ketonic carbonyl groups of nanocarbon catalysts (the backbone of nanocarbon is simplified as four benzene rings).

The rate equation based on the mechanistic interpretations of the catalytic reactions provided intrinsic rate or equilibrium constants and a quantitative analysis on the evolution and content of surface intermediates on carbon catalysts. The proposed catalytic mechanism was different from typical M-v-K or L–H process because of the unique structure of nanocarbon catalysts (lack of lattice oxygen). This is the first time that classical kinetic and mechanistic analysis based on TOF was performed in the field of carbon catalysis, which illuminated on the comprehensive structure–function relations and rational design for nonmetallic carbon catalytic systems.

In 2019, a computational investigation on the ODH mechanism of ethylbenzene catalyzed by oxidized CNTs was carried out [52]. Two different model systems (Model A and Model B) were used to simulate the catalytic oxygen-containing groups on CNT (Scheme 6.3). Model A contained two adjacent epoxy groups in axis configuration and model B consisted of a quinone functionality at the edge of the tube. DFT calculations revealed that the process could occur along three pathways. The first two bifurcate from the initial transformation of an epoxide into a hydroxyl group, which occurred via a biradical transition state and the formation of a benzyl-like intermediate, and the epoxide could further react to release styrene and form a water molecule. Alternatively, in the presence of a second adjacent epoxide, styrene was also produced without water formation along a less exothermic pathway that leaved two hydroxyl groups on the nanotube surface. Along the third pathway, two adjacent carbonyl groups (quinone functionality) also promoted

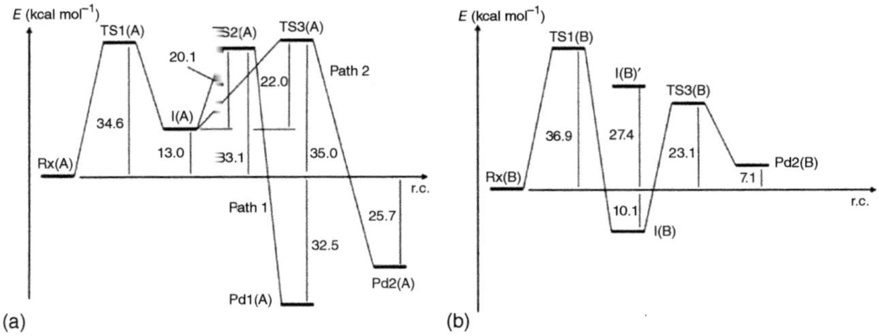

Scheme 6.3 A schematic representation of the potential energy surface for the reaction ethylbenzene → styrene obtained for (a) Model A and (b) Model B. Energy values are given in kcal mol^{-1}.

the formation of styrene, with energy barriers similar to those calculated in the presence of epoxy groups.

6.1.2 Oxidative Dehydrogenation of n-Butane

Besides ethylbenzene, ODH reactions of light alkanes were also reported using nanocarbons catalysts. Butenes, useful intermediates for the synthesis of polymers and other compounds, are synthesized traditionally by ODH of *n*-butane over complex metal oxides. In recent years, nanocarbon materials are used for ODH of butanes, but improving the butenes selectivity remains a challenge. In 2008, it was reported that CNTs with modified surface functionality efficiently catalyzed ODH of *n*-butane to butenes, especially butadiene [53]. A high selectivity of butenes was maintained for periods as long as 100 hours in low O_2/butane ratios. The CNT catalyst was neither lattice nor structural oxygen in comparison with metal oxides, but only the oxygen at active sites, and this reaction was mildly catalyzed by ketonic C=O groups and occurs via a combination of parallel and sequential oxidation steps. The results also showed that addition of a small amount of phosphorus greatly improved the selectivity by suppressing the combustion of hydrocarbons.

In 2017, a detailed analysis and evidence of the impact of electrophilic oxygen and phenol groups on the ODH of *n*-butane were reported [54]. Study results showed that electrophilic oxygen species had been identified as the main culprits or deep oxidation of *n*-butane via a clear correlation between the amount of electrophilic oxygen and the combustion reaction rate, which limited the selectivity of butenes. The amount of electrophilic oxygen on the surface of CNTs was quantified by iodometric titration method. Both reduction by $LiAlH_4$ and annealing treatment at 900 °C in nitrogen could significantly improve the selectivity of butenes by eliminating electrophilic oxygen and carboxyl groups. It was proved that phenol groups on CNTs played an essential role in catalytic performance because it could be converted to carbonyl during ODH (Scheme 6.4).

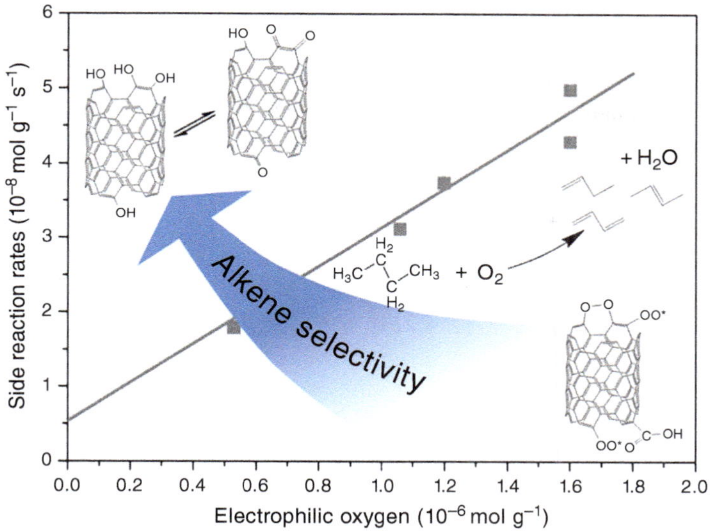

Scheme 6.4 Improving the alkene selectivity of nanocarbon-catalyzed ODH of *n*-butane by refinement of oxygen species.

6.1.3 Oxidative Dehydrogenation of Isobutane

Research on oxidative dehydrogenation of isobutane to isobutene (ODHI) has been intensified, due to its importance as raw material for the chemical, petrochemical, and polymer industries. The activated carbons are active and selective catalysts for the ODHI and the best performance was obtained when the catalyst has a high concentration of carbonyl/quinone groups on the surface.

In 2006, the performance of surface-modified activated carbons in the ODHI was tested [55]. The results showed that different treatments lead to a given catalytic performance associated with the degree of oxidation. At 648 K, isobutene yields were higher in the initial stages for the activated carbons with the highest amounts of carbonyl/quinone groups on the surface, which are considered the active sites for this reaction. As the reaction proceeds, the reactant mixture (isobutene–oxygen and argon as carrier gas) was able to introduce active sites onto the carbon surface to the point where they were stabilized for the case of treatment in H_2 and without pretreatment. For the N_2O and oxygen treatments, the active sites were rearranged and stabilized after a while.

Identification of the highly active oxygen functionalities would enable engineering of nanocarbons for ODH of alkanes. In 2014, few-layered graphenes were used as model catalysts to synthesize reduced GO samples with varying oxygen concentrations for ODH of isobutene [56]. Periodic DFT calculations were performed on graphene nanoribbon models with a variety of oxygen functionalities at the edges to calculate their thermal stability and to simulate reaction mechanisms for ODH of isobutane. The results showed that dicarbonyls at the zigzag edges and quinones at armchair edges were appropriately balanced for high activity. Both dehydrogenation and regeneration of catalytic sites were relevant at the dicarbonyls, whereas regeneration was facile compared with dehydrogenation at quinones. The catalytic mechanism involved weakly adsorbed isobutane reducing functional oxygen and leaving as isobutene, and O_2 in the feed, weakly adsorbed on the hydrogenated functionality, reacting with that hydrogen and regenerating the catalytic sites (Scheme 6.5).

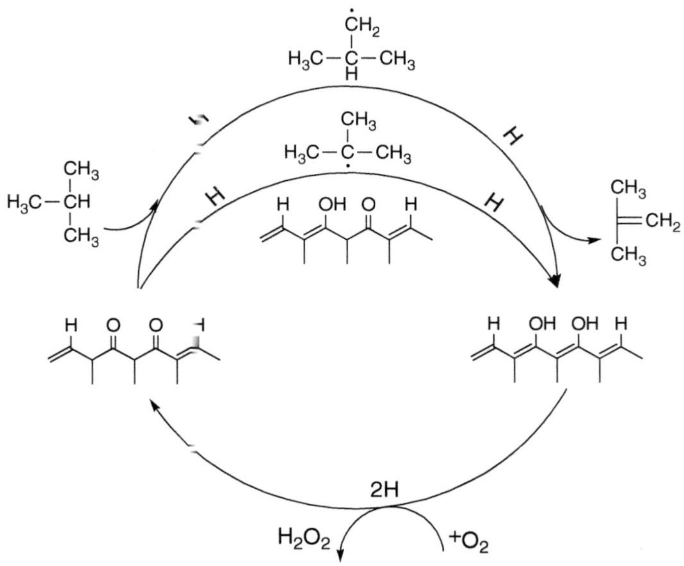

Scheme 6.5 Schematic representation of the ODH of isobutene by dicarbonyl groups on the zigzag edge.

Modifications that supercritical fluids promote in the surface chemistry of carbon materials are scarcely investigated. Due to this lack of knowledge, carbon materials exposed to supercritical fluids have not been tested in any catalytic application until 2015. The ODHI catalyzed by an activated carbon fiber cloth that previously exposed to supercritical carbon dioxide, supercritical water, and nitric acid was reported [57]. Carbonyl–quinone groups were confirmed to be the active sites for the reaction by their direct correlation with catalytic activity. The presence of acidic groups is adverse for the isobutene formation. This reaction mechanism implied that exposure to supercritical fluids constituted a suitable method to prepare efficient catalysts for isobutane ODH. At the same year, oxidized and nitrogen-doped carbon xerogels were used as catalysts for the ODH of isobutane in a fixed-bed reactor at 375 °C using a stoichiometric feed composition [58]. Carbonyl–quinone groups as the active sites and the negative effect of the carboxylic anhydride groups on the reaction were confirmed by selectively removing oxygen groups from the surface by thermal treatments performed on the oxidized samples. Good correlations were obtained between the catalytic activity and the concentration of carbonyl–quinone groups (Scheme 6.6). On the other hand, the catalytic activity increased with the nitrogen content of the carbon xerogels.

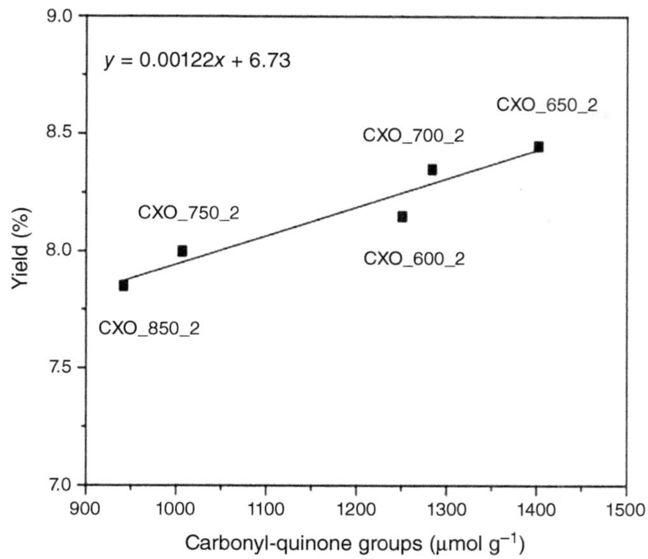

Scheme 6.6 Correlation between isobutene yield and the amounts of carbonyl–quinone groups on the surface of carbon xerogels after 65 minutes of reaction.

6.1.4 Oxidative Dehydrogenation of Propane

Graphitic carbon nitride (g-C_3N_4) is another analog of graphite in addition to h-BN, consisting of tri-*s*-triazine rings with π-conjugated planar layers. It is commercially available and can be easily fabricated. The unique structural features present outstanding properties such as amenability for chemical modification and high thermal/chemical stability against oxidation. In 2020, graphitic carbon nitride as a new kind of metal-free catalyst for exothermic oxidative dehydrogenation of propane (ODHP) provided high selectivity for propylene (74.7%) and ethylene (14.9%) at propane conversion of 12.8% [59]. The spectroscopic insights and in situ experiments confirmed the carbonyl groups modified on the graphitic carbon nitride as active sites. A feasible ODHP reaction pathway was hypothesized by DFT in which two adjacent carbonyl groups on the edge of carbon nitride could activate two C–H bonds of propane simultaneously, thus avoiding the generation of free C_3H_7* radicals and leading to the high selectivity of light olefins (Scheme 6.7).

6.2 Polymer-Catalyzed Selective Oxidation Reactions

6.2.1 Oxidative Dehydrogenation of Ethylbenzene

The organic polymers containing quinone groups which act as active sites in the dehydrogenations are considered to have a well-defined uniform surface. In 1973, the styrene synthesis from ethylbenzene over polynaphthoquinone was reported at 170–270 °C [60]. Polynaphthoquinone, a quinone-type polymer, can be obtained

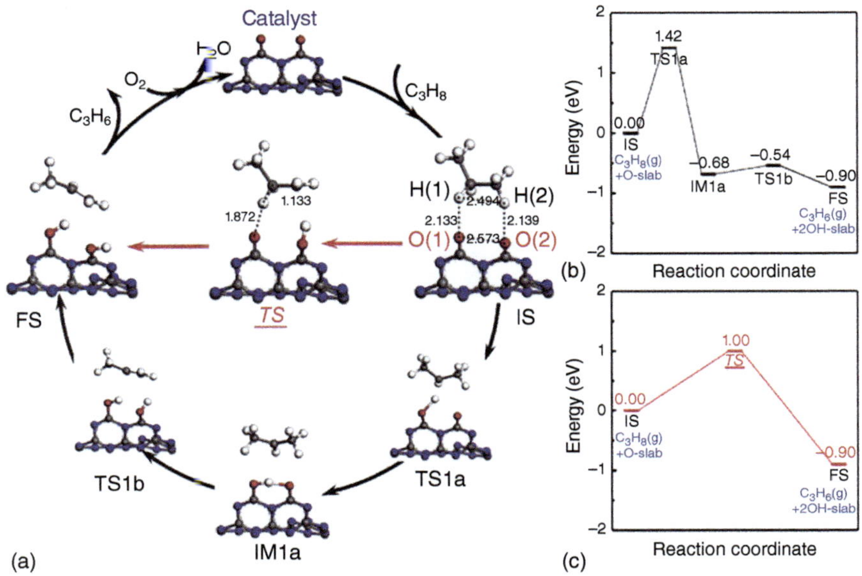

Scheme 6.7 (a) Schematic reaction cycle of the ODHP reaction using carbonyl group-modified g-C_3N_4 catalyst. (b) The energy profile of route 1 (black pathway) and (c) the energy profile of route 2 (red pathway). Notes: carbon is gray, nitrogen is blue, hydrogen is white, and oxygen is red.

by oxidation of 1,7-naphthalenediol. The mechanism showed that dehydrogenation was the hydrogen transfer reaction from ethylbenzene to the quinone monomer. The catalytic activities of various catalysts having the polynaphthoquinone was tested for styrene formation from ethylbenzene at 200 °C in a conventional flow apparatus. As a result, the $FeCl_3$-Q (mixing $FeCl_3$ with the polynaphthoquinone) catalyst showed higher catalytic activity than SiO_2-Q, Al_2O_3-Q, KOH-Q, and Pd-Q.

In 2009, macrocyclic trimer as a suitable model to study ODH catalysis under conditions being rather severe for a molecular catalyst was reported [61]. The superior activity for styrene formation could be related to the abundance of diketone groups. The results of temperature-programmed oxidation/thermogravimetric analysis revealed a weight loss less than 1% at the reaction temperature to indicate the oxidative stability of the molecule.

6.2.2 Oxidative Dehydrogenation of Heterocyclic Compounds

Generally, a homogeneous catalyst exhibits good activity and defined active sites but it is difficult to recycle. Meanwhile, a heterogeneous catalyst can easily be reused but its active site is difficult to reveal. It is interesting to bridge the gap between homogeneous and heterogeneous catalysis via controllable construction of a heterogeneous catalyst containing defined active sites. In 2018, it is reported that a molecularly defined, single-active site heterogeneous catalyst had been designed and prepared via the oxidative polymerization of maleimide derivatives

(Scheme 6.8) [62]. These polymaleimide derivatives were proved to be active catalysts for the selective oxidation of heterocyclic compounds to quinoline and indole via the recycling of –C=O and –C–OH groups, which was confirmed by tracing the reaction with GC-MS using maleimide as the catalyst and by FT-IR analysis with polymaleimide as the catalyst. These results might promote the development of heterogeneous catalysts with molecularly defined single active sites exhibiting a comparable activity to homogeneous catalysts.

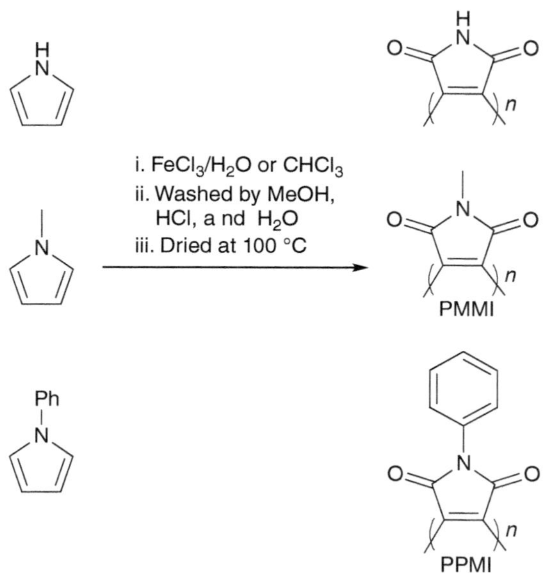

Scheme 6.8 Polycatalysts synthesized. PMI, polymaleimide; PMMI, poly-*N*-methyl-maleimide; PPMI, poly-*N*-phenyl-maleimide. Source: Zhang et al. [62].

6.3 Aldehyde/Ketone-Catalyzed Borrowing-Hydrogen Reactions

6.3.1 Dehydrative β-C-Alkylation Reaction of Methyl Carbinols with Alcohols

Alcohols are frequently used as alkylating reagents in synthesis, the reactions of alcohol-based alkylation and aerobic oxidative reactions and some efficient and selective aldehyde/ketone-catalyzed or autocatalyzed dehydrative N- and C-alkylation reactions were developed. In 2013, the efficient and green aldehyde-catalyzed transition metal-free dehydrative β-C-alkylation reaction of methyl carbinols with various benzylic and aliphatic primary alcohols was reported, which demonstrated that addition of aldehydes was indeed an effective protocol of alcohol activation, preparing the useful long-chain alcohols (Scheme 6.9) [63]. This method was preferable from a synthetic point of view because it required no expensive metal catalysts and activating ligands, merely the cheap and readily available aldehydes and base, which might in turn solve the catalyst recovery, reuse, deactivation, and metal leaching and residue problems accompanied with the metal-catalyzed methods. However, the defect of this protocol was that the aldehyde as the catalyst varied with the change of alcohol as the substrate.

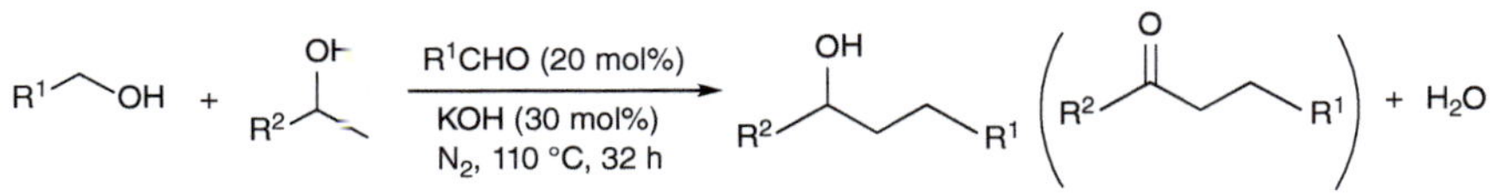

Scheme 6.9 Aldehyde-catalyzed transition metal-free dehydrative β-C-alkylation reaction of methyl carbinols with various benzylic and aliphatic primary alcohols. Source: Xu et al. [63].

6.3.2 Dehydrative α-Alkylation Reactions of Ketones with Alcohols

Direct dehydrative α-alkylation reactions of ketones with alcohols are now realized under simple, practical, and green conditions without using external catalysts. In 2014, catalyst-free autocatalyzed dehydrative α-alkylation reactions of ketones with alcohols under simple, practical, green, and one-pot conditions were reported (Scheme 6.10) [64]. This catalyst-free autocatalyzed alkylation method could afford useful alkylated ketone or alcohol products in a one-pot manner and on a large scale by C=C bond formation of the in situ generated intermediates with subsequent controllable and selective Meerwein–Ponndorf–Verley–Oppenauer-type redox processes.

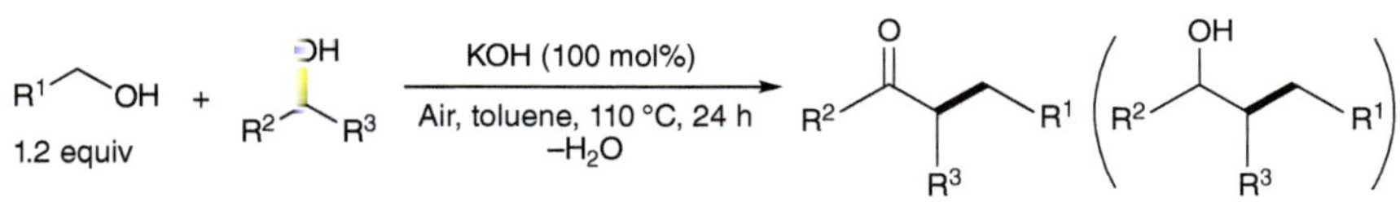

Scheme 6.10 Catalyst-free autocatalyzed dehydrative α-alkylation reactions of ketones with alcohols. Source: Xu et al. [64].

6.3.3 Dehydrative Alkylation Reactions of Fluorenes with Alcohols

Syntheses of 9-monoalkylated and 9,9-dialkylated fluorene monomers have attracted much attention in the past few decades as they are key starting materials for poly(alkylfluorene)s. With the 9-carbon position vacant for further transformations to unsymmetrical 9,9-difunctionalized fluorenes with oriented functional groups and hence oriented chemical, physical, and photoelectric properties, 9-monoalkylated fluorenes are a class of attractive building blocks. In 2017, a general and highly selective 9-monoalkylation reaction of fluorenes with alcohols using aldehydes or ketones as the catalyst and CsOH as the base was developed (Scheme 6.11) [65]. This method could tolerate a wide range of substrates such as the usual primary (hetero)benzylic alcohols, the less reactive short- to long-chain primary alcohols, the sterically more bulky and less reactive secondary benzylic, aliphatic, and cycloalkyl alcohols, as well as alcohols and fluorenes bearing the reactive fluoro, chloro, and bromo substituents. Preliminary mechanistic studies revealed that both aldehydes or ketones as the catalyst and the base additive were crucial to the reaction and the carbonyl catalysts could be quantitatively regenerated and recovered in the reaction cycles.

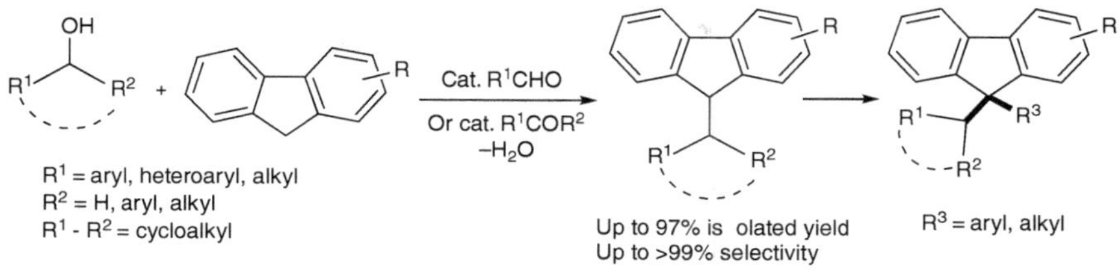

Scheme 6.11 Aldehyde/ketone-catalyzed high selective 9-monoalkylation of fluorenes with primary and secondary alcohols. Source: Chen et al. [65].

6.3.4 Dehydrative N-Alkylation Reactions of Amines with Alcohols

In contrast to the borrowing-hydrogen-type N-alkylation reactions, in which alcohols were activated by transition metal-catalyzed anaerobic dehydrogenation, the addition of external aldehydes was accidentally found to be a simple and effective protocol for alcohol activation in air. In 2013, an efficient and scalable aldehyde-catalyzed dehydrative N-alkylation method for a variety of amides, amines, and alcohols was reported, which required no transition metal catalysts, activating ligands, solvents, and inert atmosphere protection but could efficiently afford the target amide and amine derivatives on a large scale using only the economic and readily available aldehydes and bases (Scheme 6.12) [66]. However, the mechanism for the transition metal-free N-alkylation reactions was still unclear.

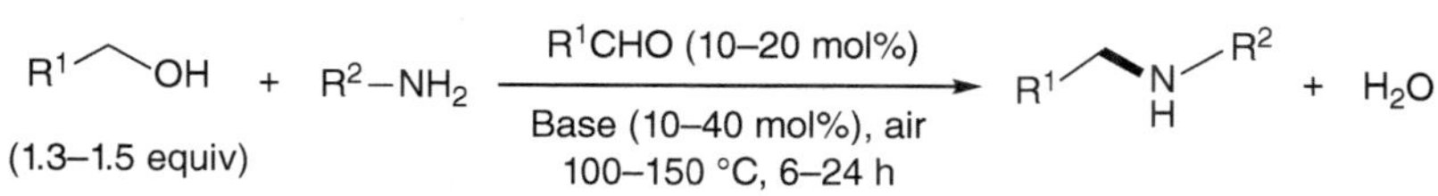

Scheme 6.12 Aldehyde-catalyzed dehydrative N-alkylation. Source: Based on Xu et al. [66].

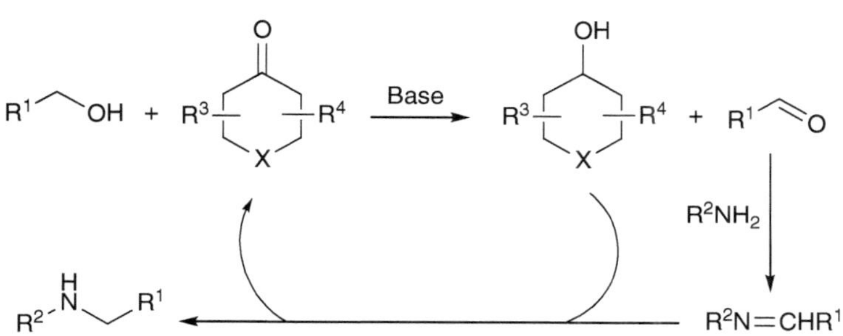

Scheme 6.13 Alcohol amination reactions mechanism.

Until 2015, an efficient molecular-defined organocatalyst was developed for the transfer-hydrogenation reactions including alcohol amination and carbonyl compound reduction [67]. Under the optimized reaction conditions, the yields to the desired products reached 98%. The reaction mechanism and kinetic study supposed that carbonyl–hydroxyl groups were the catalytically active sites, and the transfer hydrogenation reactions progress via the recycling of carbonyl and hydroxyl groups (Scheme 6.13). The catalytic process showed promise as an efficient and economic route for alcohol amination reactions.

6.4 Carbon-Catalyzed Borrowing-Hydrogen Reactions

Transition metal catalysts are essential for the reductive hydrogen atom transfer reaction, which is also known as the transfer hydrogenation reaction or the borrowing-hydrogen reaction. Graphene has been reported to be active for ethylene and nitrobenzene reductions, but carbon-based materials as catalysts for alcohol amination via the borrowing-hydrogen reaction mechanism has not been reported. In 2015, the transition metal-free reductive hydrogen atom transfer transformation was reported using a carbon material as the catalyst with alcohol amination as well as nitro compound and ketone reduction as model reactions [68]. The carbon catalyst exhibited excellent catalytic performance and good generality in the reaction of alcohol amination. The results of XPS, XRD, SEM, FT-IR, and N_2 adsorption–desorption characterizations suggested that the C=O group might be the catalytically active site (Scheme 6.14), while a high surface area was also important for achieving high activity. Moreover, the carbon catalyst was recyclable for several runs without observable deactivation. This study might provide an attractive and useful methodology for a wider range of applications for industrially important compounds and intermediates.

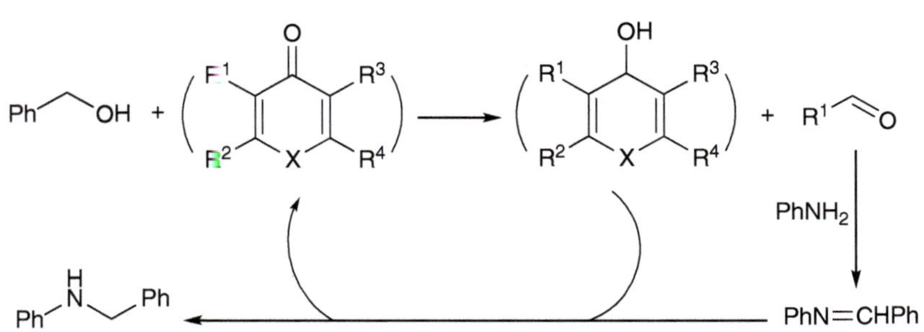

Scheme 6.14 Carbon-based materials as catalysts for alcohol amination via the borrowing-hydrogen reaction mechanism.

Part II References

1 Chebbi, A. and Carlier, P. (1996). *Atmos. Environ.* 30: 4233–4249.
2 Cohen, T. and Lipowitz, J. (1964). *J. Am. Chem. Soc.* 86: 5611–5616.
3 Gisin, B.F. and Merrifield, R.B. (1972). *J. Am. Chem. Soc.* 94: 3102–3106.
4 Lammiman, S.A. and Satchell, R.S. (1974). *J. Chem. Soc. Perkin Trans.* 2: 877–883.
5 Mittal, S., Gupta, K.S., and Gupta, Y.K. (1981). *Indian J. Chem. Sect. A* 20A: 1220–1221.
6 Srinivasan, V.S., Radlowski, C.A., and Gould, E.S. (1981). *Inorg. Chem.* 20: 3172–3175.
7 Defaye, J., Gadelle, A., and Pedersen, C. (1990). *Carbohydr. Res.* 205: 191–202.
8 Kergomard, A. and Renard, M. (1968). *Tetrahedron* 24: 6643–6651.
9 Lillford, P.J. and Satchell, D.P.N. (1968). *J. Chem. Soc. B*: 889–897.
10 Jaacks, V. and Pennewiss, H. (1969). *Makromol. Chem.* 124: 59–69.
11 Hirata, H. and Fukuzumi, K. (1974). *Nippon Kagaku Kaishi* 1974: 1929–1937.
12 Hirata, H., Tanaka, H., and Fukuzumi, K. (1975). *Nippon Kagaku Kaishi* 1975: 1004–1018.
13 Hirata, H. and Fukuzumi, K. (1976). *Yukagaku* 25: 281–286.
14 Awano, H. and Tagaki, W. (1986). *Bull. Chem. Soc. Jpn.* 59: 3117–3123.
15 Eliason, R. (1999). *Can. J. Chem.* 77: 744–751.
16 Chou, P.T., Chen, Y.C., Wei, C.Y., and Chen, W.S. (2000). *J. Am. Chem. Soc.* 122: 9322–9323.
17 Santamaría, P. and Eguiazabal, J.I. (2013). *Polym. Adv. Technol.* 24: 300–306.
18 Wang, W., Wang, X., Li, R. et al. (2009). *J. Appl. Polym. Sci.* 114: 2036–2042.
19 Bassan, A., Zou, W., Reyes, E. et al. (2005). *Angew. Chem. Int. Ed.* 44: 7028–7032.
20 Shirakawa, S. and Kobayashi, S. (2006). *Org. Lett.* 8: 4939–4342.
21 Levchik, S.V., Weil, E.D., and Lewin, M. (1999). *Polym. Int.* 48: 532–557.
22 Klausen, R.S., Kennedy, C.R., Hyde, A.M., and Jacobsen, E.N. (2017). *J. Am. Chem. Soc.* 139: 12299–12309.
23 Silva, G.D. (2010). *Angew. Chem. Int. Ed.* 49: 7523–7525.
24 Wei, Y., Lin, S., Zhang, J. et al. (2011). *Chem. Commun.* 47: 12394–12396.
25 Wei, Y., Lin, S., Xue, H. et al. (2012). *Org. Lett.* 14: 712–715.

Carbonyl Compounds: Reactants, Catalysts and Products, First Edition.
Feng Shi, Hongli Wang and Xingchao Dai.

26 Hashimoto, T., Omote, M., and Maruoka, K. (2011). *Angew. Chem. Int. Ed.* 50: 3489–3492.

27 Hashimoto, T., Galvez, A.O., and Maruoka, K. (2013). *J. Am. Chem. Soc.* 135: 17667–17670.

28 Mir, E., Mollashahi, E., and Marandi, G. (2018). *Curr. Catal.* 7: 217–223.

29 Sui, Y., Cui, P. Liu, S. et al. (2018). *Eur. J. Org. Chem.* 2018: 215–218.

30 Dabrowski, M., Lulinski, S., Madura, I. et al. (2000). *J. Organomet. Chem.* 597: 190–195.

31 Ho, H.E., Asao, N., Yamamoto, Y., and Jin, T. (2014). *Org. Lett.* 16: 4670–4673.

32 Zhou, Q., Xu, S., and Zhang, R. (2019). *Tetrahedron Lett.* 60: 734–738.

33 Mochida, I., Ogaki, M., Fujitsu, H. et al. (1983). *Fuel* 62: 867–868.

34 Ahmed, S.N., Baldwin, R., Derbyshire, F. et al. (1993). *Fuel* 72: 287–292.

35 Teng, H., Tu, Y.-T., Lai, Y.-C., and Lin, C.-C. (2001). *Carbon* 39: 575–582.

36 Szymański, G.S., Grzybek, T., and Papp, H. (2004). *Catal. Today* 90: 51–59.

37 Boyano, A., Gálvez, M.E., Moliner, R., and Lázaro, M.J. (2008). *Fuel*: 2058–2068.

38 Zhang, J., Gao, Q., Li, X.M. et al. (2017). *Phys. Chem. Chem. Phys.* 19: 22462–22471.

39 Su, C.L., Acik, M., Takai, K. et al. (2012). *Nat. Commun.* 3: 1–9.

40 Charmot, A., Chung, P.-W., and Katz, A. (2014). *ACS Sustain. Chem. Eng.* 2: 2866–2872.

41 To, A.T., Chung, P.W., and Katz, A. (2015). *Angew. Chem. Int. Ed.* 54: 11050–11053.

42 Kobayashi, H., Kaiki, H., Shrotri, A. et al. (2016). *Chem. Sci.* 7: 692–696.

43 Qi, X., Yan, L., Shen, F., and Qiu, M. (2019). *Bioresour. Technol.* 273: 687–691.

44 Wen, G., Wang, B., Wang, C. et al. (2016). *Angew. Chem. Int. Ed.* 55: 1–6.

45 Wang, B.L., Wen, G.D., and Su, D.S. (2017). *Nano Res.* 10: 2954–2965.

46 Schraut, A., Emig, G., and Sockel, H.G. (1987). *Appl. Catal.* 29: 311–326.

47 Pereira, M.F.R., Órfão, J.J.M., and Figueiredo, J.L. (1999). *Appl. Catal. A-Gen.* 184: 153–160.

48 Pereira, M.F.R., Orfão, J.J.M., and Figueiredo, J.L. (2000). *Appl. Catal. A-Gen.* 196: 43–54.

49 Zhang, J., Su, D., Zhang, A. et al. (2007). *Angew. Chem. Int. Ed.* 46: 7319–7323.

50 Zhang, J., Su, D.S., Blume, R. et al. (2010). *Angew. Chem. Int. Ed.* 49: 8640–8644.

51 Liu, W., Wang, C., Su, D., and Qi, W. (2018). *J. Catal.* 368: 1–7.

52 Marforio, T.D., Bottoni, A., Zerbetto, F., and Calvaresi, M. (2019). *ChemNanoMat* 5: 499–505.

53 Zhang, J., Liu, X., Blume, R. et al. (2008). *Science* 322: 73–77.

54 Li, J., Yu, P. Xie, J. et al. (2017). *ACS Catal.* 7: 7305–7311.

55 de Jesús Díaz Velásquez, J., Suárez, L.M.C., and Figueiredo, J.L. (2006). *Appl. Catal. A-Gen.* 311: 51–57.

56 Dathar, G.K., Tsai, Y.T., Gierszal, K. et al. (2014). *ChemSusChem* 7: 483–491.

57 Martin-Sanchez, N., Soares, O.S.G.P., Pereira, M.F.R. et al. (2015). *Appl. Catal. A-Gen.* 502: 71–77.

58 Pelech, I., Soares, O.S.G.P., Pereira, M.F.R., and Figueiredo, J.L. (2015). *Catal. Today* 249: 176–183.

59 Cao, L., Dai, P., Zhu, L. et al. (2020). *Appl. Catal. B-Environ.* 262: 1–9.
60 Iwasawa, Y., Nobe, H., and Ogasawara, S. (1973). *J. Catal.* 31: 444–449.
61 Zhang, J., Wang, X., Su, Q. et al. (2009). *J. Am. Chem. Soc.* 131: 11296–11297.
62 Zhang, Y., Pang, S., Wei, Z. et al. (2018). *Nat. Commun.* 9: 1–10.
63 Xu, Q., Chen, J., and Liu, Q. (2013). *Adv. Synth. Catal.* 355: 697–704.
64 Xu, Q., Chen, J., Tian, H. et al. (2014). *Angew. Chem. Int. Ed.* 53: 225–229.
65 Chen, J., Li, Y., Li, S. et al. (2017). *Green Chem.* 19: 623–628.
66 Xu, Q., Li, Q., Zhu, X., and Chen, J. (2013). *Adv. Synth. Catal.* 355: 73–80.
67 Dai, X., Cui, X., Deng, Y., and Shi, F. (2015). *RSC Adv.* 5: 43589–43593.
68 Yang, H., Cui, X., Dai, X. et al. (2015). *Nat. Commun.* 6: 1–11.

Part III

The Synthetic Applications of Carbonyl Compounds

7

Synthesis of Functional Molecules

7.1 Reduction of Carbonyl Compounds

7.1.1 Aldehydes and Ketones to Alcohol

The reduction of carbonyl compounds to alcohols is one of the most widely used and fundamental reactions in organic chemistry [1]. Transition metal-catalyzed hydrogenation [2, 3], biocatalytic reductions with yeast reductase [4–7] and chemical reductions by $LiAlH_4$, nickel–aluminum alloy [8], tri-*n*-butyltin hydride [9], sodium formate [10], alkyl phosphines [11], poly(methylhydrosiloxane) [12], or borane reagents such as $ZnBH_4$ [13], $NaBH_4$ [14, 15], $NaBH_3CN$ [16], BH_3–THF [17], and amine boranes [18, 19] have been utilized to accomplish the reduction of the carbonyl group (Scheme 7.1). We herein only selected some of the important examples and summarized the classical development and characteristic about reduction of aldehydes and ketones to alcohols.

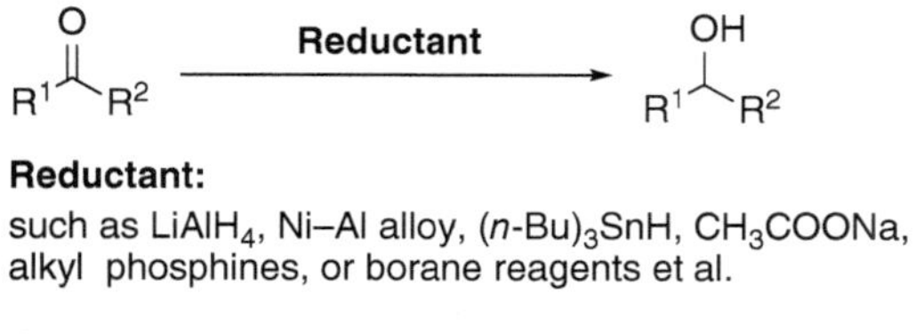

Scheme 7.1 Reduction of aldehydes and ketones to alcohols.

Reduction of aldehydes and ketones through organosilanes to alcohols in aqueous acidic media was reported by Doyle et al. in 1974 [20], and they found that both concentrated hydrochloric acid and aqueous sulfuric acid are suitable aqueous acids (Scheme 7.2). A nonreactive solvent such as acetonitrile or ethyl ether was used in this system to minimize ether formation.

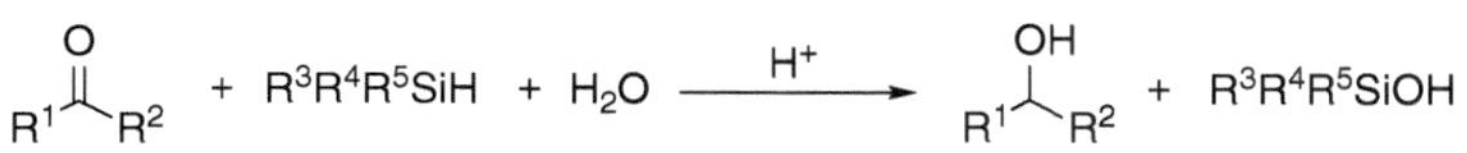

Scheme 7.2 Silane reductions of aldehydes and ketones to alcohols in acidic media.

Carbonyl Compounds: Reactants, Catalysts and Products, First Edition.
Feng Shi, Hongli Wang and Xingchao Dai.

Among the various photoreactions of carbonyl compounds, the photoreduction of aldehydes and ketones is one of the best understood and usually affords 1,2-dihydroxy compounds. Sonoda and coworkers reported the photoreduction of ketones and aldehydes with H_2Se to give alcohols in desirable yields (Scheme 7.3) [21]. And they also found that the reduction of aryl-substituted carbonyl compounds proceeded more rapidly than that of alkyl-substituted carbonyl compounds.

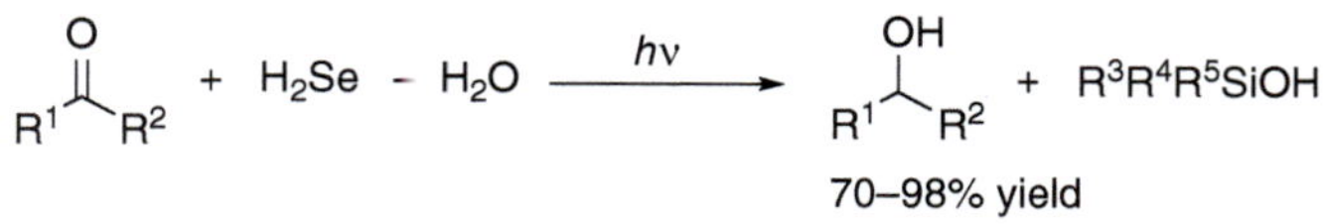

Scheme 7.3 Photoreduction of ketones and aldehydes to alcohols with H_2Se. Source: Based on Kambe et al. [21].

In 1996, Kwon and coworkers reported the application of diisobutylalkoxyalanes (DIBAL–OR) for the selective reduction of α,β-unsaturated aldehydes and ketones to the corresponding allylic alcohols (Scheme 7.4) [22]. In the process of exploring the reduction of diisobutylchloroalane (DIBAL–Cl), they found that the reagents are much milder than DIBAL–Cl and effectively reduce aldehydes and ketones under mild conditions. They attributed such a high selectivity to the cyclic mechanism in which the β-hydride of the isobutyl group is shifted to the carbonyl carbon of the substrate: the reduction is very similar to a Meerwein–Ponndorf–Verley-type process.

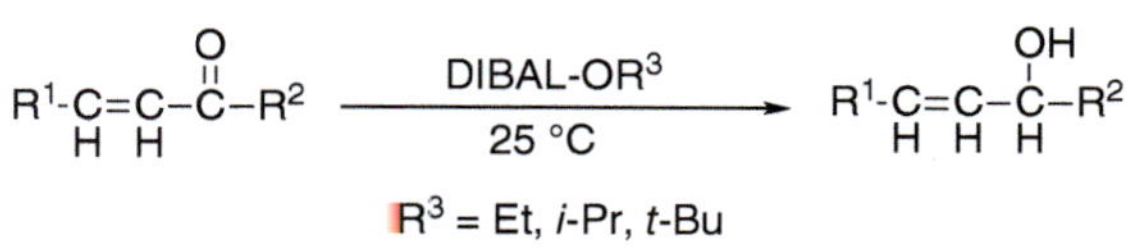

Scheme 7.4 Reduction of aldehydes and ketones with DIBAL–OR. Source: Modified from Cha et al. [22].

A simple and rapid procedure for quantitative reduction of aromatic aldehydes, ketones, diketones, and oxo aldehydes to alcohols by application of $TiCl_3/NH_3$ in aqueous methanol solution was reported by Porta and coworkers in 2002 (Scheme 7.5) [23]. The reducing system distinguishes between different classes of aldehydes and/or ketones, and many functional groups that usually do not survive under other reducing conditions are tolerated well.

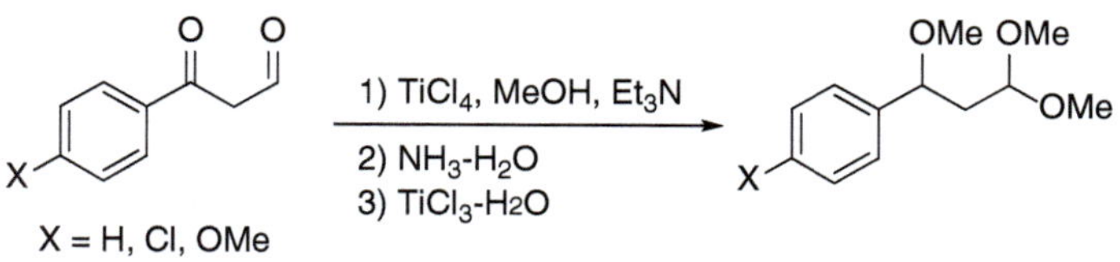

Scheme 7.5 Reduction of diketones and oxo aldehydes to alcohols by an aqueous $TiCl_3/NH_3$. Source: Modified from Clerici et al. [23].

Kaupp and coworkers found that stoichiometric molecular solid-state vibrational, solvent-free kneading, and mechanochemical ball milling of various aldehydes and ketones with sodium borohydride under mild reaction temperature could use all hydrogen atoms of the reducing agent in fast reactions [24], which provides thermally stable sodium tetraalkoxyborates in quantitative yields (Scheme 7.6). The easily isolated solids are extremely sensitive toward hydrolysis, leading to quantitative yields of the corresponding alcohols. Notably, this useful protocol is regiospecific and stereoselective and is favorable for the reduction of conjugated aldehydes and ketones to provide the corresponding allylic alcohols in quantitative yields.

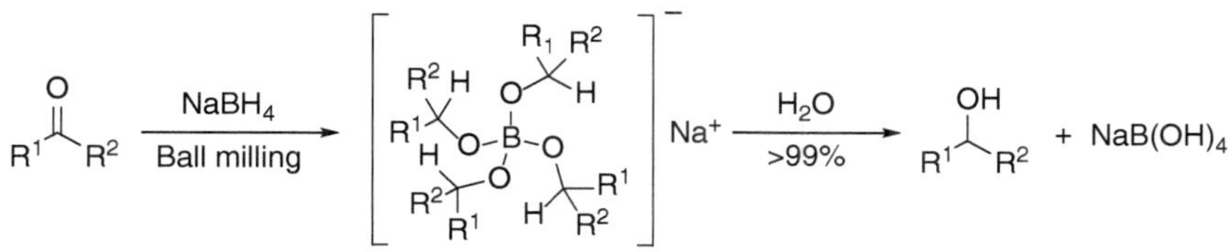

Scheme 7.6 Ball milling solvent-free reduction of carbonyl compounds via sodium tetraalkoxyborates.

Chemoselective reduction of various carbonyl compounds to alcohols with ammonia borane (AB) in water was achieved by Zhang and coworkers in 2012 in quantitative conversions and high isolated yields (Scheme 7.7) [25]. Interestingly, α- and β-keto esters are selectively reduced to corresponding hydroxyl esters, while diols are obtained when sodium borohydride was used as a reducing agent. The procedure is also compatible in the presence of a variety of base-labile protecting groups, such as acetyl, ester groups, and acid-labile protecting groups such as trityl and TBDMS groups, and others, such as the unsaturated double bond and cyano groups. Moreover, a kilo-scale reaction of methyl benzoylformate with AB was conducted in water giving methyl mandelate in 94% yield.

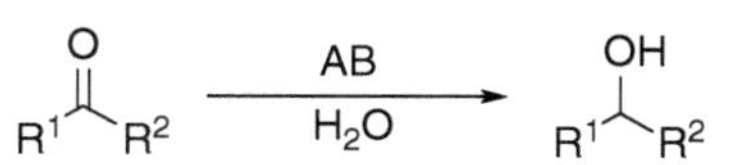

R^1 = aryl, alkyl; R^2 = H, aryl, alkyl, CO_2R, CH_2CO_2R

Scheme 7.7 Reduction of aldehydes and ketones to alcohols with ammonia borane. Source: Modified from Shi et al. [25].

7.1.2 Acids to the Alcohols and Aldehydes

7.1.2.1 To Alcohols

As early as 1905, the reduction of aromatic amino acids to the corresponding alcohols was studied by Langguth [26]. Then, in 1981, selective reduction of the mixed anhydride of 3α,7α,12α-tris(acetyloxy)cholic acid and ethyl chloroformate with sodium borohydride had been reported by Hoshita and coworkers [27].

Subsequently, Sharts developed a simple procedure for the reduction of bile acids and formylated bile acids to corresponding alcohols in good yields with excess BH_3·THF complex in THF at room temperature (Scheme 7.8) [28].

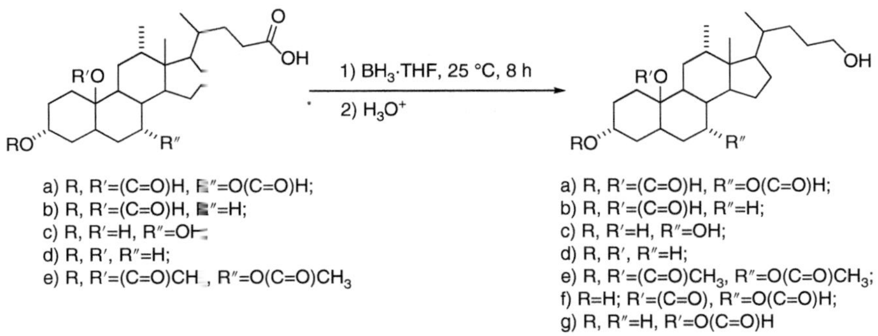

Scheme 7.8 Selective reduction of carboxylic acids to alcohols. Source: Malik and Sharts [28].

The hydrogenation of aliphatic carboxylic acid via $9Ni/Al_2O_3$ catalyst to provide desired alcohol was reported by Kalló and coworkers, and the yield of which can be increased drastically by 10 wt% In_2O_3 doping [29]. The presence of indium and nickel can effectively guide the catalytic reduction to alcohols step by step instead of shortening the decarbonylation chain. Indium oxide can be more easily reduced to metallic form at lower temperature than supported nickel oxide, but this difference is not reflected by any change of catalytic behavior over a wide range of pretreatment temperature (350 and 550 °C). It is worth mentioning that Cu and Cu_2In-type catalysts was also an effective alternative method for the reduction of biomass-derived carboxylic acids to alcohols [30].

An efficient method for the synthesis of alcohols from the corresponding carboxylic acids was described using 1-propanephosphonic acid cyclic anhydride (T_3P) as an activator and $NaBH_4$ as a reductant at 0 °C with excellent yields (up to 96% yield) and broad substrate scope (Scheme 7.9) [31]. Reduction of several alkyl/aryl carboxylic acids and N^{α}-protected amino acids/peptide acids as well as N^{β}-protected amino acids was successfully carried out to obtain corresponding alcohols in excellent yields.

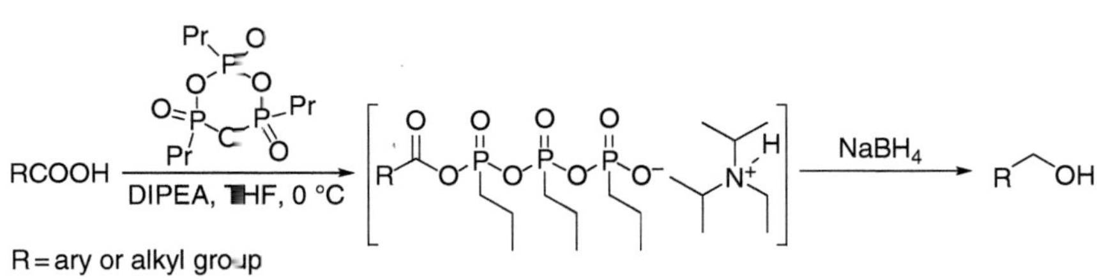

Scheme 7.9 Reduction of carboxylic acids to corresponding alcohols mediated by T_3P. Source: Nagendra et al. [31].

In 2014, Pattarawarapan and coworkers developed a simple method for $NaBH_4$ reduction of carboxylic acids to alcohols under solvent-free conditions using a combination of 2,4,6-trichloro-1,3,5-triazine (TCT) with a catalytic amount of triphenylphosphine as an acid activator (Scheme 7.10) [32]. With the 1 : 0.2 : 1.5 : 2 mole ratio of TCT : PPh_3 : K_2CO_3 : $NaBH_4$, carboxylic acids including aromatic acids, aliphatic acids, and N-protected amino acids could readily undergo reduction to give the corresponding alcohols in good to excellent yields within 10 minutes.

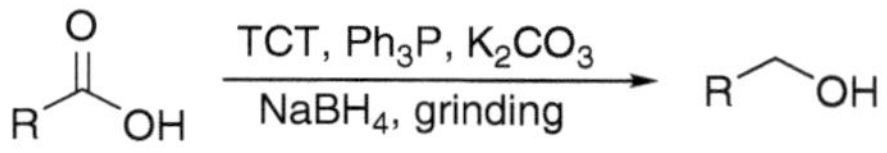

Scheme 7.10 Solvent-free reduction of carboxylic acids with $NaBH_4$ promoted by TCT–PPh_3. Source: Jaita et al. [32].

Recently, Mousavi and coworkers developed an efficient protocols for the direct reduction of carboxylic acids to corresponding alcohols in good to excellent yields using four different types of hydrogen donors, namely $NaBH_4$, HCO_2NH_4, glycerol, and *i*-PrOH in the presence of Fe_3O_4@APTMS@$ZrCp_2Clx$ (x = 0, 1, 2) as the reusable magnetic nanocatalyst (Scheme 7.11) [33]. These practical and environmentally benign strategies could be beneficial for the green and convenient reduction of carboxylic acids to alcohols.

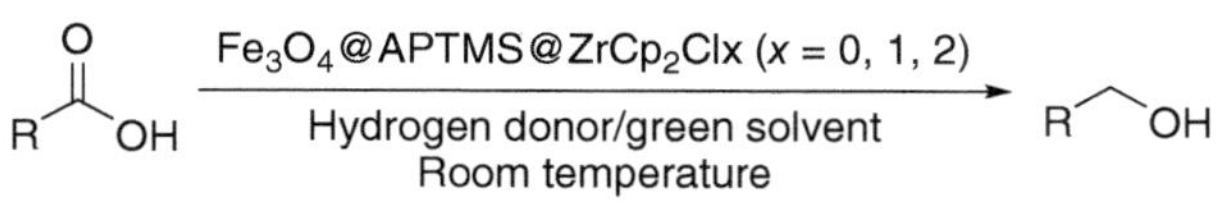

Hydrogen donor: $NaBH_4$, HCO_2NH_4, glycerol, *i*-propanol
Green solvent: H_2O:THF, H_2O:PEG-400, H_2O
R=aromatic or aliphatic

Scheme 7.11 Reduction of carboxylic acids to alcohols using four different hydrogen donors. Source: Modified from Zeynizadeh et al. [33].

7.1.2.2 To Aldehydes

In the early 1900s, electrolytical reduction for the sodium salts of benzoic and salicylic acids to the corresponding aldehydes in 30–50% yields was reported by Mettler [34]. Another contribution for the reduction of salidyl acid to a salicyl aldehyde using 2% sodium amalgam with a yield of 60% was described by Weil at the same time [35].

In 1974, bis(*N*-methylpiperazinyl)aluminum hydride (BMPA) was first prepared from the reaction of aluminum hydride with *N*-methylpiperazine to reduce aliphatic and aromatic acids to aldehydes in good yields under room temperature or reflux (Scheme 7.12) [36]. Subsequently, Wiemer and coworkers improved the synthetic method of BMPA using *N*-methylpiperazine dihydrochloride and $LiAlH_4$ as starting materials, making it much cheaper and easy to synthesis for the reduction of carboxylic acid [37].

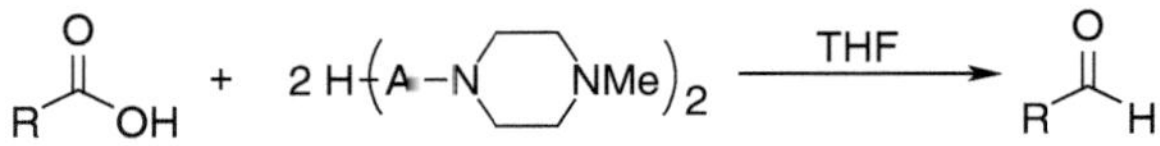

Scheme 7.12 Selective reduction of carboxylic acids to aldehydes using aminoaluminum hydride. Source: Based on Muraki and Mukaiyama [36].

In 1999, Taddei and coworkers discovered that the activated ester of *N*-Boc amino acids with 2-chloro-4,6-dimethoxy[1,3,5] triazine can be reduced into the corresponding aldehydes in good yields with H_2 over Pd/C catalyst [38]. The method does not require the use of anhydrous solvents, strong reaction conditions or dangerous or expensive reagents, and is suitable for the large-scale preparation of aldehydes. The main drawback is the necessity to monitor the reaction to prevent the formation of the alcohol. At the same time, a facile and simple method for the reduction of N-protected amino acids and peptides to the corresponding alcohols using $NaBH_4$–I_2 reductant with excellent yields and retention of optical purity was reported by Haq and coworkers [39].

Palladium catalyzed reduction of carboxylic acids to the corresponding aldehydes with hydrosilanes as reducing agent and pivalic anhydride as an indispensable reagent had been developed by Tsuji and coworkers in 2013 (Scheme 7.13) [40]. The commercially available bis(dibenzylidene-acetone)palladium, tri(*p*-tolyl) phosphane, and methylphenylsilane as a simple system realized the reduction of various aliphatic carboxylic acids as well as benzoic acids to aldehydes in good to high yields.

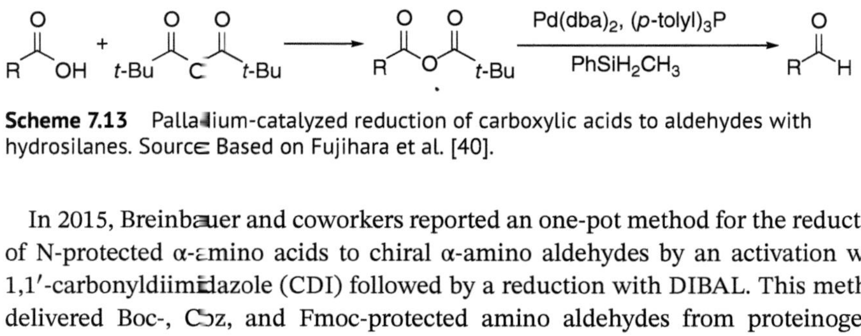

Scheme 7.13 Palladium-catalyzed reduction of carboxylic acids to aldehydes with hydrosilanes. Source: Based on Fujihara et al. [40].

In 2015, Breinbauer and coworkers reported an one-pot method for the reduction of N-protected α-amino acids to chiral α-amino aldehydes by an activation with 1,1′-carbonyldiimidazole (CDI) followed by a reduction with DIBAL. This method delivered Boc-, Cbz, and Fmoc-protected amino aldehydes from proteinogenic amino acids in good isolated yields and with complete retention of configuration (Scheme 7.14) [41]. Compared to established two-step protocols, the advantages of this method are (i) its operational simplicity, (ii) the use of inexpensive reagents, (iii) the simple extractive workup, and (iv) its short overall processing time (typically less

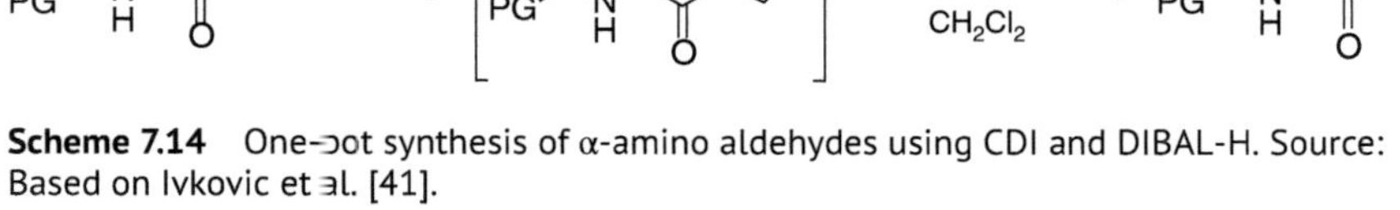

Scheme 7.14 One-pot synthesis of α-amino aldehydes using CDI and DIBAL-H. Source: Based on Ivkovic et al. [41].

than four hours) to deliver the product in high purity. While the method is excellent for proteinogenic amino acids leading to good yields and preserved stereointegrity, it has its limitations in phenylglycine and dipeptides where epimerization was observed.

7.1.3 Ester to Alcohols and Ethers

7.1.3.1 To Alcohols

In early stage, Sato et al. developed a series of insightful work for reduction of unsaturated fatty esters to corresponding alcohols by selective high-pressure catalytic hydrogen [42]. Then, Cp^2TiC1_2-catalyzed Grignard reactions with esters provided a general methodology for the reduction of esters to the corresponding primary alcohols (Scheme 7.15).

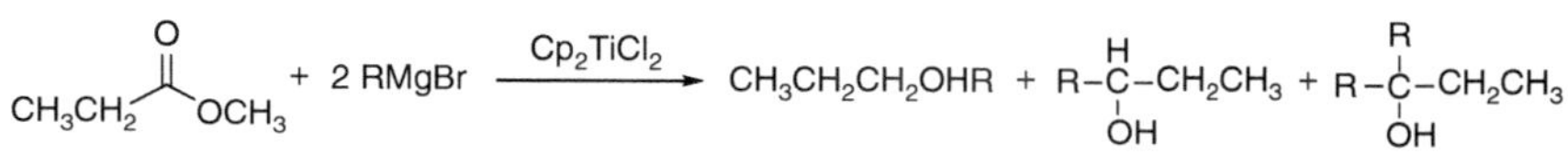

Scheme 7.15 The reduction of esters to alcohols by Cp_2TiCl_2-catalyzed Grignard reaction.

Reduction of esters to alcohols by means of sodium borohydride in polyethylene glycols was reported by Santaniello et al. in 1981 [43], and they thought that stable polymeric dialkoxyborohydrides $[BH_2(OR)_2]$– where $(OR)_2$ = $-OCH_2CH_2(OCH_2CH_2)_6CH_2CH_2O-$ were formed, which might show an higher reactivity than $NaBH_4$ itself. Except for this, a series of similar catalytic systems using borohydride as reducing agent, such as $ZnBH_4$ [5, 44], $NaBH_4$–MeOH [45, 46], and $NaBH_4$–$CeCl_3$ [47], were implemented successively, which led to the further improvement and development of the catalytic system for reduction of esters to alcohols.

In 1991, Buchwald and coworkers reported a titanium-catalyzed hydrosilylation for the conversion of esters to primary alcohols in high yields under mild conditions with silanes as the stoichiometric reductant (Scheme 7.16) [48]. This procedure is selective and might represent a safer and more convenient alternative to the use of reducing agents such as $LiAlH_4$ and DIBAL on a large scale. After this, rhodium-catalyzed reduction of carboxylic esters to corresponding alcohols in high yields using diphenylsilane as reductant at room temperature was developed [49]. For example, ethyl decanoate and ethyl phenylacetate were converted to decanol and 2-phenylethanol by $[RhCl(cod)]_2/4PPh_3$ in 98% and 92% yields, respectively.

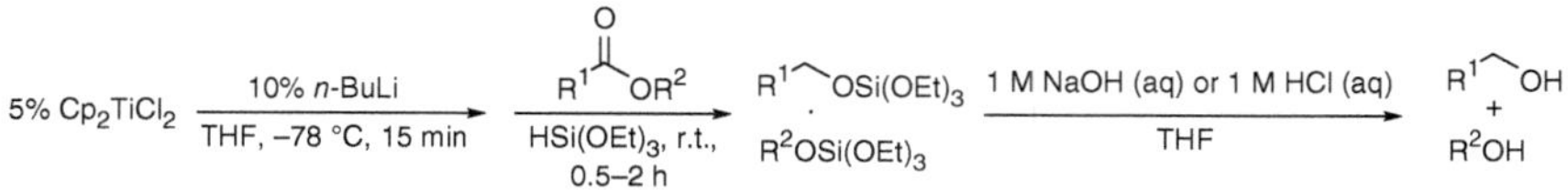

Scheme 7.16 Titanium-catalyzed reduction of esters to alcohols. Source: Berk et al. [48].

In 2007, Saudan et al. developed a reduction of aliphatic and aromatic carboxylic esters to alcohols with H_2 catalyzed by homogeneous ruthenium complexes with N and P ligands in excellent catalytic performance (turnover number (TON) ≈ 2000, turnover frequency (TOF) = 800–2000 h^{-1}) (Scheme 7.17) [50]. Under the relatively mild conditions ($p(H_2)$ = 50 bar, T = 100 °C), esters with a di- or trisubstituted C=C bond were reduced to the corresponding unsaturated alcohols with high chemoselectivity (unsaturated/saturated product >98 : 2) in over 85% yield. In addition, optically active esters could readily undergo reduction to give the corresponding alcohols without loss of their optical purity under mild (T = 80 °C) and neutral conditions. Except for this, other methodology of homogeneous transition metal catalyzed the reduction of esters to alcohols, such as Zn [51], Fe [52], and Co [53], have been developed accordingly.

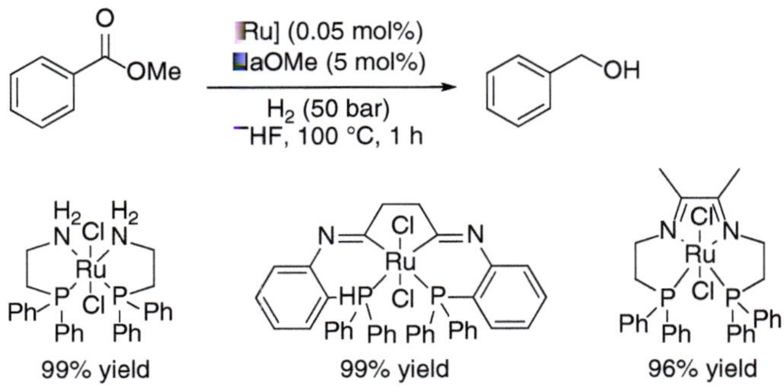

Scheme 7.17 Reduction of carboxylic esters to alcohols under the catalysis of homogeneous ruthenium complexes. Source: Saudan et al. [50].

7.1.3.2 To Ethers

The direct reduction of esters to their corresponding ethers can be achieved using the Lewis acid $BF_2OTf{\cdot}Et_2O$ generated via anionic redistribution between TMSOTf and $BF_3{\cdot}Et_2O$ with Et_3SiH as the reducing agent [54]. This procedure could obtain ethers in up to 71% isolated yield with alcohols as the only side product.

In 2008, Sakai et al. found that the $InBr_3$–Et_3SiH could promote the direct and selective reduction of esters to yield the corresponding unsymmetrical ethers [55]. This simple catalytic system appeared to be tolerant of some functional groups and was applicable to other important compounds, such as a tertiary amide and a carboxylic acid.

In 2018, Motoyama and coworkers reported that the partial reduction of carboxylic esters or lactones with 1,1,3,3-tetramethyldisiloxane (TMDS) in the presence of commercially available Pd/C catalyst and $Cu(acac)_2$ as a co-catalyst could proceed smoothly at 50 °C to afford the silyl acetal derivatives in high yields, which were easily converted to the corresponding ethers with high selectivity by treatment of silyl acetals with catalytic amounts of TMSOTf at −78 °C (Scheme 7.18) [56].

7.1.4 Amides to Amines

Amides can be reduced by $LiAlH_4$ to obtain the amines by the means of converting the amide carbonyl group into a methylene group (Scheme 7.19). This kind of

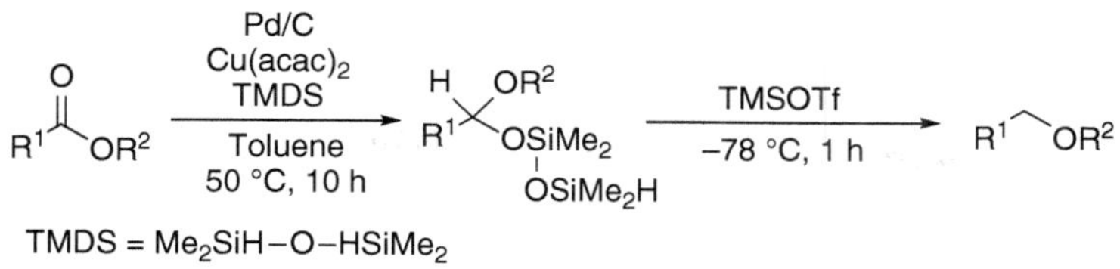

Scheme 7.18 Pd/C catalyzed reduction of carboxylic ester to ethers. Source: Based on Hosokawa et al. [56].

reaction is specific for amides and does not occur with other carboxylic acid derivatives. Notably, the isolation of nitriles in $LiAlH_4$ reductions of amides indicated that the reaction involved dehydration of amide to nitrile followed by the reduction of nitrile to amine.

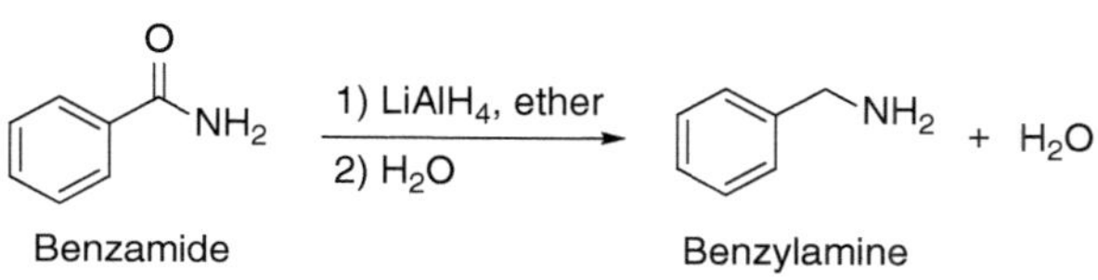

Scheme 7.19 Reduction of amides to amines.

In 1965, $NaBH_4$, as a mild reducing agent, was used to the reduction of tertiary amides to the corresponding tertiary amines in refluxing pyridine with moderate yields (Scheme 7.20) [57]. Both *N*,*N*-dimethylhydrocinnamamide and *N*,*N*-diethylbenzamide were reduced to the corresponding tertiary amines with a trace of recovery starting amides, and aromatic tertiary amide seemed to be reduced faster than aliphatic derivatives. Then, a new method for the reduction of amides to amines with $NaBH_4$–PCl_5 was reported in 1976, which involved the generation of the Vilsmeir complex of an amide/imide with PCl_5 that could react via the chlorenamine with appropriate electrophiles [58].

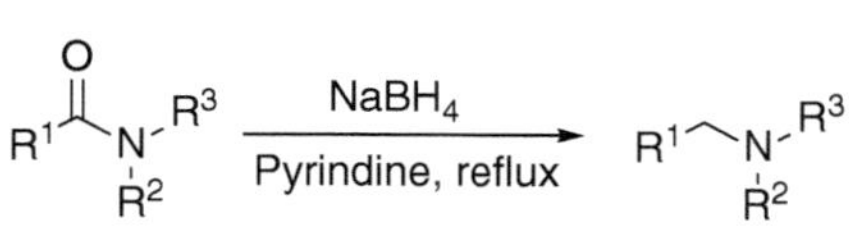

Scheme 7.20 $NaBH_4$ as the reducing agent for reduction of amides to amines. Source: Based on Yamada et al. [57].

A simple and convenient procedure for the conversion of amides to the corresponding amines with diborane as reducing agent was developed by Brown and coworker in 1973; primary, secondary, and tertiary amides proceeded smoothly [59]. The disadvantage of diborane for such reductions occurred to unsaturated derivatives, such as *N*,*N*-dimethylcinnamamide, owing to the rapid addition of diborane to the double bonds.

The reaction of $NaBH_4$ with bis(2-bromoethyl)selenium dibromide and/or diethylselenium dibromide in THF gave borane. In 1991, Akabori and coworker reported that treatment of tertiary amides with mixtures of $NaBH_4$ and either of the dibromides in THF gave the corresponding amines (Scheme 7.21) [60]. However, similar reactions with primary and secondary amides did not proceed. Furthermore, triphenylborane was also found to catalyze the reduction of tertiary amides with hydrosilanes to give amines under mild condition with high chemoselectivity in the presence of ketones, esters, and imines [61].

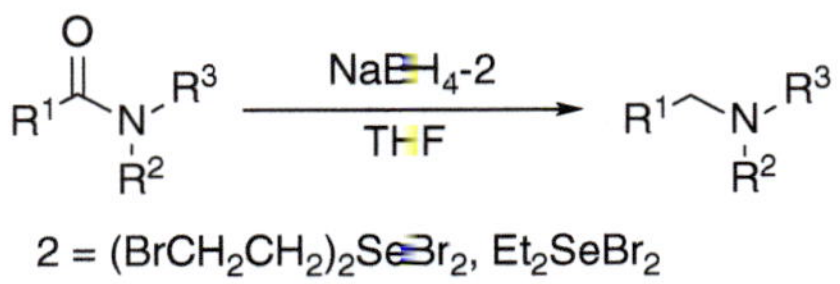

Scheme 7.21 Reduction of carboxylic amides to corresponding amines by $NaBH_4$-2. Source: Modified from Akabori et al. [60].

In 2009, Beller and coworkers reported the first iron-catalyzed reduction of secondary and tertiary amides with inexpensive polymethylhydrosiloxane (PMHS) [62]. The new protocol exhibited broad substrate scope and functional group tolerance with high selectivity, thus providing a variety of amines in good to excellent yields. And, they also established the first example of a highly chemoselective reduction of amides to the corresponding amines with silanes using inexpensive zinc catalysts under mild conditions [63]. This procedure was suitable for the catalytic hydrosilylation of primary, secondary, and tertiary amides and proceeded with excellent chemoselectivity in the presence of sensitive ester, nitro, azo, nitrile, olefins, and other functional groups.

Catalytic reduction of secondary amides to corresponding amines had been achieved using readily available iridium catalysts such as $[Ir(coe)_2Cl]_2$ with diethylsilane as reductant [64]. The stepwise reduction to secondary amine proceeded through an imine intermediate that can be isolated when 2 equiv of silane is used (Scheme 7.22). This system requires low catalyst loading and shows high efficiency (up to 1000 turnovers at room temperature with 99% conversion) and an appreciable level of functional group tolerance.

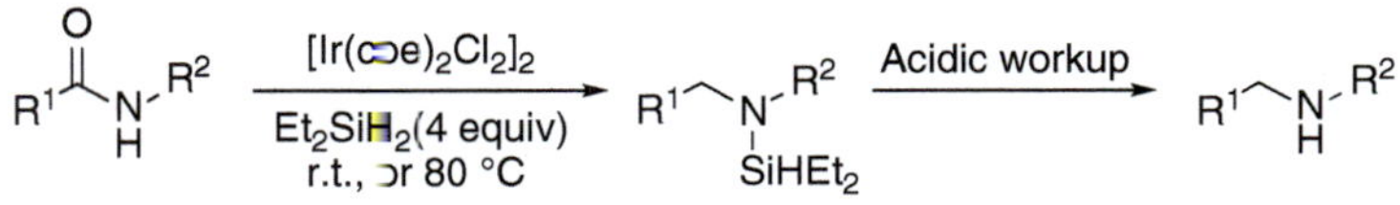

Scheme 7.22 Ir-catalyzed reduction of amides to amines.

Metal-free chemoselective hydrosilylation for the reduction of primary, secondary, and tertiary amides to the corresponding amines in the presence of boronic acids had been developed by Beller and coworkers in 2013 [65]. By applying a commercially available air- and moisture-tolerant boronic acid as the catalyst, this convenient methodology allowed a clean and straightforward synthesis of amines with good functional group tolerance (Scheme 7.23).

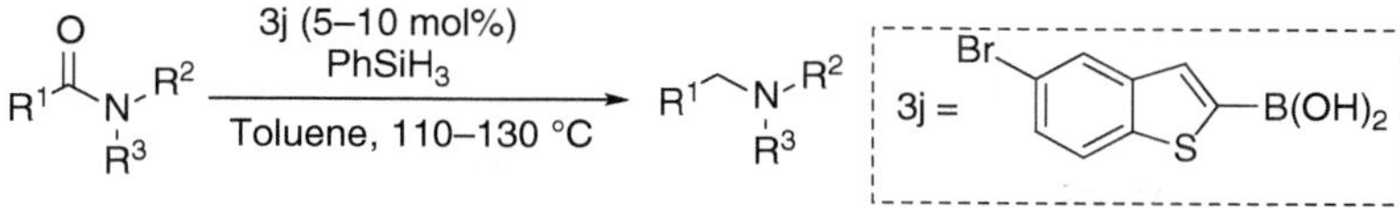

Scheme 7.23 Reduction of amides to amines by aromatic boric acid.

The catalytic hydroboration of amides for deoxygenation to amines was reported for the first time in 2015 [66]. This transformation employed an earth-abundant Mg-based catalyst, and tertiary and secondary amides could be reduced to amines at room temperature in the presence of pinacolborane (HBpin) and catalytic amounts of To^{M}MgMe (To^{M} = tris(4,4-dimethyl-2-oxazolinyl)phenylborate) (Scheme 7.24).

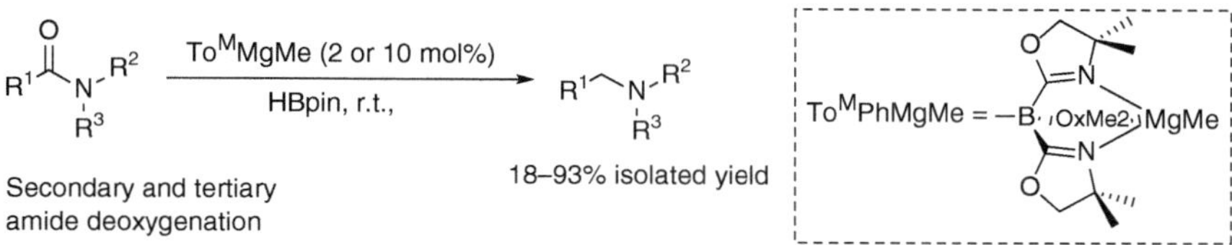

Scheme 7.24 Magnesium-catalyzed amide deoxygenation via hydroboration.

The reduction of tertiary amides to corresponding amines using $(i\text{-Bu})_2AlBH_4$ was complete within five minutes under ambient conditions with an isolated yield up to 99% by a simple acid–base extraction (Scheme 7.25) [67]. This methodology worked well for the reduction of tertiary aliphatic and aromatic amides as well as lactams to the corresponding amines without column chromatography.

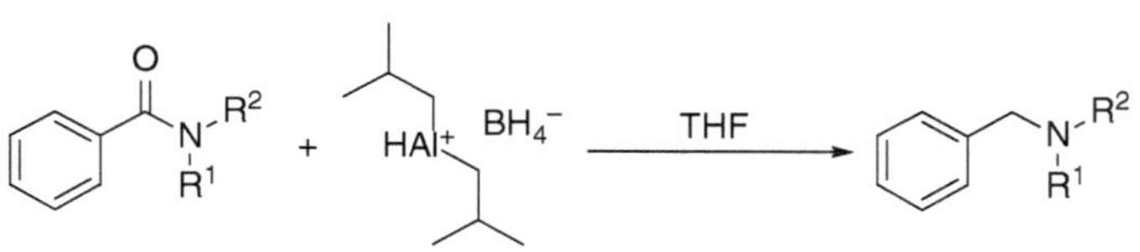

Scheme 7.25 Synthesis of amines from amide via $(i\text{-Bu})_2AlBH_4$. Source: Modified from Snelling et al. [67].

7.1.5 Clemmensen Reduction

The simple ketones and aldehydes can be converted to the corresponding alkanes upon refluxing for several hours with 40% of aqueous hydrochloric acid, amalgamated zinc (Zn/Hg), and a hydrophobic organic co-solvent such as toluene, which was developed by Clemmensen in 1913 [68]. This method of converting a carbonyl group to the corresponding methylene group is known as the Clemmensen reduction (Scheme 7.26). The original procedure is rather harsh, so not surprisingly the Clemmensen reduction of acid-sensitive substrates and polyfunctional ketones is rarely successful in yielding the expected alkanes. The Clemmensen reduction has been widely used in synthesis and several modifications were developed to improve its synthetic utility by increasing the functional group tolerance. Yamamura and coworkers have developed a mild procedure which uses organic solvents (THF, Et_2O, Ac_2O, and benzene) saturated with dry hydrogen halides (HCl or HBr) and activated zinc dust at ice-bath temperature [69–72]. Compared to the original Clemmensen procedure, these modified reductions are complete within one hour at 0 °C and are appropriate for acid and heat-sensitive compounds. Certain carbonyl compounds, however, have very low solubility in the usual solvents used for the Clemmensen reduction, so in these cases a second solvent (acetic acid, ethanol, or 1,4-dioxane) is added to the reaction mixture to increase the solubility of the substrate and allow the reduction to take place.

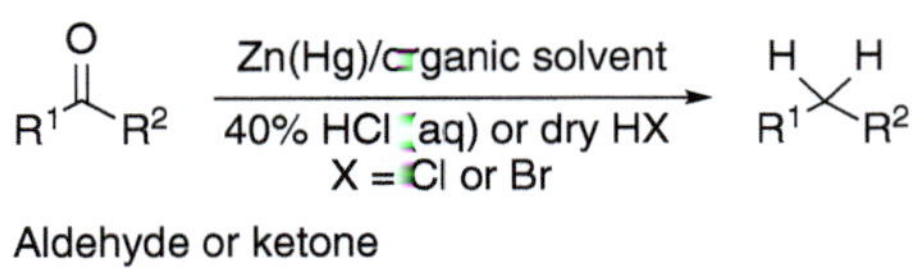

Scheme 7.26 General reaction equation of Clemmensen reduction.

During the enantioselective total synthesis of denrobatid alkaloid (−)-pumiliotoxin C by Kibayashi and coworkers, an aqueous acylnitroso Diels–Alder cycloaddition was used as the key step. In the endgame of the total synthesis, the cis-fused decahydroquinolone was subjected to the Clemmensen reduction conditions to give a 2 : 1 epimeric mixture of deoxygenated products in 57% yield. Subsequent debenzylation converted the major isomer into 5-*epi*-Pumiliotoxin C (Scheme 7.27) [73].

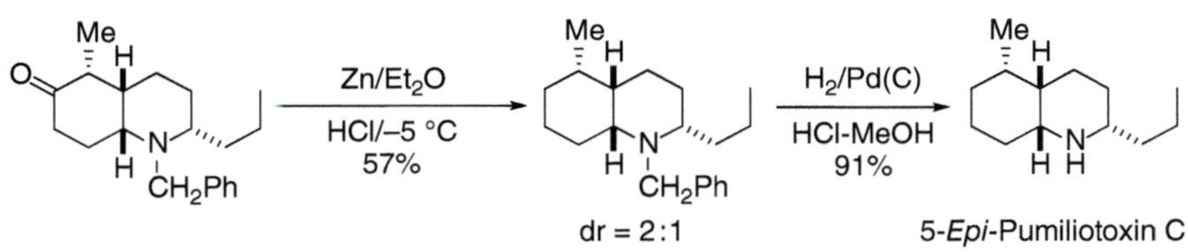

Scheme 7.27 Clemmesen reduction as an important step for the synthesis of 5-*Epi*-Pumiliotoxin C. Source: Based on Naruse et al. [73].

7.1.6 Wolff–Kishner Reduction

The Wolff–Kishner reduction is a reaction used in organic chemistry to convert carbonyl group into alkanes, originally reported by Kishner [74] in 1911 and Wolff [75] in 1912 (Scheme 7.28), and it has been applied to the total synthesis of scopadulcic acid [76] B, aspidospermidine [77, 78], and dysidiolide [79]. The Wolff–Kishner reduction is not suitable for base-sensitive substrates and can be hampered by steric hindrance surrounding the carbonyl group under certain conditions.

Later, Wolff accomplished the same result by heating an ethanol solution of semicarbazones or hydrazones in a sealed tube to 180 °C in the presence of sodium ethoxide. The method developed by Kishner had the advantage of avoiding the requirement of a sealed tube, but early variations of Wolff's procedure involved the

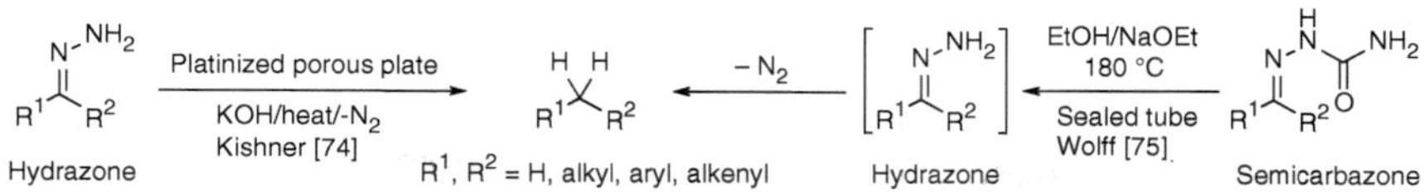

Scheme 7.28 General reaction equation of Wolff–Kishner reduction.

use of high-boiling solvents, such as triethylene glycol, to alleviate this necessity [80, 81]. Subsequently, these initial modifications were followed by many other improvements (Scheme 7.29).

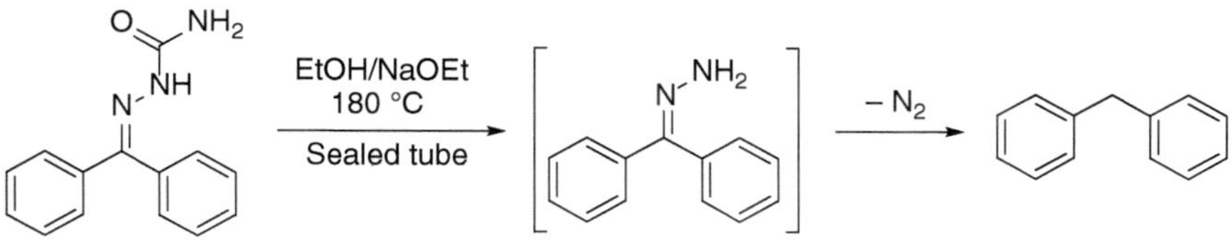

Scheme 7.29 Modification of Wolff–Kishner reduction.

In 1946, Huang-Minlon reported a modified procedure for the Wolff–Kishner reduction of ketones in which excess hydrazine and water were removed by distillation after hydrazone formation (Scheme 7.30) [82, 83]. This modification significantly reduced reaction times and improved yields. Huang-Minlon's original report described the reduction of β-(*p*-phenoxybenzoyl)propionic acid to γ-(*p*-phenoxyphenyl)butyric acid in 95% yield compared to 48% yield obtained by the traditional procedure.

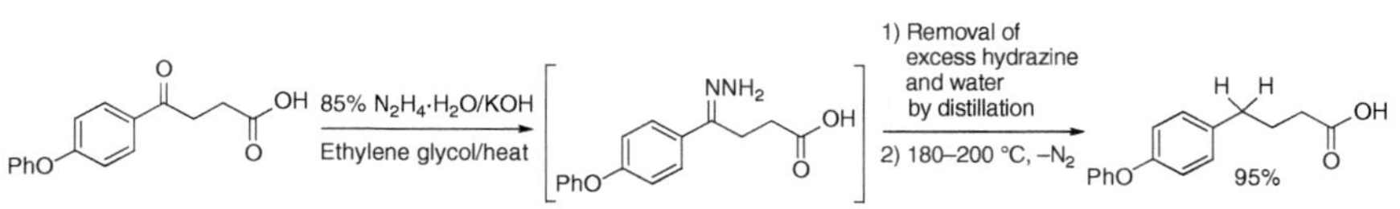

Scheme 7.30 Huang-Minlon modification of Wolff–Kishner reduction. Sources: Huang-Minlon [82], Huang-Minlon [83].

Nine years after Huang-Minlon's first modification, Barton developed a method for the reduction of sterically hindered carbonyl groups [84]. This method featured vigorous exclusion of water, higher temperatures, and longer reaction times as well as sodium in diethylene glycol instead of alkoxide base. Under these conditions, some of the problems that normally arise from hindered ketones can be alleviated, for example, the C11-carbonyl group in the steroidal compound shown below was successfully reduced under Barton's conditions while Huang-Minlon conditions failed to effect this transformation (Scheme 7.31).

The Wolff–Kishner reduction has also been used on kilogram scale for the synthesis of a functionalized imidazole product in good yield (Scheme 7.32) [85].

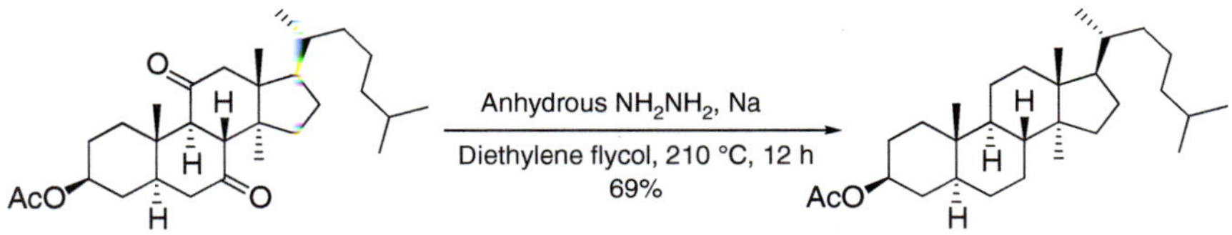

Scheme 7.31 Wolff–Kishner reduction of sterically hindered carbonyl groups.

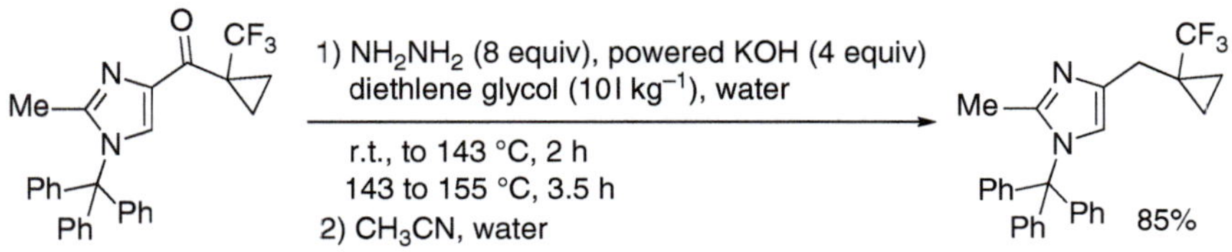

Scheme 7.32 Wolff–Kishner reduction in kilogram scale. Source: Based on Kuethe et al. [85].

7.2 Nucleophilic Addition Reactions of Aldehydes and Ketones

7.2.1 Carbon Nucleophiles

7.2.1.1 Grignard Reagent and Other Organometallic Reagents

The nucleophilic addition of Grignard reagents and organic lithium to carbonyl compounds is an important method for the preparation of alcohols (Scheme 7.33) [86]. Except for formaldehyde, other aldehydes produce secondary alcohols and ketones produce tertiary alcohols. When the groups on both sides of the carbonyl group are too large, the ketone cannot react normally. For example, when R is C_2H_5, the addition reaction proceeds normally with 80% yield. When R is *n*-propyl and *i*-propyl, the yield of addition products is 30% and 0%, respectively.

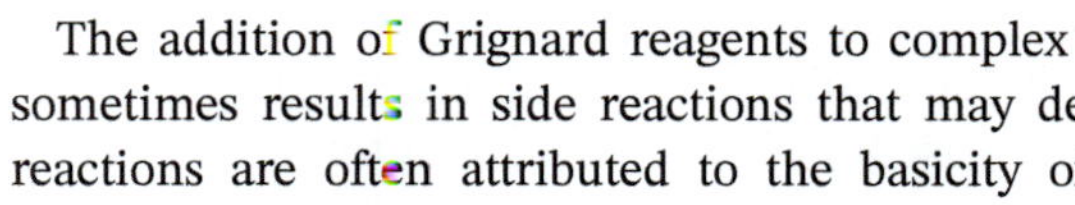

Scheme 7.33 Nucleophilic addition reaction of aldehydes and ketones by Grignard reagent. Source: Based on Cornils et al. [86].

The addition of Grignard reagents to complex molecules with carbonyl group sometimes results in side reactions that may destroy the substrate. These side reactions are often attributed to the basicity of the reagent. Therefore, more nucleophilic derivatives must be prepared. This was the case during the total synthesis of (−)-lochneridine by Kuehne and Xu, when the attempted conversion of a pentacyclic ketone to the corresponding tertiaryalcohol with ethylmagnesium bromide failed [87]. However, the formation of an organocerium reagent by adding the Grignard reagent to anhydrous $CeCl_3$ increased its nucleophilicity; therefore, the reaction afforded the desired tertiary alcohol in 73% yield with complete diastereoselection (Scheme 7.34).

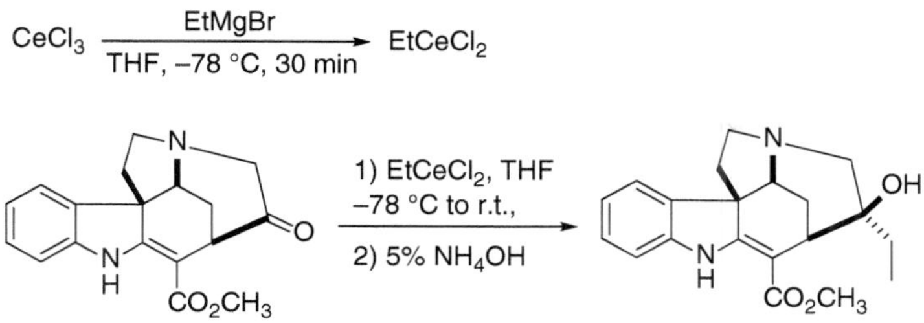

(−)-Lochneridine

Scheme 7.34 Synthesis of (−)-Lochneridine via nucleophilic addition reaction using $EtCeCl_2$.

7.2.1.2 Reformatsky Reaction

In 1887, Reformatsky reported that iodoacetic acid ethyl ester reacted with acetone to yield 3-hydroxy-3-methylbutyric acid ethyl ester in the presence of zinc metal [88]. Since this initial report, the classical Reformatsky reaction was defined as the zinc-induced reaction between a α-halo ester and an aldehyde or ketone (Scheme 7.35).

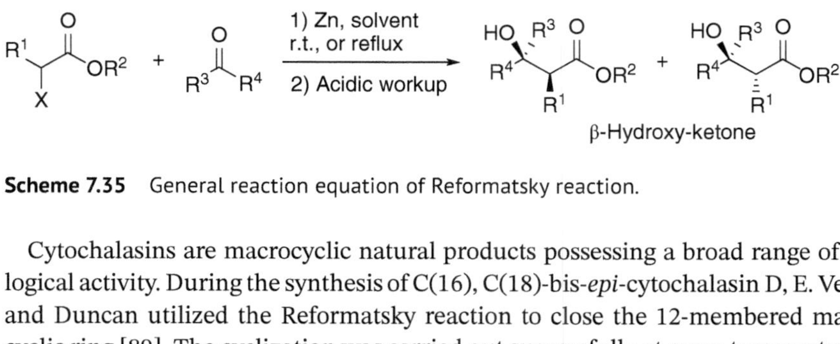

β-Hydroxy-ketone

Scheme 7.35 General reaction equation of Reformatsky reaction.

Cytochalasins are macrocyclic natural products possessing a broad range of biological activity. During the synthesis of C(16), C(18)-bis-*epi*-cytochalasin D, E. Vedejs and Duncan utilized the Reformatsky reaction to close the 12-membered macrocyclic ring [89]. The cyclization was carried out successfully at room temperature to give the 12-membered macrocyclic ring product in 67% yield (Scheme 7.36).

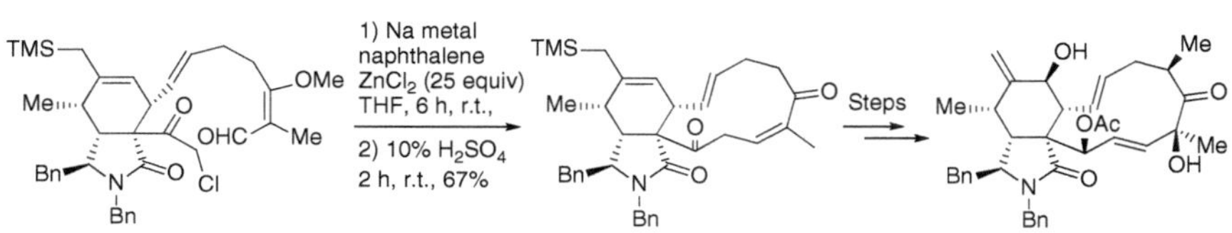

C(16), C(18)-bis-*epi*-Cytochalasin D

Scheme 7.36 Reformatsky reaction for the synthesis of C(16), C(18)-bis-*epi*-cytochalasin D intermediate.

7.2.1.3 Benzoin Condensation

Upon treating certain aromatic aldehydes or glyoxals (α-keto aldehydes) with cyanide ion (CN^-), the carbon nucleophile reagent is formed. Subsequently, benzoins (α-hydroxy-ketones or acyloins) can be produced via the addition of carbon nucleophile reagent to carbonyl group, which is called the benzoin condensation (Scheme 7.37). The condensation involves the addition of one molecule of aldehyde to the C=O group of another. One of the aldehydes serves as the donor and the other serves as the acceptor. Some aldehydes can only be donors (e.g. *p*-dimethylaminobenzaldehyde) or acceptors, so they are not able to self-condense, while other aldehydes (benzaldehyde) can perform both functions and are capable of self-condensation. Certain thiazolium salts can also catalyze the reaction in the presence of a mild base [90–92]. This version of the benzoin condensation is more synthetically useful than the original procedure because it works with enolizable and non-enolizable aldehydes and asymmetric catalysts may be used.

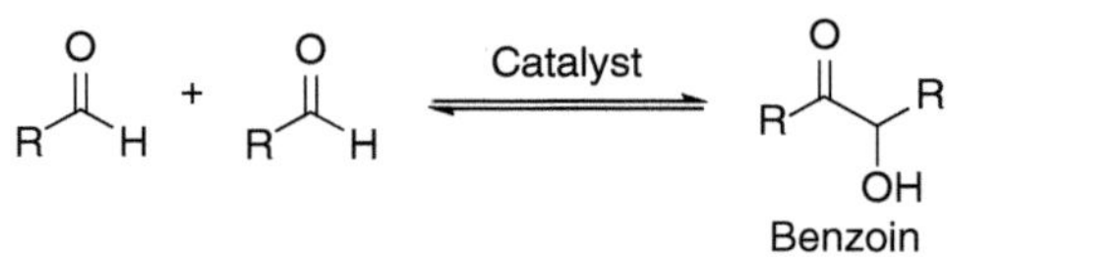

R = aryl, heteroaryl, 3° alkyl, C(=O)-alkyl;
Catalyst = NaCN, KCN, thiazolium, *N*-heterocyclic carbenes

Scheme 7.37 General reaction equation of benzoin condensation.

Miyashita et al. have developed a new method for the synthesis of ketones based on the concept that the benzoin condensation is reversible (retro-benzoin condensation) and affords the most stable product [93]. When α-benzylbenzoin was treated with KCN in DMF, the C–C bond was cleaved, resulting in the formation of deoxybenzoin and benzaldehyde (Scheme 7.38).

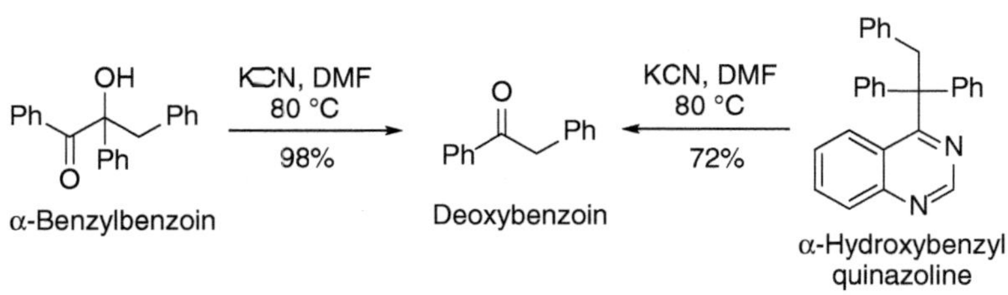

Scheme 7.38 Retro-benzoin condensation for the synthesis of ketones.

7.2.1.4 CN Group

Carbon of cyanide ion can also undergo nucleophilic addition with a variety of reactive aldehydes and ketones to produce α-hydroxynitrile, which is the raw material for the preparation of α-hydroxy-acid. The α-hydroxyl acid can further lose water to become α,β-unsaturated acid. Hydrocyanic acid is a weak acid with little dissociation. The addition of trace base can increase the concentration of CN^-, which greatly increases the reaction rate. However, the addition of acids can lead to protonation of hydrogen ions to carbonyl groups, which increase the electrophilic properties of carbon atoms on carbonyl group, but lowering the concentration of CN^-, thus reducing the reaction rate of nucleophilic addition and making the reaction difficult to occur (Scheme 7.39) [94].

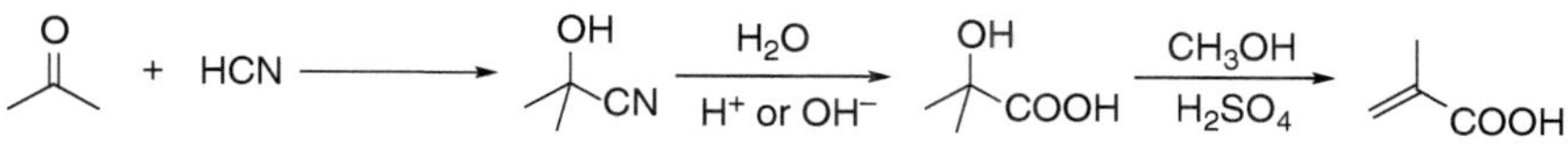

Scheme 7.39 Nucleophilic addition of CN anion to acetone. Source: Vollhardt and Schore [94].

7.2.1.5 Aromatic and Aliphatic C–H Bond

While different classes of nucleophiles have been employed, the transition-metal-catalyzed addition of C–H bonds to aldehydes and ketones has recently emerged as an efficient and powerful approach. To date, a number of powerful catalyst systems have been developed for C–H bond addition to carbonyls for the construction of oxygen-containing products [95–97].

In the seminal report on additions to activated aldehydes, Li and coworkers employed Rh(III) catalyst for the direct addition of aromatic C–H bonds to destabilized, electron-deficient aldehydes to furnish unprotected alcohols (Scheme 7.40) [98]. When several nitrogen heterocycles as effective directing groups, ethyl glyoxylate was employed to obtain the addition products in high yield. Favorable formation of alcohol products was also achieved for aromatic aldehydes bearing electron-withdrawing functionalities.

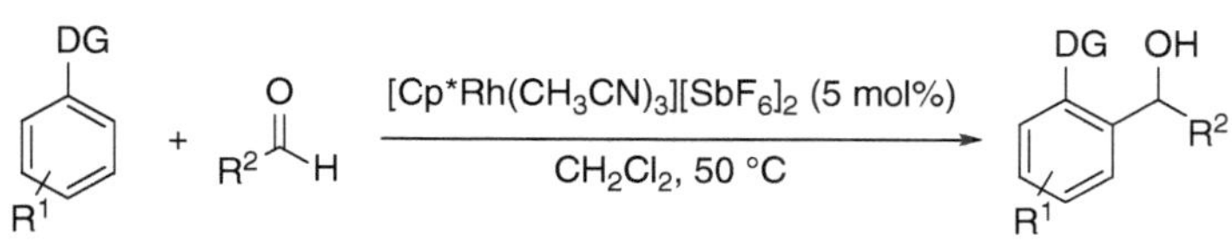

Scheme 7.40 Rh(III)-catalyzed arene C–H bond addition to electron-deficient aldehydes. Source: Modified from Yang et al. [98].

In 2007, Takai and coworkers reported the first example of Mn-catalyzed C–H bond additions to aldehydes (Scheme 7.41) [99]. When a stoichiometric amount of $[MnBr(CO)_5]$ was employed, the desired alcohol was obtained. The addition of triethylsilane proved critical to achieving a high yield under catalytic conditions by trapping alcohol intermediates through protection as silyl ethers. The substrate scope of the transformation was demonstrated with a broad range of aldehydes, including linear and branched alkyl, aromatic, and heteroaromatic derivatives.

In 2013, Yang and coworkers reported the Pd-catalyzed intermolecular addition of heteroaryl C–H bonds to the highly electron-deficient and destabilized ketone carbonyl present in isatins (Scheme 7.42) [100]. A range of electron-rich and -deficient benzoxazole coupling partners and other azole derivatives were effective.

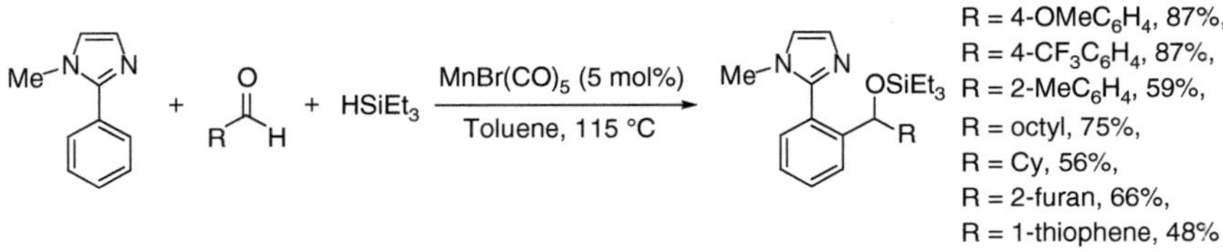

Scheme 7.41 Mn-catalyzed arene C–H addition to aldehydes followed by silylation. Source: Modified from Kuninobu et al. [99].

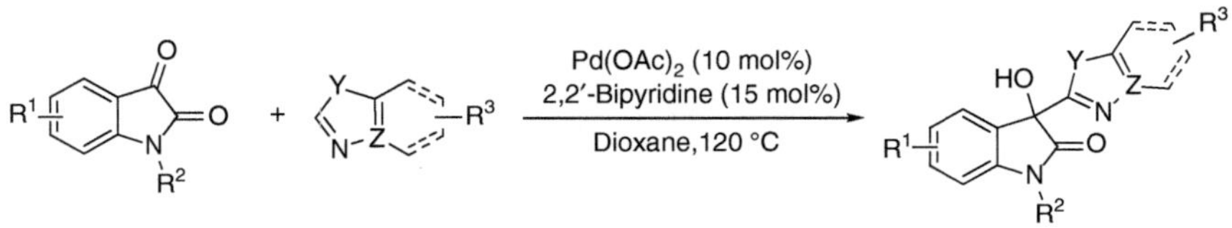

Scheme 7.42 Pd-catalyzed nucleophilic additions of heteroarenes to isatins. Source: Modified from Wang et al. [100].

In 2016, the Kim's laboratory reported the Ru(II)- or Rh(III)-catalyzed pyrimidinyl-directed Grignard-type C–H additions of *N*-pyrimidyl indolines and indoles with activated aldehydes and ketones to provide the alcohol product in moderate to good yields (Scheme 7.43) [101]. For *N*-pyrimidyl indoles, derivatives with halogens at the C4 and C5 positions were effective substrates, but methyl substitution at the C3 position resulted in only a moderate yield of the alcohol product. Moreover, pyrimidyl carbazole gave the alcohol in modest yield, and when pyrimidyl pyrrole was evaluated, it resulted in a high yield of the overaddition product. When *N*-pyrimidyl indoline was used, selective C–H functionalization at the C7 position occurred to provide the alcohol product in moderate yield.

Scheme 7.43 Ru(II)- or Rh(III)-catalyzed C–H additions of *N*-pyrimidyl indolines and indoles with activated aldehydes and ketones. Source: Based on Jo et al. [101].

Compared with the above discussed direct C(sp^2)–H addition to aldehydes and ketones, transition-metal-catalyzed direct C(sp^3)–H addition to aldehydes and ketones still represents a challenging task for chemists, and only a few studies have been reported. In 2012, an efficient Lewis acid-catalyzed addition of 2-alkylazaarenes to ethyl glyoxylate was reported by Jin et al. (Scheme 7.44) [102]. In the presence of 10 mol% of Cu(OTf)$_2$ and 5 mol% of phen, various 2-alkylquinolines and 2-alkylpyridines reacted with ethyl glyoxylate to give the desired addition products in moderate to good yields. However, the addition of 2-alkylazaarenes to ethyl pyruvate, aliphatic aldehydes, and aromatic aldehydes failed to form the desired addition products under the same reaction conditions.

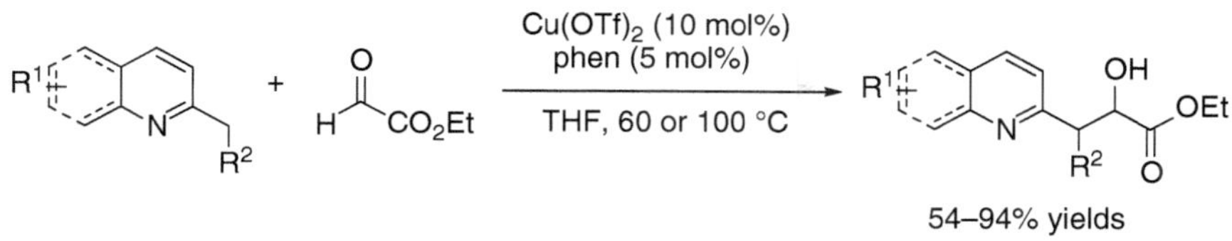

Scheme 7.44 Cu-catalyzed addition of 2-alkylazaarenes to ethyl glyoxylate. Source: Modified from Jin et al. [102].

$Yb(OTf)_3$ was found to be a good catalyst for the reaction of 2-alkylazaarenes with trifluoromethyl carbonyl compounds (Scheme 7.45) [103]. Using $Yb(OTf)_3$ as a catalyst, the direct addition of 2- or 4-methylazaarenes to isatins has been realized [104].

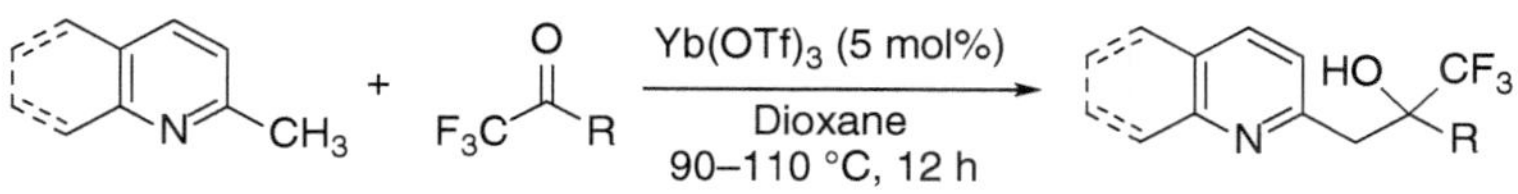

Scheme 7.45 $Yb(OTf)_3$-catalyzed addition of 2-alkylazaarenes with trifluoromethyl carbonyl compounds. Source: Modified from Graves and Shaikh [103].

An extensive study of the reactions between 2-methylazaarenes and simple aldehydes was reported by Wang and coworkers in 2014 (Scheme 7.46) [105]. They found that these reactions occurred smoothly in the presence of $LiNTf_2$ to afford 2-(pyridin-2-yl)ethanols in 20–98% yields.

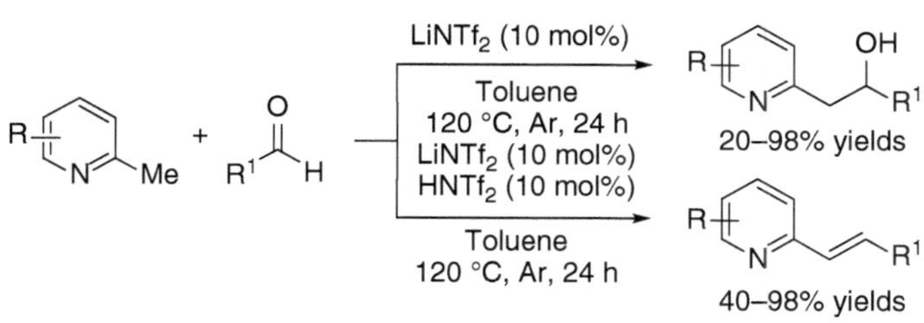

Scheme 7.46 Addition of 2-methylazaarenes to simple aldehydes using $LiNTf_2$. Source: Modified from Mao et al. [105].

7.2.2 Nitrogen Nucleophiles

The addition and elimination of aldehydes and ketones with nitrogen-containing nucleophiles are also described. They mainly react with amine (ammonia), *azo*-compound, hydroxylamine, and other derivatives to produce imine, hydrazone, oxime, etc. [94]

Ammonia or amine can undergo nucleophilic addition with the carbonyl group of aldehyde or ketone, but the product is usually unstable and proceeds to dehydrating into an imine (Schiff base) immediately by losing the hydrogen on the nitrogen. Aliphatic imines are very unstable and easy to break down. However, aromatic imines are stable and can be isolated. Imines can be hydrolyzed in dilute acids to produce the original carbonyl compounds and amines, so it is also a way to protect the carbonyl compounds. Carbonyl compounds containing α-hydrogen react with secondary amines to form enamine in a different way to lose water. Enamine has two reaction sites, namely nitrogen and carbon on double bond, which can introduce various groups to nitrogen and carbon atoms, respectively, which is very important in organic synthesis (Scheme 7.47).

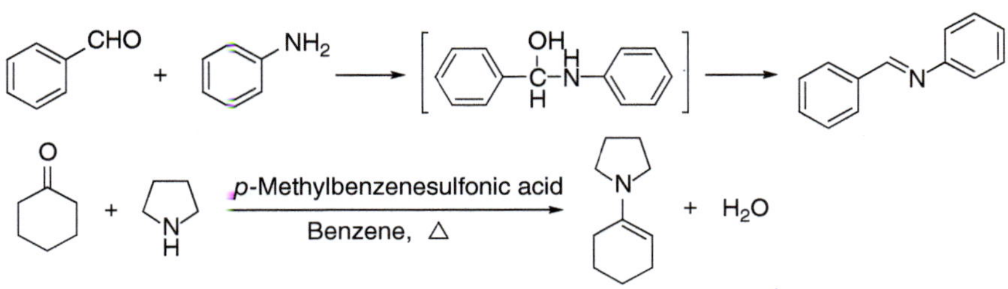

Scheme 7.47 Nucleophilic addition of amines to aldehydes or ketones.

The products obtained by the acid-catalyzed reaction of aldehydes or ketones with *azo* compounds are called hydrazone (Scheme 7.48). When the reaction takes place in a nonaqueous solvent, the concentration of hydrogen ions is very low, mainly the whole weak acid, such as acetic acid, that is because carbonyl group may be bonded to the whole weak acid in a way of hydrogen bond during the reaction, which increases the electrophilic properties of the carbonyl group and promotes the nucleophilic addition of its free amino derivatives.

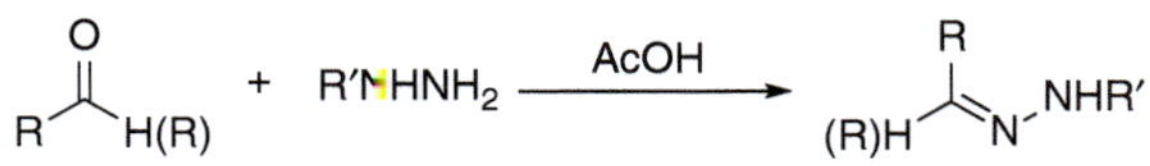

Scheme 7.48 Nucleophilic addition of *azo* compounds to aldehydes or ketones.

Aldehydes or ketones can react with hydroxylamine to form oxime. Oxime and nitrite compounds are tautomeric with the following equilibrium. Only in the absence of α-hydrogen can make nitroso compounds stable; otherwise, the equilibrium is favorable for oxime. Oxime has *Z* and *E* isomers, but it often exists in a more stable *E* configuration. For example, the *Z*-isomer of the benzaldehyde oxime dissolved in an alcohol under acid condition becomes an *E*-isomer. However,

E-isomer can be converted to *Z*-benzaldehyde oxime only under the action of light (Scheme 7.49).

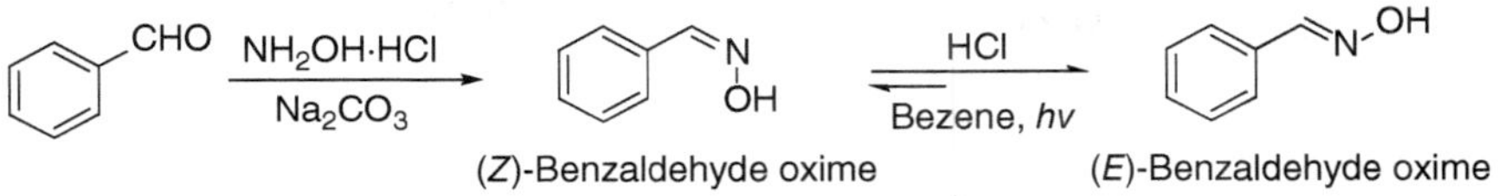

Scheme 7.49 Nucleophilic addition of hydroxylamine to aldehydes or ketones.

7.2.3 Oxygen Nucleophiles

7.2.3.1 H_2O as a Nucleophile

Water, as a nucleophile, reacts with aldehydes or ketones in acidic conditions and the resulting products are called the hydrate of aldehydes or ketones (Scheme 7.50) [94]. Since the two hydroxyl groups of the hydrate are attached to the same carbon, which is thermodynamically unstable and is prone to losing H_2O and reverting to aldehydes or ketones, that is, the addition of water to aldehydes or ketones is a reversible reaction and the equilibrium is much more preferred to the reactants. Only a few of aldehydes, such as formaldehyde, turn almost entirely into hydrates in an aqueous solution, but the hydrates cannot be separated owing to the water loss during the separation process. However, if the carbonyl group is connected to strong electron-withdrawing groups, such as Cl_3C–, RCO–, –CHO, –COOH, and FCH_2–, the stable hydrates can be formed owing to the enhancement of electrophilicity of the carbonyl group.

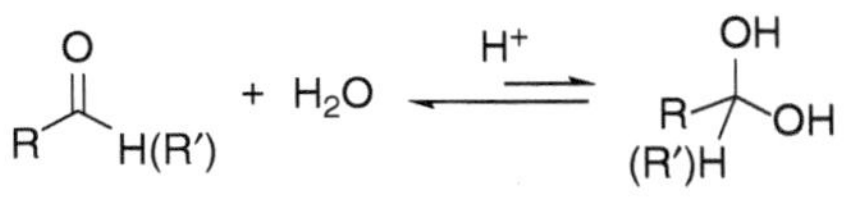

Scheme 7.50 Water as nucleophile for the addition of aldehydes or ketones.

7.2.3.2 ROH as a Nucleophile

Alcohols are also nucleophilic, can easily undergo nucleophilic addition with aldehydes or ketones to form hemiacetal or hemiketal in the presence of acid catalysts such as *p*-toluenesulfonic acid and hydrogen chloride, and further react to form acetal and ketal (Scheme 7.51) [94]. In addition, in order to increase the yield of acetal and ketal, Dean–Stark trap need to be used to continuously remove the water generated by the reaction, so that the balance can move toward the product.

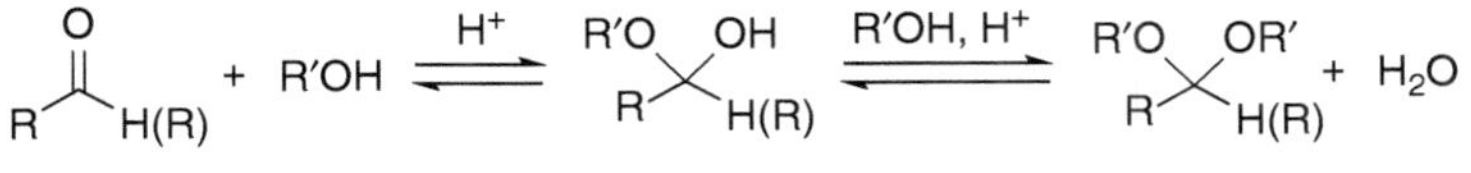

Scheme 7.51 Alcohols as nucleophile for the addition of aldehyde or ketones. Source: Vollhardt and Schore [94].

7.3 Addition Elimination Reactions of Aldehydes and Ketones

7.3.1 Aldol Reaction

The aldol reaction involves the addition of the enol/enolate of a carbonyl compound (nucleophile) to an aldehyde or ketone (electrophile). Originally, the aldol reaction was carried out with Brönsted acid [106] or Brönsted base catalysis [107, 108]. The initial product of the reaction is α-hydroxycarbonyl compound that undergoes dehydration to generate the corresponding α,β-unsaturated carbonyl compound (Scheme 7.52). The transformation takes its name from 3-hydroxybutanal, the acid-catalyzed self-condensation product of acetaldehyde, which is commonly called aldol.

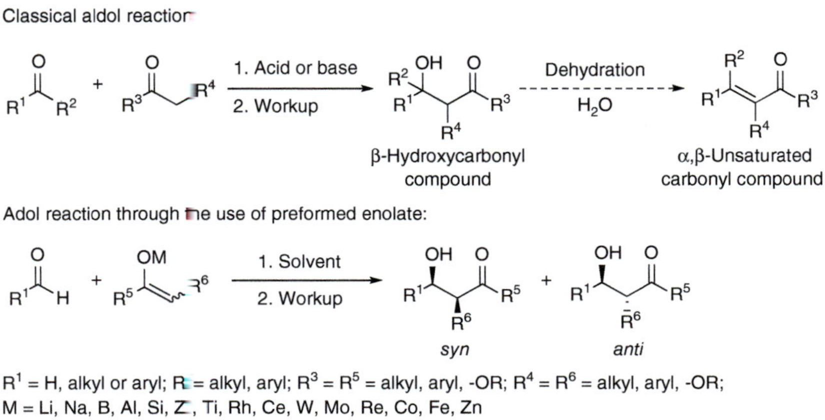

Scheme 7.52 General reaction equation of aldol condensation. Source: Kane [106], Claisen and Claparede [107], and Claisen [108].

During the total synthesis of Rhizoxin D by White et al., an asymmetric aldol reaction was utilized to achieve the coupling of two key fragments [109]. The aldol reaction of the aldehyde and the chiral enolate derived from (+)-chlorodiisopinocampheylborane afforded the product with a diastereomeric ratio of 17–20 : 1 at the C13 stereocenter (Scheme 7.53).

Shibasaki and coworkers developed a bifunctional chiral catalyst, (*S*)-LLB (L = lanthanum; LB = lithium binaphthoxide), which could be successfully applied in direct catalytic asymmetric aldol reactions [110]. An improved version of this catalyst derived from (*S*)-LLB by the addition of water and KOH was utilized in the formal total synthesis of fostriecin (Scheme 7.54) [111].

7.3.2 Perkin Reaction

The condensation of aromatic aldehydes with the anhydrides of aliphatic carboxylic acids in the presence of a weak base to afford α,β-unsaturated carboxylic acids is

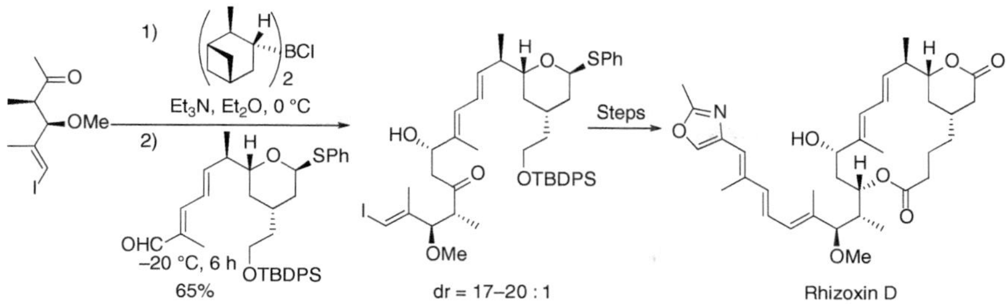

Scheme 7.53 Application of aldol reaction for the synthesis of Rhizoxin D intermediate.

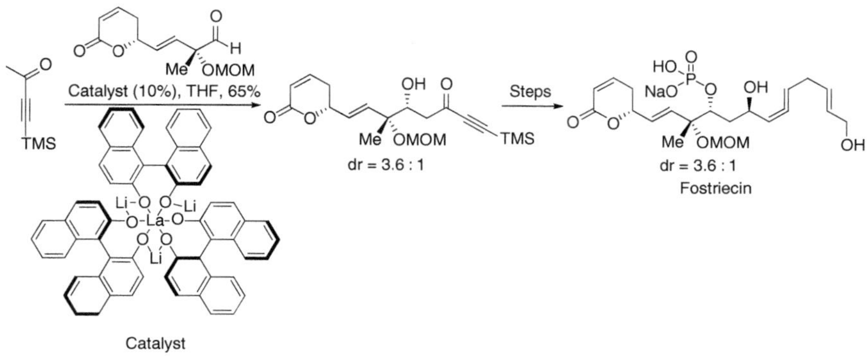

Scheme 7.54 Application of aldol reaction for the synthesis of fostriecin intermediate.

known as the Perkin reaction (or Perkin condensation). In 1868, Perkin described the one-pot synthesis of coumarin by heating the sodium salt of salicylaldehyde in acetic anhydride (Scheme 7.55) [112]. After this initial report, Perkin investigated the scope and limitation of the process and found that it was well suited for the efficient synthesis of cinnamic acids [113].

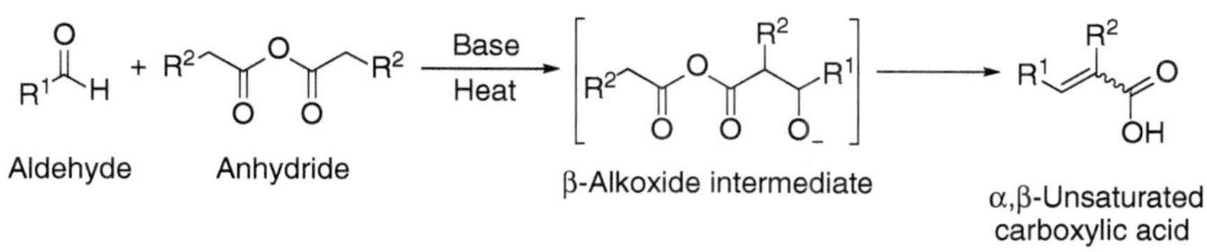

Scheme 7.55 General reaction equation of Perkin reaction. Source: Modified from Perkin [112].

The application of Perkin reaction was demonstrated by the synthesis of (*Z*)-combretastatin A-4 [114]. 3,4,5-Trimethoxyphenylacetic acid and 3-hydroxy-4-methoxbenzaldehyde were heated with triethylamine and acetic anhydride at reflux for several hours to give the α,β-unsaturated acid in good yield after the acidification and had the expected (*E*)-stereochemistry. Decarboxylation of this acid was affected by heating it with copper powder in quinoline to afford the natural product (*Z*)-combretastatin A-4 (Scheme 7.56).

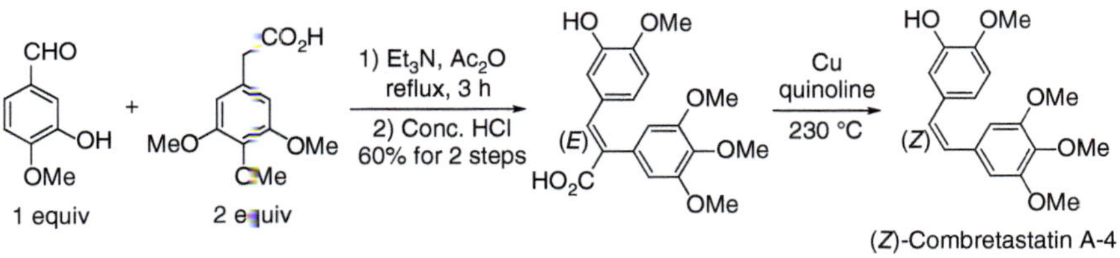

Scheme 7.56 Synthesis of (*Z*)-combretastatin A-4. Source: Modified from Baker et al. [114].

The large-scale pilot plant preparation of the chiral aminochroman antidepressant ebalzotan (also known as NAE-086) was developed by H.-J. Federsel and coworkers [115]. The backbone of the target molecule was constructed using the Perkin condensation of 2-hydroxy-6-methoxybenzaldehyde with hippuric acid under mild conditions (Scheme 7.57).

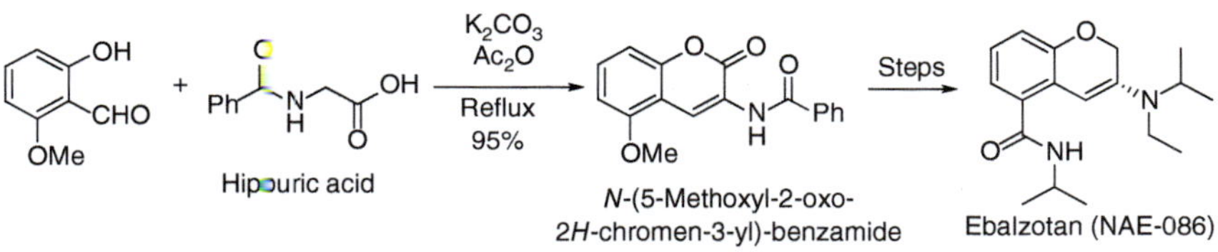

Scheme 7.57 Synthesis of ebalzotan (NAE-086).

7.3.3 Knoevenagel Condensation

The reaction of aldehydes and ketones with active methylene compounds in the presence of a weak base to afford α,β-unsaturated dicarbonyl or related compounds is known as the Knoevenagel condensation. In 1894, E. Knoevenagel reported the diethylamine-catalyzed condensation of diethyl malonate with formaldehyde in which he isolated the bis adduct (Scheme 7.58) [116]. In this reaction, the carbonyl group is an aldehyde or a ketone. The catalyst is usually a weakly basic amine. The active hydrogen component has the form Z–CH_2–Z or Z–CHR–Z, for instance diethyl malonate, Meldrum's acid, ethyl acetoacetate or malonic acid,

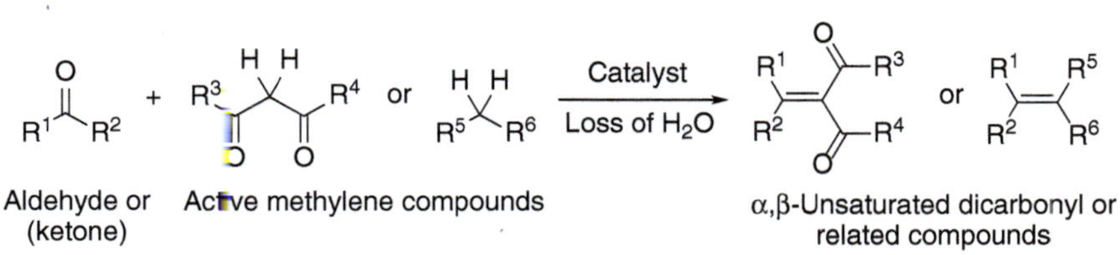

Scheme 7.58 General reaction equation of Knoevenagel condensation.

or cyanoacetic acid [117], Z–CHR^1R^2, for instance nitromethane, where Z is an electron-withdrawing functional group.

The domino Knoevenagel condensation or hetero-Diels–Alder reaction was used for the enantioselective total synthesis of the active *anti*-influenza A virus indole alkaloid hirsutine and related compounds by L.F. Tietze and Y. Zhou [118]. The Knoevenagel condensation was carried out between an enantiopure aldehyde and Meldrum's acid in the presence of ethylenediamine diacetate. The resulting highly reactive 1-oxa-1,3-butadiene underwent a hetero-Diels–Alder reaction with 4-methoxybenzyl butenyl ether ($E/Z = 1:1$) in situ. The product exhibited a 1,3-asymmetric induction greater than 20 : 1 (Scheme 7.59).

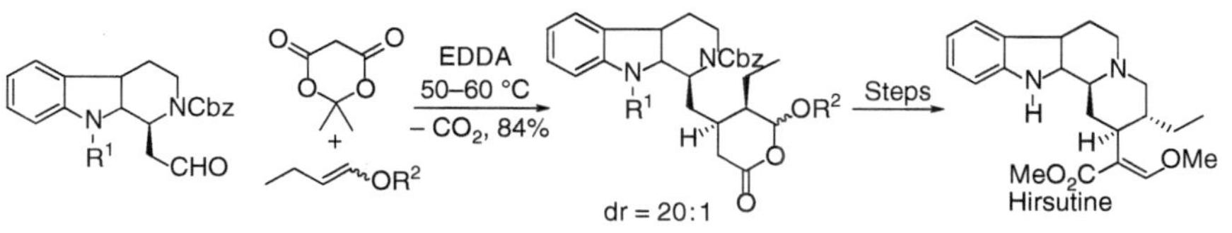

Scheme 7.59 Synthesis of hirsutine. Source: Modified from Tietze and Zhou [118].

During the total synthesis of (+)-leporin A, a tandem Knoevenagel condensation/inverse electron demand intramolecular hetero-Diels–Alder reaction was employed by B.B. Snider and Q. Lu to construct the key tricyclic intermediate [119]. The condensation of pyridone with the enantiopure acyclic aldehyde in the presence of triethylamine as catalyst afforded an intermediate that underwent a [4 + 2] cycloaddition to afford the tricyclic core of the target (Scheme 7.60).

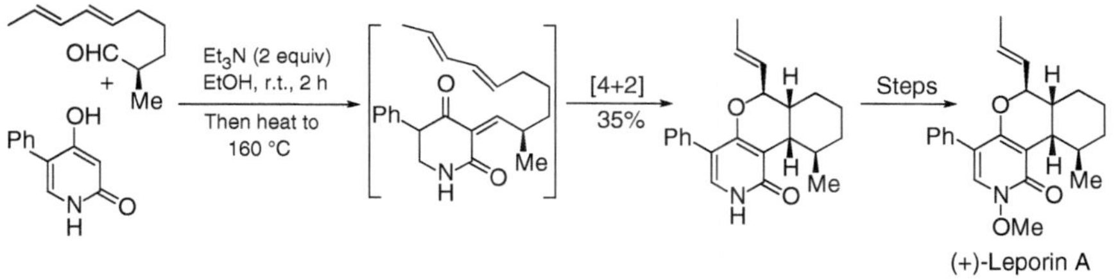

Scheme 7.60 Synthesis of (+)-leporin A.

7.4 Oxidation of Aldehydes and Ketones

7.4.1 Baeyer–Villiger Oxidation

The Baeyer–Villiger oxidation is an organic reaction that forms an ester from a ketone or a lactone from a cyclic ketone, using peroxyacids or peroxides as the oxidant. The reaction is named after Adolf von Baeyer and Victor Villiger who first reported the reaction in 1899 (Scheme 7.61) [120, 121]. Typical peracids in this oxidation include peroxybenzoic acid, *m*-chloroperoxybenzoic acid (*m*CPBA),

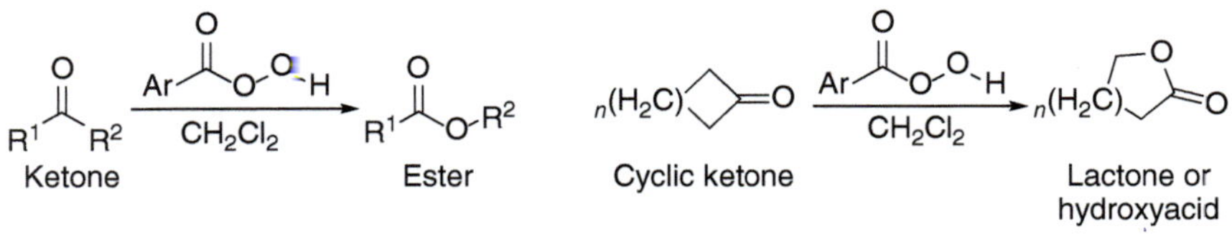

Scheme 7.61 General reaction equation of aldehydes and ketones to acids.

peroxyacetic acid, and trifluoroperoxyacetic acid. In this reaction, the more substituted group often migrates preferentially, and the migratory aptitudes of these groups are tertiary > secondary > cyclohexyl > benzyl > phenyl > primary > methyl. More importantly, both the stereochemistry and chirality of the migrating group are retained during the migration.

Investigations by Oh showed that the cycloaddition of dichloroketene to glucal followed by Baeyer–Villiger oxidation afforded a bicyclic γ-lactone, an α-D-C-glucoside, which was further transformed to a C1-methyl glucitol derivative (Scheme 7.62) [122].

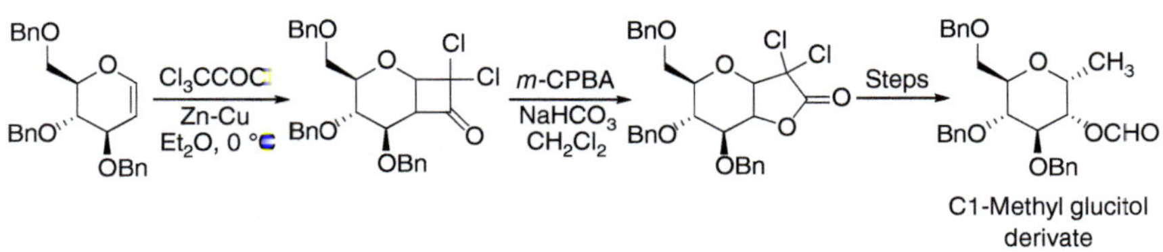

Scheme 7.62 Synthesis of C1-methyl glucitol derivatives. Source: Based on Oh [122].

In 2013, Świzdor reported the transformation of the dehydroepiandrosterone to anticancer agent testololactone by the use of a Baeyer–Villiger oxidation induced by fungus that produces Baeyer–Villiger monooxygenases (Scheme 7.63) [123].

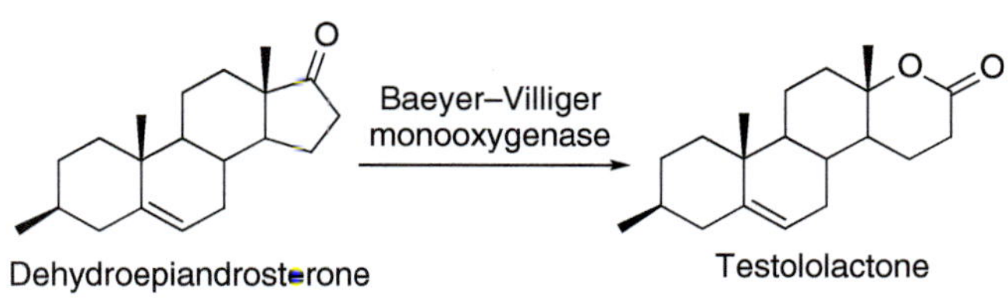

Scheme 7.63 Synthesis of testololactone. Source: Modified from Świzdor [123].

7.4.2 To Acid

In 1965, the base-catalyzed autoxidation of acetophenone and C5 to C12 cyclic aliphatic ketones was studied in hexamethylphosphoramide (HMPA) at 23.5 and 80 °C [124]. The cyclic ketones were autoxidized to their corresponding dibasic acids in moderate to excellent yields in the presence of either potassium or sodium hydroxide.

Subsequently, Corey et al. found that the cyanide-catalyzed oxidation of aldehydes in methanol can be directed to give either methyl ester (with active MnO_2) or carboxylic acid (with AgO) [125]. Using a molar ratio of oxide to aldehyde of 4 : 1 and reaction time of 14 hours, dodecanal and 3-cyclohexenylcarboxaldehyde were oxidized to the corresponding acids in 90% and 97% yields, respectively.

A variety of reliable oxidizing reagents for the conversion of "sensitive" alcohols to aldehydes or ketones are available, but for the oxidation of "sensitive" aldehydes to carboxylic acids there are only a few. In 1986, Masamune and coworkers modified the conventional catalytic system of $KMnO_4$ using a mixture of *t*-BuOH and aqueous NaH_2PO_4 as a reaction medium to make aldehydic compound with one or more protected hydroxyl groups effectively oxidized with $KMnO_4$ to the corresponding carboxylic acids, and this methodology is operational simplicity, efficiency, and selectivity (Scheme 7.64) [126].

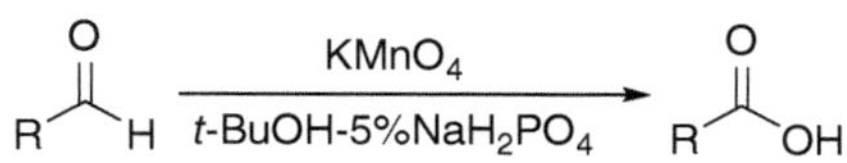

Scheme 7.64 Oxidation of aldehydes to carboxylic acids by $KMnO_4$. Source: Modified from Abiko et al. [126].

A convenient method for oxidative transformation of aromatic, heteroaromatic, and aliphatic aldehydes into carboxylic acids was presented by Mlochowski in 2000 [127]. It was based on the oxidation of aldehydes in THF using 30% of hydrogen peroxide in the presence of 5 mol% of selenium(IV) oxide.

Solid-supported reagents such as phosphate-buffered (PB) silica gel (SiO_2)-supported potassium permanganate ($KMnO_4$) and polymer-supported (PS) chlorite have been prepared and used in the conversion of aldehydes to carboxylic acids, affording products without any need for conventional workup procedures [128]. After this, Huang and coworkers prepared polystyrene-bound phenylseleninic acid, which was used as a catalyst for the oxidation of aldehydes to carboxylic acids with hydrogen peroxide. In all cases, the recovered resin can be used again with no further treatment and no discernible loss of reactivity [129].

Oxone, a potassium triple salt containing potassium peroxymonosulfate, is an effective oxidant for numerous transformations. The metal-free oxidation of dialkyl, alkyl aryl, or alkyl heteroaryl ketones or arylalkynes to the corresponding carboxylic acids is achieved using an oxidative mixture of oxone and trifluoroacetic acid (Scheme 7.65) [130]. This green method is a simple and mild protocol to obtain carboxylic derivatives in good to excellent yields.

In 2007, Sain and coworkers for the first time developed an efficient and new transition metal-free method for the oxidation of aromatic aldehydes to carboxylic acids using *N*-methylpyrrolidin-2-one hydrotribromide (MPHT) as a catalyst and hydrogen peroxide as the oxidant under mild reaction conditions [131]. The safe handling, ease of preparation, crystalline nature of the reagent, excellent yields of products, and simple workup make this synthetic tool for the oxidation of various aromatic aldehydes to acids.

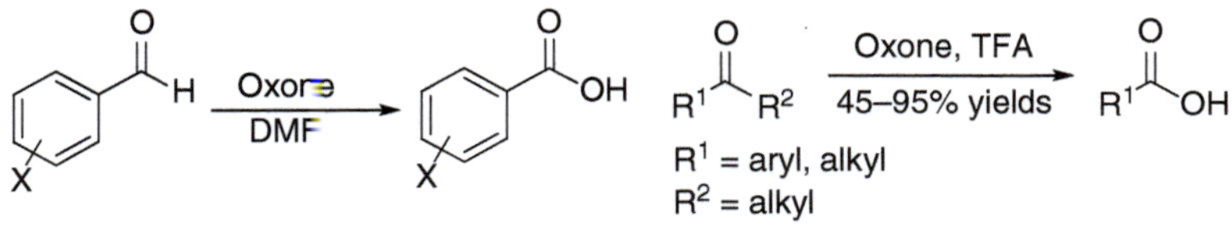

Scheme 7.65 Oxone as an oxidant for the conversion of aldehydes to acids. Source: Modified from Travis et al. [130].

Sulfonated carbon as a strong and stable solid acid catalyst exhibited excellent catalytic performance in various acid-catalyzed reactions. In 2013, Yang and coworkers studied the catalytic performance of sulfonated carbon in the oxidation of aldehydes to the corresponding carboxylic acids using 30 wt% of H_2O_2 as the oxidant [132]. The sulfonated carbon is found to be an efficient, easy recyclable heterogeneous catalyst for the oxidation of aldehydes to carboxylic acids.

The first example of homogeneous silver(I)-catalyzed aerobic oxidation of an aldehyde in water was discovered by Li and coworkers in 2015 [133]. The reaction occurred at a very mild temperature, using atmospheric oxygen as the oxidant, and this reaction can proceed with an extremely low loading of the Ag(I)catalyst (360 ppm catalyst on a gram-scale experiment). More than 50 different kinds of aldehydes were tested, and all underwent transformation to their corresponding carboxylic acids in mostly excellent to quantitative yields, indicating a good versatility and a variety of possible applications for this reaction (Scheme 7.66).

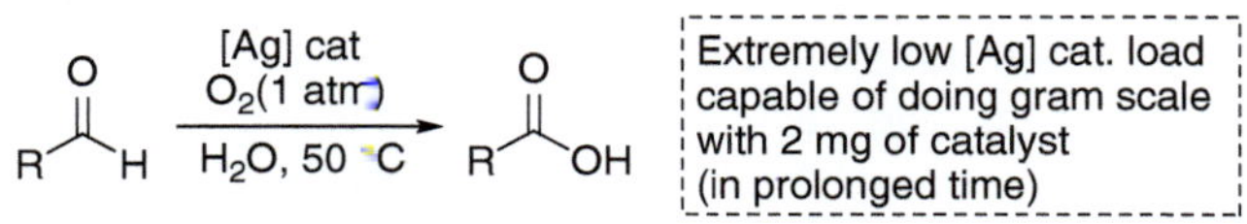

Scheme 7.66 Silver(I)-catalyzed aerobic oxidation of an aldehyde. Source: Modified from Liu et al. [133].

In 2018, Ma and coworkers developed an efficient catalytic room temperature aerobic oxidation of aldehydes to the corresponding acids [134]. With 5 mol% of $Fe(NO_3)_3·9H_2O$ as a catalyst and 1 atm of pure O_2 or oxygen in air as the oxidant, different types of aldehydes with various useful functional groups could be oxidized to acid smoothly. The reaction could be conducted easily on a 50-mmol scale with O_2 or a slow air flow (Scheme 7.67).

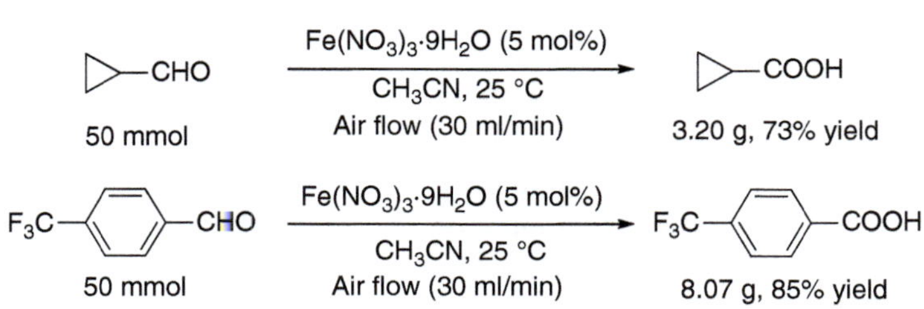

Scheme 7.67 Fe(III)-catalyzed aerobic oxidation of aldehydes to the corresponding acids.

Recently, organo-catalyzed aerobic oxidation of aldehydes to carboxylic acids in both organic solvent and water under mild conditions was developed for the first time by Kang and coworkers, using *N*-hydroxyphthalimide (NHPI) (5 mol%) as the organocatalyst, and molecular O_2 as the sole oxidant (Scheme 7.68) [135]. A wide range of carboxylic acids bearing diverse functional groups were obtained from aldehydes, even from alcohols, in high yields.

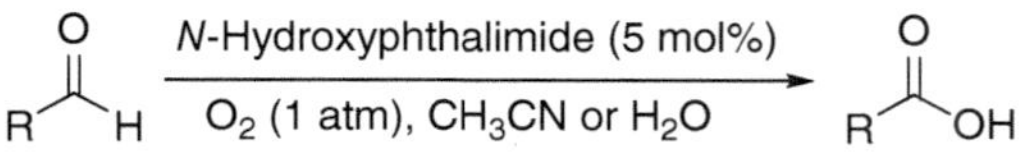

Scheme 7.68 *N*-Hydroxyphthalimide-catalyzed aerobic oxidation of aldehydes to carboxylic acids.

7.5 Wittig Reaction

The formation of carbon–carbon double bonds (olefins) from carbonyl compounds and phosphoranes (phosphorous ylides) is known as the Wittig reaction, for which he was awarded the Nobel Prize in Chemistry in 1979. In the early 1950s, Wittig and coworker investigated pentavalent phosphorous and described the reaction between methylenetriphenylphosphorane ($Ph_3P{=}CH_2$) and benzophenone, which gave 1,1-diphenylethene and triphenylphosphine oxide ($Ph_3P{=}O$) in quantitative yield [136]. Since its discovery, the Wittig reaction has become one of the most important and most effective methods for the synthesis of alkenes. The active reagent in this transformation is the phosphorus ylide, which is usually prepared from a triaryl- or trialkylphosphine and an alkyl halide (1° or 2°) followed by deprotonation with a suitable base (e.g. RLi, NaH, NaOR, etc.) (Scheme 7.69).

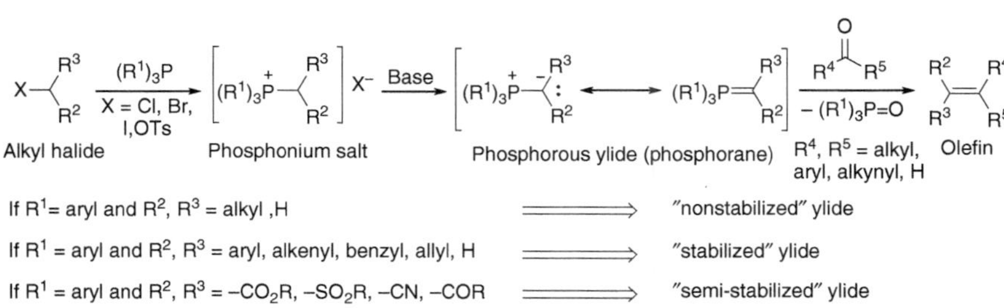

Scheme 7.69 General reaction equation of Wittig reaction. Source: Wittig and Geissler [136].

The most popular use of the Wittig reaction is for the introduction of a methylene group using $Ph_3P{=}CH_2$. Using this reagent, even a sterically hindered ketone such as Camphor can be converted to its methylene derivative. In another example, the phosphorane is produced using sodium amide as a base, and this reagent converts the aldehyde into alkene in 62% yield (Scheme 7.70) [137].

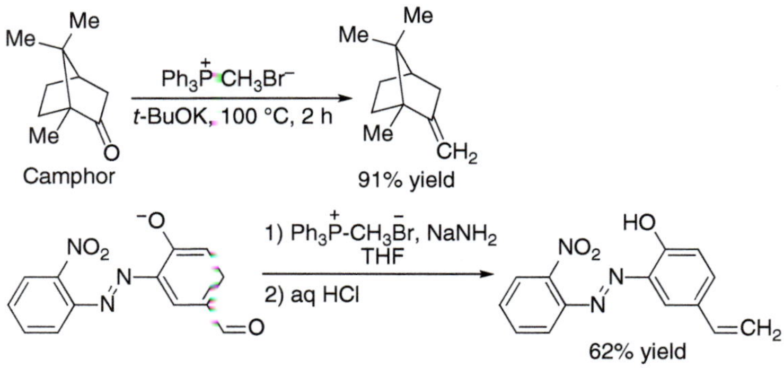

Scheme 7.70 Wittig reaction for the synthesis of methylene derivatives.

The synthesis of heterocyclic systems and naturally occurring alkaloids has been successfully accomplished by utilizing the tandem *aza*-Wittig/intramolecular cyclization (AW-IC), among which is lavendamycin [138]. Phosphazenes, in which the indole ring is linked with a flexible alkyl chain containing two carbon atoms, reacted through an *aza*-Wittig reaction with aldehydes to afford the heterocyclic compounds (Scheme 7.71) [139]. In an analogous manner, the synthesis of heteroaromatic β-carboline-fused pentacycles has been reported [140].

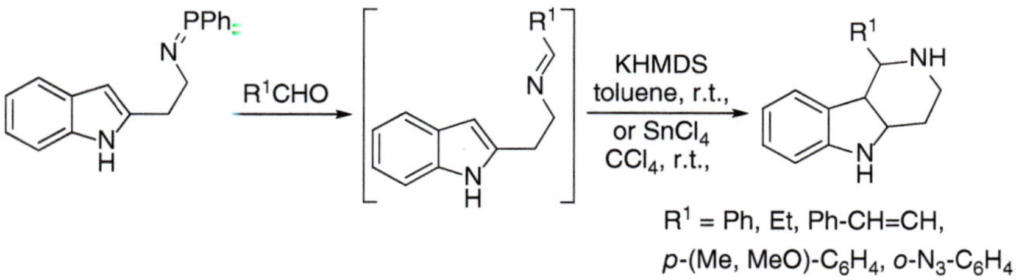

Scheme 7.71 Tandem *aza*-Wittig/intramolecular cyclization. Source: Modified from Molina et al. [139].

7.6 Reductive Amination Reaction

The preparation of amines from aldehydes and ketones is an important method in organic synthesis because of their versatile utility as intermediates for synthesis of pharmaceuticals [141] and agrochemicals [142, 143]. For this transformation, the reductive amination allows the conversion of carbonyl functionality to an amine by directly treating a mixture of the carbonyl compound and the amine with suitable reducing agent in a single operation without preformation of an intermediate imine or iminium salt. A variety of homogeneous and heterogeneous reducing reagents have been reported for this conversion.

7.6.1 Homogeneous Catalyst System

Although there are several reports describing synthesis of amines using different reductive amination procedures [144], few have dealt with anilines bearing electron-withdrawing groups [145]. As early as in 1990, Maryanoff and coworker reported that the weakly basic chloro-, nitro-, cyano-, ethoxy-, carbomyl-, and carboxy-substituted anilines react with aldehydes and ketones under reductive amination conditions, employing sodium triacetoxyborohydride ($NaH(OAc)_3$), to give excellent yields of the corresponding secondary aromatic amines (Scheme 7.72) [146].

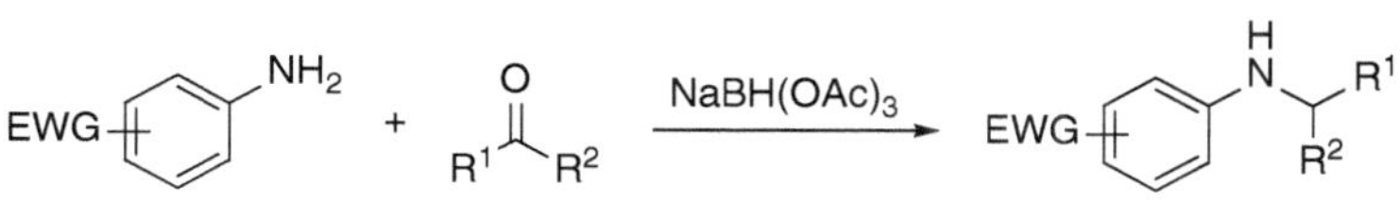

Scheme 7.72 Reductive amination of aldehydes and amines under $NaBH(OAc)_3$.

Procedures for using the mild and selective reagent have been developed for a wide variety of substrates by Shah and coworkers in 1996 [147]. The scope of the reaction includes aliphatic acyclic and cyclic ketones, aliphatic and aromatic aldehydes, and primary and secondary amines including a variety of weakly basic and nonbasic amines. The procedure is carried out effectively in the presence of acid-sensitive functional groups such as acetals and ketals; it can also be carried out in the presence of reducible functional groups such as C–C multiple bonds and cyano and nitro groups. In comparison with other reductive amination procedures such as $NaBH_3CN$/MeOH, borane pyridine, and catalytic hydrogenation, $NaBH(OAc)_3$ gave consistently higher yields and fewer side products.

In 1998, an efficient procedure using the combination of silica gel and zinc borohydride for providing the first general and efficient methodology for one-pot reductive amination of conjugated aldehydes and ketones was developed by Ranu et al. [148]. During the process, a mixture of carbonyl compound and amine was adsorbed on the surface of silica gel, which was benefit for yielding imines, thus obtaining the final amines under zinc borohydride (Scheme 7.73).

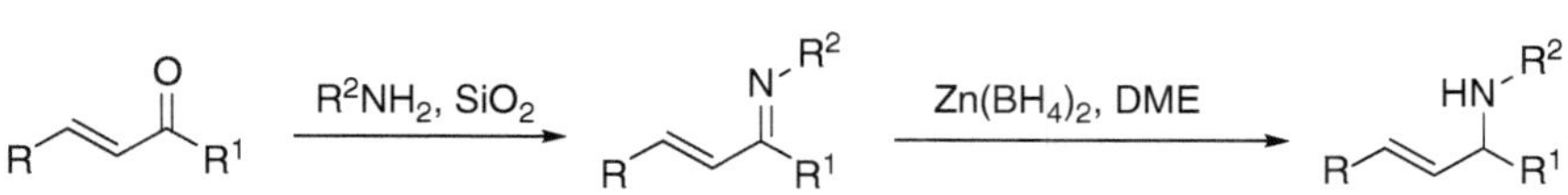

R, R^2 = alkyl or aryl; R^1 = H or alkyl

Scheme 7.73 Combination of silica gel and zinc borohydride for one-pot reductive amination. Source: Modified from Ranu et al. [148].

Two years later, Tararov et al. for the first time reported that cationic rhodium(I) complexes $[Rh(dppb)(cod)]BF_4$ [dppb = 1,4-bis(diphenylphosphino)butane, cod = cycloocta-1,5-diene] and $[Rh(dppe)(cod)]BF_4$ [dppe = 1,2-bis(diphenylphosphino) ethane] are highly efficient precatalysts in the homogeneously catalyzed reductive amination of aldehydes and ketones under room temperature with 1–50 bar H_2 as the reducing agent (Scheme 7.74) [149].

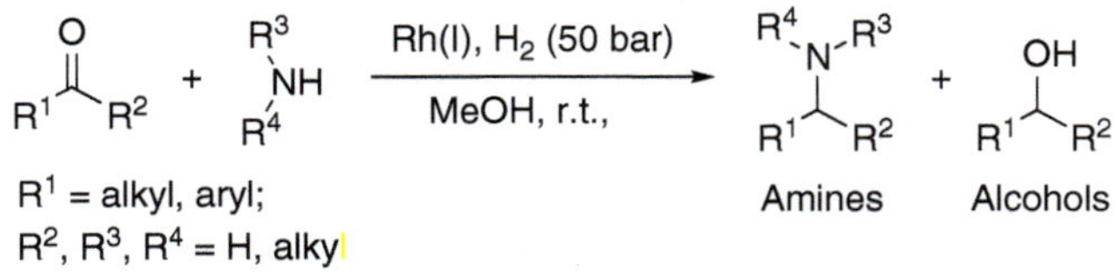

Scheme 7.74 Reductive amination of aldehydes and ketones with Rh(I) catalysts.

In 2001, Apodaca and coworker developed a simple direct reductive amination procedure which employs phenylsilane as a stoichiometric reductant and dibutyltin dichloride as a catalyst under room temperature [150]. Both anilines and dialkylamines with ketones were reductively aminated with anilines and secondary alkylamines; however, the reaction failed with primary alkylamines (Scheme 7.75). Five years later, the reductive amination of aldehydes or ketones using Ph_2SiH_2 or $PhSiH_3$ had been effectively promoted by the direct use of Bu_2SnClH–pyridine *N*-oxide as a catalyst by Baba and coworkers [151]. This method had advantages in terms of its mild conditions and wide application to various carbonyls and amines, including aliphatic examples.

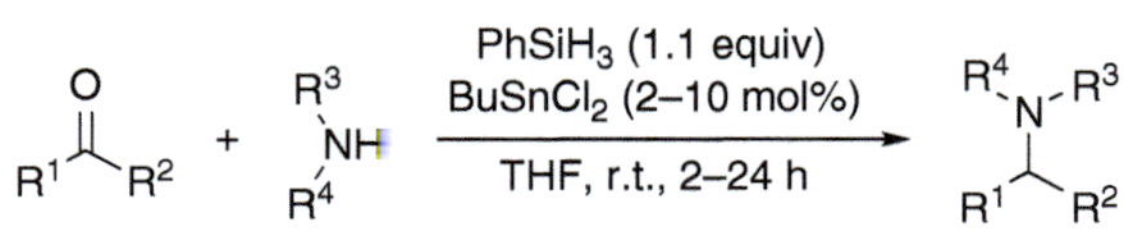

Scheme 7.75 Combination of phenylsilane and dibutyltin dichloride for reductive amination. Source: Modified from Apodaca and Xiao [150].

A simple protocol for direct reductive amination of aldehydes and ketones, including α,β-unsaturated carbonyl compounds, had been developed by Basu et al. in 2003 [152], which required potassium formate as reductant and palladium acetate as catalyst. Suitable amines included both primary and secondary aliphatic and aromatic amines (Scheme 7.76). The method was useful for preparing all classes of amines from suitable carbonyl compounds and the amines. Furthermore, the method could be of importance in view of cheap reducing agent, which decomposes to environmentally friendly chemicals.

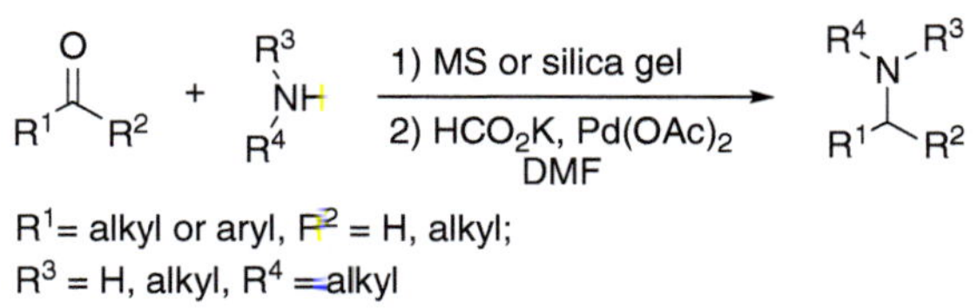

Scheme 7.76 Reductive amination using KCO_2K as reductant and $Pd(OAc)_2$ as catalyst.

In 2005, Cho et al. established a direct and indirect reductive amination of aldehydes and ketones using sodium borohydride activated by inorganic and organic solid acid, such as boric acid, *p*-toluenesulfonic acid, and benzoic acid, as reducing agents under solvent-free conditions (Scheme 7.77) [153]. Soon after, a simple and convenient procedure for direct reductive amination of aldehydes and ketones with sodium borohydride was described by Heydari et al. [154] The reaction was carried out in methanol in the presence of a catalytic amount of $H_3PW_{12}O_{40}$ (0.5 mol%).

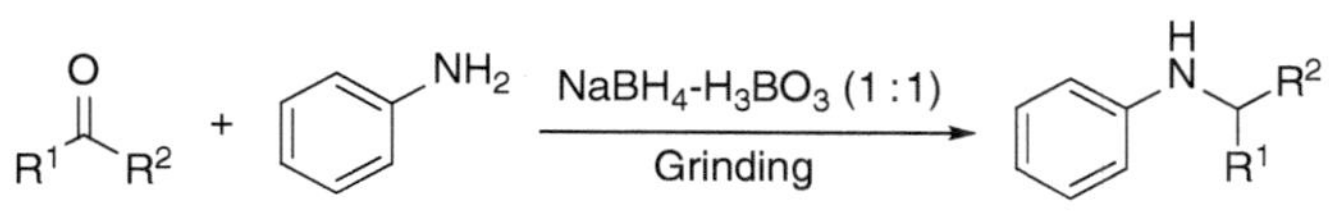

Scheme 7.77 Reductive amination of functionalized aldehydes and ketones with H_3BO_3-actived $NaBH_4$.

In 2007, Yu and coworkers found that $ZrCl_4$/Hantzsch 1,4-dihydropyridine was a selective and efficient reagent combination for the direct reductive amination [13]. Weakly basic amines such as anilines substituted by electron-withdrawing group and heteroaromatic amines can be reductively alkylated with electron-rich aldehydes and ketones under mild conditions to form the secondary amines in excellent yields (Scheme 7.78).

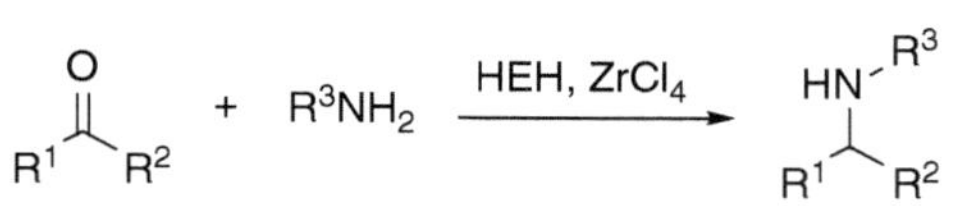

Scheme 7.78 $ZrCl_4$/HEH-mediated reductive amination of aldehydes and ketones with weakly basic amines. Source: Modified from Liu et al. [133].

In 2018, Zhou and coworkers reported a green and efficient electrocatalytic reductive amination of electron-deficient aldehydes/ketones, which is promising to be used in large-scale preparation of racemic clopidogrel [155]. A plausible mechanism involving an iminium cation intermediate was proposed to explain this reaction (Scheme 7.79).

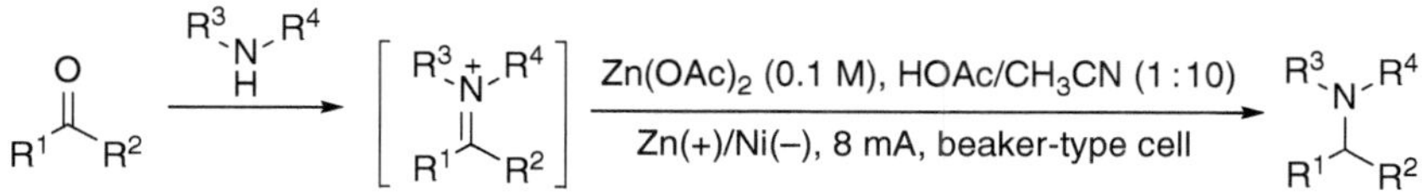

Scheme 7.79 Electrocatalytic reductive amination to tertiary amines.

Recently, Ru-catalyzed direct asymmetric reductive amination of *ortho*-OH-substituted diaryl and sterically hindered ketones with ammonium salts and H_2 was disclosed by Zhang et al. [156], offering a concise route toward the synthesis of useful chiral primary diarylmethylamines and sterically hindered benzylamines (up to 97% yield, 93–>99% ee) (Scheme 7.80).

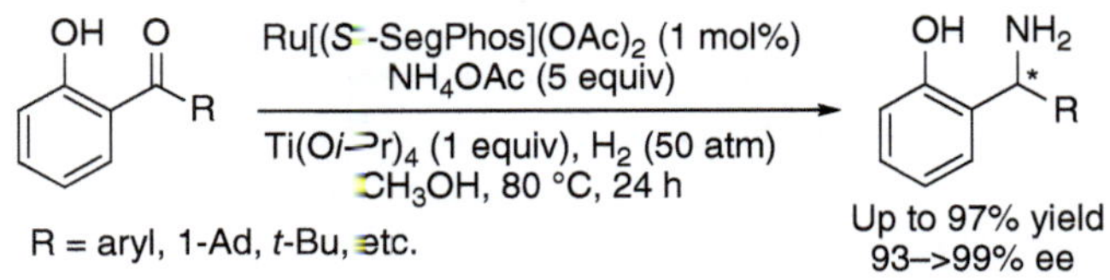

Scheme 7.80 Ru-catalyzed direct asymmetric reductive amination of ketones with ammonium salts and H_2.

7.6.2 Heterogeneous Catalyst System

Direct reductive amination of aldehydes and ketones was accomplished efficiently using high-capacity, ionene-based, polymer-supported borohydride reagent in isopropyl alcohol at reflux under neutral conditions by Tajbakhsh et al. in 2008 (Scheme 7.81) [157]. The reagent of poly-2,4-ionene borohydride (IBER) is easily prepared by mixing aqueous solution of 2,4-ionene bromide with an alkaline solution of sodium borohydride at room temperature. The generality of reaction was established using both aliphatic and aromatic aldehydes, ketones, and amines.

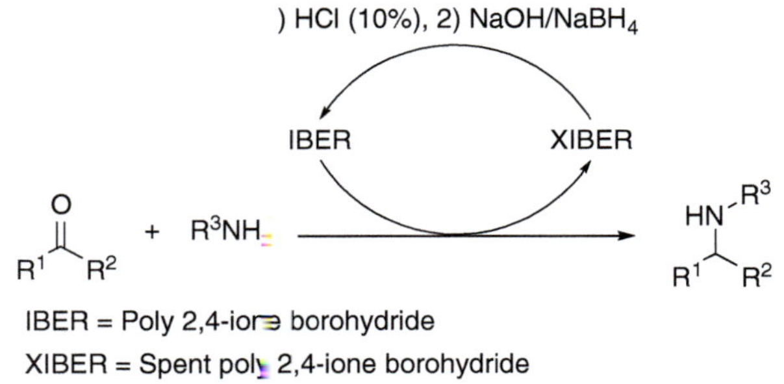

Scheme 7.81 Regeneration of poly-2,4-ionene borohydride.

In 2009, Zare and coworkers described an effective and regioselective method for the preparation of amines by direct reductive amination of different aldehydes and ketones with various amines using sodium borohydride in the presence of silica-gel-supported sulfuric acid (SSA) as a stable solid acid under heterogeneous and solvent-free conditions at room temperature (Scheme 7.82) [158]. While ago, a similar procedure was introduced for which had been carried out in solid state in the presence of catalytic amount of wet carbon-based solid acid [159].

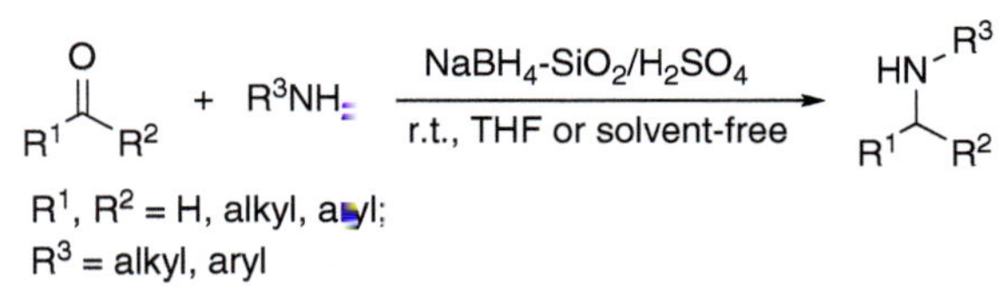

Scheme 7.82 Reductive amination of aldehydes and ketones with amines using $NaBH_4$-SiO_2/H_2SO_4. Source: Modified from Alinezhad et al. [158].

A facile and efficient process for the synthesis of dibenzylamines through direct reductive amination of aldehydes with ammonia has been developed using the unsupported ultra-thin Pt nanowire catalyst under mild reaction conditions by Qi et al. in 2012 [160]. The Pt nanowire catalyst exhibited excellent activity and selectivity in reductive amination using ammonia or ammonium acetate as the substrate and can be recycled easily (Scheme 7.83).

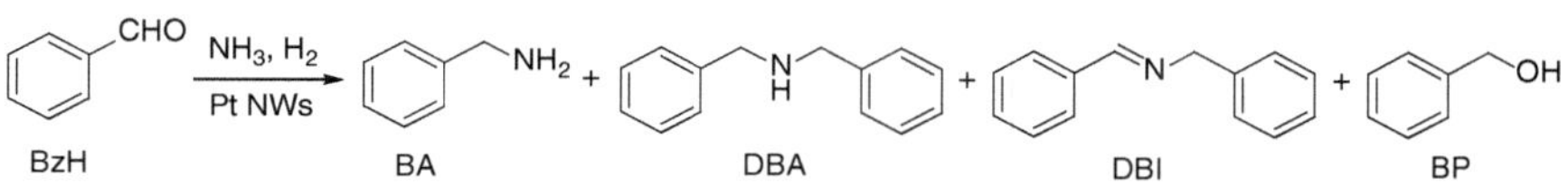

Scheme 7.83 Reductive amination of benzaldehydes with ammonia. Source: Qi et al. [160].

In 2014, Boukherroub group has developed an efficient, simple, and convenient method for the synthesis of various amine derivatives by the reductive amination of aldehydes and ketones [161]. The reductive amination of aromatic, heteroaromatic, and aliphatic aldehydes and ketones with aromatic or aliphatic amines in the presence of triethylsilane and a catalytic amount of Pd/C in ethanol afforded the corresponding amines in excellent yields (Scheme 7.84). In the same year, Nasrollahzadeh reported the synthesis and use of palladium nanoparticles as heterogeneous catalysts for the reductive amination of aldehydes and hydrogenation of unsaturated ketones with the advantages of high yields, simple methodology, and easy workup [162]. The catalyst can be recovered and reused several times without significant loss of catalytic activity.

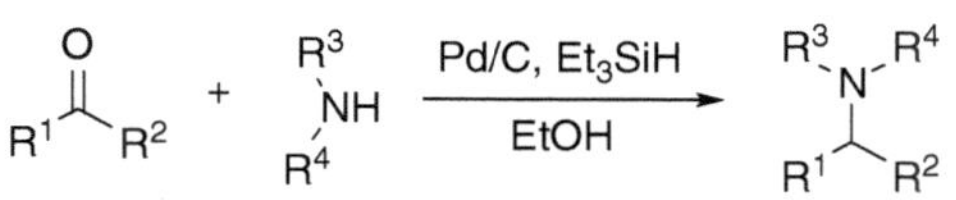

Scheme 7.84 Reductive amination of carbonyl compounds using triethylsilane and Pd/C catalyst with amines.

In 2015, Chung and coworker developed the first Co_2Rh_2 nanoparticles/charcoal-catalyzed reductive amination of aldehydes and ketones with amines using a water–gas shift reaction instead of hydrogen [163]. The reaction can be extended to the tandem reduction of aldehydes and ketones with nitroarenes (Scheme 7.85).

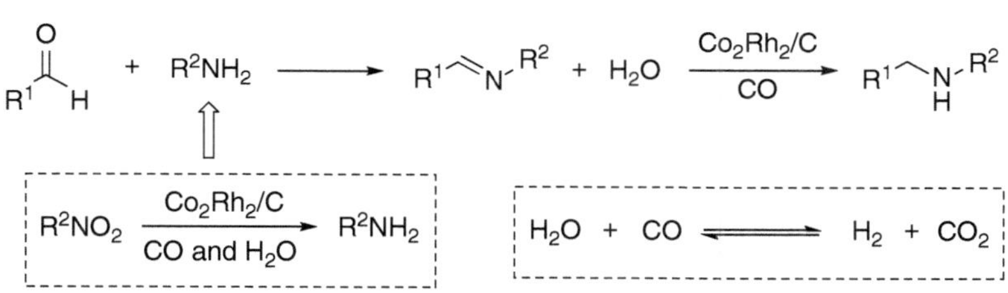

Scheme 7.85 Co_2Rh_2 nanoparticles/charcoal-catalyzed reductive amination.

The catalytic system is stable under the reaction conditions and could be reused eight times without losing any catalytic activity.

In 2016, Xu and coworkers developed an efficient reductive amination methodology, in which the heterogeneous Au/TiO_2 was used as catalyst, and environmentally friendly formic acid was used as transfer hydrogen reagent [164]. Aldehydes and ketones were converted into secondary and tertiary amines in good to excellent yields. Moreover, this gold catalyst could be easily recycled without significant loss of reactivity. Meanwhile, magnetic nanoparticle-supported phosphine gold(I) complex [Fe_3O_4@SiO_2–P–AuCl] as a highly efficient and recyclable catalyst for the direct reductive amination of aldehydes and ketones in dichloromethane at room temperature had been reported by Cai group (Scheme 7.86) [165]. The new heterogeneous gold catalyst can be easily prepared and separated from the reaction mixture by applying an external magnet and recycled at least 10 times without any loss of activity.

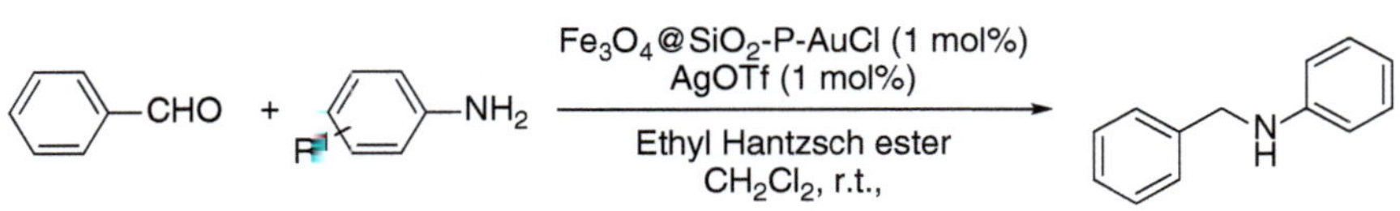

Scheme 7.86 Reductive amination of benzaldehyde with aromatic amines catalyzed by Fe_3O_4@SiO_2–P–AuCl.

Transformation of biomass into valuable nitrogen-containing compounds is highly desired, yet limited success has been achieved. In 2017, Zhang group demonstrated that Ru/ZrO_2-containing multivalence Ru association species functioned as a highly efficient catalyst for the reductive amination of a variety of biomass-based aliphatic and aromatic aldehydes/ketones in aqueous ammonia under mild reaction conditions [166]. The coexistence of Ru and RuO_2 on the surface of Ru/ZrO_2-150 furnished its both strong Lewis acid sites and metallic hydrogenation sites. RuO_2 worked as acidic promoter to facilitate the activation of carbonyl groups, while metallic Ru worked as active sites for imine hydrogenation. The cooperation between two types of active sites leaded to excellent performance for the production of primary amines. They also demonstrated that cellulose could be converted into ethanolamine by a two-step approach, with an overall yield of 10% (Scheme 7.87).

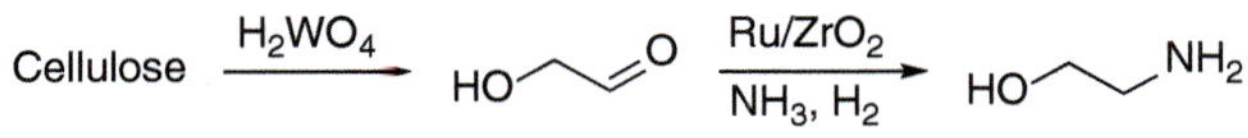

Scheme 7.87 Direct conversion of cellulose to ETA via reductive amination. Source: Modified from Yang et al. [165].

At the same time, Beller and coworkers reported that metal–organic framework (MOF)-derived cobalt nanoparticles encapsulated by a graphitic shell are broadly effective reductive amination catalysts [167]. Their convenient and practical preparation entailed template assembly of cobaltdiamine–dicarboxylic acid metal organic frameworks on carbon and subsequent pyrolysis under inert atmosphere. The resulting stable and reusable catalysts were active for synthesis of primary, secondary, tertiary, and *N*-methylamines (Scheme 7.88). The reaction couples easily accessible carbonyl compounds (aldehydes and ketones) with ammonia, amines, or nitro compounds, and molecular hydrogen under industrially viable and scalable conditions, offering cost-effective access to numerous amines, amino acid derivatives, and more complex drug targets.

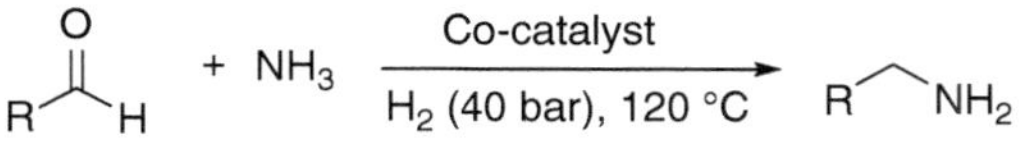

Scheme 7.88 Reductive amination using MOF-derived cobalt nanoparticles.

7.7 Hydroboration/Hydrophosphonylation/Hydrosilylation/Hydroacylation of Aldehydes and Ketones

7.7.1 Hydroboration

In 1970, Brown and Kabalka reported a oxygen-induced reaction of organoboranes with the inert α,β-unsaturated carbonyl derivatives, which was a convenient aldehyde and ketone synthesis via hydroboration [168]. Trialkylboranes, readily available via hydroboration, undergo 1,4-addition to α,β-unsaturated carbonyl compounds (Scheme 7.89). Many α,β-unsaturated carbonyl compounds, such as ethylideneacetone and crotonaldehyde, undergo a rapid reaction with the organoborane, without added catalysts. However, other α,β-unsaturated carbonyl compounds, the "inert" group, fail to react in the absence of added catalysts, such as diacyl peroxides or photochemical activation.

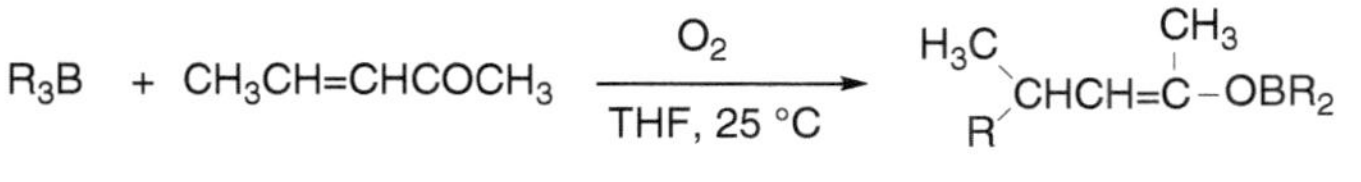

Scheme 7.89 Oxygen-induced hydroboration of organoboranes with α,β-unsaturated carbonyl derivatives.

In 2009, Clark and coworkers developed the hydroboration of aldehydes and ketones using ruthenium dimer $[2,5\text{-}Ph_2\text{-}3,4\text{-}Tol_2(\eta^5\text{-}C_4CO)Ru(CO)_2]_2$ as a catalyst at 50–70 °C [169]. The hydroboration of aryl ketones, however, required strongly electron-withdrawing substituents to induce hydroboration in reasonable reaction times (Scheme 7.90).

The easily prepared and inexpensive magnesium alkyl complex $[CH_2Mg\textit{n}\text{-}Bu]$ was applied to the efficient hydroboration of a variety of aromatic and aliphatic aldehydes and ketones under mild conditions and at low catalyst loadings, which was demonstrated by Arrowsmith et al. in 2012 [170]. Catalytic turnover most likely proceeded via the formation of a catalytically active magnesium hydride species, insertion of

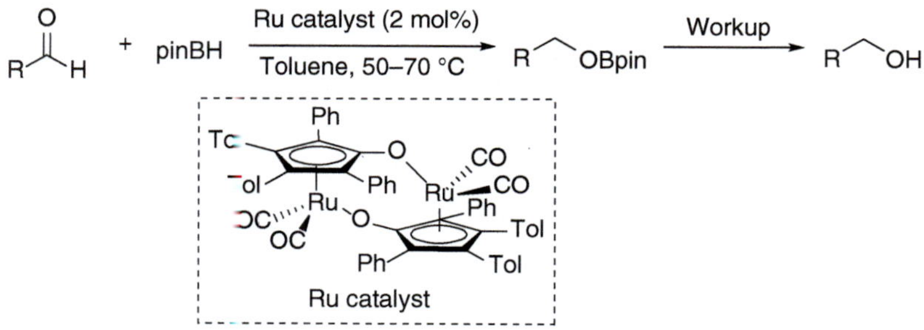

Scheme 7.90 Catalytic hydroboration of aldehydes.

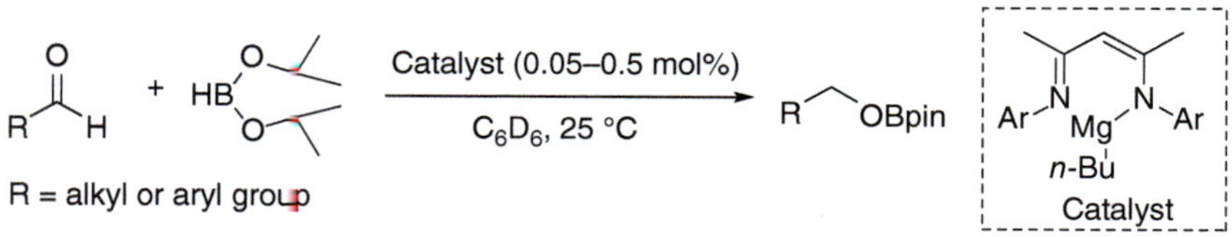

Scheme 7.91 Hydroboration of aldehydes and ketones using magnesium alkyl complex.

the carbonyl moiety into the Mg–H bond, and subsequent σ-bond metathesis with pinacolborane (Scheme 7.91).

In 2016, Mankad and coworker demonstrated that the copper carbene complex, (IPr)CuO*t*-Bu, could catalyze the hydroboration of ketones and aldehydes even at low catalyst loadings (0.1 mol%), in some cases with TOFs exceeding 6000 h^{-1} [171]. Carbonyl reduction occurred with high selectivities in the presence of other reducible functional groups, including alkenes, nitriles, esters, and alkyl chlorides (Scheme 7.92). Recently, a novel nonanuclear copper(II) complex obtained by a facile one-pot self-assembly was found to catalyze the hydroboration of ketones and aldehydes in the absence of an activator under mild, solvent-free conditions [172]. The catalyst is air- and moisture-stable, displaying high efficiency (1980 h^{-1} TOF) and chemoselectivity on aldehydes over ketones and ketones over imines.

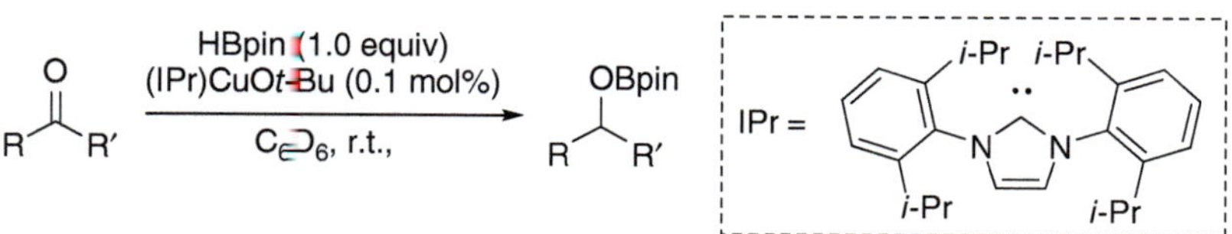

Scheme 7.92 Hydroboration of carbonyl compounds by copper carbine catalysis.

In 2017, Findlater and coworker for the first time reported an iron-catalyzed hydroboration of aldehydes and ketones under room temperature with $Fe(acac)_3$ as precatalyst [173]. The hydroboration proceeded efficiently to yield 1° and 2° alcohols; chemoselective hydroboration of aldehydes over ketones was attained under these conditions (Scheme 7.93). Recently, the in air catalytic hydroboration of ketones and aldehydes with pinacolborane by an iron(II) coordination polymer was carried out by Zhang et al. [174], which was under mild and solvent-free conditions in the presence of *t*-BuOK as an activator, achieving a high TON of up to 9500. This catalyst was also recyclable for reuse for at least five times without losing its effectiveness.

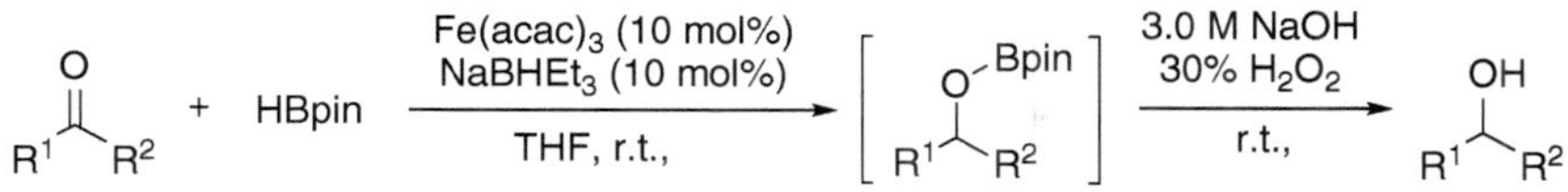

Scheme 7.93 Fe-catalyzed hydroboration of aldehydes and ketones.

An operationally convenient and general method for the hydroboration of aldehydes and ketones employing $Co(acac)_3$ as a precatalyst was reported by Findlater and coworkers in 2018 [175]. Chemoselective experiments revealed that the catalytic system was selective for aldehydes over ketones (Scheme 7.94).

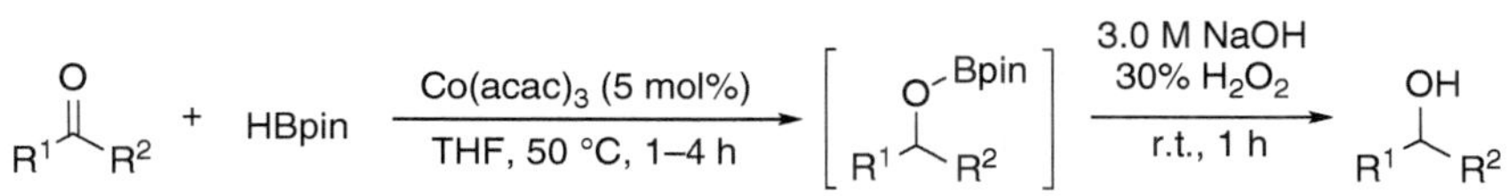

Scheme 7.94 Co-catalyzed hydroboration of aldehydes and ketones.

In addition, metal-free catalyst system has been also developed for the hydroboration of various aldehydes and ketones. Zhao and coworkers reported the first convenient, and versatile hydroboration of various aldehydes and ketones with B–H compounds using NaOH powder instead of unfriendly transition metals [176]. The reaction was conducted under mild condition with remarkable substrate tolerance, high chemoselectivity, and good yields (Scheme 7.95).

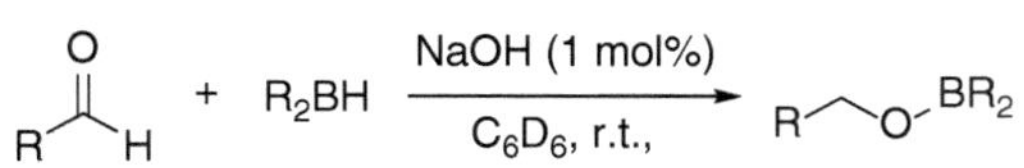

Scheme 7.95 Hydroboration of various aldehydes and ketones using NaOH as catalyst.

In 2018, the use of simple *n*-BuLi as an efficient catalyst for the selective hydroboration of aldehydes and ketones was described by Bao and coworkers for the first time (Scheme 7.96) [177].

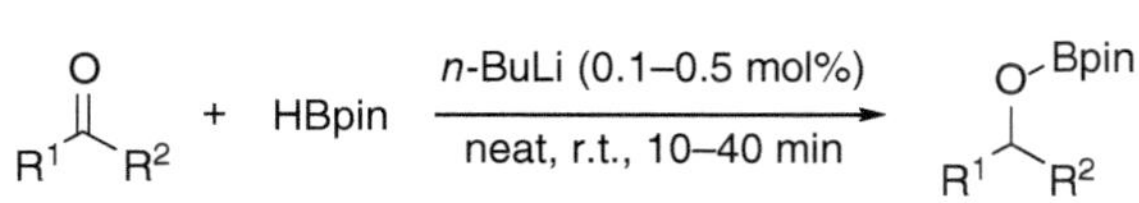

R^1 = alkyl, aryl;
R^2 = H, alkyl, aryl

Scheme 7.96 Hydroboration of various aldehydes and ketones using *n*-BuLi as catalyst.

7.7.2 Hydrophosphonylation

In 2010, Tajbakhsh and coworkers developed a method for the synthesis of α-hydroxy phosphonates by reacting aldehydes and ketones with trimethylphosphite in the presence of a catalytic amount of pyridine 2,6-dicarboxylic acid as a novel bifunctional organocatalyst in aqueous media (Scheme 7.97) [178]. The simple experimental procedure, application of an inexpensive catalyst, short reaction times, and high yields were the notable advantages of the protocol. In many cases, the products crystallize directly out of the reaction mixture.

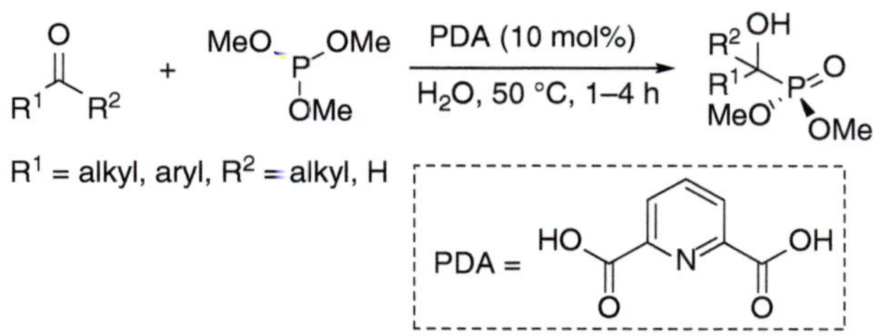

Scheme 7.97 Synthesis of α-hydroxy phosphonates from various aldehydes and ketones. Source: Modified from Jahani et al. [178].

The alkali metal salt-free dinuclear trivalent lanthanide amido complexes $(\eta^5:\eta^1:\eta^5:\eta^1\text{-Et}_8\text{-calix[4]-pyrrolyl})_2$ [179] (Ln = Nd, Sm, Gd) were prepared through the silylamine elimination reactions of calix[4]-pyrrole $[Et_2C(C_4H_2NH)]_4$ (1) with 2 equiv of $[(Me_3Si)_2N]_3Ln(\mu\text{-Cl})Li(THF)_3$ (Ln = Nd, Sm, Gd) in toluene at 110 °C [180]. Studies on the catalytic activity of the new lanthanide amido complexes revealed that these complexes can be used as efficient catalysts for the hydrophosphonylation of aldehydes and unactivated ketones, affording the products in high yields by employing a low catalyst loading (0.1 mol%) at room temperature in a short time (20 minutes) (Scheme 7.98).

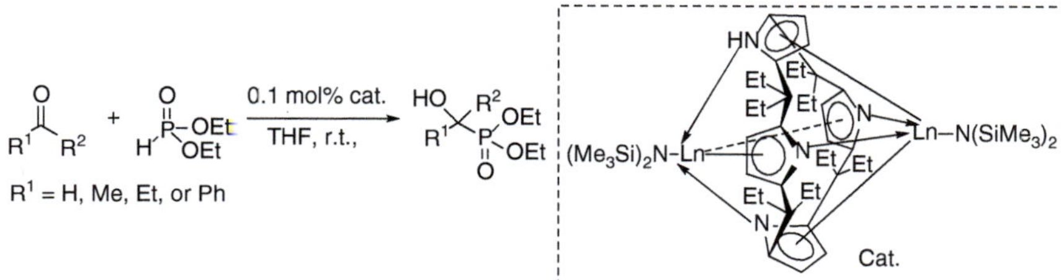

Scheme 7.98 Hydrophosphonylation of aldehydes and unactivated ketones via lanthanide amido complexes. Source: Modified from Zhou et al. [179].

A series of rare-earth metal amides supported by a cyclohexyl-linked bis(β-diketiminato) ligand were synthesized, and their catalytic activities for hydrophosphonylation of aldehydes and ketones were examed [181]. Investigation of the catalytic properties of the complexes revealed that these complexes exhibited a high catalytic activity toward the hydrophosphonylation of aldehydes and ketones in the presence of a very low loading of rare-earth metal amides (0.1–1 mol%) at room temperature in a short time (Scheme 7.99).

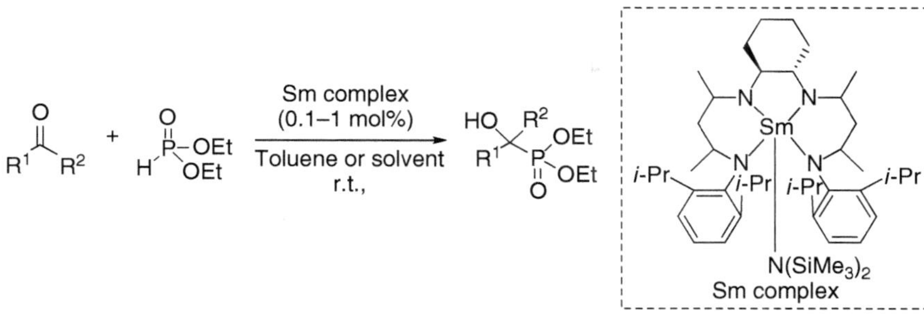

Scheme 7.99 Hydrophosphonylation of aldehydes and ketones catalyzed by Sm complex.

In 2014, Yao and coworkers for the first time found that organic alkali metal compounds served as highly efficient precatalysts for the hydrophosphonylation reactions of aldehydes and unactivated ketones with dialkyl phosphite under mild conditions [182]. The hydrophosphonylation reactions catalyzed by 0.1 mol% *n*-BuLi were completed within five minutes for a broad range of substrates and generated a series of α-hydroxy phosphonates in high yields (Scheme 7.100). Meanwhile, they realized the synthesis and characterization of amidate rare-earth metal amides and their catalytic activities toward the hydrophosphonylation reaction of aldehydes and unactivated ketones [183]. Subsequently, they also developed the synthesis of bimetallic lanthanide bis(amido) complexes stabilized by bridged bis(guanidinate) ligands and studied their catalytic activity toward the hydrophosphonylation reaction of aldehydes and ketones [184].

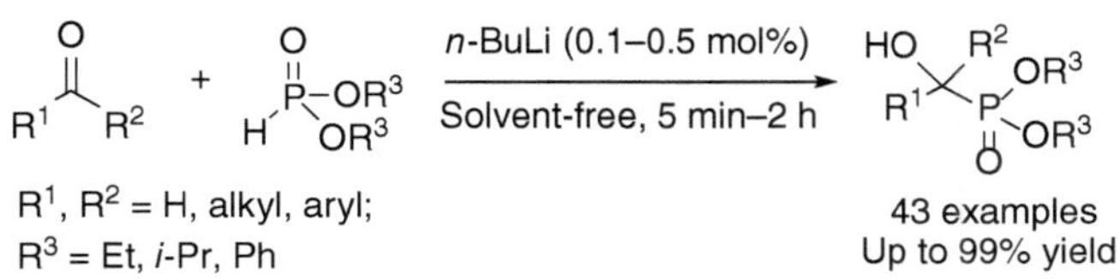

Scheme 7.100 Hydrophosphonylation of aldehydes and ketones catalyzed by *n*-BuLi.

7.7.3 Hydrosilylation Reactions

As early as 1979, the hydrosilylation of saturated and α,β-unsaturated carbonyl compounds by heterogeneous catalysis without solvent and in the presence of salts (KCO_2H, *o*-$Ph(CO_2K)_2$, KF, CsF) was carried out with good yields by Corriu and coworkers [185]. CsF is a very efficient salt for the hydrosilylation of aldehydes and ketones: for instance, α-$NpSiH_3$ reacted quantitatively at room temperature with benzophenone giving α-NpSi(OCHPh$_2$)$_3$. It allowed the isomerization of allylsilyl ethers in silyl enol ethers, providing that the transferring hydrogen is benzylic. Also, in 1989, Lukevics and coworkers reported that the hydrosilylation of aldehydes and ketones with dimethylphenylsilane occurring readily in dichloromethane at room temperature in the presence of catalytic amounts of cesium fluoride and 18-crown-6 to afford silyl ethers of the corresponding aryl and hetaryl carbinols in good yields [186].

In 2001, Lipshutz et al. reported a method for the hydrosilylation of aldehydes and ketones, which can be used to arrive at silylated alcohols bearing any of several commonly used silicon protecting groups [187]. The reaction conditions are very mild, reagents are inexpensive, and the amount of $[Ph_3P(CuH)]_6$ required could be as little as 0.1 mol% in the case of aldehydes (Scheme 7.101). Also, copper(I) complexes featuring *N*-heterocyclic carbenes (NHCs) had been synthesized and tested in the hydrosilylation of functionalized and/or sterically demanding ketones and aldehydes with TONs up to 1000 by Matt and Brenner in 2015 [188].

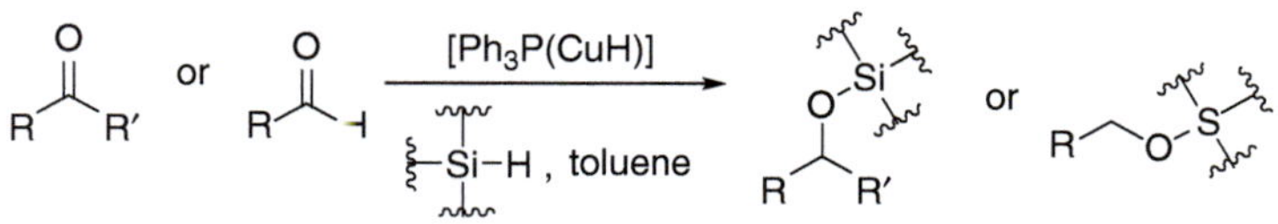

Scheme 7.101 Hydrosilylation of aldehydes and ketones catalyzed by $[Ph_3P(CuH)]_6$.

High valent molybdenum-dioxo complex $[MoO_2Cl_2]$ catalyzed the addition of dimethylphenylsilane to aldehydes and ketones to afford the corresponding dimethylphenylsilyl ethers in quantitative yield was reported by Royo and coworkers in 2005 and 2006 [189, 190].

Chemoselective hydrosilylation of aldehydes and ketones in the presence of other functional groups catalyzed by nickel PCP–pincer hydride complexes was disclosed by Guan and coworkers in 2009 [191], where the insertion of carbonyl groups into nickel–hydrogen bonds had been directly observed (Scheme 7.102).

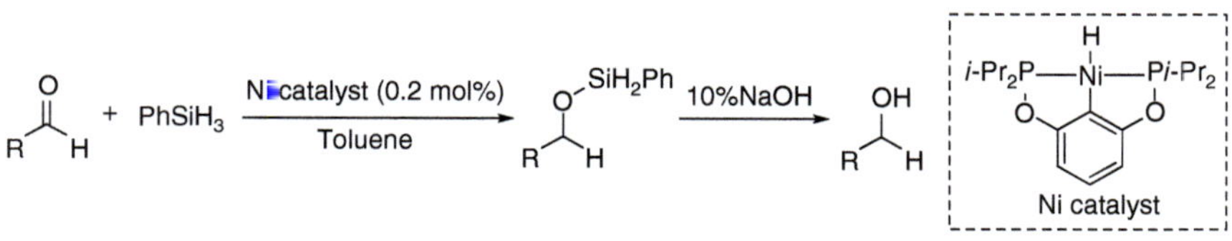

Scheme 7.102 Hydrosilylation of aldehydes and ketones catalyzed by Ni PCP–pincer hydride complexes.

In 2011, Du and coworkers for the first time demonstrated that the air-stable, cationic Ru(VI)N-salen was an efficient catalyst for the hydrosilylation of ketones and aldehydes using phenylsilane as a reductant in the presence of a tertiary silane [192]. A variety of ketones and aldehydes were reduced to alcohols with good to high isolated yields (Scheme 7.103).

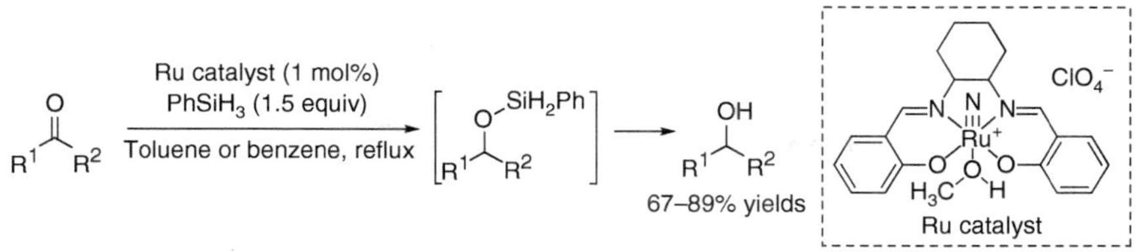

Scheme 7.103 Ru-catalyzed hydrosilylation of ketones and aldehydes.

In 2012, Sudalai and coworkers discovered that $Pd(OAc)_2$ was a highly effective catalyst for the selective hydrosilylation of aryl ketones and aldehydes in DMF at 25 °C using triethylsilane as hydride source [193]. However, the corresponding benzylic alcohols could be obtained in excellent yields when the reaction was performed in DMF/H_2O (4 : 1) as a solvent system (Scheme 7.104).

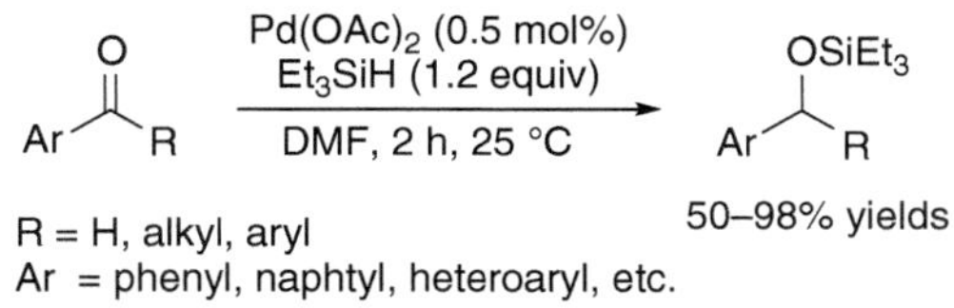

Scheme 7.104 Pd-catalyzed hydrosilylation of aryl ketones and aldehydes.

In 2015, the group of Gade presented a new iron alkoxide complex bearing a chiral bis(oxazolinylmethylidene)isoindoline ligand able to catalyze the hydrosilylation of a wide range of ketones with high enantiomeric excesses and excellent conversions [194]. Reaction conditions were optimized for the use of a 5 mol% of iron catalyst and 2 equiv of $(EtO)_2MeSiH$ in toluene solution, adding the silane at −78 °C and then allowing the mixture to warm to room temperature (Scheme 7.105). Most of the substrates were reduced above 95% conversion after six hours and the enantiomeric excesses surpassed 90% ee in most cases.

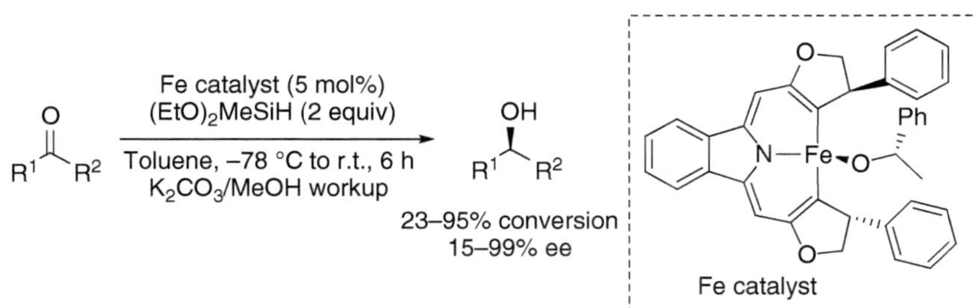

Scheme 7.105 Iron-catalyzed enantioselective hydrosilylation of ketones.

In the same year, Parkin group demonstrated that the zinc hydride complex, [κ^3-Tptm]ZnH, was an effective catalyst for the multiple insertion of carbonyl compounds into the Si–H bonds of $PhSiH_3$ and Ph_2SiH_2 [195]. Multiple insertion of prochiral ketones resulted in the formation of diastereomeric mixtures of alkoxysilanes that can be identified by NMR spectroscopy. Since the alkoxysilanes can also be obtained via the zinc-catalyzed dehydrocoupling of alcohols with silanes, the reactions with (*R*)-(+)-1-phenylethanol and (*S*)-(−)-1-phenylethanol provided a means of identifying the diastereomers that result from the hydrosilylation of acetophenone (Scheme 7.106).

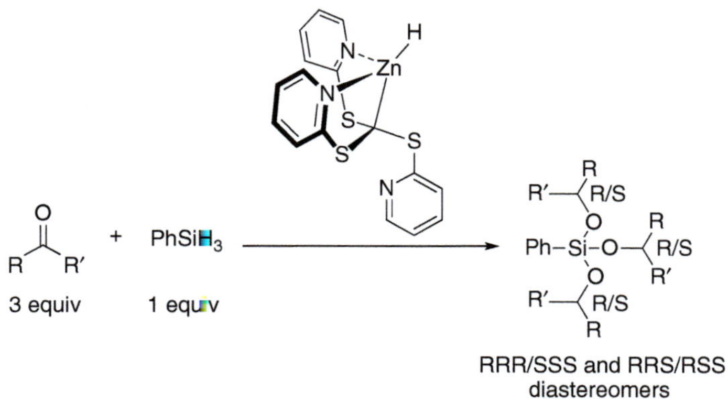

Scheme 7.106 Hydrosilylation of aldehydes and ketones catalyzed by zinc hydride complex.

7.7.4 Hydroacylation Reactions

In 2006, Scheidt and coworker demonstrated that hydroacylations of activated ketones can be catalyzed by NHCs to form related ester or alcohols [196]. This process occurs via separate reduction and acylation steps in which the organocatalyst is responsible for two key bond-forming steps of the catalytic cycle (Scheme 7.107).

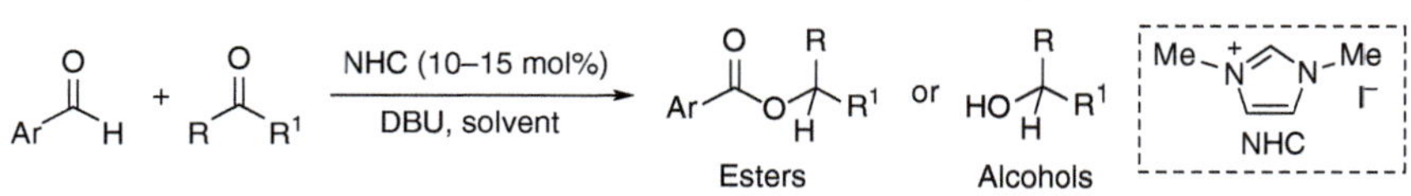

Scheme 7.107 NHC-catalyzed hydroacylation of activated ketones.

In 2009, Dong and coworkers reported an atom economical approach to phthalides by the enantioselective intramolecular hydroacylation of ketones in the presence of $[Rh(cod)Cl]_2$ [197]. They found that the appropriate choice of counterion was crucial in suppressing decarbonylation and controlling enantioselectivity (Scheme 7.108).

In 2013, an enantioselective hydroacylation of ketones using Noyori's transfer hydrogenation catalyst to form lactones from keto alcohols was developed by Dong's group [198]. The alcohol is oxidized in situ to an aldehyde, obviating the need to prepare sensitive keto aldehyde substrates (Scheme 7.109).

In 2014, Yoshikai and coworker reported that cobalt–chiral diphosphine catalytic systems promote intramolecular hydroacylation reactions of 2-acylbenzaldehydes

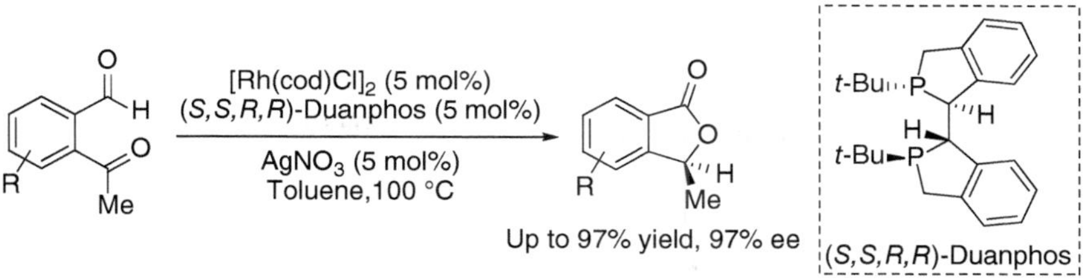

Scheme 7.108 Rh-catalyzed enantioselective hydroacylation of ketones.

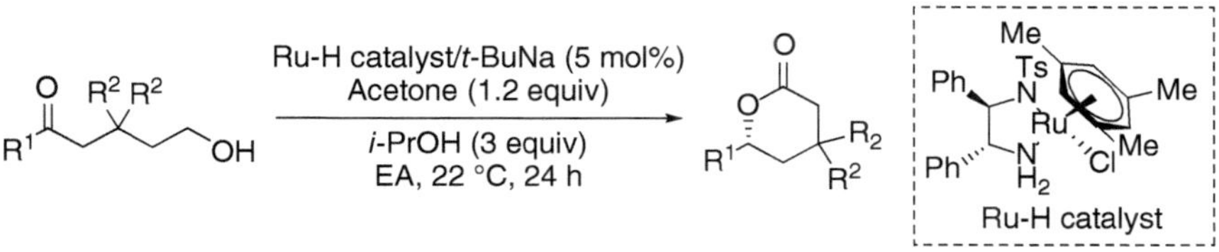

Scheme 7.109 Enantioselective hydroacylation of 1,5-keto alcohols.

and 2-alkenylbenzaldehydes to afford phthalide and indanone derivatives, respectively, in moderate to good yields with high enantioselectivities (Scheme 7.110) [199]. The ketone hydroacylation did not exhibit a significant H/D kinetic isotope effect (KIE) with respect to the aldehyde C–H bond, indicating that C–H activation would not be involved in the rate-limiting step.

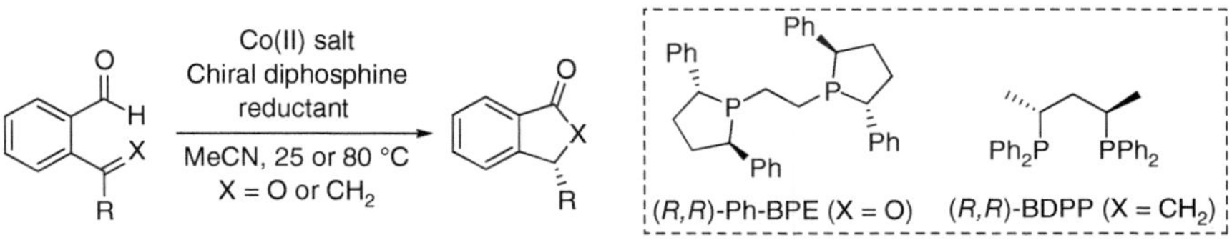

Scheme 7.110 Enantioselective intramolecular hydroacylation of ketone with olefin.

In 2018, Zhang's group developed a cobalt-catalyzed intermolecular hydroacylation of aldehydes to obtain esters with moderate to good yield (Scheme 7.111) [200]. The key feature in this catalytic system is the application of the Lewis basic amine which can deprotonate a hydridocobalt intermediate or activate the Co–H bond, thereby facilitating the C=O insertion step.

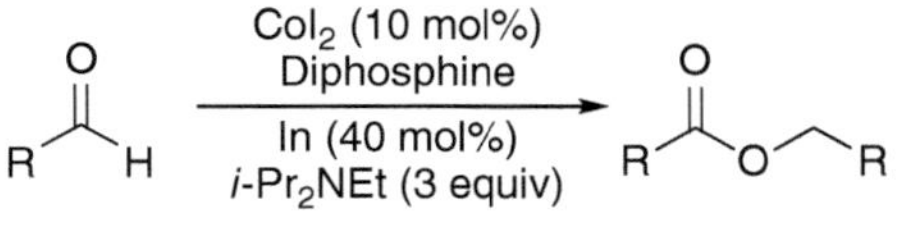

Scheme 7.111 Co-catalyzed intermolecular hydroacylation of aldehydes. Source: Modified from Li et al. [200].

7.8 Oxidative Cross-Coupling Reaction of Aldehydes

7.8.1 Homogeneous Catalyst System

As early as in 1970, Leffingw exploited MnO_2 as an oxidant to realize the oxidative coupling of α-substituted aliphatic aldehydes, which readily undergo α-hydrogen abstraction and subsequent formation of C–C and C–O dimerization products (Scheme 7.112) [201]. The results of preliminary work indicated that certain aldehydes are in fact easily dimerized by a number of common oxidizing agents, and the products are stable and easily isolable materials of considerable synthetic interest. Later, a similar work with respect to oxidative coupling of aldehydes and the rearrangement of dioxa-1,5-hexadienes was reported by Lewis and coworker [202].

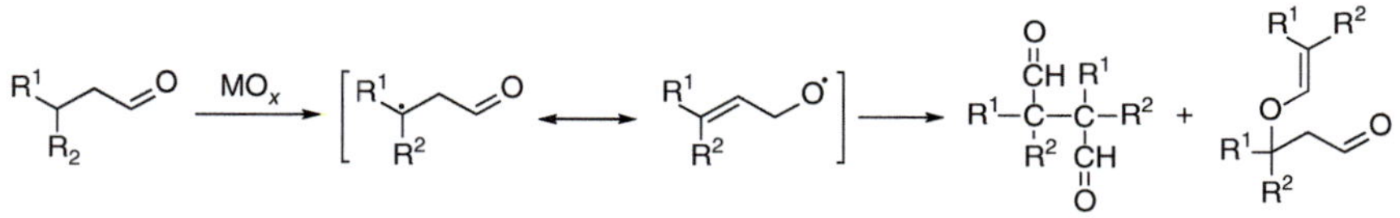

Scheme 7.112 MnO_2 as an oxidant for the oxidative coupling of α-substituted aliphatic aldehydes. Source: Modified from Leffingw [201].

In 2010, a one-pot direct synthesis of aryl α-ketoamides by a $ZnCl_2$-promoted formal oxidative coupling of aromatic aldehydes and isocyanides has been developed under mild conditions using both *N*-methylhydroxylamine and acetic acid as shuttle molecules, affording the aryl α-ketoamides in moderate to good yields (Scheme 7.113) [203].

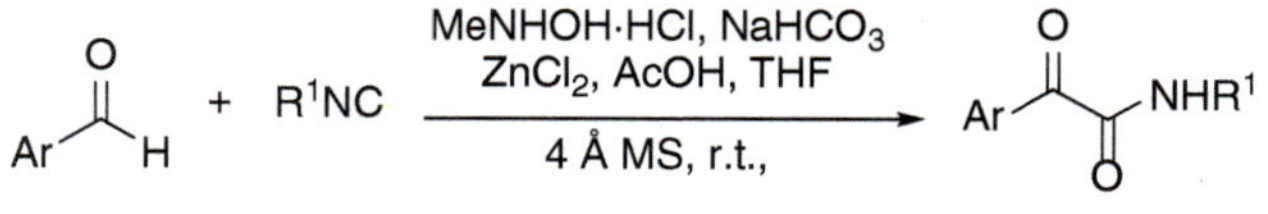

Scheme 7.113 Synthesis of aryl α-ketoamides via a coupling of an aromatic aldehyde with isocyanide. Source: Bowma et al. [203].

In 2011, Kwong and coworkers succeeded in showing aromatic and aliphatic aldehydes that can act as the acyl sources in the oxidative coupling of two C–H bonds [204]. In the presence of a $Pd(TFA)_2$ catalyst and *tert*-butylhydroperoxide at 90 °C in general, an array of *ortho*-acylacetanilides can be afforded in good yields (Scheme 7.114).

In 2012, Patel and coworkers developed a new and efficient catalytic approach for the synthesis of benzylic ester from aldehydes and alkylbenzenes [205]. This reaction proceeded via benzylic $C(sp^3)$–H bond activation of alkylbenzenes in the presence of an inexpensive copper catalyst and *t*-butyl hydroperoxide (TBHP) combination (Scheme 7.115). Exclusive formation of monoesters was observed for polyalkylated benzenes indicating a high degree of selectivity. This methodology provides an avenue to the synthesis of various unsymmetrical benzylic esters from

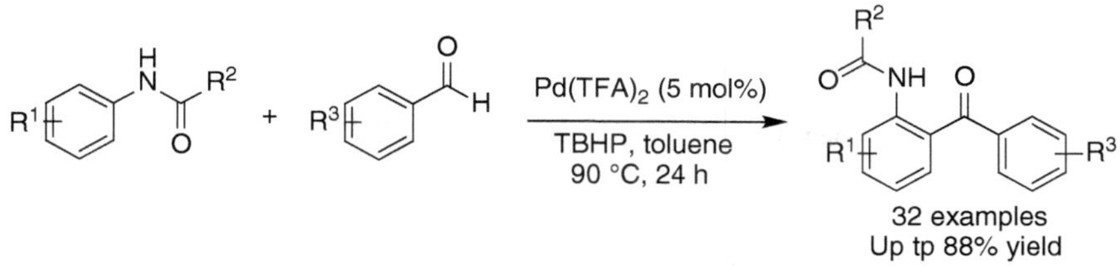

Scheme 7.114 Pd-catalyzed oxidative cross-coupling reaction of aldehydes. Source: Wu et al. [204].

diverse alkylbenzenes and aldehydes. Subsequently, different types of oxidative cross-coupling related to a variety of aromatic aldehydes with or without α-carbonyl group in the presence of Cu homogeneous catalyst were also achieved [206–214].

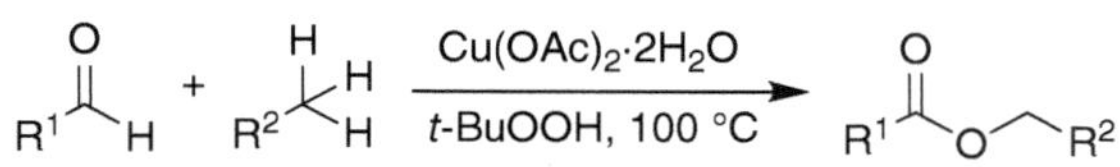

Scheme 7.115 Cu-catalyzed oxidative esterification of aldehydes with alkylbenzenes.

In 2013, Seayad and coworkers developed a novel oxidative coupling reaction of aldehydes for the synthesis of C3-susbtituted phthalides by Rh(III)-amine dual catalysis [215]. This methodology was applied for the homo- and hetero-coupling of aldehydes forming various functionalized aryl and alkyl phthalide derivatives in moderate to high yields (Scheme 7.116).

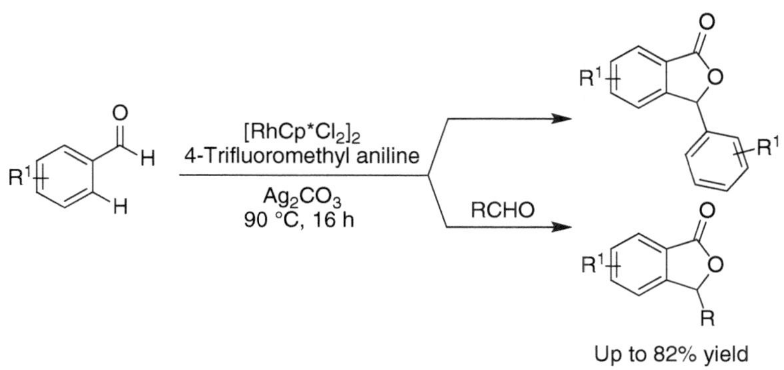

Scheme 7.116 Rh(III)-amine dual catalysis for the oxidative coupling of aldehydes.

In the same year, Soeta et al. achieved the oxidative coupling reaction of aldehydes with *N,N′*-disubstituted carbodiimides catalyzed by NHC under aerobic conditions (Scheme 7.117) [216]. This reaction gives the corresponding *N*-acylurea derivatives in good to high yields. Various kinds of aldehydes including aliphatic ones and carbodiimides are applicable to this reaction.

In 2015, Lu and coworkers developed a base-catalyzed, one-pot synthetic procedure for the oxidative coupling of styrene derivatives with aldehydes in the absence

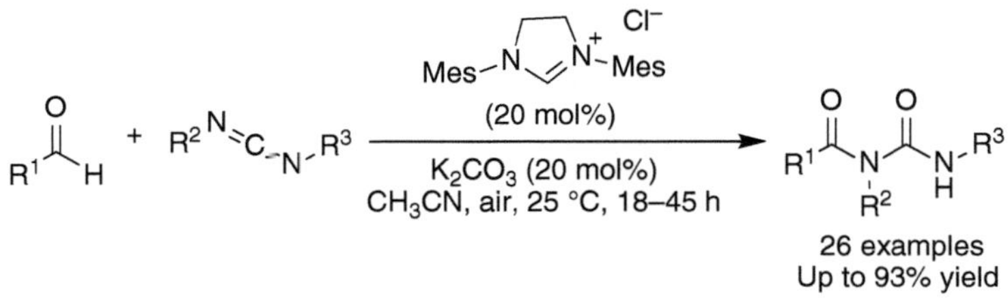

Scheme 7.117 Oxidative coupling reaction of aldehydes with *N*,*N*′-disubstituted carbodiimides.

of transition metal species (Scheme 7.118) [217]. They also determined that various functional groups could be substituted on both the aldehyde and styrene. This work demonstrates a facile method for the synthesis of α,β-epoxy ketones.

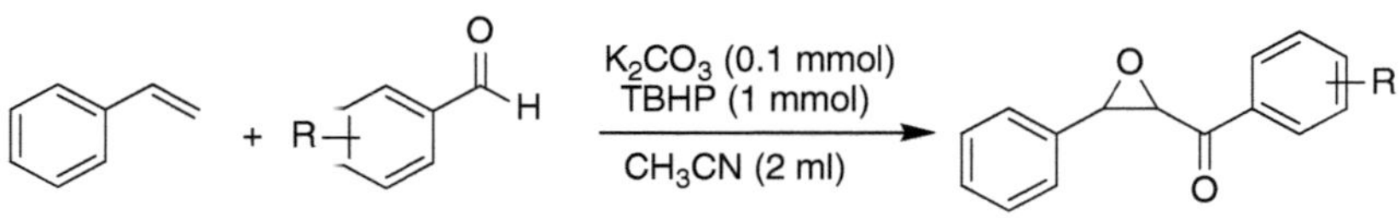

Scheme 7.118 Oxidative coupling of styrene derivatives with aldehydes. Source: Ke et al. [217].

In 2017, Xie and coworkers developed an efficient approach for the synthesis of *N*-hydroxyimide esters via cross-dehydrogenative coupling (CDC) of aldehydes with NHPI under oxidative catalyst-free conditions (Scheme 7.119) [218]. Notably, the use of molecular oxygen as the terminal oxidant and simple purification process made this method environmentally friendly and practical.

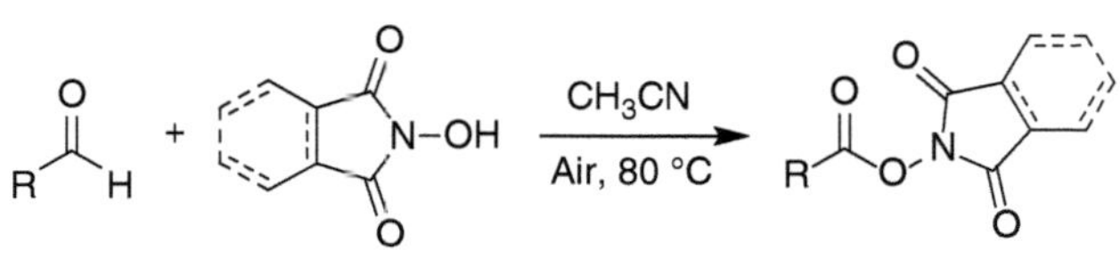

Scheme 7.119 Cross-dehydrogenative coupling of aldehydes with *N*-hydroxyphthalimide. Source: Modified from Xu et al. [218].

7.8.2 Heterogeneous Catalyst System

In 2012, Riisager and coworkers demonstrated that one-pot synthesis of amides by aerobic oxidative coupling of alcohols or aldehydes with amines via intermediate formation of methyl esters was highly efficient and selective when using a catalytic system comprised of supported gold nanoparticles and added base in methanol (Scheme 7.120) [219]. A variety of aldehydes, alcohols, and amines undergo smoothly with excellent yield.

In 2014, Zhang and coworkers reported that a La–Mg composite oxide modified extra-large mesoporous FDU-12 supported Au–Ni bimetallic catalyst exhibited

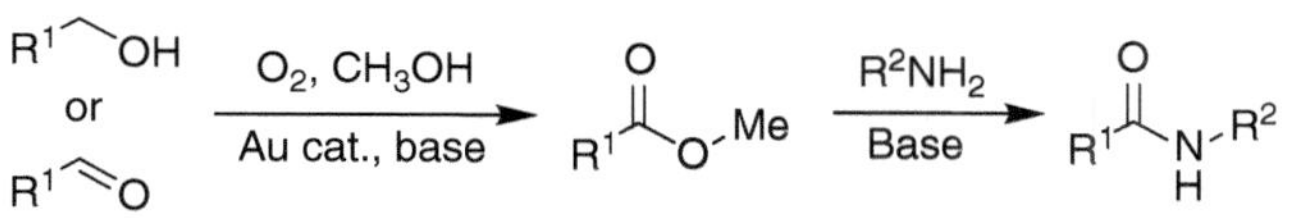

Scheme 7.120 Synthesis of amides by aerobic oxidative coupling of alcohols or aldehydes with amines. Source: Modified from Kegns et al. [219].

excellent activity and stability for oxidative coupling of aldehydes (including benzyl and aliphatic aldehydes) with methanol to directly produce methyl esters (Scheme 7.121) [220]. No deactivation was observed after recycling for six times.

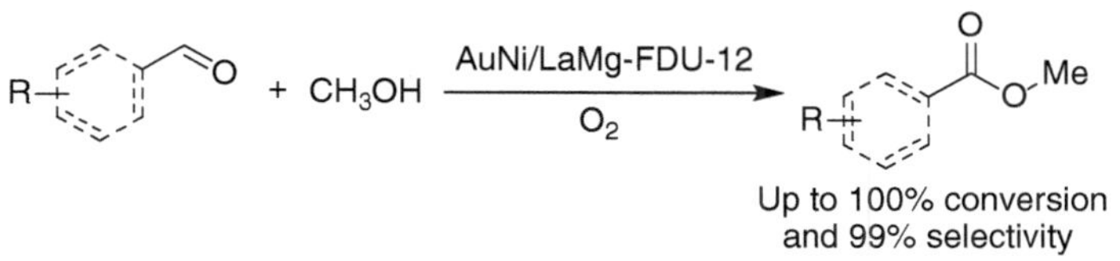

Scheme 7.121 Oxidative coupling of aldehydes with methanol to produce methyl esters. Source: Han et al. [220].

Meanwhile, Corma and coworkers shown that copper MOFs containing aza-heterocyclic ligands, viz., pyrimidine [Cu(2-pymo)$_2$] and imidazole [Cu(im)$_2$], are active, stable, and reusable catalysts for oxidative C–O coupling reactions by direct C–H activation of formamides, aldehydes, and ethers [2]. The materials used in this study also clearly outperform the previously reported [Cu$_2$(BPDC)$_2$(BPY)] MOF as a catalyst, attaining more than a sixfold increment of the initial reaction rate (in terms of TOF). Later, oxidative CDC of amines and α-carbonyl aldehydes to form α-ketoamides using air oxidant over heterogeneous Cu-MOF-74 catalyst was also developed by Phan and coworkers (Scheme 7.122) [221]. High yields were achieved in the presence of a catalytic amount of the Cu-MOF without any added ligand and base. Furthermore, leaching test indicated no contribution from homogeneous-leached species and the catalyst could be recovered and reused several times without a significant degradation in catalytic activity.

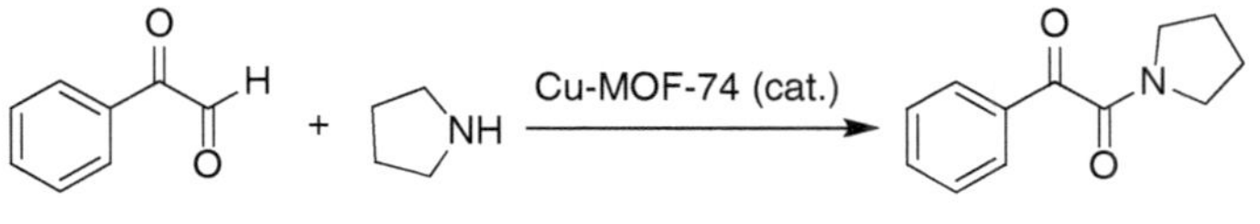

Scheme 7.122 The oxidative coupling of pyrrolidine and phenylglyoxal using Cu-MOF-74 catalyst. Source: Modified from Thanh et al. [221].

Co-METS-10 and Ni-METS-10 catalysts showed high activity in the oxidative coupling reaction of styrenes with benzaldehydes through one-pot synthetic procedure [222]. In addition, Co-METS-10 exhibits the highest product selectivity to form α,β-epoxy ketones (Scheme 7.123). Co (or Ni) METS-10 has bifunctional characteristic with transition metal and strong basicity sites. The highly dispersed Co and Ni species facilitate the generation of the alkyloxy and alkylperoxy radicals from *t*-BuOOH, leading to the occurrence of radical addition for the alkenes with aldehyde and alkylperoxy to form β-peroxy ketones. Meanwhile, the basic sites on Co (or Ni) METS-10 catalyst favor the transformation from intermediate β-peroxy ketone to α,β-epoxy ketones, resulting in the enhanced catalytic activity.

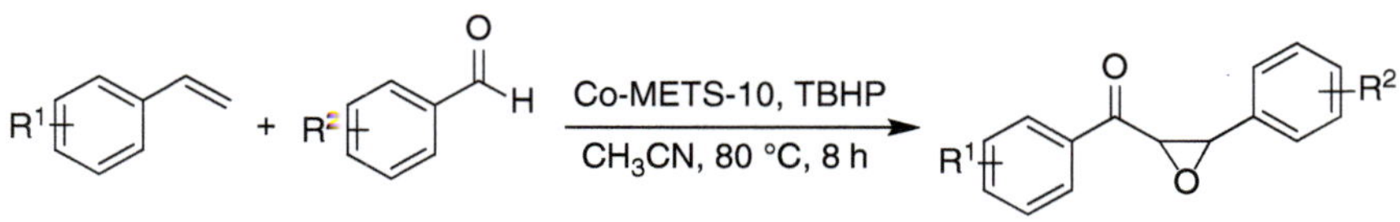

Scheme 7.123 The oxidative coupling of styrenes with benzaldehydes catalyzed by Co-METS-10. Source: Modified from Hu et al. [222].

Recently, Yang group developed a reusable heterogeneous non-precious iron nanocomposite comprising metallic Fe–Fe_3C nanoparticles and Fe–Nx sites on *N*-doped porous carbon, which allows for highly efficient synthesis of quinolines and quinazolinones via oxidative coupling of amines and aldehydes using H_2O_2 as the oxidant in aqueous solution under mild conditions [223]. A set of quinazolines and quinazolinones were synthesized in high yields with a broad substrate scope and good tolerance of functional groups (Scheme 7.124). Characterization and control experiments disclosed that a synergistic effect between the metallic Fe nanoparticles and built-in Fe–Nx sites is primarily responsible for the outstanding catalytic performance. Furthermore, the iron nanocomposite could be readily recovered for successive use without appreciable loss in catalytic activity and selectivity.

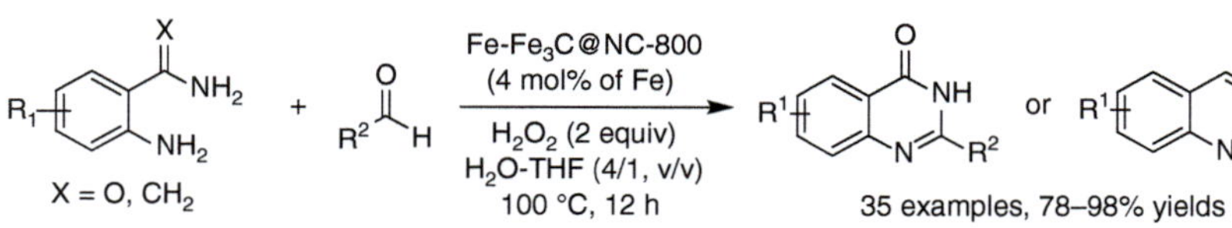

Scheme 7.124 Oxidative coupling of amines and aldehydes catalyzed by Fe-Fe_3C@NC-800.

7.9 Reductive Coupling Reactions of Aldehydes

In 1970s, Rautenstrauch for the first time reported the reductive coupling of simple aliphatic aldehydes using metallic lithium [224]. Subsequently, reductive coupling of aromatic aldehydes with $Fe(CO)_5$ or $Fe_3(CO)_{12}$ in pyridine was also achieved by Otsuji and coworkers in 1980, giving the corresponding 1,2-diaryl-l,2-ethanediols as major products in good yields (Scheme 7.125) [225]. A reactive species of this reaction is octacarbonyl diferrate $[Fe_2\,(CO)_8]^-$.

In 1999, Periasamy et al. demonstrated that aromatic aldehydes could be converted to the corresponding diols using the low valent titanium species generated by the reaction of $TiC1_4$ with triethylamine in moderate to excellent yield (Scheme 7.126) [226].

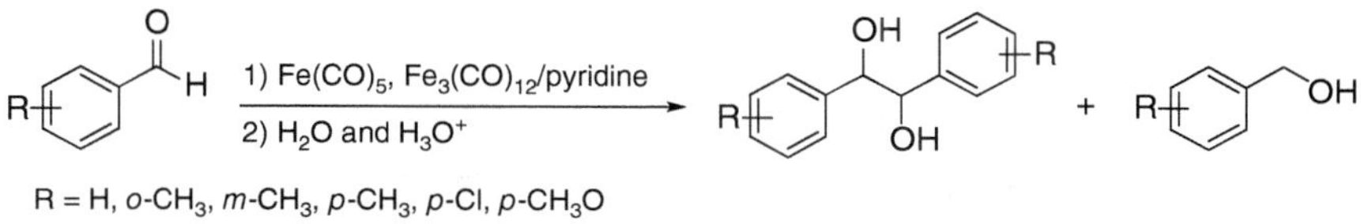

Scheme 7.125 Fe-catalyzed reductive coupling of aromatic aldehydes.

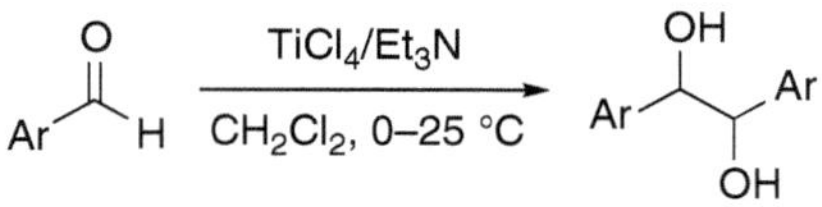

Scheme 7.126 Conversion of aromatic aldehydes to 1,2-diols using $TiCl_4/Et_3N$. Source: Modified from Periasamy et al. [226].

Sunlight is a cheap, efficient, and environmentally friendly natural resource. In 2003, reductive coupling of aromatic aldehydes and ketones leading to 1,2-diols has been achieved with high yield in isopropanol in sunlight (Scheme 7.127) [227]. The important advantages offered by this procedure include higher yield, operational simplicity, low cost, minority of byproduct, and no pollution. Then, in 2015, Rueping and coworkers reported a photoredox-catalyzed reductive coupling of aldehydes and ketones to pinacols with visible light and using nickel complex as a photocatalyst [228]. Their synthetic utility was demonstrated by the reductive dimerization of aldehydes and ketones under mild reaction conditions, with broad functional group tolerance and without additives. Recently, reductive cross-coupling of aldehydes and imines mediated by visible light photoredox catalysis was also developed by Walsh et al. [229]

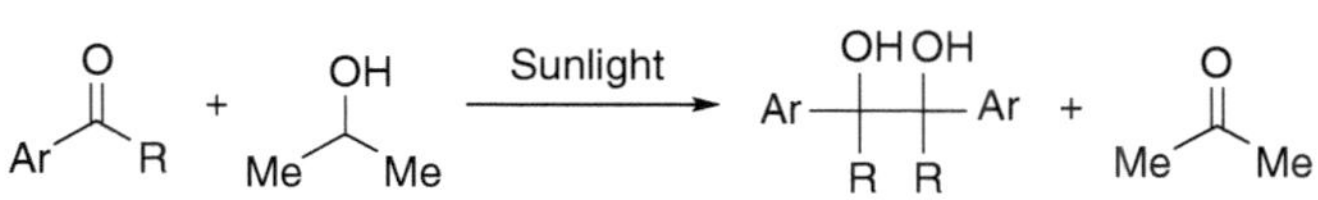

Scheme 7.127 Reductive coupling of aromatic aldehydes and ketones in sunlight. Source: Modified from Li et al. [227].

In 2004, ligand-dependent scope and divergent mechanistic behavior in nickel-catalyzed reductive couplings of aldehydes and alkynes was studied by Montgomery and coworkers (Scheme 7.128) [230]. Later, they reported several nickel-catalyzed reductive couplings of aldehyde and alkyne about its regiocontrol and mechanical study [231–233].

In 2009, the intermolecular reductive coupling reaction of cyclopent-2-enone and aromatic aldehydes was realized by Nishiyama using chiral rhodium-(bisoxazolinyl) phenyl catalyst, Rh(Phebox-Ph)$(OAc)_2(H_2O)$, with diphenylmethylsilane as a hydride donor to give the corresponding β-hydroxyketones in high antiselectivity (up to 96%) with high enantioselectivity (up to 93%) (Scheme 7.129) [234]. Recently, conversion of aldehydes to branched or linear ketones via regiodivergent

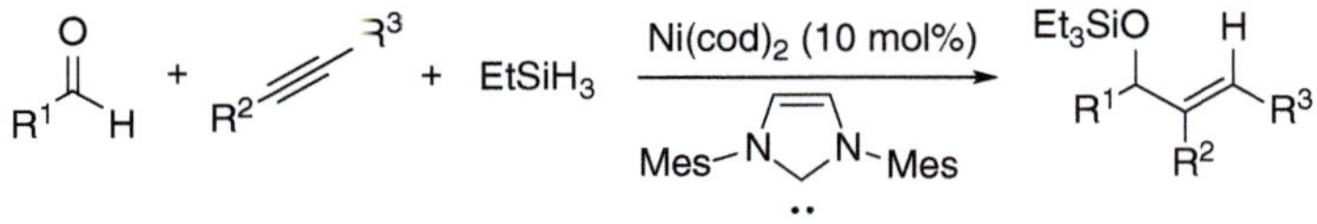

Scheme 7.128 Ni-catalyzed reductive couplings of aldehydes and alkynes. Source: Modified from Mahandru et al. [230].

rhodium-catalyzed reductive coupling-redox isomerization mediated by formate was also developed by Krische and coworkers [235].

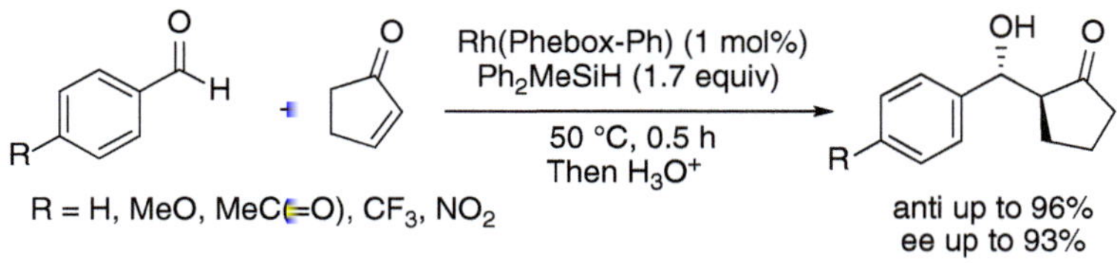

Scheme 7.129 Rh-catalyzed reductive coupling of cyclopent-2-enone and aromatic aldehydes. Source: Shiomi et al. [234].

In 2015, Fuchs and coworkers reported the successful reductive coupling of an aldehyde on a Au(111) surface leading to the formation of a PPV derivative (Scheme 7.130) [236]. Scanning tunneling microscopy and photoelectron spectroscopy experiments confirmed oxygen dissociation and its desorption from the surface.

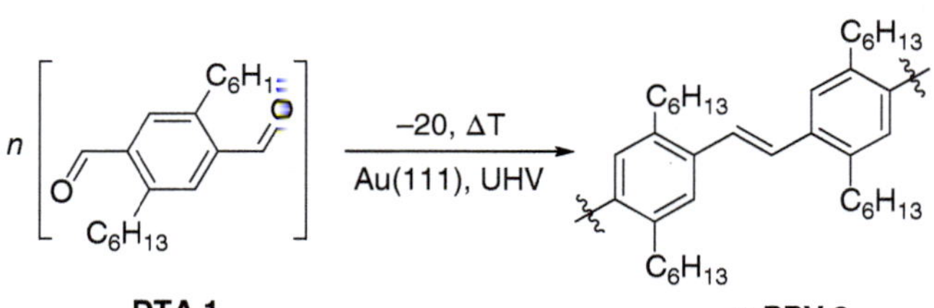

DTA 1 | p-PPV 2

Scheme 7.130 Synthesis of *p*-PPV-2 through reductive coupling of aldehyde. Source: Modified from Arado et al. [236].

Umpolung strategies of aldehydes and imines have dramatically expanded the scope of carbonyl and iminyl chemistry by facilitating reactions with non-nucleophilic reagents. In 2017, Ngai and coworkers reported the first example of visible light photoredox-catalyzed β-selective reductive coupling of alkenylpyridines with carbonyl or iminyl derivatives with the aid of a Lewis acid co-catalyst (Scheme 7.131) [237]. Their process tolerates complex molecular scaffolds (e.g. sugar, natural product, and peptide derivatives) and is applicable to the preparation of compounds containing a broad range of heterocyclic moieties.

Scheme 7.131 Reductive coupling of aldehydes with 4-vinylpyridine catalyzed by Ru. Source: Lee et al. [237].

A pyridine-boryl radical-promoted metal-free reductive coupling reaction of aldehydes with 1,1-diarylethylenes was established by Li group in 2018 with a broad substrate scope and good functional compatibility (Scheme 7.132) [238]. Density functional theory calculations and control experiments suggested that the ketyl radical from the addition of the pyridineboryl radical to aldehydes is the key intermediate for this C–C bond formation reaction.

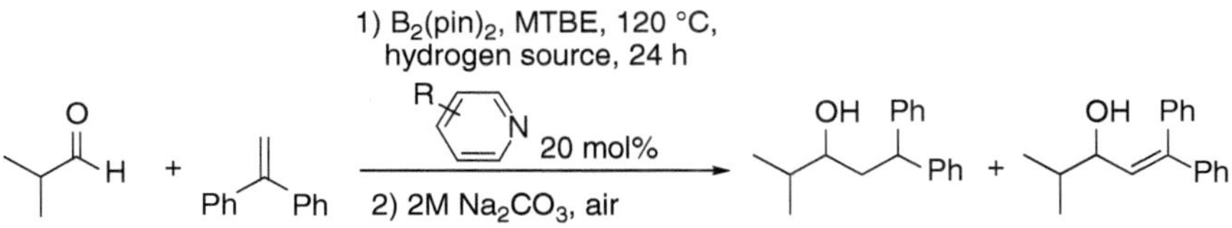

Scheme 7.132 Metal-free reductive coupling reaction of aldehydes with 1,1-diarylethylenes. Source: Cao et al. [238].

In the same year, a novel route to synthesize unsymmetrically *N,N*-disubstituted formamides was reported by Liu and coworkers [239], which was achieved via reductive coupling of primary amine and aldehyde with CO_2/H_2 over a cobalt-based catalytic system composed of CoF_2, $P(CH_2CH_2PPh_2)_3$ and K_2CO_3 (Scheme 7.133).

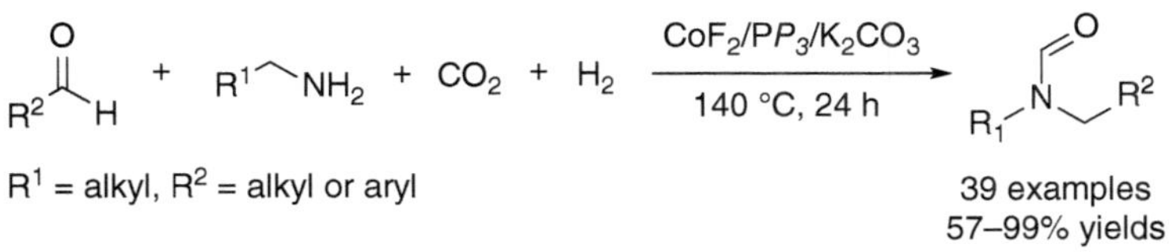

Scheme 7.133 Reductive coupling of primary amine and aldehyde with CO_2/H_2. Source: Ke et al. [239].

Recently, Ni-catalyzed branch-selective reductive coupling of aldehydes with 1,3-dienes has been developed under visible light photoredox dual catalysis [240]. Coupling of widely available 1,3-dienes with (hetero)aryl and alkyl aldehydes afforded homoallylic alcohols with broad scope and good functionality tolerance under the synergistic catalysis (Scheme 7.134). Hantzsch ester is used as the hydrogen radical source to oxidize low-valent nickel salt affording Ni–H species.

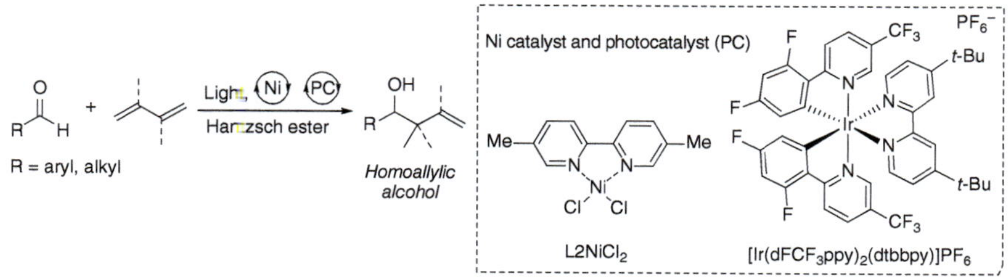

Scheme 7.134 Photoredox Ni-catalyzed branch-selective reductive coupling of aldehydes with 1,3-dienes.

7.10 Reaction of Acids as Starting Materials

7.10.1 Esterification Reactions

Perhaps the most useful reaction of carboxylic acids is their conversion into esters by reaction with an alcohol by the substitution of –OH by –OR and this is called the Fischer esterification reaction, the simplest method involves heating the carboxylic acid with an acid catalyst in an alcohol solvent (Scheme 7.135). This reaction is an equilibrium, all steps in the Fischer esterification reaction are reversible, and the position of the equilibrium can be driven either forward or backward depending on the reaction conditions. Ester formation is favored when alcohol is used as the solvent, but carboxylic acid is favored when the solvent is water. Intramolecular esterification of γ- and δ-hydroxy carboxylic acids forms five- and six-membered lactones. Esterification of a carboxylic acid occurs in the presence of acid but not in the presence of base. Base removes a proton from the carboxylic acid, forming an electron-rich carboxylate anion, which does not react with an electron-rich nucleophile [94].

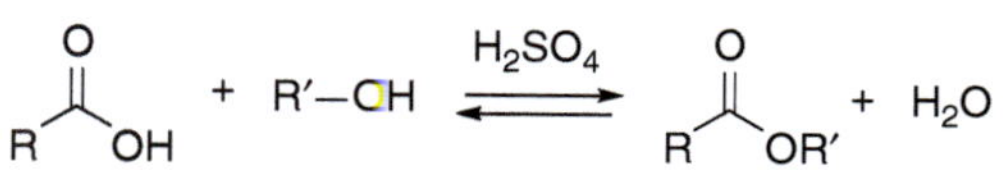

Scheme 7.135 Esterification of carboxylic acids with alcohols.

7.10.2 Amidation Reactions

The direct conversion of a carboxylic acid to an amide with NH_3 or an amine is very difficult, even though a more reactive acyl compound is being transformed into a less reactive one (Scheme 7.136). The problem is that carboxylic acids are strong organic

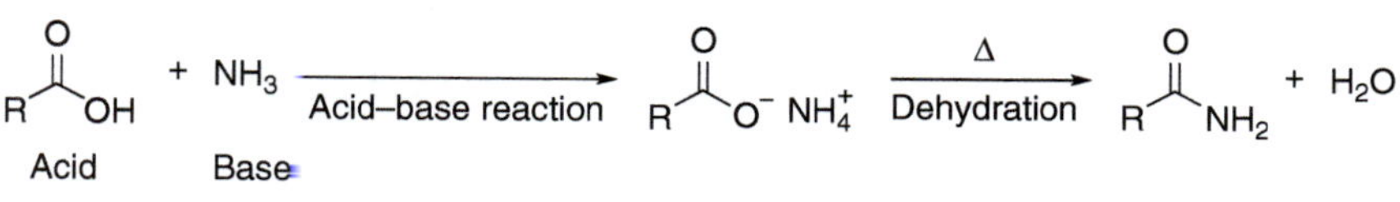

Scheme 7.136 Amidation of carboxylic acids.

acids and NH_3 and amines are bases, so they undergo an acid–base reaction to form an ammonium salt before any nucleophilic substitution occurs [94].

Heating at high temperature (>100 °C) dehydrates the resulting ammonium salt of the carboxylate anion to form an amide, though the yield can be low. Therefore, the overall conversion of RCOOH to $RCONH_2$ requires two steps: acid–base reaction of RCOOH with NH_3 to form an ammonium salt; dehydration at high temperature (>100 °C) (Scheme 7.137).

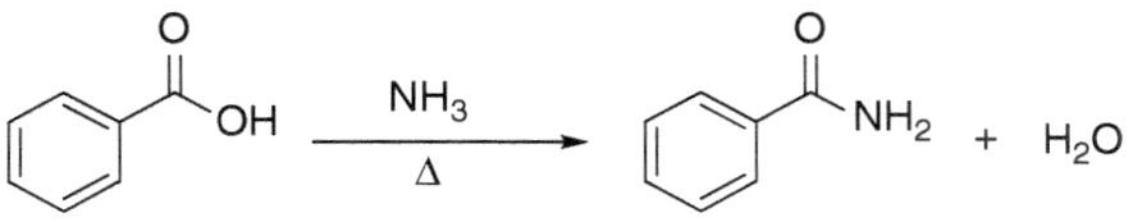

Scheme 7.137 Amidation of aromatic acids.

A carboxylic acid and an amine readily react to form an amide in the presence of a dehydratingreagent, dicyclohexylcarbodiimide (DCC), which is converted to the by-product dicyclohexylurea in the course of the reaction (Scheme 7.138). The dicyclohexylurea by-product is formed by adding the elements of H_2O to DCC. DCC promotes amide formation by converting the carboxy –OH group into a better leaving group.

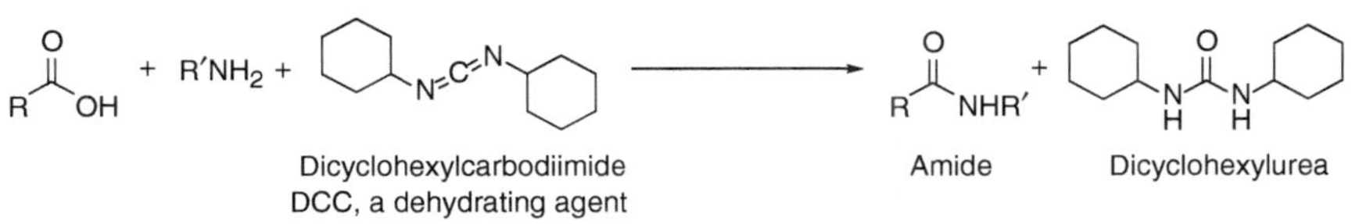

Scheme 7.138 Synthesis of amide.

7.10.3 Decarboxylation Coupling Reactions

Carboxylic acids are commercially available in large structural diversities at low cost, and they can be readily prepared by means of numerous well-established methods [241]. One of the most fascinating transformations of carboxylic acids is decarboxylation [242, 243], in which the C–CO_2H group undergoes C–C bond cleavage to liberate one equivalent of carbon dioxide along with the generation of an active carbon species (e.g. a carbon radical or a C–M species), then react with other carbon reagent to afford the desired products.

The oldest decarboxylation reaction is the Kolbe electrolysis developed in 1848 [244–246]. Here, radicals are generated electrochemically from carboxylic acids and dimerize (Scheme 7.139). Therefore, the scope of the reactions is quite limited and cross-couplings are hard to achieve via electrolysis. In addition, preparative electrolysis requires specialized laboratory equipment. Nevertheless, electrochemical methods allow decarboxylative reactions and are applied in cross-coupling [247–249].

In 2002, Myers et al. reported a novel palladium-catalyzed decarboxylative coupling of arene carboxylic acids with olefinic substrates for the synthesis of vinylarenes (Scheme 7.140) [250]. The methodology is applicable to heteroaromatic

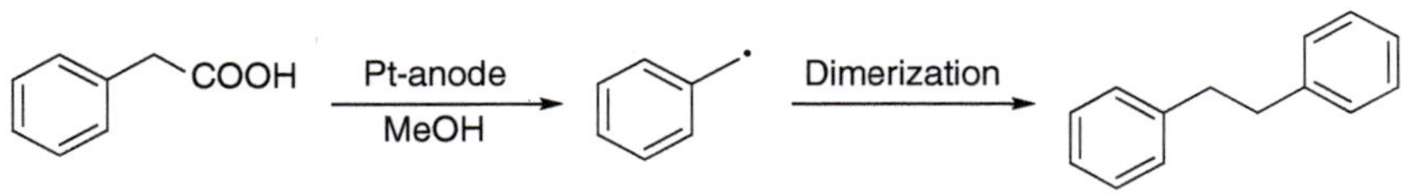

Scheme 7.139 The classic Kolbe reaction: electronic chemical decarboxylative dimeration of phenylacetic acid.

2-carboxylic acids or aromatic carboxylic acids bearing a substituent in the *ortho* position. The scope with regard to the olefinic substrate includes styrene, acrylates, (*E*)-ethyl crotonate, and cyclohexenone.

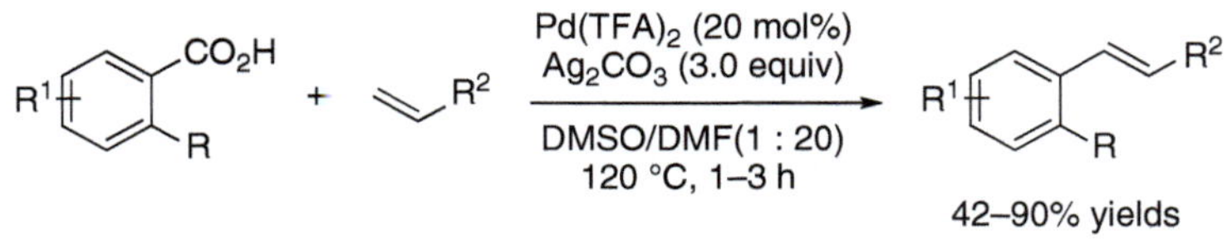

Scheme 7.140 Decarboxylation Heck reactions using a silver salt as an oxidant.

In 2006, an efficient cross-coupling protocol for the large-scale synthesis of biaryls using 2 mol% $Pd(acac)_2$/1.2 equiv $CuCO_3$/KF or 0.5 mol% $Pd(acac)_2$/1 mol% CuI/K_2CO_3 catalytic system was developed by Goossen et al. (Scheme 7.141) [251]. In contrast to traditional cross-couplings requiring the prior preparation of organometallic reagents, the application of a copper catalyst to generate the carbon nucleophiles in situ via decarboxylation of arylcarboxylic acid salts was achieved. The practicability of this reaction is demonstrated by their broad substrate scope and excellent yields (up to 99% yield), one of which is an intermediate in the large-scale production of the agricultural fungicide Boscalid.

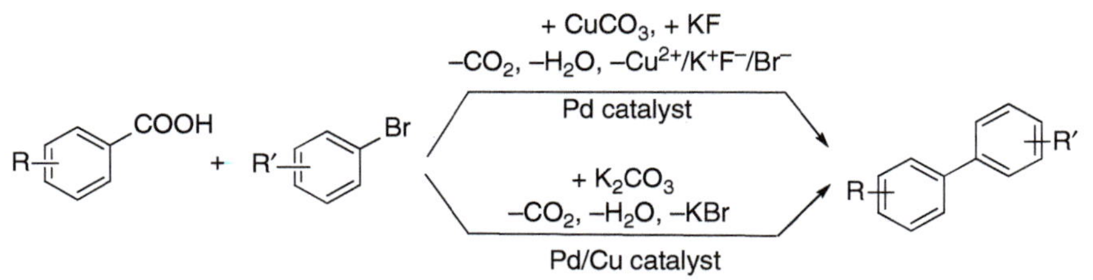

Scheme 7.141 Biaryl synthesis with excess copper and catalytic copper. Source: Modified from Goossen et al. [251].

An example regarding the decarboxylative C(sp)–H bond functionalization was disclosed by Liang, Li, and coworkers in 2009 [252]. In this protocol, readily available α-amino acids reacted with terminal alkynes catalyzed by a copper(I) bromide to give rise to various nitrogen-containing compounds in moderate to good yields, except for the noncyclic amino acid that showed poor reactivity (Scheme 7.142).

In the same year, Yu et al. demonstrated that aryl acylperoxides are surrogates to benzoic acids for the decarboxylative arylation of aromatic C–H bonds

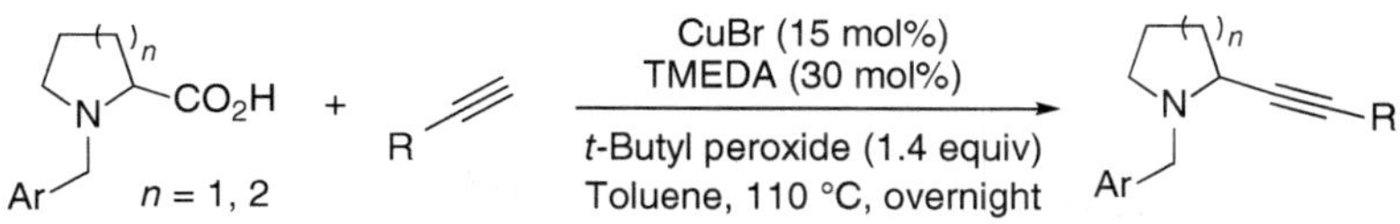

Scheme 7.142 Cu-catalyzed decarboxylative alkynylation. Source: Modified from Bi et al. [252].

(Scheme 7.143) [253]. In Yu's work, a wide range of arenes showed good reactivity toward the Pd-catalyzed decarboxylative arylation reactions, which were aided by various directing groups, such as pyridine, oxazole, and oxime. The authors also figured out that the aryl acylperoxides with bulky or electrondonating groups were poor substrates. For example, reaction of 4-methoxybenzoic peroxyanhydride with 2-phenylpyridine afforded an arylcarboxylation product in 90% yield.

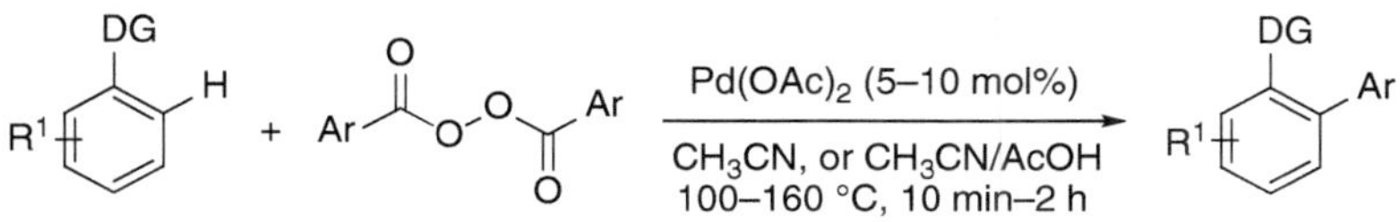

Scheme 7.143 Pd-catalyzed decarbonylative C–H arylation with acylperoxides as arylating reagents.

In 2009, Duan et al. reported the first Pd-catalyzed decarboxylative coupling of ortho-substituted aryl carboxylic acids with thiols for the synthesis of aryl sulfides (Scheme 7.144) [254]. When Y is $-NO_2$ group, the products were obtained as mixtures of nitrobenzene and aminobenzene sulfides. However, these reactions have several limitations, such as harsh reaction conditions, high temperature, strong bases, and the need for toxic polar solvents. In addition, an electron-withdrawing group on the aromatic carboxylic acid is required to afford good yields.

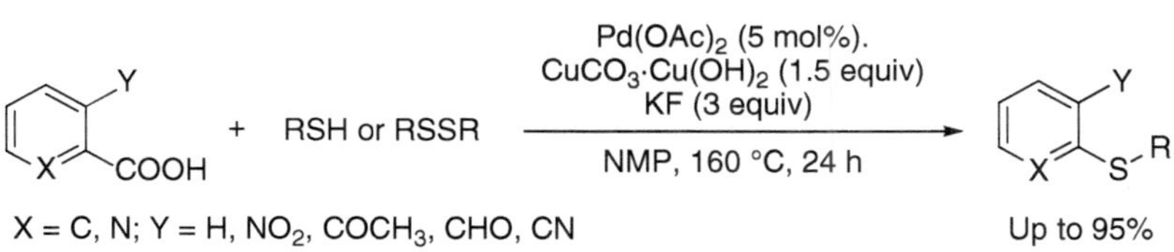

Scheme 7.144 Pd-catalyzed decarboxylative coupling of ortho-substituted aryl carboxylic acids with thiols. Source: Modified from Duan et al. [254].

In 2010, Ge and coworkers developed a protocol to realize the *ortho* C–H bond acylation of anilides with α-oxocarboxylic acids in the presence of $Pd(TFA)_2$ and $(NH_4)_2S_2O_8$ (Scheme 7.145) [255]. An obvious feature of this method is the mild reaction conditions. In this case, a variety of useful functionalities, such as bromide, nitro, chloro, and trifluoromethyl groups, are compatible. Both phenylglyoxylic acids and aliphatic α-oxocarboxylic acids are suitable reagents for the acylation reactions. A similar method for the *ortho* C–H bond acylation of acetanilides was reported by Saxena and coworkers afterward, in which the iron peroxo complex Fe(III)EDTA-H_2O_2 was employed as the oxidant [256].

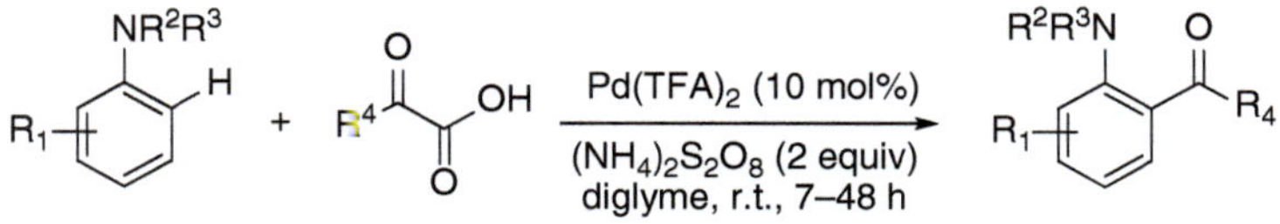

Scheme 7.145 Pd-catalyzed decarboxylative acylation between anilides and α-oxocarboxylic acids. Source: Modified from Fang et al. [255].

In 2013, Wang and coworkers reported a $CuBr_2$-catalyzed decarboxylative acylation reaction between formamides and α-oxocarboxylic acids, resulting in α-ketoamides, which are core structures for many natural products and pharmaceuticals (Scheme 7.146) [257]. By using $CuBr_2$ as the catalyst and di-*tert*-butyl peroxide (DTBP) as the oxidant, the reaction enjoys a good substrate scope; both *N*-alkylformamides and *N*,*N*-dialkylformamides were suitable substrates.

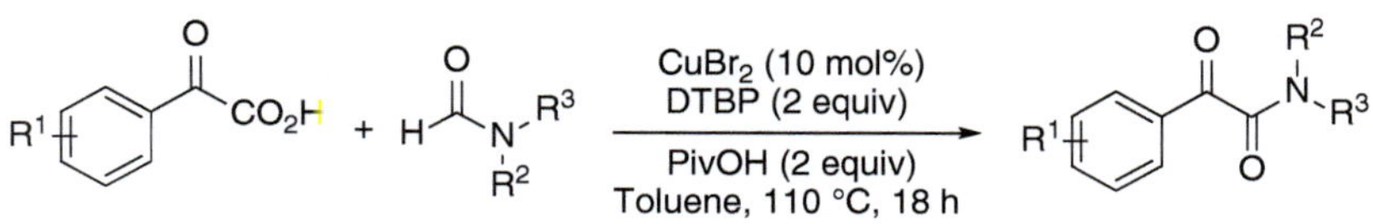

Scheme 7.146 Decarboxylative cross-coupling of foramides and α-oxocarboxylic acids. Source: Modified from Li et al. [257].

At the end of 2014, Lei and coworkers developed the first visible-light-induced decarboxylative amidation of α-oxocarboxylic acids using $[Ru(phen)_3]Cl_2$ as the catalyst, thus providing a novel and efficient route to various amides in good yields (Scheme 7.147) [258]. Feature of this work was the use of O_2 as a green terminal oxidant. Different aryl-, heteroaryl-, or alkyl-substituted aketo acids could be successfully used in this transformation. Importantly, the method can be successfully applied to the efficient synthesis of nitrogen-containing heterocyclic compounds when ortho-substituted anilines are involved.

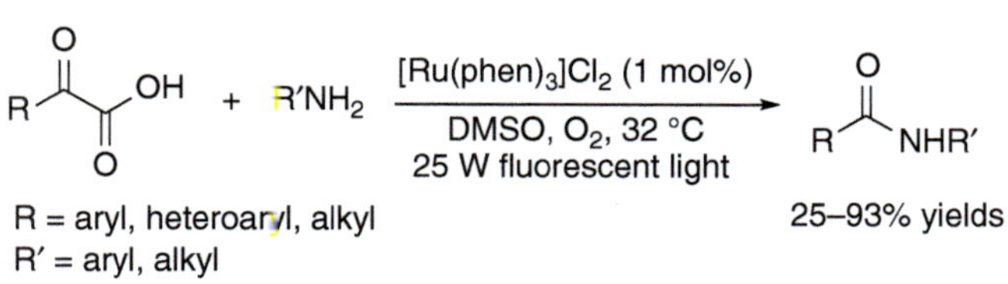

Scheme 7.147 Ru-catalyzed visible-light-induced decarboxylative amidation of α-oxocarboxylic acids. Source: Modified from Liu et al. [258].

In 2014, the Wen group reported one example of nickel- and manganese-catalyzed decarboxylative cross-coupling of arylvinyl carboxylic acids and cyclic ethers

(Scheme 7.148) [259]. The reaction gave different products when different catalysts were used: manganese acetate produced the normal decarboxylative cross-coupling products, while nickel resulted in only phenylethanone derivatives. In most cases, both directions enjoyed moderate to good yields of products. However, the carboxylic acid substrates were limited to arylvinyl carboxylic acids.

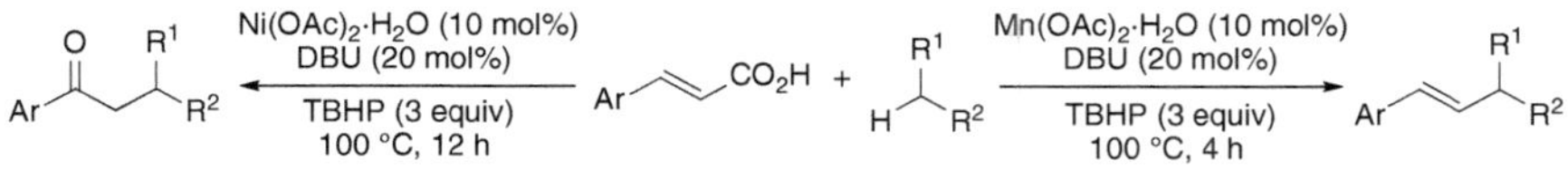

Scheme 7.148 Selective Ni- and Mn-catalyzed decarboxylative cross-coupling of arylvinyl carboxylic acids with cyclic ethers. Source: Modified from Zhang et al. [259].

Meanwhile, Su and coworkers developed a methodology that realized the direct arylation of pyridines with benzoic acids catalyzed by a silver salt for the first time (Scheme 7.149) [260]. Depending upon the silver salts used (e.g. $AgNO_3$, AgTFA, and Ag_2SO_4), various benzoic acids reacted with pyridine and its derivatives smoothly.

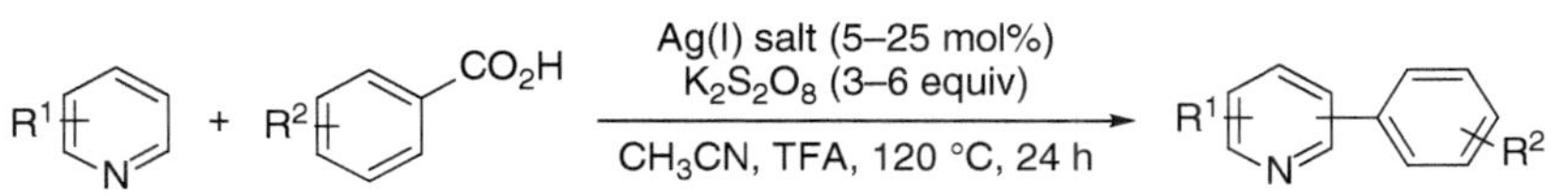

Scheme 7.149 Ag-catalyzed C–H arylation of pyridines. Source: Modified from Kan et al. [260].

In 2015, a visible-light-mediated decarboxylative alkynylation reaction using BI–alkyne [261, 262] as electrophilic alkynylating agent was described (Scheme 7.150) [263]. Remarkably, in the presence of 60 bar carbon monoxide, this reaction becomes a decarboxylative carbonylative alkynylation reaction and affords valuable ynones in good yields. Importantly, the synthetic application of the method was elegantly demonstrated by the preparation of naturally occurring ursolic acid and the synthesis of oxazolidinones.

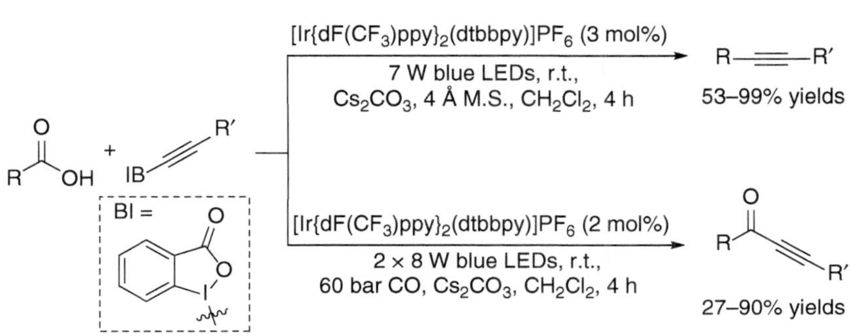

Scheme 7.150 Visible-light-mediated decarboxylative alkynylation reaction catalyzed by iridium. Source: Modified from Zhou et al. [263].

In the same year, Pd/Cu catalytic systems were developed especially by Goossen and coworkers [264]. They improved the efficiency of the reaction and expanded the substrate scope to aryl halides, heteroaryl halides, and sulfonates on the one side and to non-activated aromatics without *ortho* substituent as well as heteroaromatic acids on the other side. (Scheme 7.151)

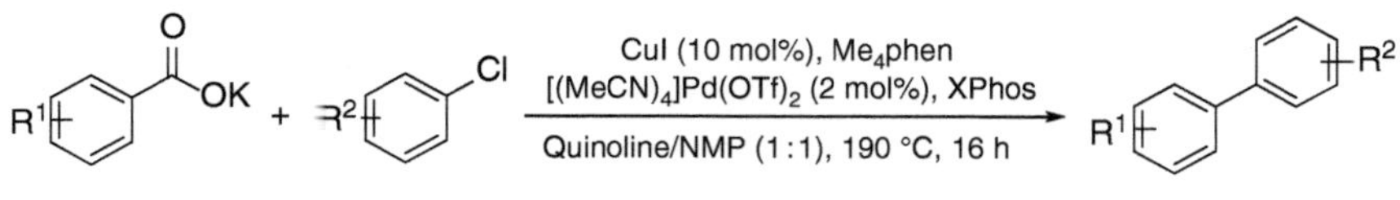

Scheme 7.151 Cu/Pd-catalyzed decarboxylative arylation of aromatic acids without activating ortho substituent. Source: Modified from Tang et al. [264].

A photocatalytic protocol was developed by König group using the organic dye Eosin Y (10 mol%) and visible light (535 nm) for decarboxylative alkylation with broad substrate (amino acids, α-oxy acids, and fatty acids) (Scheme 7.152) [265]. The carboxylic acids are activated by esterification to *N*-(acyloxy)phthalimides and then reductively cleaved upon irradiation with green light. This generates alkyl radicals, which undergo cross-coupling with electron-deficient alkenes.

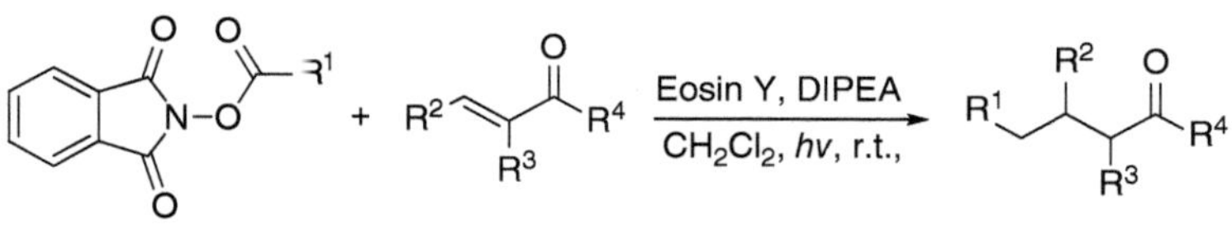

Scheme 7.152 Decarboxylative alkylation using Eosin Y and visible light.

In 2016, Wan, Hao, and coworkers reported a Ag-catalyzed decarboxylative cyclization method for the synthesis of gem-difluoromethylenated phenanthridines (Scheme 7.153) [266]. Under the relatively gentle reaction conditions, various difluoroaetates, including thiophene-derived and sterically hindered ones, were reacted with biarylisonitriles to afford the corresponding products in moderate to good yields. Nevertheless, the reactions were limited to aryl difluoroacetates. For example, potassium trifluoroacetate showed no reactivity to the reaction.

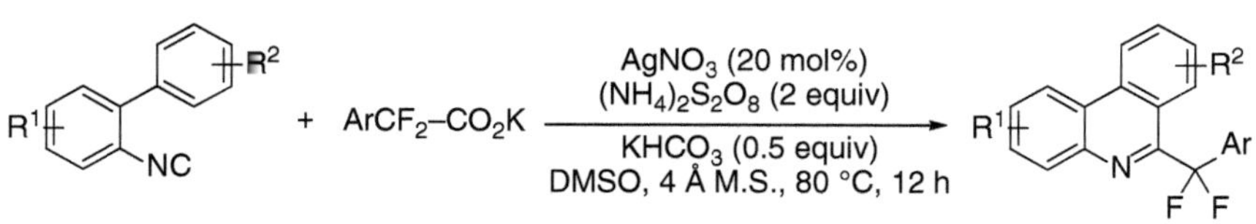

Scheme 7.153 Synthesis of gem-difluoromethylenated phenanthridines. Source: Modified from Wan et al. [266].

Most photoredox catalysts in current use are precious metal complexes or synthetically elaborate organic dyes, the cost of which can impede their application for large-scale industrial processes. In 2019, photocatalytic decarboxylative alkylation mediated by triphenylphosphine and sodium iodide under 456-nanometer irradiation was reported by Fu group in moderate to excellent yields, which can catalyze the alkylation of silyl enol ethers by decarboxylative coupling with redox-active esters (Scheme 7.154) [267]. Importantly, this photoredox system can catalyzes Minisci-type alkylation of *N*-heterocycles and operate in tandem with chiral phosphoric acids to achieve high enantioselective products.

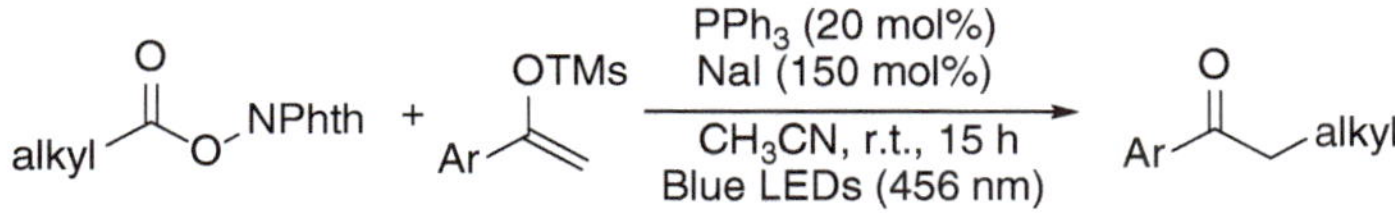

Scheme 7.154 Photocatalytic decarboxylative alkylation mediated by PPh_3 and NaI. Source: Modified from Fu et al. [267].

Recently, Aggarwal group reported the reaction of metallaphotoredox-catalyzed decarboxylative conjunctive cross-coupling of vinyl boronic esters with carboxylic acids and aryl iodides, which was developed for the synthesis of complex alkyl boronic esters from feedstock materials (Scheme 7.155) [268]. The reaction proceeded under mild metallaphotoredox conditions and involves an unprecedented decarboxylative radical addition/cross-coupling cascade of vinyl boronic esters. Excellent functional group tolerance is displayed, and application of a range of carboxylic acids, namely secondary α-amino acids, and aryl iodides provides efficient access to highly functionalized alkyl boronic esters. The decarboxylative conjunctive cross-coupling was also applied to the synthesis of sedum alkaloids.

Scheme 7.155 Ni-catalyzed decarboxylative conjunctive cross-coupling. Source: Mega et al. [268].

7.11 Reaction of Esters as Starting Materials

7.11.1 Hydrolysis Reaction

Hydrolysis reaction of esters can be divided into two types according to different catalyst or promoter (acid or base). The hydrolysis of esters in aqueous acid is a reversible equilibrium reaction that is driven to the right using a large excess of water (Scheme 7.156) [269].

Esters are hydrolyzed in aqueous base to form carboxylate anions. Basic hydrolysis of an ester is called saponification (Scheme 7.157).

Soap is prepared by the basic hydrolysis or saponification of a triacylglycerol. Heating an animal fat or vegetable oil with aqueous base hydrolyzes the three esters

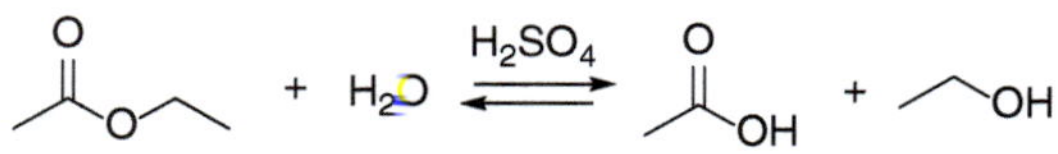

Scheme 7.156 Hydrolysis of esters under acid condition. Source: Modified from Paula [269].

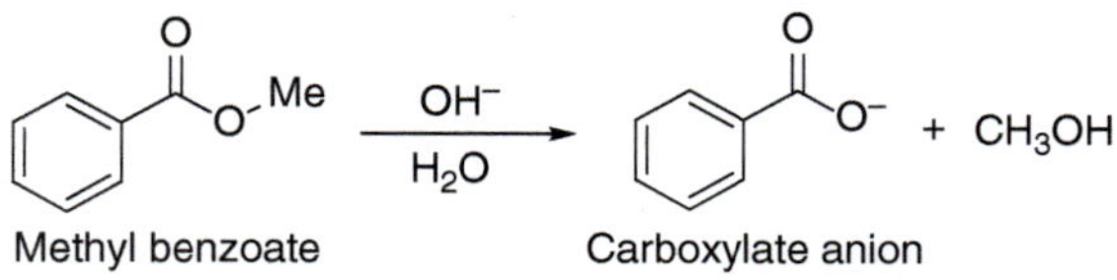

Scheme 7.157 Hydrolysis of methyl benzoate under base condition.

to form glycerol and sodium salts of three fatty acids. These carboxylate salts are soaps, which clean away dirt because of their two structurally different regions. The nonpolar tail dissolves grease and oil and the polar head makes it soluble in water (Scheme 7.158). Most triacylglycerols have two or three different R groups in their hydrocarbon chains, so soaps are usually mixtures of two or three different carboxylate salts.

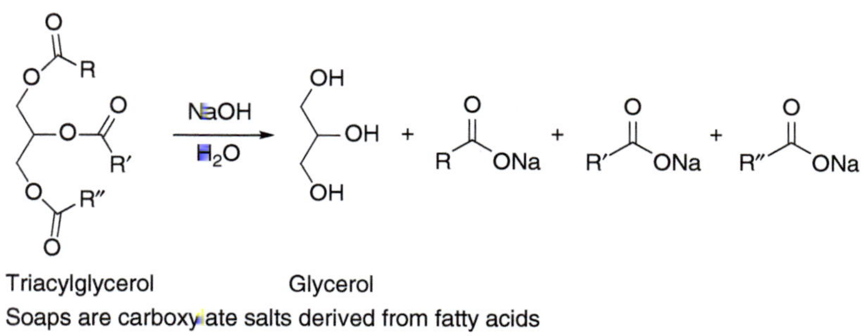

Scheme 7.158 Basic hydrolysis of a triacylglycerol.

7.11.2 Transesterification Reaction

Transesterification, the reaction of an ester with an alcohol, is catalyzed by acid. The mechanism for acid-catalyzed transesterification is identical to the mechanism for acid-catalyzed ester hydrolysis, except that the nucleophile is ROH rather than H_2O (Scheme 7.159). Both the hydrolysis and the transesterification of an ester are very slow reactions because water and alcohols are poor nucleophiles and the –OR group of an ester is a poor leaving group. Therefore, these reactions are always catalyzed

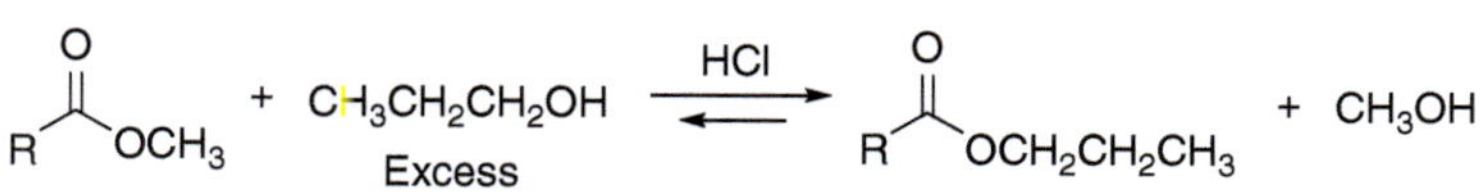

Scheme 7.159 Transesterification reaction.

when carried out in the laboratory. Both the hydrolysis and transesterification of an ester can be catalyzed by acids. The rate of hydrolysis can also be increased by hydroxide ion and the rate of alcoholysis can be increased by the conjugate base (–OR) of the reactant alcohol.

Prostaglandins have several different physiological functions. One is to stimulate inflammation and another to induce fever. Prostaglandin synthase is composed of two enzymes. One of them, cyclooxygenase, has a CH_2OH group at its active site that is necessary for enzymatic activity. When the CH_2OH group reacts with aspirin in a transesterification reaction, the enzyme is inactivated. This prevents prostaglandins from being synthesized, so inflammation is suppressed and fever is reduced. Notice that the carboxyl group of aspirin is a basic catalyst. It removes a proton from the CH_2OH group, which makes it a better nucleophile. This is why aspirin is maximally active in its basic form (Scheme 7.160).

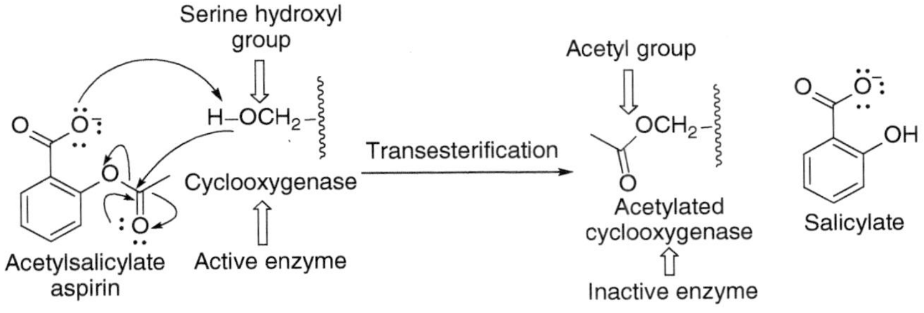

Scheme 7.160 Mechanism of aspirin for inhibition of fever.

7.11.3 Aminolysis Reaction

Esters undergo nucleophilic addition–elimination at their acyl carbon atoms when they are treated with ammonia or with primary and secondary amines, which is called ammonolysis reaction of carboxylic acids (Scheme 7.161). The amine must be primary (RNH_2) or secondary (R_2NH). Tertiary amines (R_3N) cannot form amides because they have no proton on nitrogen that can be replaced by an acyl group. These reactions take place much more slowly than those of acyl chlorides and anhydrides, but they can still be synthetically useful.

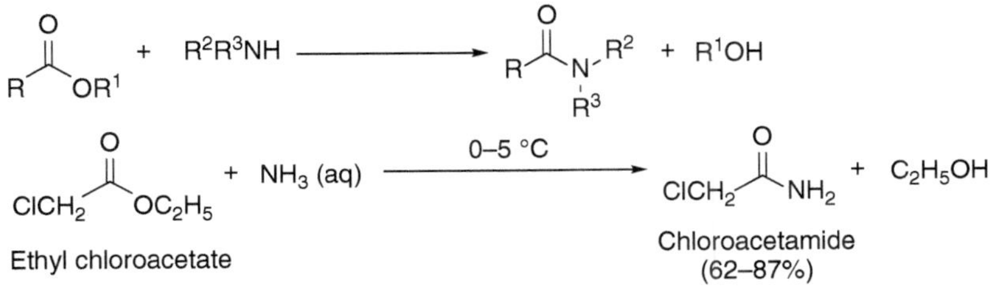

Scheme 7.161 Aminolysis reaction.

7.12 Reaction of Amides as Starting Materials

7.12.1 Hydrolysis Reaction

Because amides have the poorest leaving group of all the carboxylic acid derivatives, they are the least reactive. Under strenuous reaction conditions, amides are hydrolyzed in acid or base to form carboxylic acids or carboxylate anions (Scheme 7.162). In acid, the amine by-product is protonated as an ammonium ion, whereas in base, a neutral amine is formed [269].

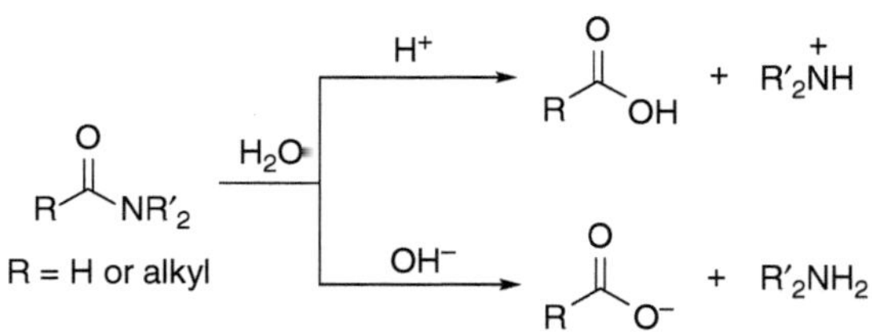

Scheme 7.162 Hydrolysis of amides.

7.12.2 Alcoholysis Reaction

Amides react with alcohols in the presence of an acid and under heat condition to form esters is called alcoholysis reaction of carboxylic amides. The mechanism for this reaction is exactly the same as the mechanism for the acid-catalyzed transesterification of an ester (Scheme 7.163).

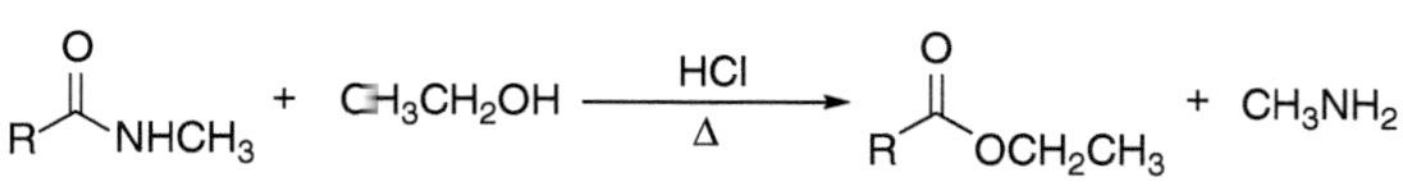

Scheme 7.163 Alcoholysis reaction of amides.

The antibiotic activity of penicillin results from its ability to acylate (put an acyl group on) a CH_2OH group of an enzyme that has a role in the synthesis of bacterial cell walls (Scheme 7.164). Acylation occurs by a nucleophilic acyl substitution

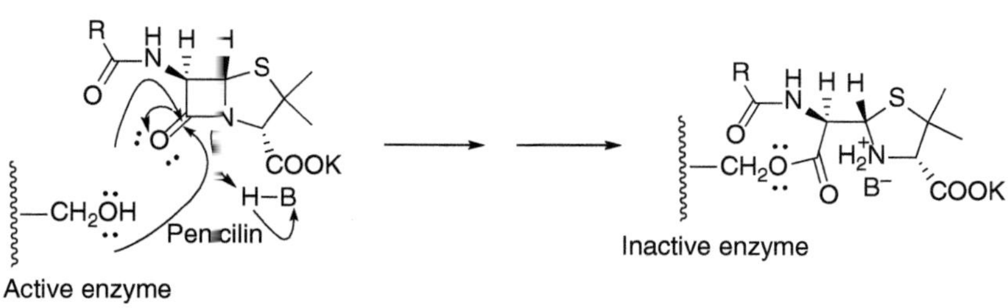

Scheme 7.164 Mechanism of penicillin for the inhibition of bacterium.

reaction: the CH_2OH group adds to the carbonyl carbon of the β-lactam of penicillin, forming a tetrahedral intermediate. The four-membered ring amide is more reactive than a noncyclic amide because when the π bond reforms, the strain in the four-membered ring is released when the amino group is eliminated.

8

Synthesis of Functional Materials

8.1 Polyamides

The polyamide (PA) is a polymeric substance, generally linear, with recurring amide groups as an integral part of the polymeric chain. PAs, and the other polymers based on the linkages of Scheme 8.1, are among the class of polymers generally formed by the mechanism of "step-growth," in which reaction takes place between multifunctional (usually bifunctional) molecules and each new bond is created as a separate step. PAs have been known since the 1930s – the dawn of the synthetic polymer age and they constitute very important part of polymeric materials [270, 271]. The outstanding properties in terms of mechanical strength, flexibility, toughness, and resistance make these polymers especially interesting for diverse and special components for automotives, aircrafts, fibers, appliances, and engineering components as well as in the medical sector [272].

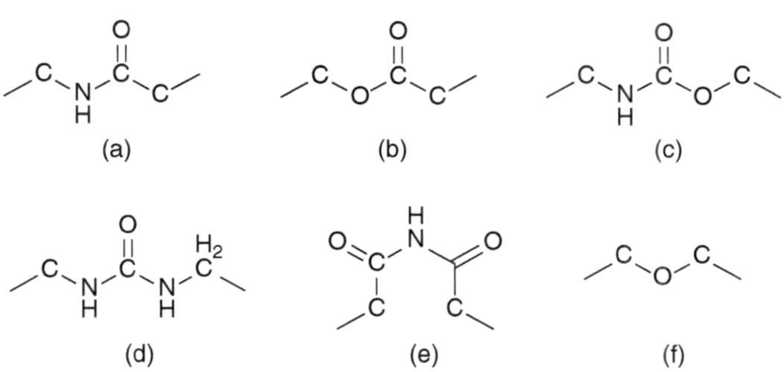

Scheme 8.1 Common polymer linkages (a: amide; b: ester; c: urethane; d: urea; e: imide; f: ether).

PAs are categorized into three groups according to the types of monomers:

1) Aliphatic polyamides
2) Aromatic polyamides
3) Long-chain semiaromatic polyamides

Carbonyl Compounds: Reactants, Catalysts and Products, First Edition.
Feng Shi, Hongli Wang and Xingchao Dai.

8.1.1 Aliphatic Polyamides

Polymers prepared with only straight-chain monomers are the original and dominant group. The nylon 6 and nylon 66, i.e. PA 6 and PA 66, together constitute about 90% of the world's overall PA usage and 99% of PA fibers [273]. Along with PA 6, other commercially significant members of the single-monomer nylon family are PA 7, PA 11, and PA 12. PA 7 makes a good fiber [274] and has lower equilibrium content of troublesome residual monomer and higher melting point (230–235 °C) than nylon 6, but it has seen limited commercial use [275]. Table 8.1 shows, via the melting temperatures, that a long-chain frequency matters a great deal to the extent and coherence of the crystalline region. Amide frequency is generally defined as 100 times the ratio of amide links to total long-chain atoms.

The two original PAs, nylon 66 and nylon 6, defined by the US Federal Trade Commission as any PA having less than 85% of the amide groups directly connected to two aromatic groups, were invented in the 1930s, but they originated from the concept of natural polymers developed by Herman Staudinger in the 1920s that large molecules comprised repeating small units or segments [276]. Inspired by this, Wallace Carothers reasoned that synthetic polymers could be made by the chemical joining of small molecules, and he eventually chose diamines and diacids as the molecules, reacting them with each other to produce amide linkages and large molecules consisting of 100 and more repeat units. The reaction is shown in Scheme 8.2.

A lot of different diacid and diamine were tested and finally the combination of hexamethylenediamine (HMD) and adipic acid was chosen, as shown in Scheme 8.3. The material was first made on 28 February 1935, and the first commercial plant for fiber started up in Seaford Delaware in December 1939 from this first PA called nylon 66 (PA 66) because of the six carbons in the diamine and diacid, respectively [277]. Concurrently, DuPont was also manufacturing nylon 66 in the form of bristles and monofilaments (e.g. fishing line) and as a molding resin [278].

On 29 January 1938, a polymer that could be drawn into a textile fiber of good properties and could also be molded into a strong rod was made by the acid hydrolysis of pure caprolactam. This polymer is called PA 6 because it was made from a single starting material possessing six carbon atoms. Like PA 66, PA 6 was immediately used as a textile fiber and as a monofilament for brushes. The key difference in

Table 8.1 Melting point: straight-chain aliphatic AB polyamides.

Polyamide	Amide frequency	Melting point (°C)
PA 6	14.3	225
PA 11	8.3	183
PA 12	7.7	180
PA 22	4.3	145

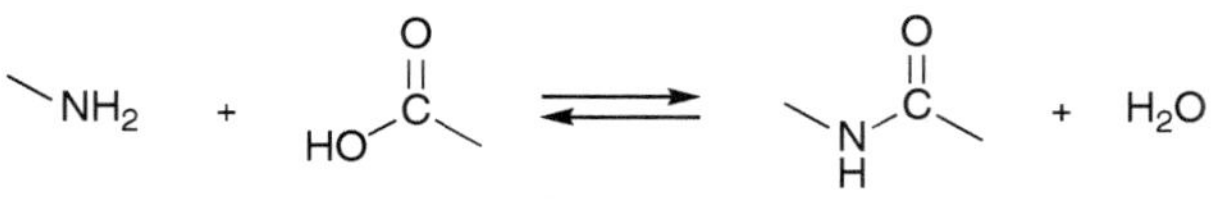

Scheme 8.2 Direct formation of amide linkage.

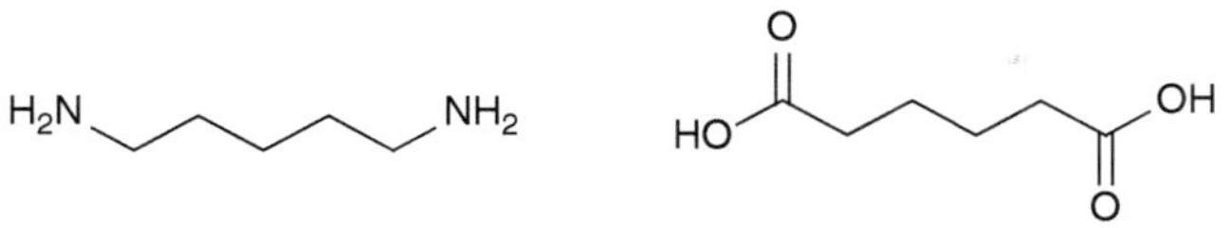

Scheme 8.3 Hexamethylenediamine (HMD) and adipic acid.

chemistry of formation of the two polymers – the difference which avoided a conflict of patents – is that PA 66 is formed by the direct amidation, whereas PA 6 depends on transamidation, as shown in Scheme 8.4. Much of the chain growth is achieved by the direct reaction of chain amine ends with the amide link contained in the caprolactam molecule [270].

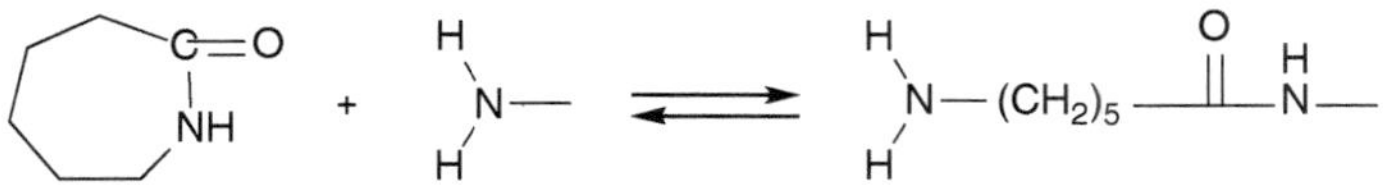

Scheme 8.4 Growth by transamidation in PA 6.

Aliphatic PAs, such as poly(e-caprolactam) (nylon 6) and poly(hexamethylene adipamide) (nylon 66), are a class of engineering thermoplastics, which play key roles in industrial and commercial applications because of their high tensile strength, excellent chemical resistance, fine abrasion, and easily processing characteristics. However, these aliphatic PAs also have some disadvantages such as high moisture absorption, and poor dimensional stability and thermal properties. Their applications are limited to those involving exposure to temperatures of <100 °C due to the low glass transition temperature of about 50 °C. In order to improve the heat resistance and strength of PAs, aromatic rings were incorporated into their backbone [279].

8.1.2 Aromatic Polyamides

Aromatic PAs, i.e. aramids, stand for synthetic PAs comprising >85% by amide groups (–CO–NH–) bound directly to two aromatic rings. These polymers are categorized as high-performance materials because of their outstanding mechanical strength and superior high thermal resistance. They are spun into fibers for advanced fabrics such as advanced sport and work protective clothing, bullet-proof body armor, advanced composites in armament and aerospace industries, in composites as asbestos substitutes, and high-temperature insulation paper, to name but a few [280].

The aramids structure is based on the rigid aromatic amide linkage, which is responsible for the outstanding properties of these materials. Highly directional and efficient interchain hydrogen bonds led to materials with high crystallization tendency and extremely high cohesive energy density. At the same time, they are also responsible for the insolubility of the wholly aromatic PAs, a drawback that impairs the expansion of the application field of these materials, where the improvement of the solubility is a topic of research interest [281].

Poly(*p*-benzamide), an all-*para*-oriented aramid, an extremely rigid rod-like macromolecule, was commercialized about 50 years ago, for a short period of time, under the tradename "Fiber B®," but with important transformation impairments, it was replaced on the market in the 1970 decade by poly(*p*-phenyleneterephthalamide) (PPTA), the unique all-*para*-oriented aramid commercialized nowadays. Besides, poly(*m*-phenylene isophthalamide) (MPIA), a less rigid and easily transformable aramid with impressive thermal resistance, only slightly underperforming the mechanical behavior of PPTA, was marketed in 1967. PPTA and MPIA were discovered and initially commercialized under the tradenames of "Kevlar®" and "Nomex®" by DuPont [281]. The structures and some of the aramid producers are shown in Table 8.2.

Aramids are synthesized at a lab scale by two procedures: low- and high-temperature solution methods. The commercial aramids PPTA and MPIA are prepared by the low-temperature solution method using *p*-phenylenediamine (PPD) and terephthaloyl dichloride (TPC), or *m*-phenylenediamine (MPD) and isophthaloyl dichloride (IPC) as the starting materials, a polar aprotic solvent such as *N*-methyl-2-pyrrolidone (NMP) or *N*,*N*-dimethylacetamide (DMA) as the reaction medium and salts such as LiCl and/or $CaCl_2$ as solubility promoters. Aramids can be also produced from aromatic diacids by the high-temperature solution method, which was reported by García et al. [280, 282, 283]. However, it is impossible to produce them by melting process due to their high glass transition and melting temperatures. The semiaromatic PA containing aliphatic chain can modify their processability [279].

Aromatic PAs are currently being studied as functional and cutting-edge materials for better thermally and mechanically resistant polymers, sensing and extraction applications, films and membranes for transport applications, optically active materials, and materials with electromechanical or electrochromic properties.

8.1.3 Long-Chain Semiaromatic Polyamides

In this group of PAs, an aromatic monomer connected to one or more aliphatic (usually straight-chain) monomers or, in some cases, a pair of aromatic monomers of same functionality (i.e. both diamines or both diacids) connected to one or more aliphatic monomers of opposite functionality was considered. This group of PAs includes, but is not restricted to, the polyphthalamides (PPA). The four represent substances are 1,3- and 1,4-phenylenediamine and 1,3- and 1,4-phthalic acid. Besides, there are a host of other more complex substances that satisfy the definition and that have been used in making PAs, at least experimentally [270].

Table 8.2 Structure, aramid type, brand names, and company of commercial aramids.

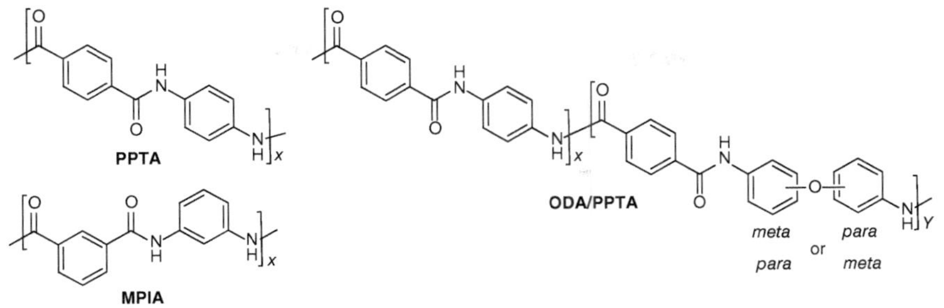

Aramid type			Brand names	Company
PPTA	MPIA	ODA/PPTA		
x			Kevlar®	DuPont, Wilmington, DE, United States
	x		Nomex®	
x			Twaron®	
	x		Teijinconex®	Teijin, Arnhem, the Netherlands
	x	x	Technora®	
x			Heracron®	Kolon Industries, Seoul, Korea
x			Alkex®	Hyosung, Seoul, Kore
	x		*Meta* Aramid Yarn & Thread®	SRO Group, Heatherbrae, New South Wales, Australia
x			Taparan® *Para*-aramid	Yantai Tayho Advanced Materials Co., Ltd, Yantai, Shandong, China
	x		Newstar® *Meta*-Aramid	
	x		Arawin®	Toray Chemical Korea Inc., Seoul, Korea
	x		Metaone	Huvis, Seoul, Korea

The simplest form of PPA is formed by a single straight-chain aliphatic diamine and a single aromatic acid, terephthalic acid (TPA), or isophthalic acid (IPA). The series of diamines, from 2-carbon to 13-carbon, with TPA has been studied by various workers [279, 284–287]. Melting and glass transition temperatures of these polymers are shown in Scheme 8.5. The melting data show the odd/even difference that is common to polymers with straight-chain monomers. The glass transition data reflect the differences in measurement method and conditions of the different sources. The high melting points of polymers containing a diamine with less than nine carbons would make them difficult to melt process without excessive degradation.

Nylon 9T is more tractable and a new PPA "based on long-chain diamines" obtained from the castor bean plant, probably 1,10-decanediamine, has been introduced [288]. These higher diamines have low moisture absorption and stable

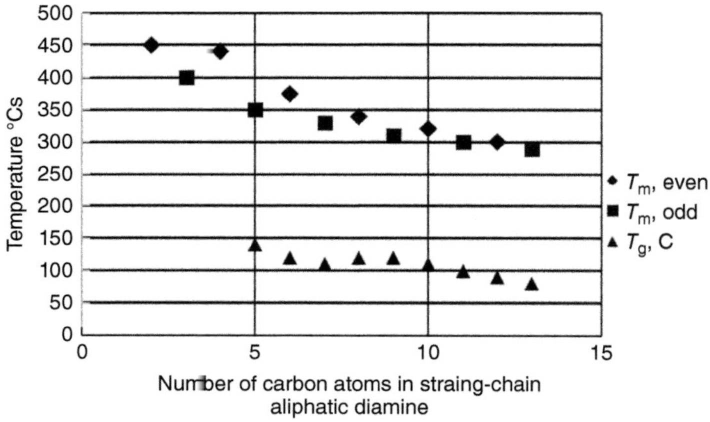

Scheme 8.5 Melting points and glass transition temperatures of TPA-based polyphthalamides.

shape and glass transition temperature. The semiaromatic polymers containing 10, 11, 12, and 13-carbon straight-chain diamines [285] and PA 10T [287, 289] have been studied, and fibers from PA 6T has been also spun [290]. However, compared with PA 66, these new polymers were slower to crystallize, so that there was further crystallization in the solid phase and also ambiguity in crystal habit.

A series of PAs, such as 2I, 3I, 4I, 6I, and 10I from straight-chain diamines with 2, 3, 4, 6, and 10 carbons and IPA, have been made and their melting points were measured (Table 8.3) [291]. Two observations may be made: (i) the melting temperatures of the IPA-based polymers are lower than those of the corresponding TPA-based polymer, reflecting a less organized crystalline structure and (ii) each IPA-based polymer has a melting temperatures range of considerable width, reflecting uncertainty in the extent of crystallinity in the measured samples.

Table 8.3 Melting behavior of isophthalic-based polyphthalamides.

Number of carbon atoms in straight-chain Aliphatic diamine, *n*	Melting (or softening) point of nylon *n*I (°C)	Melting point of corresponding TPA-based polymer (°C)
2	292–310	455
3	240–288	399
4	230–245	433
6	170–230	370.5
10	166–194	313

Source: Modified from Gorton [291].

The PPAs, often referred to as high-temperature nylons, are an important part of the nylon plastic family. They have extended the reach of the PAs into difficult applications, especially in the automotive and electronic industries. TPA is the key monomer, which is found in all the various PPA formulations. IPA plays a supporting role, helping to reduce melting point while at the same time contributing its bulkiness to the immobilization of amorphous regions and the maintenance of high glass transition temperature [270].

8.2 Phenol Formaldehyde Resins

Phenol formaldehyde resins (phenolic resins, PF), the first thermosetting plastics, are considered to be the first truly synthetic commercially available plastic resins [292]. Work in the area of phenols and formaldehydes could be traced back to prior to the twentieth century [119, 293]. The first commercially available phenolic resin, called Laccain, was introduced by Knop and Scheib [293] in 1902 as a substitute for shellac, but it was not successful in commerce [294]. The use of phenolic resins was popularized by the "heat and pressure" patents [294–297] of Dr. Leo H. Baekeland in 1907, who is known as the "father of phenolic resins." Today, some of the most popular phenolic resins bear the trade name "Bakelite" in reference to the company (General Bake-lite Company) that he built in 1910. Phenolic resins are widely used in many areas, such as insulation, electrical devices, automotive parts, and adhesives. In 1993, its consumption volume reached 3.07 billion pounds, ranking second only to polyurethane (PU) (3.476 billion pounds) among thermosets.

Phenolic resins as general-purpose thermosets were generally prepared by the polycondensation reaction between phenols and formaldehyde solutions. The three major raw materials are phenol, formaldehyde, and hexamethylene tetramine (HEXA) [298]. Initially, phenols condense with formaldehyde in the presence of either acid or alkali to form a methylolphenol or phenolic alcohol, and then dimethylolphenol. The initial attack may be at the second, fourth, or sixth positions. The second stage of the reaction involves methylol groups with other available phenol or methylol phenol, leading first to the formation of linear polymers [299] and then to the formation of hard-cured, highly branched structures.

Phenolic resins are categorized into two groups according to the types of catalysts:

8.2.1 Novolac Resins

Novolac resins are obtained with acid catalysis such as oxalic acid, sulfuric acid, hydrochloric acid, formic acid, and aromatic sulfuric acids [300, 301]. Sulfuric and oxalic acids are the two most commonly used acids. The gel point of the cure is deliberately delayed using a phenol formaldehyde feedstock generally with a phenol formaldehyde ratio of 1 : 0.8 in the first stage (pre-polymerization) [302].

Polymerization is carried out by heating the mixture for two to four hours at reflux, and water was removed at 160 °C. Then, the resultant low-molecular-weight molten intermediate is cooled and the glassy material is carefully crushed and blended with HEXA (in powder form) to produce a molding compound. The HEXA is the second part or hardener, and hence, the blended resins are referred to as two-step resins or novolac phenolic resins. Upon heating the novolac resins to about 165 °C in a mold, the HEXA decomposes to provide the formaldehyde necessary for the final curing. A molar ratio of final working phenol formaldehyde is 1 : 1.5 considering the addition of HEXA.

A novolac resin is incapable of condensing with other novolac molecules on heating without hardening agents due to no reactive methylol groups in its molecules. To complete resinification, further formaldehyde is added to cross-link the novolac resin. Phenolic rings are considerably less active as nucleophilic centers at acidic condition due to hydroxyl and ring protonation (Scheme 8.6a). However, the aldehyde is activated by protonation, compensating for this reduction in potential reactivity because the protonated aldehyde is a more effective electrophile (Scheme 8.6b) [303].

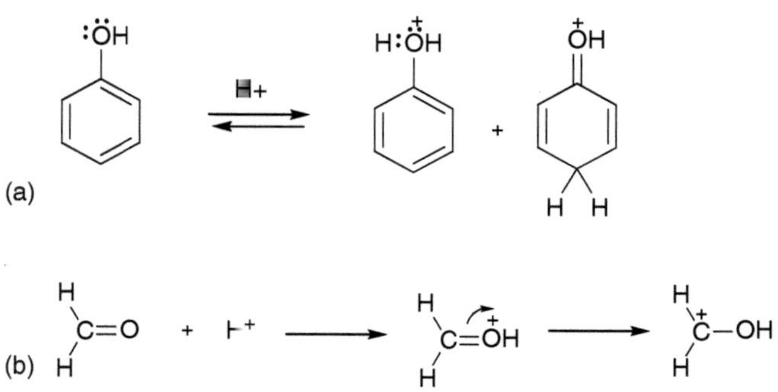

Scheme 8.6 (a) Phenolic rings as nucleophilic centers at acidic condition; (b) aldehyde protonation. Source (b) Based on Gardziella et al. [303].

8.2.2 Resole Resins

Different from novolac resins, resole phenolic resins are produced in the presence of an alkaline catalyst such as ammonia, sodium carbonate, or sodium hydroxide at 100 °C for about one hour. The molar ratio of phenol and formalin is usual 1 : 1–1 : 1.5.

Polymerization is stopped short of the gel point by cooling to provide an intermediate resole phenol formaldehyde resin. If a solid product is desired, the intermediate is dried by healing under a vacuum for three to four hours to prevent heat hardening. Resole phenolic resin is a water-soluble methylol (-CH_2OH) bearing thermoplastic, and the curing process to the final thermoset material can be initiated by just heating the resole in a mold above its gel point. The formed resole resins contain reactive methylol and hydroxyl groups, so they form larger molecules with methylene cross-links without the addition of a curing agent when heated. The resinification reaction of phenol formaldehyde resin is a typical polycondensation reaction since water is produced as a by-product [303].

Resols are obtained with alkaline catalysis and an excess of formaldehyde (Scheme 8.7) [303]. The resol molecules condense under heating to form large molecules without the addition of a hardener due to the existence of reactive methylol groups. The ionization of the phenol strengthens its function as nucleophiles but does not affect the activity of the aldehyde.

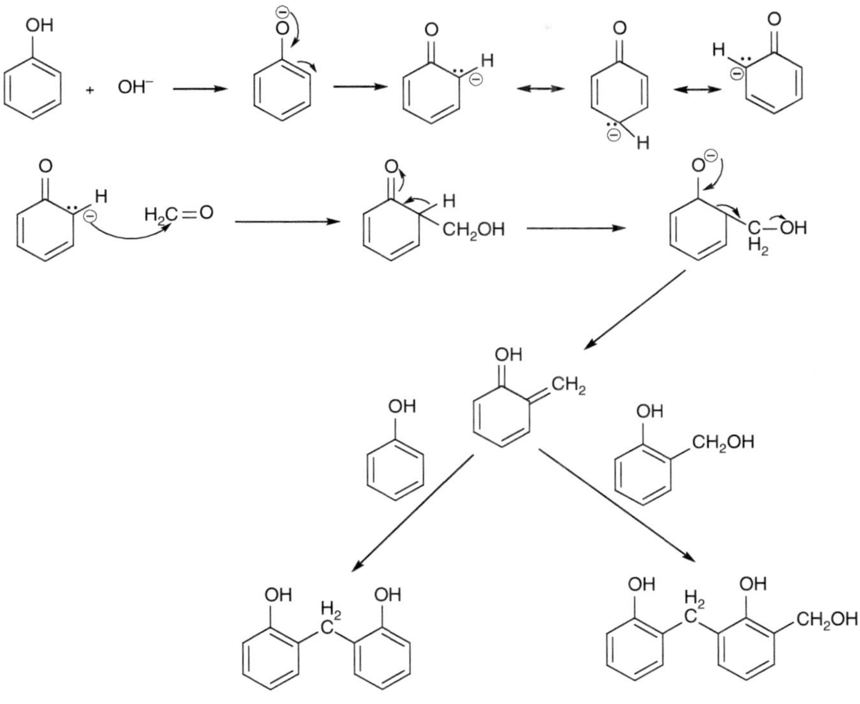

Scheme 8.7 Resols are obtained with alkaline catalysis and an excess of formaldehyde. Source: Based on Gardziella et al. [303].

The differences between resoles and novolac phenolic resins are shown in Table 8.4. It is divided into seven parts, such as catalyst, category of resins, shelf life, cure, dimensionally stable, application, and physical form.

Phenolic resins possess these prerequisites to a considerable extent. Although non-modified, inorganically filled phenolics are already considered temperature-resistant, their thermo-oxidative resistance can be further improved by chemical modification [302, 304]. The common methods to improve the thermo-oxidative resistance of phenolic resins are as following: (i) etherification or esterification of the phenolic hydroxyl group; (ii) complex formation with polyvalent elements (Ca, Mg, Zn, Cd, and others); and (iii) replacement of the methylene-linking group by heteroatoms (O, S, N, Si, and others).

Phenolic resins are mainly applied in areas of adhesives and bonding (these being dominated by wood adhesives, 97%; foundry binders, 3%), laminates, molded parts, and coatings. Wood adhesives and laminates are part of the adhesive and bonding

Table 8.4 The differences between resoles and novolac phenolic resins.

Entry	Resoles	Novolac phenolic resins
Catalyst	Alkaline	Acidic
Category of resins	Methylol-bearing	Non-methylol-bearing
Shelf life	Less than one year	Infinite
Cure	Split off water	Give off ammonia
Dimensionally stable	Novolac resins are twice more dimensionally stable than resoles	
Application	Casting and bonding resins	Molding compounds
Physical form	Liquids	Solids

market, but the large volume usage of phenolic resins in wood adhesives (in 2004 69% of the phenolic resin market) and laminates (7%) warrant their consideration as distinct areas of application of phenolic resins. Phenolic resins for insulation also reached 13% of the market in 2004 [292].

8.3 Polyurethanes

PUs, which can be easily prepared by a simple polyaddition reaction of polyol, isocyanate, and a chain extender, are an important class of polymers and exhibit an exceptionally versatile range of properties and applications [305–307]. PUs are characterized by the presence of urethane linkage (NHCOO) formed typically through the reaction of a diisocyanate and a glycol (Scheme 8.8). Often, ester, ether, urea, and aromatic rings are also present along with the urethane linkages in PU backbone.

Scheme 8.8 Schematic representation of the reaction of a diisocyanate and a glycol. Source: Based on Gardziella et al. [303].

PU was firstly discovered in 1947 by Otto Bayer and his coworkers at the laboratories of I.G. Farben in Leverkusen, Germany, and they synthesized polyurethane Perlon U by 1,6-hexamethylene diisocyanate (HDI) and 1,4-butanediol (BDO) [116]. The initial works focused on PU products from aliphatic diisocyanate and diamine till the PU products from an aliphatic diisocyanate and glycol exhibited interesting properties. Polyisocyanates became commercially available in 1952 soon after the commercial scale production of PU was from toluene diisocyanate (TDI) and polyester polyols (PEPs) [308]. In the years of 1952–1954, different polyester-polyisocyanate systems were developed by Bayer.

PEPs were gradually replaced by polyether polyols due to their several advantages such as low cost, ease of handling, and improved hydrolytic stability. Poly(tetramethylene ether) glycol (PTMG) is the first commercially available polyether polyol, which was introduced by DuPont in 1956 by polymerizing tetrahydrofuran. Later, in 1957, BASF and Dow Chemical produced polyalkylene glycols. Besides, DuPont produced a Spandex fiber called Lycra based on PTMG and 4,4′-diphenylmethane diisocyanate (MDI) and ethylene diamine. With the decades, PU evolved from flexible PU foams (1960) to rigid PU foams (polyisocyanurate foams-1967) as several blowing agents, polyether polyols, and polymeric isocyanate such as poly methylene diphenyl diisocyanate (PMDI) became available. These PMDI-based PU foams showed good thermal resistance and flame retardance [309].

In 1969, PU reaction injection molding (PU RIM) technology was introduced which further advanced into reinforced reaction injection molding (RRIM) producing high-performance PU material that in 1983 yielded the first plastic body automobile in the United States. In 1990s, due to the rising awareness toward the hazards of chloro-4 PU alkanes as blowing agents (Montreal protocol, 1987), several other blowing agents (e.g. carbon dioxide, pentane, 1,1,1,2-tetrafluoroethane, and 1,1,1,3,3-pentafluoropropane) outpoured in the market. At the same time, two-pack PU, PU polyurea spray coating technology came into foreplay, which had significant advantages of being moisture-insensitive with the fast reactivity. Then blossomed the strategy of the utilization of vegetable oil based polyols for the development of PU [305]. Today, the world of PU has come a long way from PU hybrids, PU composites, non-isocyanate PU, with versatile applications in diverse fields. PU has attracted wide attention due to their simple synthesis and application protocol, simple (few) basic reactants, and superior properties of the final product.

Isocyanates are essential components for PU synthesis, which are di- or polyfunctional isocyanates containing two or more –NCO groups per molecule and can be aliphatic, cycloaliphatic, polycyclic, or aromatic in nature. Scheme 8.9 shows examples of some common isocyanates.

In addition to bearing plurality of hydroxyl groups, polyols substances may also contain ester, ether, amide, acrylic, metal, metalloid, and other functionalities. For example, PEPs consist of ester and hydroxylic groups in one backbone.

Along with a polyol and an isocyanate, some additives including catalysts, chain extenders, cross-linkers, fillers, moisture scavengers, colorants, and others may also be required during PU production, primarily to control the reaction, modify the reaction conditions, and also to finish the final product.

PUs have excellent abrasion resistance and some properties of both rubber and plastics [309–311]. Their properties varies in numerous ways according to their demand. For example, in order to form branched or cross-linked polymers, the functionality of the hydroxyl compound as well as the isocyanate can be increased to three or more. A unique feature of PU is that a wide variety of structural changes could be brought about with different hydroxy compounds and isocyanates, leading to a wide spectrum of properties in PUs.

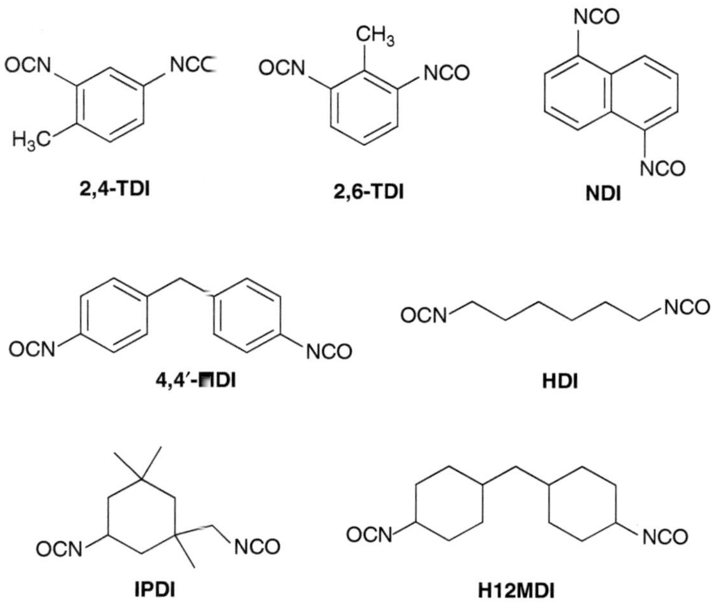

Scheme 8.9 Common isocyanates.

Originally, there are three states of matter: solid, liquid, and gas. The emergence of liquid crystal has been considered as one of the major breakthrough in polymer science [312]. Liquid crystal can be defined as an intermediate of solid (crystal) and liquid [313] where the molecules have the capabilities to flow like a liquid (mobility) as well as possessing the degree of order like solid [314]. Liquid crystalline polymers are synthesized from specific monomers containing mesogenic groups. The mesogenic moieties are contained in the main chain or in the side chains as pendant groups [315]. Thermotropic liquid crystalline PUs continue to draw attention due to their wide applications. Initial research on liquid crystalline polymers was focused on main chain thermotropic polyesters and lyotropic PAs [316]. Although PUs contribute to a broad spectrum of applications, liquid crystalline behavior in these materials is a recent development.

The preparation of liquid crystalline PU is performed essentially in two pathways Scheme 8.10: (i) a mesogenic diisocyanate with an aliphatic diol and (ii) a mesogenic diol with an aliphatic or aromatic diisocyanate. The latter is the main method, where the diol can be partially replaced by a polyol (flexible spacer). The polyaddition is carried out at a NCO/OH ratio of 1 : 1 in highly polar anhydrous solvents (dimethylformamide, dimethyl sulfoxide, chloroform, or dioxane) with or without catalysts such as dibutyltin dilaurate [312].

Properties of liquid crystalline PUs are investigated by spectrometric methods and the molecular weight distribution measurements, which do not differ from the chemical composition studies performed using common PUs. Besides, there are specific investigation methods for liquid crystalline PUs, among which the

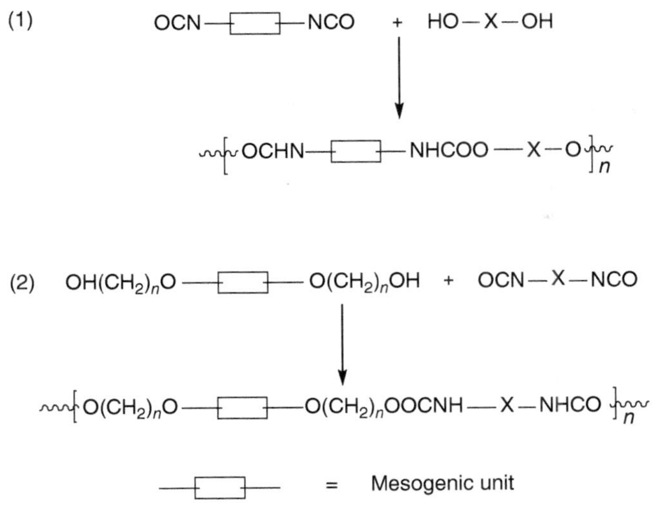

Scheme 8.10 Two synthetic pathways for the preparation of liquid crystalline polyurethanes. Reaction of (1) a mesogenic diisocyanate with an aliphatic diol and (2) a mesogenic diol with aromatic or aliphatic diols.

most important one is the determination of transition temperatures (melting and isotropization transition). For that purpose, differential scanning calorimetry (DSC), which can give the transition temperatures as well as the enthalpy values of the phase transitions, and polarization microscopy are used. In addition to the thermal transitions, morphology of liquid crystalline PUs is studied by polarizing microscopy equipped with a hot stage and by X-ray diffraction [312].

As commodity products, PUs have achieved a certain establishment status in academic science, but the activity in PU science shows no sign of abating due to its high potential for design and innovation. Initial PU consisted of foams and fibers, while today, colossal changes have happened in the world of PU. Attention is focused on green PU such as from non-isocyanate technologies, from biobased polyols and isocyanates, PU hybrids, PU composites, and so on. PU finds versatile applications as foams, adhesive, surface coatings, and sealants, to name a few.

8.4 Polyesters

As one of the most important polymers, polyesters are widely used in modern life, ranging from bottles for carbonated soft drinks and water to fibers for shirts and other apparel [317, 318] and to base for photographic film and recording tape. Household tradenames, such as Dacron®, Fortrel®, Terylene®, and Mylar®, reflect the high versatility and importance of polyesters [319].

Linear polyesters were first synthesized by Kroschwitz [320]. The first synthetic fiber was drawn from the molten polyester of trimethylene glycol and hexadecamethylene dicarboxylic acid. However, it had a low melting point and poor hydrolytic stability. Later, poly(ethylene terephthalate) (PET) with high melting

(265 °C) and good hydrolytic stability was discovered by Kroschwitz [320]. Subsequently, many other aromatic polyesters have been also synthesized. Among these polyesters, PET and poly(butylene terephthalate) (PBT) have been produced commercially for more than 50 years, and poly(trimethylene terephthalate) (PTT) was commercialized very recently.

The traditional way to synthesize polyesters has been by polycondensation using diols and a diacid (or an acid derivative) or from a hydroxy acid. The simplest polyesters are produced by the polycondensation reaction of a glycol (or dialcohol) with a difunctional carboxylic acid (or diacid). Hundreds of polyesters have been produced due to the myriad of combinations of dialcohols and diacids [319, 321]. Polyesters, especially PET, were manufactured as industrial products by ICI (UK, 1949) and Du Pont (USA, 1953) soon after the technology of manufacturing was developed by Whinfield and Dickson in 1946 [320]. PET, which is manufactured from ethylene glycol (EG) and TPA or dimethyl terephthalate (DMT), is widely used to produce synthetic fibers, films, beverage bottles, and molded plastic parts, because of its good physical properties [322]. Preparation of PET contains two steps: the first step, prepolymerization forming bis-(2-hydroxyethyl) terephthalate (BHET), the precursor for the second step, melt polycondensation (Scheme 8.11).

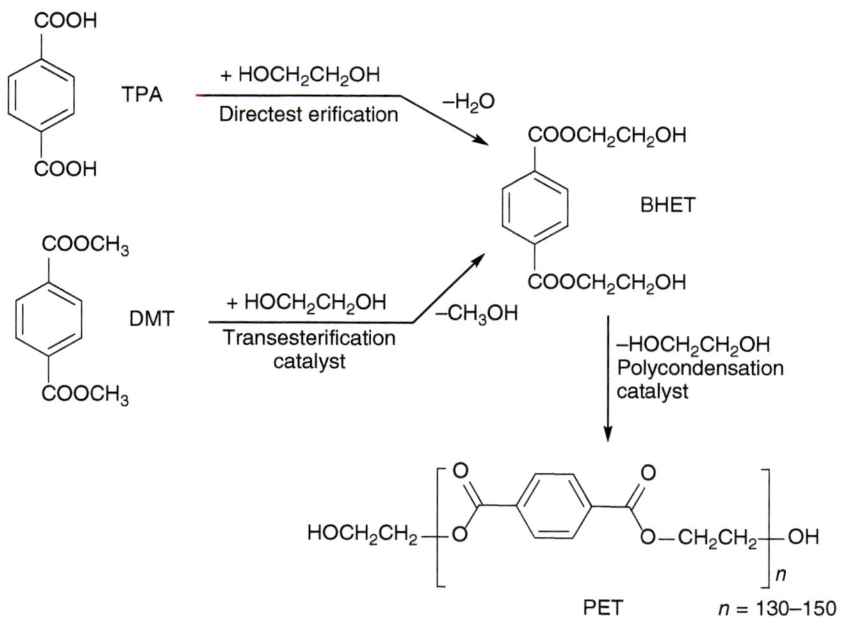

Scheme 8.11 Schemes for the PET polymerization process.

However, this method suffers from some shortcomings such as high temperature, long reaction times, removal of reaction by-products, and a precise stoichiometric balance between reactive acid and hydroxy groups. High conversion is desirable to get polymer of sufficiently high molecular masses and good mechanical properties.

However, it is very difficult to achieve a high degree of polymerization by this method because of side reactions and the volatilization of monomers, which leads to a stoichiometric imbalance of reactants.

One of the major challenges in polymer syntheses is to use renewable resources, and concomitantly to approach convenient synthetic pathways avoiding hazard and toxic substances [323]. Thus, sustainable catalytic approaches in polymer chemistry became an emerging research area with a tremendous impact on environmental and even economic issues [16], which draws the attention toward the possibility to synthesize polyesters through the use of green catalytic systems, mild polymerization conditions, and/or atom efficient approaches.

One example in polyester chemistry is the progressive replacement of the highly efficient (poly)transesterification antimonycatalysts, which were conventionally used in PET production [324], by other less toxic catalysts such as the titanium-based Lewis acid catalysts and other metal ones. For example, several furanic-based polyesters were synthesized by transesterification followed by polytransesterification using titanium(IV) tetraisopropoxide, titanium(IV) butoxide, tin(II) 2-ethyl hexanoate, or zirconium(IV) butoxide [8, 325, 326].

However, the use of these less aggressive catalysts in polyester synthesis is also hampered by the requirements for very high temperatures and reduced pressure to achieve high conversion yields and molecular weights. One approach to partially circumvent these hard conditions both applied to aromatic and aliphatic polyester syntheses is ring-opening polymerization (ROP) [322], which occurs at atmospheric pressure.

Another one, applied to the polycondensation (or just condensation) of aliphatic monomers, is to attain the (poly)condensation at the interface of the emulsion in water in the presence of a Brønsted acid surfactant catalyst [327, 328], in which the low reaction temperature is usually adopted, avoiding undesirable side reactions. Also, the aqueous media is obviously a safe, environmentally benign, and cheap solvent. One example [329] is the polycondensation of renewable-based long-chain aliphatic suberin monomers, which was conducted in water at 80 °C, for different reaction times, and using *p*-dodecylbenzenesulfonic acid as a surfactant/catalyst. The polyester products were, however, isolated in moderate yields (between 6% and 49%).

An inspiring strategy to polyester synthesis is definitely enzymatic polymerization, which is a highly selective reaction, occurs under mild and neutral conditions, at low temperature and in a quantitative conversion [16]. Hence, they sound the ideal conditions for any chemical reaction and are perfectly in tune with the "green chemistry" concept. However, these enzymatic-catalyzed reactions are not yet considered economically competitive [330] due to the high costs of the enzyme and the long reaction times. Hence, industry and scientists have to gain balance to find cheap and viable processes to enable enzyme use at a large scale in polyester synthesis.

Other promising approaches comprise the use of ionic liquids as reaction/catalyst media [331], or click chemistry [332] in polyester synthesis, but these approaches are still limited in some cases by the use of non-ecofriendly conditions.

In this regard, the use of biomass in the chemical industry is already a feasible reality, although on a small scale. Nevertheless, the continuous skyrocketing consumer demand for advanced plastics will definitely contribute to the steady conversion from petrochemical-based polyesters to sustainable and renewable-based polyesters.

Therefore, the development of sustainable polyesters from renewable resources will definitely continue to flourish since they are serious candidates to replace oil-based polymers. Notwithstanding the promising aspect of investigations related to polyesters from renewable resources, great challenges remain and there are plenty of possibilities in the future for innovation and environmentally friendlier polyesters [321].

Part III References

1 Maerten, E., Sauthier, M., Mortreux, A., and Castanet, Y. (2007). *Tetrahedron* 63: 682–689.

2 Luz, I., Corma, A., Llabres, I., and Xamena, F.X. (2014). *Catal. Sci. Technol.* 4: 1829–1836.

3 Šterk, D., Stephan, M.S., and Mohar, B. (2004). *Tetrahedron Lett.* 45: 535–537.

4 Deetz, J.S., Luehr, C.A., and Vallee, B.L. (1984). *Biochemistry* 23: 6822–6828.

5 Rodríguez, S., Kayser, M., and Stewart, J.D. (1999). *Org. Lett.* 1: 1153–1155.

6 Kaluzna, I.A., Matsuda, T., Sewell, A.K., and Stewart, J.D. (2004). *J. Am. Chem. Soc.* 126: 12827–12832.

7 Kaluzna, I.A., Feske, B.D., Wittayanan, W. et al. (2005). *J. Organomet. Chem.* 70: 342–345.

8 Ma, J., Yu, X., Xu, J., and Pang, Y. (2012). *Polymer* 53: 4145–4151.

9 Ma, S.M., Wu, B., and Zhao, S.M. (2003). *Org. Lett.* 5: 4429–4432.

10 Ma, W., Li, X., Yang, J. et al. (2006). *Synthesis* 2006: 2489–2492.

11 Zhang, W. and Shi, M. (2006). *Chem. Commun.*: 1218–1220.

12 Maerten, E., Hassouna, F., Couve-Bonnaire, S. et al. (2003). *Synlett*: 1874–1876.

13 Liu, Z.G., Li, N., Yang, L. et al. (2007). *Chin. Chem. Lett.* 18: 458–460.

14 Lledo, A., Fuster, A., Reves, M. et al. (2013). *Chem. Commun.* 49: 3055–3057.

15 Kim, J., De Castro, K.A., Lim, M., and Rhee, H. (2010). *Tetrahedron* 66: 3995–4001.

16 Loos, K. (2011). *Biocatalysis in Polymer Chemistry*. Weinheim, New Jersey: Wiley.

17 Lu, W., Yamaoka, Y., Taniguchi, Y. et al. (1999). *J. Organomet. Chem.* 580: 290–294.

18 Luo, J., Yu, H., Wang, H. et al. (2014). *Chem. Eng. J.* 240: 434–442.

19 Luo, R., Zhou, X., Fang, Y., and Ji, H. (2015). *Carbon* 82: 1–11.

20 Doyle, M.P., Debruyn, D.J., Donnelly, S.J. et al. (1974). *J. Organomet. Chem.* 39: 2740–2747.

21 Kambe, N., Kondo, K., Murai, S., and Sonoda, N. (1980). *Angew. Chem. Int. Ed.* 19: 1008–1009.

22 Cha, J.S., Kwon, O.O., and Kwon, S.Y. (1996). *Org. Prep. Proced. Int.* 28: 355–359.

Carbonyl Compounds: Reactants, Catalysts and Products, First Edition.
Feng Shi, Hongli Wang and Xingchao Dai.

23 Clerici, A., Pastori, N., and Porta, O. (2002). *Eur. J. Org. Chem.* 2002: 3326–3335.

24 Naimi-Jamal, M.R., Mokhtari, J., Dekamin, M.G., and Kaupp, G. (2009). *Eur. J. Org. Chem.* 2009: 3567–3572.

25 Shi, L., Liu, Y., Liu, Q. et al. (2012). *Green Chem.* 14: 1372–1375.

26 Langguth, S. (1905). *Ber. Dtsch. Chem. Ges.* 38: 2062–2064.

27 Kuramoto, T., Kihira, K., Matsumoto, N., and Hoshita, T. (1981). *Chem. Pharm. Bull.* 29: 1136–1139.

28 Malik, A.A. and Sharts, C.M. (1987). *Org. Prep. Proced. Int.* 19: 1–7.

29 Onyestyák, G., Harnos, S., and Kalló, D. (2011). *Catal. Commun.* 16: 184–188.

30 Harnos, S., Onyestyák, G., and Kalló, D. (2013). *Microporous Mesoporous Mater.* 167: 109–116.

31 Nagendra, G., Madhu, C., Vishwanatha, T.M., and Sureshbabu, V.V. (2012). *Tetrahedron Lett.* 53: 5059–5063.

32 Jaita, S., Kaewkum, P., Duangkamol, C. et al. (2014). *RSC Adv.* 4: 46947–46950.

33 Zeynizadeh, B., Sepehraddin, F., and Mousavi, H. (2019). *Ind. Eng. Chem. Res.* 58: 16379–16388.

34 Mettler, C. (1908). *Ber. Dtsch. Chem. Ges.* 41: 4148–4150.

35 Weil, H. (1908). *Ber. Dtsch. Chem. Ges.* 41: 4147–4148.

36 Muraki, M. and Mukaiyama, T. (1974). *Chem. Lett.* 3: 1447–1450.

37 Hubert, T.D., Eyman, D.P., and Wiemer, D.F. (1984). *J. Organomet. Chem.* 49: 2279–2281.

38 Falorni, M., Giacomelli, G., Porcheddu, A., and Taddei, M. (1999). *J. Organomet. Chem.* 64: 8962–8964.

39 Naqvi, T., Bhattacharya, M., and Haq, W. (1999). *J. Chem. Res.*: 424–425.

40 Fujihara, T., Cong, C., Terao, J., and Tsuji, Y. (2013). *Adv. Synth. Catal.* 355: 3420–3424.

41 Ivkovic, J., Lembacher-Fadum, C., and Breinbauer, R. (2015). *Org. Biomol. Chem.* 13: 10456–10460.

42 Sato, F., Jinbo, T., and Sato, M. (1980). *Tetrahedron Lett.* 21: 2175–2178.

43 Santaniello, E., Ferraboschi, P., and Sozzani, P. (1981). *J. Organomet. Chem.* 46: 4584–4585.

44 Narasimhan, S., Madhavan, S., and Prasad, K.G. (1997). *Synth. Commun.* 27: 385–390.

45 da Costa, J.C.S., Pais, K.C., Fernandes, E.L. et al. (2006). *ARKIVOC* 1: 128–133.

46 Boechat, N., da Costa, J.C.S., de Souza Mendonça, J. et al. (2004). *Tetrahedron Lett.* 45: 6021–6022.

47 Xu, Y. and Wei, Y. (2010). *Synth. Commun.* 40: 3423–3429.

48 Berk, S.C., Kreutzer, K.A., and Buchwald, S.L. (1991). *J. Am. Chem. Soc.* 113: 5093–5095.

49 Ohta, T., Kamiya, M., Kusui, K. et al. (1999). *Tetrahedron Lett.* 40: 6963–6966.

50 Saudan, L.A., Saudan, C.M., Debieux, C., and Wyss, P. (2007). *Angew. Chem. Int. Ed.* 46: 7473–7476.

51 Das, S., Moller, K., Junge, K., and Beller, M. (2011). *Chem. Eur. J.* 17: 7414–7417.

52 Junge, K., Wendt, B., Zhou, S., and Beller, M. (2013). *Eur. J. Org. Chem.* 2013: 2061–2065.

53 Rysak, V., Descamps-Mandine, A., Simon, P. et al. (2018). *Catal. Sci. Technol.* 8: 3504–3512.

54 Morra, N.A. and Pagenkopf, B.L. (2008). *Synthesis* 2008: 511–514.

55 Sakai, N., Moriya, T., Fujii, K., and Konakahara, T. (2008). *Synthesis* 2008: 3533–3536.

56 Hosokawa, S., Toya, M., Noda, A. et al. (2018). *ChemistrySelect* 3: 2958–2961.

57 Yamada, S.I., Kikugawa, Y., and Ikegami, S. (1965). *Chem. Pharm. Bull.* 13: 394–396.

58 Attaurrahman, Basha, A., Waheed, N., and Ahmed, S. (1976). *Tetrahedron Lett.* 17: 219–222.

59 Brown, H.C. and Heim, P. (1973). *J. Org. Chem.* 38: 912–916.

60 Akabori, S. and Takanohashi, Y. (1991). *J. Chem. Soc. Perkin Trans.* 1: 479–482.

61 Mukherjee, D., Shirase, S., Mashima, K., and Okuda, J. (2016). *Angew. Chem. Int. Ed.* 55: 13326–13329.

62 Zhou, S., Junge, K., Addis, D. et al. (2009). *Angew. Chem. Int. Ed.* 48: 9507–9510.

63 Das, S., Addis, D., Zhou, S. et al. (2010). *J. Am. Chem. Soc.* 132: 1770–1771.

64 Cheng, C. and Brookhart, M. (2012). *J. Am. Chem. Soc.* 134: 11304–11307.

65 Li, Y., Molina de La Torre, J.A., Grabow, K. et al. (2013). *Angew. Chem. Int. Ed.* 52: 11577–11580.

66 Lampland, N.L., Hovey, M., Mukherjee, D., and Sadow, A.D. (2015). *ACS Catal.* 5: 4219–4226.

67 Snelling, R.A., Amberchan, G., Resendez, A. et al. (2017). *Tetrahedron Lett.* 58: 4073–4077.

68 Clemmensen, E. (1913). *Chem. Ber.* 46: 1837–1843.

69 Toda, M., Hirata, Y., and Yamamura, S. (1969). *J. Chem. Soc. D*: 919–920.

70 Toda, M., Hayashi, M., Hirata, Y., and Yamamura, S. (1972). *Bull. Chem. Soc. Jpn.* 45: 264–266.

71 Yamamura, S. and Hirata, Y. (1968). *J. Chem. Soc. C*: 2887–2889.

72 Yamamura, S., Ueda, S., and Hirata, Y. (1967). *Chem. Commun.*: 1049–1050.

73 Naruse, M., Aoyagi, S., and Kibayashi, C. (1994). *Tetrahedron Lett.* 35: 9213–9216.

74 Kishner, N. (1911). *Russ. Phys. Chem. Soc.* 43: 582–583.

75 Wolff, L. (1912). *Justus Liebigs Ann. Chem.* 394: 86–108.

76 Overman, L.E., Ricca, D.J., and Tran, V.D. (1993). *J. Am. Chem. Soc.* 115: 2042–2044.

77 Marino, J.P., Rubio, M.B., Cao, G., and De Dios, A. (2002). *J. Am. Chem. Soc.* 124: 13398–13399.

78 Kawano, M., Kiuchi, T., Negishi, S. et al. (2013). *Angew. Chem. Int. Ed.* 52: 906–910.

79 Miyaoka, H., Kajiwara, Y., Hara, Y., and Yamada, Y. (2001). *J. Organomet. Chem.* 66: 1429–1435.

80 Herr, C.H., Whitmore, F.C., and Schiessler, R.W. (1945). *J. Am. Chem. Soc.* 67: 2061–2063.

81 Soffer, M.D., Soffer, M.B., and Sherk, K.W. (1945). *Justus Liebigs Ann. Chem.* 67: 1435–1436.

82 Huang-Minlon (1946). *J. Am. Chem. Soc.* 68: 2487–2488.

83 Huang-Minlon (1949). *J. Am. Chem. Soc.* 71: 3301–3303.

84 Osdene, T.S., Timmis, G.M., Maguire, M.H. et al. (1955). *J. Chem. Soc.*: 2038–2056.

85 Kuethe, J.T., Childers, K.G., Peng, Z. et al. (2009). *Org. Process. Res. Dev.* 13: 576–580.

86 Cornils, B., Herrmann, W.A., Muhler, M., and Wong, C.-H. (2007). *Catalysis from A to Z: A Concise Encyclopedia*. Weinheim: Wiley-VCH.

87 Kuehne, M.E. and Xu, F. (1998). *J. Organomet. Chem.* 63: 9434–9439.

88 Reformatsky, S. (1887). *Ber. Dtsch. Chem. Ges.* 20: 1210–1211.

89 Vedejs, E. and Duncan, S.M. (2000). *J. Organomet. Chem.* 65: 6073–6081.

90 Knight, R.L. and Leeper, F.J. (1998). *J. Chem. Soc. Perkin Trans.* 1: 1891–1894.

91 Davis, J.H. and Forrester, K.J. (1999). *Tetrahedron Lett.* 40: 1621–1622.

92 White, M.J. and Leeper, F.J. (2001). *J. Organomet. Chem.* 66: 5124–5131.

93 Miyashita, A., Suzuki, Y., Okumura, Y. et al. (1998). *Chem. Pharm. Bull.* 46: 6–11.

94 Vollhardt, K.P.C. and Schore, N.E. (2018). *Organic Chemistry: Structure and Function*. New York, USA: W. H. Freeman and Company.

95 Yan, G., Wu, X., and Yang, M. (2013). *Org. Biomol. Chem.* 11: 5558–5578.

96 Yang, L. and Huang, H. (2015). *Chem. Rev.* 115: 3468–3517.

97 Zhang, X.-S., Chen, K., and Shi, Z.-J. (2014). *Chem. Sci.* 5: 2146–2159.

98 Yang, L., Correia, C.A., and Li, C.-J. (2011). *Adv. Synth. Catal.* 353: 1269–1273.

99 Kuninobu, Y., Nishina, Y., Takeuchi, T., and Takai, K. (2007). *Angew. Chem. Int. Ed.* 46: 6518–6520.

100 Wang, G.W., Zhou, A.X., Wang, J.J. et al. (2013). *Org. Lett.* 15: 5270–5273.

101 Jo, H., Park, J., Choi, M. et al. (2016). *Adv. Synth. Catal.* 358: 2714–2720.

102 Jin, J.-J., Niu, H.-Y., Qu, G.-R. et al. (2012). *RSC Adv.* 2: 5968–5971.

103 Graves, V.B. and Shaikh, A. (2013). *Tetrahedron Lett.* 54: 695–698.

104 Niu, R., Yang, S., Xiao, J. et al. (2012). *Chin. J. Catal.* 33: 1636–1641.

105 Mao, D., Hong, G., Wu, S. et al. (2014). *Eur. J. Org. Chem.* 2014: 3009–3019.

106 Kane, R. (1838). *J. Prakt. Chem.* 15: 129–130.

107 Claisen, L. and Claparede, A. (1881). *Ber. Dtsch. Chem. Ges.* 14: 349–353.

108 Claisen, L. (1887). *Ber. Dtsch. Chem. Ges.* 20: 655–657.

109 White, J.D., Blakemore, P.R., Green, N.J. et al. (2002). *J. Organomet. Chem.* 67: 7750–7760.

110 Yoshikawa, N., Yamada, Y.M.A., Das, J. et al. (1999). *J. Am. Chem. Soc.* 121: 4168–4178.

111 Fujii, K., Maki, K., Kanai, M., and Shibasaki, M. (2003). *Org. Lett.* 5: 733–736.

112 Perkin, W.H. (1868). *J. Chem. Soc.* 21: 181–186.

113 Perkin, W.H. (1877). *J. Chem. Soc.* 31: 388–427.

114 Baker, E.C., Hendriksen, D.E., and Eisenberg, R. (1980). *J. Am. Chem. Soc.* 102: 1020–1027.

115 Federsel, H.-J. (2000). *Org. Process Res. Dev.* 4: 362–369.

116 Knoevenagel, E. (1894). *Ber.* 27: 2345–2346.

117 Becke, F., Fleig, H., and Päßler, P. (1971). *Liebigs Ann. Chem.* 749: 198–201.

118 Tietze, L.F. and Zhou, Y. (1999). *Angew. Chem. Int. Ed.* 38: 2045–2047.

119 Snider, B.B. and Lu, Q. (1996). *J. Org. Chem.* 61: 2839–2844.

120 Bayer, A. (1872). *Ber. Dtsch. Chem. Ges.* 5: 1094–1100.

121 Baeyer, A. and Villiger, V. (1899). *Ber. Dtsch. Chem. Ges.* 32: 3625–3633.

122 Oh, J. (1997). *Tetrahedron Lett.* 38: 3249–3250.

123 Świzdor, A. (2013). *Molecules* 18: 13812–13822.

124 Wallace, T.J., Pobiner, H., and Schriesc, A. (1965). *J. Organomet. Chem.* 30: 3768–3771.

125 Corey, E.J., Gilman, N.W., and Ganem, B.E. (1968). *J. Am. Chem. Soc.* 90: 5616–5617.

126 Abiko, A., Roberts, J.C., Takemasa, T., and Masamune, S. (1986). *Tetrahedron Lett.* 27: 4537–4540.

127 Brzaszcz, M., Kloc, K., Maposah, M., and Mlochowski, J. (2000). *Synth. Commun.* 30: 4425–4434.

128 Takemoto, T., Yasuda, K., and Ley, S.V. (2001). *Synlett*: 1555–1556.

129 Qian, H., Shao, L.X., and Huang, X. (2002). *J. Chem. Res.*: 514–515.

130 Travis, B.R., Sivakumar, M., Hollist, G.O., and Borhan, B. (2003). *Org. Lett.* 5: 1031–1034.

131 Joseph, J.K., Jain, S.L., and Sain, B. (2007). *Catal. Commun.* 8: 83–87.

132 Zhou, L.P., Dong, B.B., Tang, S. et al. (2013). *J. Energy Chem.* 22: 659–664.

133 Liu, M.X., Wang, H.N., Zeng, H.Y., and Li, C.J. (2015). *Sci. Adv.* 1: 1–9.

134 Jiang, X., Zhai, Y., Chen, J. et al. (2018). *Chin. J. Chem.* 36: 15–19.

135 Dai, P.F., Qu, J.P., and Kang, Y.B. (2019). *Org. Lett.* 21: 1393–1396.

136 Wittig, G. and Geissler, G. (1953). *Liebigs Ann. Chem.* 580: 44–57.

137 Murphy, P.J. and Brennan, J. (1988). *Chem. Soc. Rev.* 17: 1–30.

138 Molina, P., Murcia, F., and Fresneda, P.M. (1994). *Tetrahedron Lett.* 35: 1453–1456.

139 Molina, P., Alcántara, J., and López-Leonardo, C. (1996). *Tetrahedron* 52: 5833–5584.

140 Csányi, D., Hajós, G., Riedl, Z. et al. (2000). *Bioorg. Med. Chem. Lett.* 10: 1767–1769.

141 Samuelsson, G. (1992). *Drugs of Natural Origin*. Stockholm: Swedish Pharmaceutical.

142 Lebaron, H.M., Mcfarland, J.E., and Simoneaux, B.J. (1988). *Herbicides: Chemistry, Degradation, and Mode of Action* (eds. P.C. Kearney and D.D. Kaufman), 335. New York, ; Chapter 7: Marcel Dekker.

143 Sharp, D.B. (1988). *Herbicides: Chemistry, Degradation, and Mode of Action* (eds. P.C. Kearney and D.D. Kaufman), 301. New York, ; Chapter 7: Marcel Dekker.

144 Lane, C.F. (1975). *Synthesis*: 135–146.

145 Borch, R.F. and Hassid, A.I. (1972). *J. Organomet. Chem.* 37: 1673–1674.

146 Abdelmagid, A.F. and Maryanoff, C.A. (1990). *Synlett*: 537–539.

147 AbdelMagid, A.F., Carson, K.G., Harris, B.D. et al. (1996). *J. Organomet. Chem.* 61: 3849–3862.

148 Ranu, B.C., Majee, A., and Sarkar, A. (1998). *J. Organomet. Chem.* 63: 370–373.

149 Tararov, V.I., Kadyrov, R., Riermeier, T.H., and Borner, A. (2000). *Chem. Commun.*: 1867–1868.

150 Apodaca, R. and Xiao, W. (2001). *Can. J. Chem.* 3: 1745–1748.

151 Kato, H., Shibata, I., Yasaka, Y. et al. (2006). *Chem. Commun.*: 4189–4191.

152 Basu, B., Jha, S., Bhuiyan, M.H., and Das, P. (2003). *Synlett*: 555–557.

153 Cho, B.T. and Kang, S.K. (2005). *Tetrahedron* 61: 5725–5734.

154 Heydari, A., Khaksar, S., Akbari, J. et al. (2007). *Tetrahedron Lett.* 48: 1135–1138.

155 Zhang, Q., Zhu, W., Yao, J. et al. (2018). *Org. Biomol. Chem.* 16: 8462–8466.

156 Hu, L.A., Zhang, Y., Zhang, Q.W. et al. (2020). *Angew. Chem. Int. Ed.* 59: 5321–5325.

157 Tajbakhsh, M., Lakouraj, M.M., and Mahalli, M.S. (2008). *Synth. Commun.* 38: 1976–1983.

158 Alinezhad, H., Tajbakhsh, M., and Zare, M. (2009). *Synth. Commun.* 39: 2907–2916.

159 Shokrolahi, A., Zali, A., and Keshavarz, M.H. (2011). *Green Chem. Lett. Rev.* 4: 195–203.

160 Qi, F., Hu, L., Lu, S. et al. (2012). *Chem. Commun.* 48: 9631–9633.

161 Mirza-Aghayan, M., Tavana, M.M., Rahimifard, M., and Boukherroub, R. (2014). *Appl. Organomet. Chem.* 28: 113–115.

162 Nasrollahzadeh, M. (2014). *New J. Chem.* 38: 5544–5550.

163 Park, J.W. and Chung, Y.K. (2015). *ACS Catal.* 5: 4846–4850.

164 Liang, S., Monsen, P., Hammond, G.B., and Xu, B. (2016). *Org. Chem. Front.* 3: 505–509.

165 Yang, W., Wei, L., Yi, F., and Cai, M. (2016). *Catal. Sci. Technol.* 6: 4554–4564.

166 Liang, G., Wang, A., Li, L. et al. (2017). *Angew. Chem. Int. Ed.* 56: 3050–3054.

167 Jagadeesh, R.V., Murugesan, K., Alshammari, A.S. et al. (2017). *Science* 358: 326–322.

168 Brown, H.C. and Kabalka, G.W. (1970). *J. Am. Chem. Soc.* 92: 714–716.

169 Koren-Selfridge, L., Londino, H.N., Vellucci, J.K. et al. (2009). *Organometallics* 28: 2085–2090.

170 Arrowsmith, M., Hadlington, T.J., Hill, M.S., and Kociok-Kohn, G. (2012). *Chem. Commun.* 48: 4567–4569.

171 Bagherzadeh, S. and Mankad, N.P. (2016). *Chem. Commun.* 52: 3844–3846.

172 Zeng, H.S., Wu, J., Li, S.H. et al. (2019). *Org. Lett.* 21: 401–406.

173 Tamang, S.R. and Findlater, M. (2017). *J. Organomet. Chem.* 82: 12857–12862.

174 Zhang, G.Q., Cheng, J., Davis, K. et al. (2019). *Green Chem.* 21: 1114–1121.

175 Tamang, S.R., Bedi, D., Shafiei-Haghighi, S. et al. (2018). *Org. Lett.* 20: 6695–6700.

176 Wu, Y.L., Shan, C.K., Ying, J.X. et al. (2017). *Green Chem.* 19: 4169–4175.

177 Zhu, Z.Y., Wu, X.L., Xu, X.J. et al. (2018). *J. Organomet. Chem.* 83: 10677–10683.

178 Jahani, F., Zamenian, B., Khaksar, S., and Tajbakhsh, M. (2010). *Synthesis* 2010: 3315–3318.

179 Zhou, S.L., Wu, Z.S., Rong, J.W. et al. (2012). *Chem. Eur. J.* 18: 2653–2659.

180 Zhou, S.L., Wang, H.Y., Ping, J. et al. (2012). *Organometallics* 31: 1696–1702.

181 Miao, H., Zhou, S.L., Wang, S.W. et al. (2013). *Sci. China Chem.* 56: 329–336.

182 Liu, C.W., Zhang, Y., Qian, Q.Q. et al. (2014). *Org. Lett.* 16: 6172–6175.

183 Zhao, L., Ding, H., Zhao, B. et al. (2014). *Polyhedron* 83: 50–59.

184 Nie, K., Liu, C.W., Zhang, Y., and Yao, Y.M. (2015). *Sci. China Chem.* 58: 1451–1460.

185 Boyer, J., Corriu, R.J.P., Perz, R., and Reye, C. (1979). *J. Organomet. Chem.* 172: 143–152.

186 Goldberg, Y., Abele, E., Shymanska, M., and Lukevics, E. (1989). *J. Organomet. Chem.* 372: C9–C11.

187 Lipshutz, B.H., Chrisman, W., and Noson, K. (2001). *J. Organomet. Chem.* 624: 367–371.

188 Teci, M., Lentz, N., Brenner, E. et al. (2015). *Dalton Trans.* 44: 13991–13998.

189 Fernandes, A.C., Fernandes, R., Romao, C.C., and Royo, B. (2005). *Chem. Commun.*: 213–214.

190 Reis, P.M., Romao, C.C., and Royo, B. (2006). *Dalton Trans.*: 1842–1846.

191 Chakraborty, S., Krause, J.A., and Guan, H. (2009). *Organometallics* 28: 582–586.

192 Truong, T.V., Kastl, E.A., and Du, G. (2011). *Tetrahedron Lett.* 52: 1670–1672.

193 Chouthaiwale, P.V., Rawat, V., and Sudalai, A. (2012). *Tetrahedron Lett.* 53: 148–150.

194 Bleith, T., Wadepohl, H., and Gade, L.H. (2015). *J. Am. Chem. Soc.* 137: 2456–2459.

195 Sattler, W., Ruccolo, S., Chaijan, M.R. et al. (2015). *Organometallics* 34: 4717–4731.

196 Chan, A. and Scheidt, K.A. (2006). *J. Am. Chem. Soc.* 128: 4558–4559.

197 Phan, D.H.T., Kim, B., and Dong, V.M. (2009). *J. Am. Chem. Soc.* 131: 15608–15609.

198 Murphy, S.K. and Dong, V.M. (2013). *J. Am. Chem. Soc.* 135: 5553–5556.

199 Yang, J. and Yoshikai, N. (2014). *J. Am. Chem. Soc.* 136: 16748–16751.

200 Li, T., Xu, B.H., Zhu, D.P. et al. (2018). *Org. Chem. Front.* 5: 1933–1939.

201 Leffingw, J.C. (1970). *Chem. Commun.*: 357–358.

202 Leavell, K.H. and Lewis, E.S. (1972). *J. Organomet. Chem.* 37: 918–919.

203 Bouma, M., Masson, G., and Zhu, J. (2010). *J. Organomet. Chem.* 75: 2748–2751.

204 Wu, Y., Li, B., Mao, F. et al. (2011). *Org. Lett.* 13: 3258–3261.

205 Rout, S.K., Guin, S., Ghara, K.K. et al. (2012). *Org. Lett.* 14: 3982–3985.

206 Zhang, C., Zong, X., Zhang, L., and Jiao, N. (2012). *Org. Lett.* 14: 3280–3283.

207 Barve, B.D., Wu, Y.-C., El-Shazly, M. et al. (2014). *J. Organomet. Chem.* 79: 3206–3214.

208 Gong, W., Xu, L., Ji, T. et al. (2014). *RSC Adv.* 4: 6854–6857.

209 Wang, Q., Zheng, H., Chai, W. et al. (2014). *Org. Biomol. Chem.* 12: 6549–6553.

210 Zhang, C. and Jiao, N. (2014). *Org. Chem. Front.* 1: 109–112.

211 Saberi, D., Shojaeyan, F., and Niknam, K. (2016). *Tetrahedron Lett.* 57: 566–569.

212 Nandi, G.C. and Raju, C. (2017). *Org. Biomol. Chem.* 15: 2334–2339.

213 Wang, C.-S., Roisnel, T., Dixneuf, P.H., and Soule, J.-F. (2017). *Org. Lett.* 19: 6720–6723.

214 Zhou, B., Guo, S., Fang, Z. et al. (2019). *Tetrahedron Lett.* 60: 150914–150917.

215 Tan, P.W., Juwaini, N.A.B., and Seayad, J. (2013). *Org. Lett.* 15: 5166–5169.

216 Soeta, T., Tabatake, Y., Fujinami, S., and Ukaji, Y. (2013). *Org. Lett.* 15: 2088–2091.

217 Ke, Q., Zhang, B., Hu, B. et al. (2015). *Chem. Commun.* 51: 1012–1015.

218 Xu, X., Li, P., Huang, Y. et al. (2017). *Tetrahedron Lett.* 58: 1742–1746.

219 Kegns, S., Mielby, J., Mentzel, U.V. et al. (2012). *Chem. Commun.* 48: 2427–2429.

220 Han, J., Zhang, S., Zhang, J., and Yan, R. (2014). *RSC Adv.* 4: 58769–58772.

221 Thanh, T., Dang, G.H., Tran, N.V. et al. (2015). *J. Mol. Catal. A Chem.* 409: 110–116.

222 Hu, J., Xia, F., Yang, F. et al. (2017). *RSC Adv.* 7: 41204–41209.

223 Ma, Z., Song, T., Yuan, Y., and Yang, Y. (2019). *Chem. Sci.* 10: 10283–10289.

224 Rautenstrauch, V. (1975). *Synthesis* 1975: 787–788.

225 Ito, K., Nakanishi, S., and Otsuji, Y. (1980). *Chem. Lett.*: 1141–1144.

226 Periasamy, M., Srinivas, G., Karunakar, G.V., and Bharathi, P. (1999). *Tetrahedron Lett.* 40: 7577–7580.

227 Li, J.T., Yang, J.H., Han, J.F., and Li, T.S. (2003). *Green Chem.* 5: 433–435.

228 Nakajima, M., Fava, E., Loescher, S. et al. (2015). *Angew. Chem. Int. Ed.* 54: 8828–8832.

229 Wang, R., Ma, M., Gong, X. et al. (2019). *Org. Lett.* 21: 27–31.

230 Mahandru, G.M., Liu, G., and Montgomery, J. (2004). *J. Am. Chem. Soc.* 126: 3698–3699.

231 Malik, H.A., Sormunen, G.J., and Montgomery, J. (2010). *J. Am. Chem. Soc.* 132: 6304–6305.

232 Liu, P., Montgomery, J., and Houk, K.N. (2011). *J. Am. Chem. Soc.* 133: 6956–6959.

233 Haynes, M.T. II, Liu, P., Baxter, R.D. et al. (2014). *J. Am. Chem. Soc.* 136: 17495–17504.

234 Shiomi, T., Adachi, T., Ito, J., and Nishiyama, H. (2009). *Org. Lett.* 11: 1011–1014.

235 Swyka, R.A., Shuler, W.G., Spinello, B.J. et al. (2019). *J. Am. Chem. Soc.* 141: 6864–6868.

236 Arado, O.D., Moenig, H., Franke, J.-H. et al. (2015). *Chem. Commun.* 51: 4887–4890.

237 Lee, K.N., Lei, Z., and Ngai, M.-Y. (2017). *J. Am. Chem. Soc.* 139: 5003–5006.

238 Cao, J., Wang, G., Gao, L. et al. (2018). *Chem. Sci.* 9: 3664–3671.

239 Ke, Z., Yang, Z., Liu, Z. et al. (2018). *Org. Lett.* 20: 6622–6626.

240 Li, Y.-L., Li, W.-D., Gu, Z.-Y. et al. (2019). *ACS Catal.* 10: 1528–1534.
241 Hegedus, L.S. and Wade, L. (1977). *Compendium of Organic Synthetic Methods*, vol. 3, 8–32. New York: Wiley.
242 Goossen, L.J., Rodriguez, N., and Goossen, K. (2008). *Angew. Chem. Int. Ed.* 47: 3100–3120.
243 Rodriguez, N. and Goossen, L.J. (2011). *Chem. Soc. Rev.* 40: 5030–5048.
244 Kolbe, H. (1848). *Liebigs Ann. Chem.* 64: 339–341.
245 Fichter, F. and Stenzl, H. (1939). *Helv. Chim. Acta* 22: 970–978.
246 Wilshire, J. (1963). *Aust. J. Chem.* 16: 432–439.
247 Adamek, J., Mazurkiewicz, R., Pazdzierniok-Holewa, A. et al. (2014). *J. Organomet. Chem.* 79: 2765–2770.
248 Mazurkiewicz, R., Adamek, J., Pazdzierniok-Holewa, A. et al. (2012). *J. Organomet. Chem.* 77: 1952–1960.
249 Wang, H.-B. and Huang, J.-M. (2016). *Adv. Synth. Catal.* 358: 1975–1981.
250 Myers, A.G., Tanaka, D., and Mannion, M.R. (2002). *J. Am. Chem. Soc.* 124: 11250–11251.
251 Goossen, L.J., Deng, G., and Levy, L.M. (2006). *Science* 313: 662–664.
252 Bi, H.P., Zhao, L., Liang, Y.M., and Li, C.J. (2009). *Angew. Chem. Int. Ed.* 48: 792–795.
253 Yu, W.Y., Sit, W.N., Zhou, Z.Y., and Chan, A.S.C. (2009). *Org. Lett.* 11: 3174–3177.
254 Duan, Z., Ranjit, S., Zhang, P., and Liu, X. (2009). *Chem. Eur. J.* 15: 3666–3669.
255 Fang, P., Li, M., and Ge, H. (2010). *J. Am. Chem. Soc.* 132: 11898–11899.
256 Sharma, S., Khan, I.A., and Saxena, A.K. (2013). *Adv. Synth. Catal.* 355: 673–678.
257 Li, D., Wang, M., Liu, J. et al. (2013). *Chem. Commun.* 49: 3640–3642.
258 Liu, J., Liu, Q., Yi, H. et al. (2014). *Angew. Chem. Int. Ed.* 53: 502–506.
259 Zhang, J.X., Wang, Y.J., Zhang, W. et al. (2014). *Sci. Rep.* 4: 7446–7450.
260 Kan, J., Huang, S., Lin, J. et al. (2015). *Angew. Chem. Int. Ed.* 54: 2199–2203.
261 Merritt, E.A. and Olofsson, B. (2011). *Eur. J. Org. Chem.* 2011: 3690–3694.
262 Bouma, M.J. and Olofsson, B. (2012). *Chem. Eur. J.* 18: 14242–14245.
263 Zhou, Q.-Q., Guo, W., Ding, W. et al. (2015). *Angew. Chem. Int. Ed.* 54: 11196–11199.
264 Tang, J., Biafora, A., and Goossen, L.J. (2015). *Angew. Chem. Int. Ed.* 54: 13130–13133.
265 Schwarz, J. and König, B. (2016). *Green Chem.* 18: 4743–4749.
266 Wan, W., Ma, G., Li, J. et al. (2016). *Chem. Commun.* 52: 1598–1601.
267 Fu, M.C., Shang, R., Zhao, B. et al. (2019). *Science* 363: 1429–1434.
268 Mega, R.S., Duong, V.K., Noble, A., and Aggarwal, V.K. (2020). *Angew. Chem. Int. Ed.* 59: 4375–4379.
269 Paula, Y.B. (2017). *Organic Chemistry*. San Francisco, USA: Pearson Education, Inc.
270 Marchildon, K. (2011). *Macromol. React. Eng.* 5: 22–54.
271 Jahnke, T.S. (1996). *J. Am. Chem. Soc.* 118: 8186–8186.
272 Winnacker, M. (2017). *Biomater. Sci.* 5: 1230–1235.

273 Zimmerman, J (1990). *Polyamides*, vol. 11 (ed. J.I. Kroschwitz), 315–380. Weinheim, Germany: Wiley-VCH.

274 Horn, C., Freure, B., Vineyard, H., and Decker, H. (1963). *J. Appl. Polym. Sci.* 7: 887–896.

275 Fukada, E. (2006). *IEEE Trans. Dielectr. Electr. Insul.* 13: 1110–1119.

276 Staudinger, H. (1920). *Ber. Dtsch. Chem. Ges.* 53: 1073–1085.

277 Bolton, E.K. (1942). *J. Soc. Dye. Colour.* 58: 71–71.

278 Hounshell, D.A., Smith, J.K., and Smith, J.V. (1989). *Science and Corporate Strategy: Du Pont R and D, 1902–1980*, vol. 13. Cambridge, England: Cambridge University Press.

279 Wang, W., Wang, X., Li, R. et al. (2009). *J. Appl. Polym. Sci.* 114: 2036–2042.

280 García, J.M., García, F.C., Serna, F., and de la Peña, J.L. (2010). *Prog. Polym. Sci.* 35: 623–636.

281 Reglero Ruiz, J.A., Trigo-Lopez, M., Garcia, F.C., and Garcia, J.M. (2017). *Polymers* 9: 1–44.

282 Olabisi, O. and Adewale, K. (2016). *Handbook of Thermoplastics*. Boca Raton, Florida: CRC Press.

283 Trigo-López, M., Estévez, P., San-José, N. et al. (2009). *Recent Pat. Mater. Sci.* 2: 190–208.

284 Novitsky, T.F., Lange, C.A., Mathias, L.J. et al. (2010). *Polymer* 51: 2417–2425.

285 Wang, W. and Zhang, Y. (2009). *Express Polym. Lett.* 3: 470–476.

286 Morgan, P. and Kwolek, S. (1975). *Macromolecules* 8: 104–111.

287 Jackson, J. (1969). *Polymer* 10: 159–165.

288 Hofmann, A., Fritz, H.-G., and Kaiser, R. (2002). Google Patents: The United States Patent and Trademark Office, WO02072669A1.

289 Liu, H., Yang, G., He, A., and Wu, M. (2004). *J. Appl. Polym. Sci.* 94: 819–826.

290 Quynn, R.G. (1970). *Text. Res. J.* 40: 677–682.

291 Gorton, B. (1965). *J. Appl. Polym. Sci.* 9: 3753–3758.

292 Dodiuk, H. and Goodman, S.H. (1999). *Handbook of Thermoset Plastics*. Amsterdam: Elsevier.

293 Knop, A. and Scheib, W. (1979). *Chemistry and Application of Phenolic Resins*, vol. 3. Berlin, Germany: Springer.

294 Arno, G., Louis, A.P., and Andre, K. (2000). *Phenolic Resins: Chemistry, Applications, Standardization, Safety and Ecology*. Berlin Heidelberg: Springer-Verlag.

295 Baekeland, L.H. (1935). *Ind. Eng. Chem.* 27: 538–543.

296 Baekeland, L.H. (1915). Google Patents, US1156452A.

297 Baekeland, L.H. (1909). Google Patents, US942699A.

298 Pilato, L. (2010). *Phenolic Resins: A Century of Progress*. Berlin Heidelberg: Springer-Verlag.

299 Pizzi, A. (1986). *Wood Adhesives: Chemistry and Technology*. Routledge.

300 Pickrell, J.A., Mokler, B.V., Griffis, L.C. et al. (1983). *Environ. Sci. Technol.* 17: 753–757.

301 Babich, H. and Davis, D. (1981). *Regul. Toxicol. Pharmacol.* 1: 90–109.

302 Knop, A. and Pilato, L.A. (2013). *Phenolic Resins Chemistry, Applications and Performance*. New York: Springer Science & Business Media.

303 Gardziella, A., Pilato, L.A., and Knop, A. (2013). *Phenolic Resins Chemistry, Applications, Standardization, Safety and Ecology*. New York: Springer Science & Business Media.

304 Li, C., Wang, W., Mu, Y. et al. (2017). *J. Polym. Environ.* 26: 1297–1309.

305 Sawpan, M.A. (2018). *J. Polym. Res.* 25: 1–15.

306 Zhang, C., Garrison, T.F., Madbouly, S.A., and Kessler, M.R. (2017). *Prog. Polym. Sci.* 71: 91–143.

307 Dieterich, D. and Uhlig, K. (2000). *J. Appl. Polym. Sci.* 65: 1217–1225.

308 Ionescu, M. (2005). *Chemistry and Technology of Polyols for Polyurethanes*. Shrewsbury, UK: Smithers Rapra Technology.

309 Zafar, F. and Sharmin, E. (2012). *Polyurethane*. Norderstedt, Germany: Books on Demand.

310 Frisch, K.C. (1991). *J. Polym. Sci. A Polym. Chem.* 29: 1834–1835.

311 Saunders, J.H. and Frisch, K.C. (1964). *Polyurethanes: Chemistry and Technology: Chemistry*, vol. 16. Interscience Publishers.

312 Padmavathy, T. and Srinivasan, K.S.V. (2003). *J. Macromol. Sci. C* 43: 45–85.

313 Knight, D.P. and Vollrath, F. (2002). *Philos. Trans. R. Soc. B.* 357: 155–163.

314 Dolden, J. and Alder, P. (1998). *High Perform. Polym.* 10: 249–272.

315 Donald, A.M., Windle, A.H., and Hanna, S. (2006). *Liquid Crystalline Polymers*. Cambridgeshire, England: Cambridge University.

316 Demus, D., Goodby, J.W., Gray, G.W. et al. (2011). *Handbook of Liquid Crystals, Volume 2A: Low Molecular Weight Liquid Crystals I: Calamitic Liquid Crystals*. Hoboken, New Jersey: Wiley.

317 Cowie, J.M.G. and Arrighi, V. (2007). *Polymers: Chemistry and Physics of Modern Materials*. Boca Raton, Florida, USA: CRC press.

318 Edlund, U. and Albertsson, A.-C. (2003). *Adv. Drug Deliv. Rev.* 55: 585–609.

319 Scheirs, J. and Long, T.E. (2005). *Modern Polyesters: Chemistry and Technology of Polyesters and Copolyesters*. Hoboken, New Jersey: Wiley.

320 Kroschwitz, J.I. (1990). *Encyclopedia of Polymer Science and Engineering*, vol. 17. New York: Wiley.

321 Vilela, C., Sousa, A.F., Fonseca, A.C. et al. (2014). *Polym. Chem.* 5: 3119–3141.

322 Pang, K., Kotek, R., and Tonelli, A. (2006). *Prog. Polym. Sci.* 31: 1009–1037.

323 Mathers, R.T. and Meier, M.A. (2011). *Green Polymerization Methods: Renewable Starting Materials, Catalysis and Waste Reduction*. Hoboken, New Jersey: Wiley.

324 Thiele, U. (2004). *Chem. Fibers Int.* 54: 162–163.

325 Ma, J., Pang, Y., Wang, M. et al. (2012). *J. Mater. Chem.* 22: 3457–3461.

326 Jiang, M., Liu, Q., Zhang, Q. et al. (2012). *J. Polym. Sci. A Polym. Chem.* 50: 1026–1036.

327 Takasu, A., Takemoto, A., and Hirabayashi, T. (2006). *Biomacromolecules* 7: 6–9.

328 Manabe, K., Iimura, S., Sun, X.M., and Kobayashi, S. (2002). *J. Am. Chem. Soc.* 124: 11971–11978.

329 Sousa, A.F., Gandini, A., Silvestre, A.J., and Pascoal Neto, C. (2008). *ChemSusChem* 1: 1020–1025.

330 Vouyiouka, S.N., Topakas, E., Katsini, A. et al. (2013). *Macromol. Mater. Eng.* 298: 679–689.

331 Dukuzeyezu, E.-M., Lefebvre, H., Tessier, M., and Fradet, A. (2010). *Polymer* 51: 1218–1221.

332 Xiong, X.Q. and Xu, Y.H. (2010). *Polym. Bull.* 65: 455–463.

333 Shi, F. and Xia, C. (2019). *Chin. J. Catal.* 40: 43–50.

334 Dai, X., Yuan, H., Huang, Y. et al. (2018). *Sci. Sin. Chim.* 48: 1568.

335 Matsinha, L.C., Siangwata, S., Smith, G.S., and Makhubela, B.C.E. (2019). *Catal. Rev.* 61: 111–133.

336 Li, C., Wang, W., Yan, L., and Ding, Y. (2018). *Front. Chem. Sci. Eng.* 12: 113–123.

337 De, C., Saha, R., Ghosh, S.K. et al. (2013). *Res. Chem. Intermed.* 39: 3463–3474.

338 Franke, R., Selent, D., and Börner, A. (2012). *Chem. Rev.* 112: 5675–5732.

339 Breit, B. and Seiche, W. (2001). *Synthesis* 2001: 0001–0036.

340 Beller, M., Cornils, B., Frohning, C.D., and Kohlpaintner, C.W. (1995). *J. Mol. Catal. A: Chem.* 104: 17–85.

341 Botteghi, C., Ganzerla, R., Lenarda, M., and Moretti, G. (1987). *J. Mol. Catal.* 40: 129–182.

342 Liu, X.-F., Li, X.-Y., and He, L.-N. (2019). *Eur. J. Org. Chem.* 2019: 2437–2447.

343 Zhang, Y., Dai, X., Wang, H., and Shi, F. (2018). *Acta Phys.-Chim. Sin.* 34: 845–857.

344 Yuan, G., Qi, C., Wu, W., and Jiang, H. (2017). *Curr. Opin. Green Sustain. Chem.* 3: 22–27.

345 Tlili, A., Blondiaux, E., Frogneux, X., and Cantat, T. (2015). *Green Chem.* 17: 157–168.

346 Kondratenko, E.V., Mul, G., Baltrusaitis, J. et al. (2013). *Energy Environ. Sci.* 6: 3112–3135.

347 Dai, X., Adomeit, S., Rabeah, J. et al. (2019). *Angew. Chem. Int. Ed.* 58: 5851–5855.

348 Dai, X., Rabeah, J., Yuan, H. et al. (2016). *ChemSusChem* 9: 3133–3138.

349 Allen, C.L. and Williams, J.M.J. (2011). *Chem. Soc. Rev.* 40: 3405–3415.

350 Valeur, E. and Bradley, M. (2009). *Chem. Soc. Rev.* 38: 606–631.

9

Conclusion and Perspectives

9.1 Conclusion

Synthesis of carbonyl-containing molecules has always been an intense topic in the chemistry due to its wide applications in the synthetic, catalytic, pharmaceutical, and material chemistry [333]. From a point view of chemical reaction, the carbonyl-containing molecules can behave as both the starting material and product and catalyst. In other words, these molecules can bridge the whole catalytic process. Here, we briefly present the application of the carbonyl-containing molecules as the reactants, products, and catalysts (Figure 9.1) [334–341]. As the reactants, the first function of carbonyl molecules is as carbonyl source to synthesize valuable carbonyl-containing compounds. According to the number of carbon atoms in the molecules and the carbonyl building way, the widely used molecules as carbonyl sources can be divided into several groups: (i) C_1 carbonyl molecules such as CO, CO_2, HCHO, and HCOOH; (ii) non-C_1 carbonyl molecules such as formates, formamides, acyl chloride, carbonyl metal; (iii) carbonyl molecule precursors, namely being able to provide carbonyl sources by in situ activation way, such as primary alcohols, chloroform, oxiranes, and biomass [342–350]. These basic molecules can participate in many different reactions such as hydroformylation, alkoxycarbonylation, aminocarbonylation, the Pauson–Khand reaction, and N-formylation as well as the carbonylation of methanol, epoxides, aldehydes, diazoalkanes, and other molecules, and various kinds of carbonyl-containing chemicals such as aldehydes, ketones, carboxylic acid, esters, amides, and ureas can be synthesized via these aforementioned transformations. In addition to the direct use, these synthesized molecules containing carbonyl can be further converted to provide all kinds of functional molecules with special structures such as alcohols, aldehydes, ketones, carboxylic acid, amines, and materials such as polyamides, phenol–formaldehyde resin, polyurethanes, and polyesters.

The further transformation mainly involved the oxidation, reduction, addition, elimination, and coupling, and they are usually performed by applying different transition metal catalysts including homogeneous and heterogeneous at proper temperature and/or under certain pressure of oxidized or reduced active gas. In addition to behaving as the reactants, some carbonyl-containing molecules can be also applied as the active species to catalyze some transformations. Considering the

Carbonyl Compounds: Reactants, Catalysts and Products, First Edition.
Feng Shi, Hongli Wang and Xingchao Dai.

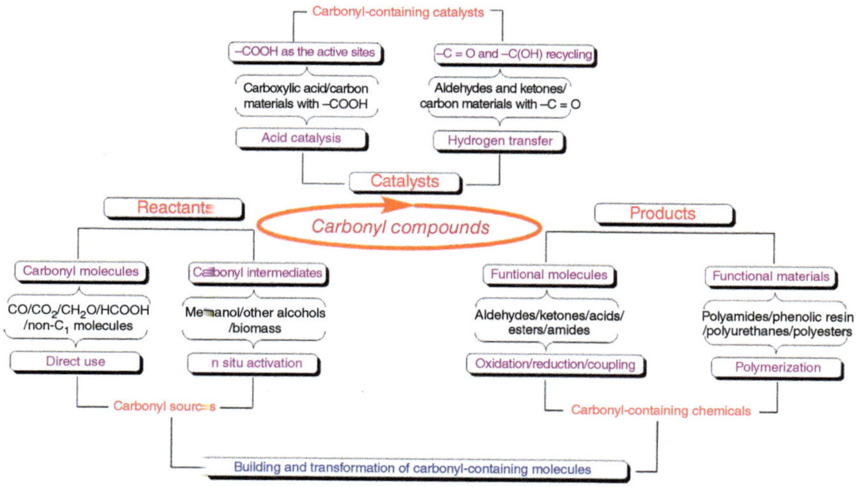

Figure 9.1 Carbonyl compounds as the molecules bridging reactants, catalysts, and products.

multifunctionalities of carbonyl-containing molecules in the catalytic processes, here, we only focus on discussing those transformations catalyzed by the molecules containing –COOH and the recycling of carbonyl and hydroxyl groups. Among them, the molecules containing –COOH mainly refer to the free carboxylic acid and some carbon materials containing –COOH, and the carbonyl and hydroxyl groups recycling is usually based on the free aldehydes/ketones as well as some carbon materials with aldehyde/ketone groups. Considering the considerable acidity of –COOH group, the molecules containing –COOH enable to catalyze many reactions catalyzed by the acid. The carbonyl and hydroxyl groups recycling is usually applied in the selective oxidation and the borrowing-hydrogen reactions due to the excellent performance of aldehyde/ketone groups in extracting and deliver H.

9.2 Perspectives

The carbonyl chemistry is an important component in catalysis chemistry, and the synthesis, activation, and transformation of carbonyl-containing molecules are the core content of the carbonyl chemistry. As one of the most widely used methods to obtain the desired bulk and fine chemicals as well as functional materials, the future development of carbonyl chemistry will mainly aim at building highly efficient and robust catalytic systems and developing new transformations. The concrete points are as follows.

(1) Development of new catalytic systems for CO activation and transformation. Although CO as the C_1 synthon has always been denounced due to the high toxicity, a fact has to be admitted is that the research of new carbonylation reaction based on CO will continue for a long time and be at the forefront of catalytic

carbonylation. For hydroformylation as the most important industrial processes for oxo synthesis, the future attention will focus on further improving the performance of the existing system by introducing the new concept and designing new ligand.

(2) Development of new catalytic systems based on non-CO C_1 molecules containing carbonyl. In order to avoid the use of toxic CO, developing new carbonyl sources as partial replacement and effective supplement of CO in some specific reaction systems will be an efficient means. CO_2, HCHO, and HCOOH as carbonyl sources to synthesize valuable carbonyl-containing chemicals have attracted a lot of attention and some progresses have been made; however, these investigations still stay at the early stage on the whole and more efforts need to be devoted.

(3) Development of new catalytic systems based on in situ building of carbonyl sources. In addition to the direct use of carbonyl-containing molecules as carbonyl sources, methanol and even methane can be also used as the carbonyl sources by in situ activation way. The former has been successfully applied for synthesis of some carbonyl-containing chemicals such as formamides; however, the investigation still stays at the lab stage. For the latter, the current study focuses on converting it into valuable C_1 molecules such as CO and methanol; however, the required harsh reaction conditions and poor efficiency of the existing systems limited their further development. Therefore, the development of new catalytic systems with higher activity is highly desirable.

(4) Development of renewable carbonyl sources based on biomass-based polyols. Compared with traditional fossil-derived hydrocarbons, biomass-based polyols possess rich oxygen atoms and can be expressed using general formula $(CO)_x(H_2)_y$, which means that these molecules might be potential natural syngas source if the C–C bond in polyol molecules can be controllably cut. The recently reported works have confirmed this point; however, the selective cleavage of C–C bond is difficult. Therefore, the key to synthesis of carbonyl-containing chemicals with renewable biomass-based molecules as carbonyl sources is the selective cleavage of C–C bond to in situ provide carbonyl sources by rational design of catalysts.

(5) The asymmetric carbonylation of olefins, alkyne, alcohol, halohydrocarbon, and ether is one of effective methods for preparing chiral aldehydes, esters, and polymer. These chiral molecules are highly valuable building blocks for the synthesis of a plethora of fine chemicals, pharmaceutical products, and function materials. Nevertheless, the development of asymmetric carbonylation with simultaneous precise control of the chemo-, regio-, and enantioselectivity of the reaction is still a challenge. Further research in this area is expected to be the development of more efficient chiral catalysts and the extension of substrate scope. In addition, detailed mechanistic investigations are likely to be beneficial for the design and discovery of new generation catalysts for asymmetric carbonylation. The asymmetric carbonylation will continue to be an exciting and promising area of carbonylation reaction.

(6) Establishment of heterogeneous catalytic systems for carbonyl chemistry. Heterogeneous carbonylation has been a long-term challenge in the carbonyl chemistry. Compared to homogeneous carbonylation catalysts widely investigated, the heterogeneous catalytic materials, especially with good performance for catalytic carbonylation, are much scarce. The loss of catalyst active component leading to activity drop and even disappearance in recycling test widely existed in the reported heterogeneous catalytic systems. Therefore, by rational design of catalytic material structure to enhance the catalyst stability, the improvement of catalytic reaction process to inhibit or solve the problem of catalyst active component leaching will be the future research focus, which is also the key to the successful application of heterogeneous catalyst system in the industry.

Index

Carbonyl Compounds: Reactants, Catalysts and Products, First Edition.
Feng Shi, Hongli Wang and Xingchao Dai.